Distillation: Fundamentals and Principles

Distillation: Fundamentals and Principles

Edited by

Andrzej Górak

Laboratory of Fluid Separations
Department of Biochemical and
Chemical Engineering
TU Dortmund University
Emil-Figge-Str. 70
D-44227 Dortmund

Eva Sorensen

Department of Chemical Engineering
UCL, Torrington Place,
UK-London WC1E 7JE

AMSTERDAM • BOSTON • HEIDELBERG • LONDON
NEW YORK • OXFORD • PARIS • SAN DIEGO
SAN FRANCISCO • SINGAPORE • SYDNEY • TOKYO
Academic Press is an imprint of Elsevier

Academic Press is an imprint of Elsevier
32 Jamestown Road, London NW1 7BY, UK
225 Wyman Street, Waltham, MA 02451, USA
525 B Street, Suite 1800, San Diego, CA 92101-4495, USA
The Boulevard, Langford Lane, Kidlington, Oxford OX5 1GB, UK

Library of Congress Cataloging-in-Publication Data
Application submitted

British Library Cataloguing in Publication Data
A catalogue record for this book is available from the British Library

ISBN: 978-0-12-386547-2

For information on all Academic Press publications
visit our web site at store.elsevier.com

This book has been manufactured using Print On Demand technology. Each copy is produced
to order and is limited to black ink. The online version of this book will show color figures
where appropriate.

Working together
to grow libraries in
developing countries

www.elsevier.com • www.bookaid.org

Bottom left and top right photographs on cover courtesy of FRI and J. Montz GmbH

Contents

Preface to the Distillation Collection

For more than 5,000 years distillation has been used as a method for separating binary and multicomponent liquid mixtures into pure components. Even today, it belongs to the most commonly applied separation technologies and is used at such a large scale worldwide that it is responsible for up to 50% of both capital and operating costs in industrial processes. It moreover absorbs about 50% of the total process energy used by the chemical and petroleum refining industries every year. Given that the chemical industry consumed 19% of the entire energy in Europe (2009), distillation is *the* big driver of overall energy consumption.

Although distillation is considered the most mature and best-understood separation technology, knowledge on its manifold aspects is distributed unevenly among different textbooks and manuals. Engineers, by contrast, often wish for just one reference book in which the most relevant information is presented in a condensed and accessible form. *Distillation* aims at filling this gap by offering a succinct overview of distillation fundamentals, equipment, and applications. Students, academics, and practitioners will find in *Distillation* a helpful summary of pertinent methods and techniques and will thus be able to quickly resolve any problems in the field of distillation.

This book provides a comprehensive and thorough introduction into all aspects of distillation, covering distillation history, fundamentals of thermodynamics, hydrodynamics, mass transfer, energy considerations, conceptual process design, modeling, optimization and control, different column internals, special cases of distillation, troubleshooting, and the most important applications in various industrial branches, including biotechnological processes.

Distillation forms part of the "Handbook of Separation Sciences" series and is available as a paper book and as an e-book, thus catering to the diverging needs of different readers. It is divided into three volumes: "Fundamentals and Principles" (Editors A. Górak and E. Sorensen), "Equipment and processes" (Editors A. Górak and Ž. Olujić), and "Operation and applications" (Editors A. Górak and H. Schoenmakers). Each volume contains chapters written by individual authors with acclaimed expertise in their fields. In addition to that, readers will find cross-references to other chapters, which allow them to gain an extensive overview of state-of-the-art technologies and various research perspectives. Helpful suggestions for further reading conclude each chapter.

A comprehensive and complex publication such as *Distillation* is impossible to complete without the support of an entire team whose enduring help I wish to acknowledge. In particular, I wish to express my heartfelt gratitude to the 42 leading world experts from the academia and industry who contributed to the chapters of this book. I thank the co-editors of the three volumes of *Distillation*—Dr Eva Sorensen,

UCL, Dr Žarko Olujić, Delft University of Technology, and Dr Hartmut Schoen-makers, former member of BASF SE, Ludwigshafen—for their knowledgeable input and expertise, unremitting patience, and continuous encouragement. The invaluable editorial assistance of Dipl.-Ing. Johannes Holtbrügge during the entire editorial process is also greatly acknowledged.

Editorial assistance of Vera Krüger is also appreciated. I thank the Elsevier team Jill Cetel, Beth Campbell and Mohanambal Natarajan for their support and valuable help through the whole editing process.

Dr Andrzej Górak
TU Dortmund University

Preface to *Distillation: Fundamentals and Principles*

This is the first book in a three-volume series covering all aspects of *Distillation*. This volume focuses on the fundamental principles of distillation with particular emphasis on practical understanding of design and operation. The chapters are written by different authors and the approach, depth, and extent of subject matter coverage may therefore differ from chapter to chapter, however, together they represent a comprehensive overview of the current state of the art.

The first chapter traces the historical development of distillation from the first applications over 5000 years ago, via the medieval period, and the nineteenth-century industrial developments, to contemporary applications with emphasis on the applications and equipment which led to our current technology. A prerequisite for the design of distillation columns is knowledge of vapor−liquid equilibrium (VLE) and of mass transfer phenomena. Chapter 2 considers thermodynamic models for the prediction of VLEs, and the conditions for the occurrence of azeotropes. In Chapter 3, an account is given of the fundamental principles of mass transfer including diffusion, mass transfer coefficients, and mass transfer of both binary and multicomponent mixtures in both tray and packed columns.

Chapter 4 sets out the fundamental principles of binary distillation including simple calculation and analysis methods. This is followed by an account of batch distillation in Chapter 5, giving an overview of the fundamentals of batch distillation, including different operating modes, alternative column configurations and more complex batch distillation processes. Chapter 6 considers energy-efficient distillation design and operation, including columns operating both above and below ambient temperatures. Various advanced and complex distillation column configurations are also introduced.

Chapters 7 and 8 consider design of distillation processes. Chapter 7 describes the conceptual design of zeotropic multicolumn distillation configurations. A computationally efficient mathematical framework is described that synthesizes configurations that use $n-1$ distillation columns for separating a zeotropic mixture into n product streams. Chapter 8 turns the attention to azeotropic systems and describes a systematic framework for their conceptual design, considering and comparing different approaches. Several shortcut methods are presented, followed by an account of rigorous optimization, and their applicability for design is discussed.

Aspects of design, analysis, and application of hybrid distillation schemes are covered in Chapter 9. These hybrid distillation schemes become necessary when separation tasks such as separation of azeotropic or close-boiling mixtures cannot be achieved in a single conventional distillation column. Chapter 10 presents an overview of modeling methods covering both simplified and rigorous models. Conceptual features are highlighted and basic equations are shown for both equilibrium- and nonequilibrium-based approaches. In addition to classical distillation, modeling of related and more complex processes is also discussed. Finally, Chapter 11 presents an overview of the main advances in optimization of zeotropic systems, ranging from systems using only conventional columns, to fully thermally coupled systems, with main focus on mathematical programming approaches for design.

I would like to thank all the authors for their contributions and assiduous efforts in making this the most comprehensive account of distillation fundamentals available to date.

Dr Eva Sorensen
UCL

List of Contributors

Rakesh Agrawal
School of Chemical Engineering, Purdue University, West Lafayette, IN, USA

Deenesh K. Babi
Department of Chemical and Biochemical Engineering, Technical University of Denmark, Lyngby, Denmark

Sergei Blagov
BASF SE, GT/SI, Ludwigshafen, Germany

José A. Caballero
Department of Chemical Engineering, University of Alicante, Alicante, Spain

Rafiqul Gani
Department of Chemical and Biochemical Engineering, Technical University of Denmark, Lyngby, Denmark

Jürgen Gmehling
University of Oldenburg, Industrial Chemistry, Oldenburg, Germany

Ignacio E. Grossmann
Department of Chemical Engineering, Carnegie Mellon University, Pittsburgh, PA, USA

Andreas Harwardt
AVT-Process Systems Engineering, RWTH Aachen University, Aachen, Germany

Megan Jobson
School of Chemical Engineering and Analytical Science, The University of Manchester, Manchester, UK

Eugeny Y. Kenig
Department of Mechanical Engineering, University of Paderborn, Pohlweg 55, Paderborn, Germany

Michael Kleiber
Hattersheim, Germany

Norbert Kockmann
Laboratory of Equipment Design, Department of Biochemical and Chemical Engineering, TU Dortmund University, Dortmund, Germany

Hendrik A. Kooijman
Clarkson University; Potsdam, NY, USA

Wolfgang Marquardt
AVT-Process Systems Engineering, RWTH Aachen University, Aachen, Germany

Vishesh H. Shah
Engineering and Process Sciences Laboratory, The Dow Chemical Company, Midland, MI, USA

Mirko Skiborowski
AVT-Process Systems Engineering, RWTH Aachen University, Aachen, Germany

Eva Sorensen
Department of Chemical Engineering, UCL, London, UK

Ross Taylor
Clarkson University; Potsdam, NY, USA

List of Symbols and Abbreviations

Latin symbols

Symbol	Explanation	Unit	Chapter
A	Matrix of constant coefficients		11
A_{AB}	Coefficient in van Laar's equation	–	4
A_c	column cross section area	m^2	10
A_e	Effective adsorption factor (Group methods)	–	11
A_{ht}	Heat transfer area	m^2	6
A_i	Coefficient in Antoine's equation for component i	Pa, bar	4
A_z	Azeotropic composition	mol/mol	5
a	Attractive parameter in cubic equations of state	Jm^3/mol^2	2
a^l	specific vapor-liquid interfacial area	m^2/m^3	10
a_i	Activity of component i	–	2
B	Bottom stream flow rate	kmol/s, kg/s	4, 11
B	Matrix of coefficients		11
B_i	Coefficient in Antoine's equation for component i	Pa K, bar °C, Pa °C,...	4
BT	Total molar bottom stream flow rate	kmol/s	11
b	Co-volume in cubic equations of state	cm^3/mol, m^3/mol	2
b	Content dependent Binary variable Vector of constant coefficients		11
C_A	Price of distillate product	\$, €	5
C_C	Unit costs for cooling in the condenser	\$/kW	11
C_F	Price of feed	\$, €	5
$C_{fix,k}$	Annualized fixed charge cost of column k	\$/a	11
C_H	Unit costs for heating in the reboiler	\$/kW	11
C_i	Coefficient in Antoine's equation for component i	K, °C	4
C_k	Annualized cost of column k	\$/a	11

(continued)

— cont'd

Symbol	Explanation	Unit	Chapter
c	Parameter in the VTPR equation of state	m^3/mol	2
c	Constant/vector of constant coefficients		11
c	molar concentration	mol/m^3	10
c_p	Molar heat capacity at constant pressure	J/(mol K)	2
c_{pL}	Liquid heat capacity at constant pressure	J/K	4
c_{pLi}	Liquid heat capacity of component i at constant pressure	J/K	4
c_{pV}	Vapor heat capacity at constant pressure	J/K	4
d	generalized driving force, Eq. (10-30)	1/m	10
D	Distillate flow rate	mol/s, kmol/s, kg/s	4, 5, 8, 10, 11
D_{AB}	binary diffusion coefficient	m^2/s	10
D_{ax}	axial dispersion coefficient	m^2/s	10
D_i	effective diffusion coefficient of component i	m^2/s	10
D_{ij}	Driving Force		9
$Đ_{ij}$	Maxwell-Stefan diffusion coefficient	m^2/s	10
DT	Total molar distillate flow rate	kmol/s	11
E	Entrainer flow rate	kmol/s	5, 8
E_{MV}	Murphree vapor efficiency	–	4
E_O	Overall tray/stage efficiency	–	4
E_{OL}	Murphree efficiency for the liquid phase	–	10
E_{OV}	Murphree efficiency for the vapor phase	–	10
Ex	Exergy	W, W/K	6
F	Feed flow rate	mol/s, kmol/s, kg/s	4, 5, 8, 10, 11
FT	Total molar feed flow rate	kmol/s	11
f	Vaporized fraction of the feed	–	4
f(·)	Scalar function		11
f_i	Fugacity of component i	Pa	2
$f_{obj}(·)$	Objective function		2, 5
G^E	Total excess Gibbs energy	J	2, 8
G_{ij}	NRTL parameter		2
g(·)	Equality constraints Scalar/vector functions		5 11
g^E	Molar excess Gibbs energy	J/mol	2
$\triangle g_{ij}$	Interaction parameter between component i and j	K	2

—cont'd

Symbol	Explanation	Unit	Chapter
H	Enthalpy	W, W/K	6
h	partial molar enthalpy	J/mol	10
H_A	Amount of distillate	mole	5
H_F	Amount of feed	mole	5
$H_{i,j}$	Henry constant of component i in j	Pa	2
$\triangle H_{vap}$	Molar heat of vaporization	J/mol	4
h	Specific enthalpy	J/kg, J/mol	4, 8, 11
$h(\cdot)$	Inequality constraints		5
	Scalar/vector functions		11
Δhv	Enthalpy of vaporization	J/mol	2
h^E	Molar excess enthalpy	J/mol	2
HETP	height equivalent to a theoretical plate	m	10
HTU	height of a transfer unit	m	10
H_c	column height	m	10
J	diffusion flux	mol/(m^2 s)	10
J	column vector consisting of J_i	mol/(m^2 s)	10
K_i	Chemical equilibrium constant of component i	–	2
K_i	Distribution coefficient/K-factor of component i	–	2, 4, 8, 11
K_k	Relation between feed and heat flow (Andrecovich & Westerberg model)	–	11
k_{ij}	Binary parameter in cubic equations of state	–	2
K^{eq}	vapor-liquid equilibrium constant	–	10
K_{OL}	overal mass transfer coefficient in terms of the liquid phase	mol/(m^2 s)	10
K_{OV}	overal mass transfer coefficient in terms of the vapor phase	mol/(m^2 s)	10
L	Liquid flow rate	kmol/s, kg/s	4, 8, 11
Ld	Reflux flow rate returned to the column	kmol/s	11
L_R	Liquid flow rate in a rectifying column section	kmol/s	11
L_S	Liquid flow rate in a stripping column section	kmol/s	11
l	axial coordinate directed from column top to bottom	m	10
L	liquid molar flow rate	mol/s	10
M	Scalar (Big M parameter)		11
m	slope of the operating line, Eq. (10-A2)	–	10

(continued)

—cont'd

Symbol	Explanation	Unit	Chapter
N	*Content dependent* Number of trays/stages Number of components	–	4, 6, 8, 11
N_R	Number of trays/stages in the rectifying section	–	4, 8, 11
N_S	Number of trays/stages in the stripping section	–	4, 8, 11
NT	Total number of trays in column section	–	4
NF	Feed tray location	mole	4
n	Number of moles		2
n_C	Number of components	–	8
n	number of mixture components	–	10
N	Content dependent molar flux	mol/(m^2 s)	10
NTU	number of transfer units	–	10
O_i	Offcuts	–	5
P	Hourly profitability	profit/hr	5
P_i	Product cuts	–	5
Poy_i	Poynting factor of component i		2
p	Pressure	Pa (or bar, atm)	2, 4, 8, 11
$p_{0,i}^{LV}$	Vapor pressure of component i	Pa	4
pb_i	Individual flow rate of the bottom product	kmol/s, kg/s	11
p_i	Partial pressure of component i	Pa (or bar, atm)	4
pt_i	Individual flow rate of the top product	kmol/s, kg/s	11
Q_{EX}	Exchanged heat	kJ	11
$\dot{Q}$	Duty	W	4, 5, 6, 8, 11
$\dot{Q}_{flash}$	Energy added or removed in the flash drum	W	4
$\dot{Q}_{ht}$	Heat transfer duty	W	6
$\dot{Q}_{hx}$	Energy added or removed in the heat exchanger	W	4
$\dot{Q}_k$	Heat duty for column k	W	11
Q_B / Q_R	reboiler/boilup heat duty	–	8
q	Energy to convert one mol of feed to saturated vapor at dew point divided by the molar heat of vaporization	–	4
q_{LS}	Liquid fraction of a stream	mol/mol	11
q	liquid molar fraction in the feed stream	–	10
Q	heat flux	W/m^2	10
R	Gas constant	J/mol·K	2

—*cont'd*

Symbol	Explanation	Unit	Chapter
R	Boolean variable		11
R_{min}	Reflux ratio	–	4, 5, 6, 8, 11
r	*Content dependent* Binary variable Interest rate Scalar function		11
rec	Component recovery		11
$\Re$	gas constant	8.3144 J/(mol K)	10
R_b	reboil ratio	–	10
R_f	reflux ratio	–	10
S	Stripping factor	–	11
S	Entropy	W, W/K	6
S	Side stream flow rate	kmol/s, kg/s	4
S_e	Effective stripping factor (Group methods)	–	11
S_{ij}	Selectivity	–	2, 8
s	Renewal frequency, parameter of the surface renewal model	1/s	10
T	Absolute temperature	K	2, 4, 6, 8, 11
T_{bp}	Boiling temperature	K, (°C)	4, 5
T_0	Reference temperature	K	6
$T_{min,ea}$	Minimum exchanger approach temperature	K	11
T_{st}	Steam temperature	K	11
ΔT_{lm}	Logarithmic mean temperature difference	K	6
t	Time	s	5, 10
t_f	Total operating/final time	s	5
t_e	exposure time, parameter of the penetration model	s	10
T	temperature	K	10
U	Overall heat transfer coefficient	W/m²K	6
U	Big M Parameter		11
u	Control variables		5
u_L	liquid-phase velocity	m/s	10
U	length-specific molar holdup	mol/m	10
V	Vapor flow rate	kmol/s, kg/s	4, 8, 11
V_R	Vapor flow rate in a rectifying column section	kmol/s	11
Vr	Reboil flow rate returned to the column	kmol/s	11

(*continued*)

—cont'd

Symbol	Explanation	Unit	Chapter
V_S	Vapor flow rate in a stripping column section	kmol/s	11
v	Molar volume	cm^3/mol, m^3/mol	2
v	Design variables		5
V	vapor molar flow rate	mol/s	10
W	Still holdup in differential distillation	mole	4, 5
W	Boolean variable		11
W_{ideal}	Ideal compression power demand	W	6
WC	Boolean variable (determine if a condenser exists)		11
WR	Boolean variable (determine if a reboiler exists)		11
w	Binary variable	–	11
X	Parameter in the Gilliland graphical correlation		11
x	Vector of real variables		11
x	liquid mole fraction	mol/mol	10
x	liquid-phase composition vector	mol/mol	10
x_i	Mole fraction of component i in the liquid phase	mol/mol	2, 4, 5, 8, 11
Y	*Content dependent* Parameter in the Gilliland graphical correlation Boolean variable		11
y	Algebraic variable		5, 11
y_i	Mole fraction of component i in the vapor phase	mol/mol	2, 4, 5, 8, 11
y	vapor mole fraction	mol/mol	10
y	vapor-phase composition vector	mol/mol	10
Z	*Content dependent* Boolean variable Objective variable in optimization problems	–	11
z	Boolean variable		11
z	Compressibility factor	–	2
z_i	Mole fraction of component i	mol/mol	2, 4, 8, 11
z	film coordinate; transformed liquid-phase concentration, Eq. (10-A11)	m –	10
z	liquid-phase composition vector consisting of z_i	–	10

Greek Symbols

Symbol	Explanation	Unit	Chapter
α	Relative volatility	–	2, 4, 5, 10, 11
$\boldsymbol{\alpha}$	relative volatility vector consisting of α_i	–	10
α^T	heat transfer coefficient	W/(m^2 K)	10
α_{ij}	Non-randomness parameter in the NRTL equation ($\alpha_{ij} = \alpha_{ji}$)	–	2
β_k	Size factor for column k (Andrecovich & Westerberg model)		11
β, β_{ik}	binary mass transfer coefficient	mol/(m^2 s)	10
$[\beta]$	matrix of mass transfer coefficients	mol/(m^2 s)	10
γ_i	Activity coefficient of component i	–	2, 4
γ	component net interstage flow, Eq. (10-A3)	–	10
$\boldsymbol{\gamma}$	component net interstage flow vector consisting of γ_i	–	10
Γ	Group activity coefficient	–	2
$[\Gamma]$	matrix of thermodynamic correction factors	–	10
δ	film thickness	m	10
ζ	Split fraction	–	11
λ	Eigenvalue	–	8
ρ	Density	kg/m, mol/m^3	2
ν_k	Number of structural groups in the mixture	–	2
$\nu_k^{(i)}$	Number of structural groups in pure solvent	–	2
ϕ	Liquid phase ratio	–	8
ϕ_A	Recovery factor for absorption section	–	11
ϕ_i	Recovery factor for component i	–	11
ϕ_R	Underwood root	–	11
ϕ_S	Recovery factor for stripping section	–	11
ϕ	Fugacity coefficient	–	2
φ_{FB}	Fischer–Burmeister function		11
ϕ	volumetric holdup	m^3/m^3	10
μ	chemical potential	J/mol	10
θ	root of Underwood's equation	–	10
$\boldsymbol{\theta}$	vector of roots of Underwood's equation	–	10
τ	transformed time parameter, Eq. (10-1)	–	10
ψ	Recovery fraction	–	11
ϑ	Temperature	°C	2
τ	Dimensionless time	–	8
τ_{ij}	NRTL parameter	–	2
ξ	Dimensionless time scale	–	2
$\Omega(\cdot)$	Boolean function		11
ω	Acentric factor	–	2

Subscripts

Symbol	Explanation	Chapter
ave	Average	4
az	Azeotropic point	2
B	Bottom stream	4, 8, 11
bot	Bottom section (aggregated models)	11
C, cond	Condenser	4, 6, 8, 11
c	Critical point	2
cool	Cooling	6, 11
D	Distillate	4, 5, 11
dp	Dew point	4
evap	Evaporation	6
F	Feed	4, 5, 8, 11
f	Formation reaction	2
heat	Heating utility	11
HK	Heavy key	11
HP	High pressure	5
hx	Heat exchanger	4
in	Inlet	4
int	Internal	4
i, j, k	Component, tray position, iteration	2, 11
i,j,k	component, reaction or stage indices	10
k	Number of heterogeneous trays	8
L	Liquid	2, 4, 8, 11
L	liquid phase	10
LK	Light key	11
LP	Low pressure	5
min	Minimum/minimal	4, 6, 8, 11
mb	Top part of the bottom section (aggregated models)	11
mt	Bottom part of the top section (aggregated models)	11
N	Stage/tray number, component in a mixture	11
opt	Optimal	4
r	Chemical reaction	2
R, reb	Reboiler	4, 5, 8, 11
S	State	11
S	Side stream	4
s	Position in column	11

(*continued*)

—cont'd

Symbol	Explanation	Chapter
st	Steam	11
t	total	10
top	Top section (aggregated models)	11
V	Vapor	2, 4, 8, 11
V	vapor phase	10
w	Still	4, 5
0	Initial	4, 5, 11

Superscripts

Symbol	Explanation	Chapter
—	Partial molar property	2
∞	At infinite dilution	2
*	equilibrium	10
av	average	10
b	bottom product	10
B	bulk phase	10
C	Combinatorial part	2
d	distillate product	10
f	column feed stream	10
I	phase interface	10
o	Standard state	2
r	Reduced property	2
r	rectifying section	10
R	Residual part	2
S	Solvent free basis	2
s	At saturation	2
s	stripping section	10
spec	Specification	5

Abbreviations

Abbreviation	Explanation	Chapter
BM	Big-M	11
BVM	Boundary value method	8, 11
CAMD	Computer-aided molecular design	8
CAMD	Computer-Aided Molecular Design	9
CC	Composition control	5
CDR	Continuous distillation region	8
CDRM	Continuous distillation region method	8
CH	Convex hull reformulation	11
CHR	Convex hull reformulation	11
CMO	Constant molar overflow	8
CNF	Conjunctive normal form	11
CRV	Constant relative volatility	8
DAE	Differential algebraic equation	5
DB	Distillation boundary	8
DBB	Disjunctive branch and bound	11
DDB	Dortmund Data Bank	2
DIPPR	Design Institute for Physical Property Data	2
DL	Distillation line	8
DRD	Distillation region diagram	1
EC	Eigenvalue criterion	8
ECP	Extended cutting plane	11
EMAT	Exchanger minimum approach temperature	11
EOS	Equation of state	2, 8
FAM	Feed angle method	8
FPM	Feed pinch method	8
FRI	Fractionation Research Inc.	1
FTC	Fully thermally coupled	11
FUG	Fenske-underwood-gilliland	4, 11
GBD	Generalized benders decomposition	11
GDP	Generalized disjunctive programing	11
GM	Group method	11
CFG	Computational fluid dynamics	1
HDS	Hybrid Distillation Scheme	9
HETP	Height equivalent to a theoretical plate	1, 4
HI	Heat integration	8
HIAG	Holzindustrie AG	2
HIDiC	Heat integrated distillation columns	6, 11
HP	High-pressure	5, 6

HTU	Height of transfer unit	1
ICAS	Integrated Computer-Aided System	9
L	Liquid	2, 11
LBOA	Logic based outer approximation	11
LC	Level control	5
LP	Linear programing	11
LP	Low-pressure	5, 6
LLE	Liquid–liquid equilibrium	2
MAC	Minimum angle criterion	8
MED	Minimum energy demand	8
MESH	Mass balances, equilibrium equations, molar fraction summation and enthalpy (H) balances	8, 11
MIDO	Mixed-integer dynamic optimization	5
MILP	Mixed-integer linear programming	11
MINDLP	Mixed-integer nonlinear dynamic optimization problem	5
MINLP	Mixed-integer nonlinear programming	8, 11
MP	Medium-pressure	6
MSF	Multi stage flash	4
MTN	Minimum number of trays	8
M-HDS	Membrane-based Hybrid Distillation Scheme	9
NLP	Nonlinear programing	5, 8, 11
NRTL	Nonrandom two liquid	2
OA	Outer approximation	11
OTOE	One task one equipment	11
PDB	Pinch distillation boundary	8
PR	Peng–Robinson	2
P-HDS	Pressure-Swing Hybrid Distillation Scheme	9
RB	Rectification body	8
RBM	Rectification body method	8, 11
RC	Residue curve	8
RCM	Residue curve map	1, 8
RD	Reactive Distillation	9
RDM	Reversible distillation model	11
R-HDS	Reactive agent-based Hybrid Distillation Scheme	9
SDB	Simple distillation boundary	8
SEN	State equipment network	11
SLE	Solid–liquid equilibrium	2
SRK	Soave–Redlich–Kwong	2
SR-MINLP	Successive relaxed MINLP	11
S-HDS	Solvent-based Hybrid Distillation Scheme	9
SSLM	Shortest stripping line method	8

(continued)

—cont'd

Abbreviation	Explanation	Chapter
STN	State task network	11
TAC	Total annual costs	8, 11
TC	Temperature control	5
TRB	Total reflux boundary	8
UNIFAC	Universal quasi-chemical theory functional group activity coefficients	2, 8
VHP	Very high pressure	6
VLE	Vapor–liquid equilibrium	2, 8
VLLE	Vapor–liquid–liquid equilibrium	2, 8
VRC	Vapor recompression cycle	11
VTE	Variable task equipment	11
VTPR	Volume translated Peng–Robinson	2
ZVC	Zero volume criterion	8

Abbreviations of chemical compounds

Abbreviation	Explanation	Chapter
CFC	Chlorofluorocarbon	6
DMSO	Dimethyl sulfoxide	8
HCFC	Hydrochlorofluorocarbon	6
HFC	Hydrofluorocarbon	6
HOAc	Acetic acid	9
H_2O	Water	9
NFM	*N*-formylmorpholine	2
NMP	*N*-methylpyrrolidone	2
MeOAc	Methyl-acetate	9
MeOH	Methanol	9
MTBE	Methyl-tert-butylether	1

History of Distillation

Norbert Kockmann

Laboratory of Equipment Design, Department of Biochemical and Chemical Engineering,
TU Dortmund University, Dortmund, Germany

CHAPTER OUTLINE

1.1 Introduction

The history of technical developments must also include social, cultural, and political perspectives. To reconstruct history from the current point of view means that we have to rely on arguable data and information. Findings on which we want to build a certain argumentation need an imaginative interpretation. The result is often not a factual, engineering-like argument presented with precision and reliability.

Distillation is a well-defined separation unit consisting of the partial evaporation of a liquid mixture and successive condensation, with a composition that differs from that of evaporation. The word *distillation* derives from the Latin verb *destillare*, meaning to drop down or to trickle down. Distillation had a broader meaning in ancient and medieval times because nearly all purification and separation operations were subsumed under the term *distillation*, such as filtration, crystallization, extraction, sublimation, or mechanical pressing of oil. This becomes evident because in earlier times there was no clear understanding of heat or the consistency of materials. Here, no further treatment will be presented on the history of alchemy, since chemistry and alchemy were not fully distinguishable until the modern age.

Distillation: Fundamentals and Principles. http://dx.doi.org/10.1016/B978-0-12-386547-2.00001-6

The equipment used for distillation flourished in Alexandria during the Roman Empire, and the apparatus did not change much until the sixteenth century. With the increased knowledge made possible by the invention of printing and with the larger demand for distilled products such as concentrated alcoholic or mineral acids, various stills thrived and were placed partly into industrial production. French scientists, English industrialists, and German craftsmen brought the equipment to the lab and fostered its industrial application. The modern era with its development of high-tech information has made possible a much wider picture and large-scale global development.

Wherever possible, we rely on the primary literature. To get a wider picture, however, it is also necessary to consult the secondary literature. Robert Forbes (1943) [1] gives an extensive illustration of the history of the art of distillation until 1840 and the death of Cellier-Blumenthal, one of the most gifted designers of distillation columns. Ludwig Deibele [2] presents an intensive treatment of the development of distillation until the end of the nineteenth century, with only a short look into the twentieth century. In 1935, A.J.V. Underwood wrote a little book on the historical development of distillation plants [3]. However, developments over the last 100 years are nearly completely missing and will be described in this contribution. In earlier times, developments have been presented in journal review articles and conference contributions. Hence, we have found descriptions in encyclopedias quite helpful (Ullmann's [4]) or handbooks (Perry's [5]).

The text follows the historical timeline, though not always rigidly so. Often product or equipment lines give a better understanding of innovative developments. Special emphasis is placed on the innovation process and its main drivers and motivations.

Two hundred years ago, Jean-Baptiste Cellier-Blumenthal invented the first continuously working distillation column in France and patented it in 1813. With this hallmark of distillation, we will celebrate this new chapter in the historical development of distillation.

1.2 From neolithic times to alexandria (3500 BC–AD 700)

The first civilizations started in Mesopotamia, Egypt, Syria, and China and spread from there. We cannot directly conclude from ancient texts what the ancient civilizations knew about the processing of foods or pharmaceutical products such as ointments, balsam, tinctures, or creams. Often priests and temple servants used distillation devices and kept their recipes secret. Hermann Schelenz [6] maintained that distillation was invented by the Persians, who used this process to produce rose water, rose oil, and other perfumes. He further stated that distillation was derived from dry distillation of wood for turpentine and wood tar. Together with Edmund von Lippmann [7], Schelenz found that the Egyptian *Ebers* papyrus (1550 BC) on medical issues already described the distillation of essential oils from herbs.

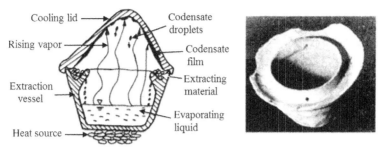

FIGURE 1.1

Extraction pot with evaporating liquid, condensing vapor, and extraction material in the chamfer of the pot [8]. The lid is sealed against the pot to avoid vapor and condensate losses.

Around 3500 BC the Sumerians were the first to apply evaporation and condensation of a liquid to refine a substance for extracting essential oils from herbs [8]; see Figure 1.1. Many of the pots and stills shown on the right side of the figure were found in excavations undertaken 250 km north of Baghdad in Iraq [9]. The liquid in the still evaporates by gentle heating from below and condenses at the colder cap. Droplets run down to be collected in the ring, where organic material such as leaves and herbs are extracted by the liquid. The typical dimensions of the earthenware pot are approximately 50 cm in diameter and 25–50 cm in height [9]. From fermented substances, alcohol may also rise and condense to help extract the ingredients, which were often essential oils or fragrances. This distillation under reflux and extraction is still employed today with the lab apparatus called the Soxhlet extractor.

Forbes discusses the early appearance of distillation. Because of his narrow definition of distillation, he concludes that the Alexandrian chemists were the first to develop and apply distillation to refine essential oils, rose water, wood turpentine, and other substances. Asphalt was found naturally and distilled to more viscous tar for ship- or house-building. Hence, both light and heavier fractions were the first products of ancient distillation processes.

Chinese distillation activities from ancient times have also been reported [10]. Early types of pots similar to those seen in Figure 1.1 were also found in China around 2000 BC. Kettles were found dating from 1000 BC, which indicate distillation operations. Joseph Needham [10] illustrates the development of the distillation apparatus in a genealogical tree (see Figure 1.2). The first primitive stills were earthenware pots heated from below with a lid (1), through which the vapor could condense. Further developments led to the device shown in Figure 1.1 with the internal collection of the lighter fraction (Figure 1.2a,b). Besides distillation, these rims could also serve as water sealants for anaerobic fermentation of foodstuffs, such as cabbage to produce sauerkraut.

Two mayor drawbacks of the early stills can be observed with this design: The condensate is collected in the warm part of the still and the internal volume is

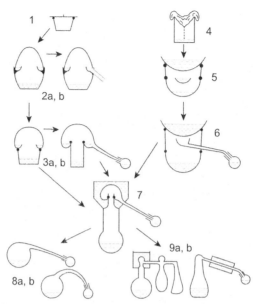

FIGURE 1.2 Illustration of Early Distillation Still Development

Adopted from Ref. [10].

limited, which may lead to unwanted reflux of the condensate back to the still. This is now the view of the modern engineer, who has a trained view of technical drawbacks. We should not forget the limits of knowledge at that time and must not judge based on our current views. It was quite difficult to remove the entire product from the gutter. An outlet tube was added (2b) to increase the capacity and performance, probably during the Hellenic era. Similar devices have been found in excavations of medieval apothecaries in Europe.

The next important step was to move the gutter to the colder cap (3a). The combination with the outlet tube formed the typical Hellenic still, in which the gutter (3b) was provided at one point leading off to a receiver. Descriptions of this type are found in Alexandrian manuscripts, with some variations of one and more outlets or different heating and cooling measures (see Figure 1.3(a)). Similar devices in the Far Eastern and Western civilizations indicate the existence of an information and equipment exchange over ancient routes, such as the Silk Road or the Indian Ocean merchants. Figure 1.3(b) depicts an Arabian still for rose water, with eight stills in parallel and central heating.

Typically, heating was performed with hot water or a sand bath, warm dung, or the sun. Cooling was usually performed by air or wet linen-covered sheets. This development can also be seen on the right side of Figure 1.2, where other types of stills are depicted. All have in common an additional cooling by water. The top vessel is covered by fleece or a ball of floss above the liquid to be distilled (4).

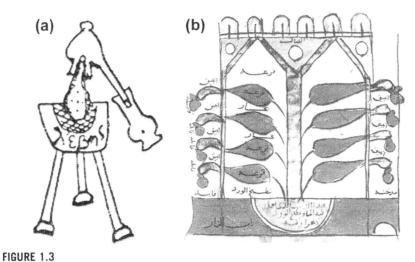

FIGURE 1.3

(a) Still of Democritos, also called the alembic of Synesios, λεβηζ: lebes, which originally meant kettle, in which water was boiled [1], fourth century AD [11]. (b) A distillation plant in Damascus consisting of multiple units for producing rose water, thirteenth century [1].

Both Pliny the Elder and Dioscorides described the use of this device for production of turpentine or on ships to obtain potable water from sea water [11].

Separation of the cooling water and the condensing vapor would lead to the next setup. This can be found, for example, as Mongolian stills (5). The collecting cup was held centrally within the still body in a variety of ways [10]. The Chinese still added a side tube to the cup and receiver. Around the eighth century AD, the Mongols and Chinese possessed sufficient knowledge to distill fermented horse milk for liquor called karakumyss. The alcohol content was controlled by the amount of cooling water added. The Chinese used fermented rice, millet seed, or barley to produce liquors (6). Another interesting technique from China and the Far East involves the production of camphor—the dry distillation of herbs and the sublimation of crystals in the cold part, a technique that is still used in this form today [10].

The "Moor's head" helmet or still top, in which cooling water surrounds the Hellenic annular rim and side tube (7), is shown in the middle part of Figure 1.2. This type was typical in medieval times and remained the favored technique until the beginning of the nineteenth century. The bottom part of Figure 1.2 describes further development steps; starting from the left: retort with cooled receiver (8a,b) deriving from the Gandharan (Persian-Indian) tradition and in China in the Ming period (fourteenth—seventeenth century) [10]; followed by a dephlegmator in medieval Europe (9a); a second vessel connects the cooled still-head and the receiver to condense the less volatile fractions and separate components of the distillate. On the rightmost side is a cooling condenser applied to the side tube of the still (9b), with no additional cooling at the head.

A dephlegmator is used for partial condensation and reflux of a high-boiling component ("phlegma" means stays calm and at the bottom), which is concentrated as the bottom product. The light boiling product, the spirit, is enriched in the dephlegmator and condensed at the end or in the receiver.

One of the oldest graphical images of distillation equipment is depicted in Figure 1.3(a) and originated from the Alexandrian philosophy schools in the first centuries AD. The still is typical for the Hellenic-Egyptian period in Alexandria and remained in use from AD 100 to 900, which overlaps with the Arabian period (AD 700–1600). The apparatus consists of four elements that are also used today: head (helmet—*alembicum*), receiver (vessel—*recaptaculum*), still ("cucumber"—*Cucurbita*), and sand or a water bath on a tripod. The denotation of the *Cucurbita* for the boiling vessel may originate from the usage of hollow cucumbers or pumpkins in early times [11]. Some elements such as the sand bath or the helmet are still found in current equipment. The materials employed vary from earthenware, glass (mainly for the helmet and receiver), and metal such as bronze, tin, or copper (mainly for the *cucurbit* or larger apparatus).

Forbes [1] divided the Alexandrian alchemists into three groups:

1. The followers of Democritus the Younger (second century AD), who focus on metal treatment and coloring, including Zosimos (around AD 350–420).
2. The school of Maria the Jewess and Comarius, including Hermes Trismegistus and Cleopatra the alchemist, who frequently employed an apparatus for distillation and sublimation.
3. Fragments of writings from Pamnenes and others, which, unfortunately, are too small to be classified.

The major achievements in this period, according to Forbes [1], were the discovery and elaboration of distillation and sublimation, and probably extraction as well; the design of prototypes of our present chemical apparatus; and the collection of new facts on the properties of materials. Nevertheless, the early activities have some limitations, such as the ancients' lack of knowledge about solvents (besides water) and mineral acids (acetic acid was probably the strongest acid). Additionally, the sealing of the equipment was not perfect, and so lighter fractions could easily escape. Sealing was done by linen sheets (or clay) but did not allow higher pressures. Metal was used only exceptionally; glass and earthenware were dominant until modern industrial applications.

1.3 The alembic, the arabs, and albertus magnus (AD 700–1450)

The Arabs adopted the technology of the Alexandrian and Syrian chemists, mainly for the production of perfume, rose water, and oil, as well as for medical substances. Geber, the author of the most important chemical book in the Arab period, recommended glass due to its nonporous surface and wide chemical resistance. He probably

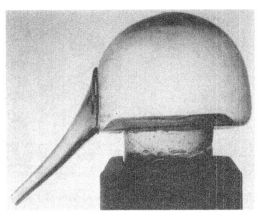

FIGURE 1.4 Glass Still-Head (Alembic) from Alexandria, Egypt (Fifth−Eighth Centuries AD) [10]

lived in the second half of the eighth century, and his Arabic name was Abū Mūsā Dschābir ibn Hayyān. His work was translated into Latin (*Summa Perfectionis*) and copied by many authors until the fifteenth century. Forbes [1] does not accept this theory; instead he argues that two separate series of legends from southern Spain and Syria were interwoven to make the figure of Geber. The Arabs did not contribute any further developments to the equipment and used the traditional distillation equipment mainly for rose water, perfumes, and medical substances. A glass alembic from this period found in Alexandria is displayed in Figure 1.4.

Nevertheless, Arab knowledge spread to the southern part of Europe. Centers of Arabic influence included Salerno (AD 900−1100), Cremona, Seville, Venice, and Murano. Venice and Murano became famous for their skilled glass crafts and trade with the Levant and Egypt. Salerno was famous for its medical school in the eleventh century (see Figure 1.5). Barthelomaeus of Salerno is said to have written a booklet on distilled waters in the first half of the twelfth century [1].

Albert the Great (Albertus Magnus, 1193−1280; Cologne) worked on the distillation of wine and other spirits. He considered distillation a most important method in alchemy: "the alchemist requires two or three rooms exclusively devoted to sublimations, solutions and distillations" [1]. Albert or one of his pupils describes two methods of gaining alcohol (*aqua ardens*). First, "When wine is sublimed like rosewater a light inflammable liquid is obtained." The second method sounds much like a chemical recipe: "Take thick, strong and old black wine, in one quart throw quicklime (calcium oxide), powdered sulfur, good quality tartar (calcium or potassium tartrate) and white common salt, all well pulverized, then put them together in a well-luted (sealed) cucurbit with alembic; you will distill from it aqua ardens which should be kept in a glass vessel" (translated by Forbes [1]). Here, too, glass is the material of choice, which is also recommended by Arnaldus Villanova (1233−1312), the Spanish alchemist. With alcohol distillation, water cooling of the alembic was introduced in Europe. Both developments led to a wide use of

(a)   (b)

FIGURE 1.5 **Medieval Book Miniatures from the Medical School of Salerno:** *The Articella*

(a) Physician with a flask containing a liquid, possibly comparing it to pictures or descriptions in a book. (b) A physician displaying a flask to a student [12].

Courtesy of the National Library of Medicine.

alcohol in medicine and pharmacy in the fourteenth and fifteenth centuries. During that period, the Great Plague swept across Europe (especially beginning in 1347), wiping out one-third of the entire population of the continent. Medicines of all kind were needed, and alcohol served as a solvent as well as an active agent (aqua vitae) for disinfection or circulation stabilization.

Around 1300 the distillation processes were classified according to two methods: per ascensum and per descensum. The first method means rising vapor, while the second describes downward-flowing vapor. Both methods and descriptions were accepted into the eighteenth century, but the per-descensum method is now rarely applied. It was mainly used for the dry distillation of solid substances such as woods, barks, and herbs. The ascending method follows the vapor path and is realized in nearly all current processes.

1.4 Printed books and the rise of science (1450−1650)

The invention of printed books with movable letters at the end of the fifteenth century led to a wider dissemination of knowledge and further promotion of inventions. With this invention, effective and rapid multiplication and distribution of knowledge was now possible. The printing press also led to the rise of the vernacular languages such as French, English, or German. Until then, the scientific (and communication) language had been Latin, which only a few people understood. Now that more people could read and understand the new methods, the apparatus and the recipes also began to be developed. People with more technical backgrounds now had the chance to access traditional knowledge which hitherto had been exclusive to the monasteries and noble courts. The main authors on distillation with publication dates were Michael Puff von Schrick (1481) [13],

FIGURE 1.6 "Rosenhut" Still for Alcohol Distillation, Puff von Schrick [13]

Please note the pieces in front of the oven, the flasks on the shelf, and other details.

Hieronymus Brunschwygk (1500) [14], Philip Ulsted (1526) [15], and Walter Ryff (1545) [16]. These books provided knowledge of equipment and design details, as well as many recipes to produce a multitude of "distilled waters" from plants and animals.

Puff von Schrick [13] wrote a small booklet (under 30 pages) on medical liquids and tinctures for the heart, stomach, liver, head, and eyes. In a later edition, an illustration of an herb witch in front of a Rosenhut still appears on the first pages (see Figure 1.6). She holds a bellows in her hands to heat up the still. On the floor in front of her are scattered pieces of coal and herbs. The board behind her contains several flasks, which were probably used as receivers. The pure air cooling of the head limits the size of the still and its throughput.

Obviously, many pharmacists and apothecaries used advanced methods of distillation to concentrate herb extracts or to mix alcohol with other ingredients. They put the apparatus directly in herb gardens, as can be seen in the title figure from Brunschwygk [14] (see Figure 1.7).

FIGURE 1.7 Title Page of Brunschwygk's Book on the Art of Distillation [14]

Displaying an herb garden with two stills.

Two different distillation setups are shown in the corners: a Rosenhut still with a man filling an empty flask in the top left corner and with a man operating an alembic still in the opposite corner. In the top right corner, a man is drinking from a small bottle, probably the distilled product; note his legs and satisfied face. The lower part of the illustration shows a garden with many different plants, harvested by two women and a man, probably the farmer with a hoe in his hand. On the lower right side, a young man with a floral wreath in his hair, like the young women harvesting, is heating a distillation vessel with an alembic on top. While heating with his right hand, he is testing the temperature of the upper part of the cucurbit with his fingertips.

Brunschwygk's book consists of two parts: the first 60 pages are on distillation technology, and the second part, with more than 400 pages, contains a detailed description of many herbs and other materials to be distilled. This part can be treated as an encyclopedia of the pharmaceutical knowledge of the time. The medical indications of Brunschwygk are compared with the modern standards in the work of Heike Will in 2006 [17]. This knowledge, together with other contemporary works, is collected in the work of Pfeiffer [18]. He describes the multitude of equipment from 1500 to 1900 in pharmaceutical applications, classified in alembic, Rosenhut, retort, and receiver. He rebuilt and tested some of the old apparatus, together with their setup and related recipes with herbs and other materials. His experiments with modern analytics gave interesting insights into the quality of distillates depending on the form and material of the stills. High-quality distillates were gained from stills with low thermal stress and rapid product release in alembics and in retorts with straight, downward-bowed tubes. A common practice at that time was the long-lasting distillation and reflux in a closed vessel, called circulation of the distillate. This operation sometimes took days and often led to the loss of the low-boiling components or decomposition of heat-sensitive products.

Ulsted's book [15] was written in Latin in 1526 and contains many pictures similar to Brunschwygk's. Another important book on the art of distillation was published by Ryff in 1545 [16], which treats mainly the medical applications of distillates. Multistep distillation was already described by Conrad Gesner [19] in 1555. Georg Agricola, in his famous book *De Re Metallica* [20] in 1556, described the distillation and concentration of mineral acids such as sulfuric or nitric acid. These acids were used for the processing of ores and the separation and purification of metals. Due to the large amount needed, the stills were arranged in parallel around the heating source, called *Fauler Heinz* in German.

The various books published on the art of distillation led to the rapid development of the apparatus and its variations. The innovations focused on heating and cooling methods, which are critical for performance. The early books of the sixteenth century mention three different heating methods: the air bath (or direct heating), the water bath, and the sand and ash bath. The first method was used in larger ovens, with many stills positioned around the heat source, as depicted in Figure 1.4 (right) or in Figure 1.8 in the background. The ring-wise arrangement was sometimes added to a terrace-like cone with the fire in the middle. Here, a smaller glass still could be arranged in parallel to increase the throughput or to distill different products. The heating was controlled by air drafts in the chimney. A central funnel was useful for coal refilling in continuous distillation over several days. Sometimes quite strange heat sources were used, such as a trough of horse dung combined with quicklime or fermenting fruit waste. In hot climate zones, hollow mirrors bundled the sunlight, which was focused on the still. This was often used on ships to produce drinking water, where an open fire was too dangerous.

The (al-)chemical laboratory in Figure 1.8 depicts many unit operations such as the mechanical treatment of seeds (oil press), the distillation oven with Rosenhut stills, as well as mortars and pans to mill and mix solid material. In the background is a large still with four visible alembics operated by two persons: one is filling coal in the center funnel and the second is refilling one still with some liquid. The center

FIGURE 1.8 Jan van der Straat, Laboratory of Grand Duke Francesco I (1570), with at Least Seven Distillation Units [21]

With kind permission of Springer Science + Business Media.

of the images is occupied by a technically advanced still with spiral dephlegmator, intercooler (ring) with water filling, second serpentine dephlegmator, alembic, and a large receiver from blown glass. The still has a long inlet tube, closed by a tiny plug. On the right side is a tiny valve whose function is not quite clear. Most of the materials seem to be glass or earthenware, while the still with the hottest part could be made from copper. These devices could tolerate higher temperatures and contain larger volumes due to their higher mechanical strength.

Unfortunately, copper tends to spoil the product and may even release toxic materials. Hence, copper stills were often internally lined with tin for an inert surface [13]. On the positive side, the catalytic effect of copper can decompose traces of sulfuric compounds and give a better taste of the distillate. Johannes Krafft (1519–1585) warned against using copper equipment in acetic acid distillation. The Parisian medic Ambroise Paré (1510–1590) noticed that alcohols had a milky color when distilled in lead equipment [11].

In Figure 1.8 (right-hand side, rear), another still with alembic and glass receiver is visible. Behind this still in the background, probably in another room, a man is standing on a ladder and reaching a bottle downward. The ladder leans against caseboards, indicating a kind of early high-rack warehouse for the storage of chemicals.

Already during the dry distillation of wood, it was found that duration and heating strength result in different distillates [11]. In glass stills, it was observed that the residue needs higher temperature to separate the last components. This "fractional" distillation was employed in the serial switching of three or four receivers, which were heated again with milder conditions. The typical products were different tar fractions from wood or essential oils with different molecular weights. Wood tar distillation is an old process with niche applications: The heavy fraction of resinous wood was used for sealing ships and roofs, windows, textiles for weaving, shoe making, and other functions. The coking of wood gives not only charcoal, but also tar, pitch, and gaseous products. The gas is condensed to tar and oil, which are collected and further refined. The oil produced was split into light oil and turpentine (turps) and used for machine oil, for example. A quite modern setup is given in Ref. [22] with a three-vessel setup for wood acetic acid.

Proper operation requires the tight sealing of the distillation parts. This sealing was often done by a cement or filler made from eggwhite and wheat or clay mixed with glass powder, iron cuttings, or lead glance. The cement was brought onto linen sheets and pressed into the gaps and joints. The first attempts at thermal insulation were reported by Raimond Lullus (1235–1315), who clad the glass vessels with clay mixed with hair. Clay with wood chips had a similar effect.

The first steam distillation was probably performed by Claude Dariot (1533–1594), who heated the vessel part between bottom and head. Direct steam distillation was probably done quite early by Chinese chemists starting in the seventh century AD [10]. The Chinese used a water bath and led the steam through a bamboo grid carrying the distilled goods. The steam dissolves lighter fractions and is condensed in a cold vessel containing the dissolved material. Johann Wecker (1574) placed three and more (up to 13 estimated) stills over a large fire-heated vessel with boiling water. The rising steam gently heats the cucurbit. Johann Glauber (1604–1668) already separated the heating source from the still, as it is standard in modern, continuously working distillation columns. He is considered one of the founders of modern chemistry, especially for its technical realization.

Cooling is the second part of the distillation process dealing with heat transfer. From the beginning, air cooling was the standard method, and remains in use today for smaller distillation columns. Air cooling limits the size, throughput, and selectivity of the process. In the Far East from time immemorial, cooling was assisted by a water jacket on top of the still-head. When it reached a high temperature, it was replaced by fresh, cooler water. Later, continuously running cooling water was used for cooling either the head or the outlet pipe of the still. The first setup is also called Moor's head (see Figure 1.2) and can be found in many contemporary illustrations. Water cooling of a spiral glass tube was invented around 1250 in Florence and is still in use today. The local invention did not mean that the technique was used throughout Europe. Often, these details were kept secret and were not spread any further, especially not to competitors. According to Brunschwygk [14], there were several different cooling methods ranging from pure air cooling by an enlarged area of Rosenhut head or a water drum with the outlet tube going through. The central still in Figure 1.8 has a partly water-cooled setup: The ring in the middle can be filled with water and lead to a partial reflux of higher condensing components. Beginning in the sixteenth century, the first stills were equipped with continuous water cooling. Control of the head temperature by different cooling techniques was important for the performance and selectivity of the distillation. Besides the cooling medium, the form of the riser tube was varied with a zigzag or spiral shape. Nicolas Lemery (1645−1715) described the riser tube as follows: "The tube must be of sufficient length that only the subtle material arrives to the head. The serpentines serve to refine the spirit, because the 'phlegma' is not able to enter the head and drops back to the ground" [11]. The first countercurrent flow for cooling was described in 1770 by Jean Pissonier.

1.5 From laboratory to industry (1650−1800)

The 1650−1800 period is characterized by the installation of complete laboratories for chemical investigations and large production of alcohol and mineral acids. In 1595, Andreas Libau (Libavius) described in Latin the general setup of a chemical laboratory [23]. He distinguished the rooms according to their functions and chemical operations, and he used a tube system for "tapped water". He also categorized the distillation equipment according to heating, cooling, and vapor flow direction (ascending or descending). He still adhered to some alchemical traditions; for example, he described the distillation of acetic acid with circulation for hours and later separation, although he knew that during acetic acid distillation, first the water fraction evaporates and only the last third can be used [11].

Figure 1.9 shows the typical late medieval distillation setup, which was standard in France, Germany, and England. The letters in the left-hand image signified the following: A, the head (alembic); B, the sealing and connection tube to the still; C, the still; D, the heated water bath, steam, or sand; E, the oven brickwork with fuel windows; F, the steam outlet of the water bath in D; and G, the receiver. Elsholtz wrote his book in Latin [25], and he presented many illustrations of herbs and trees. Only one alembic is depicted in his book (shown in Figure 1.9, in the (b) panel). The right-hand side of the figure shows a cut through the alembic with a fire-heated stirred vessel [26] from the eighteenth century. The outlet tube goes to the left side and is cooled in a water drum (not shown here). The devices were easy to disassemble for filling and emptying as well as for cleaning and maintenance. Typical dimensions are for a height of 0.9−1.2 m and a diameter of 45−75 cm. Iron, lead, and copper became more popular for distillation equipment, especially for large devices due to their higher mechanical strength and better handling capability. Friedrich Henglein [27] showed a similar device for mobile distillation of spirits (Slivovitz) in the mid-twentieth century in Yugoslavia [27].

During the sixteenth century, scientific methods were introduced in laboratories with standard equipment. In his book *Novum Organum* (1620) [28], Francis Bacon

(a)   **(b)**   **(c)**

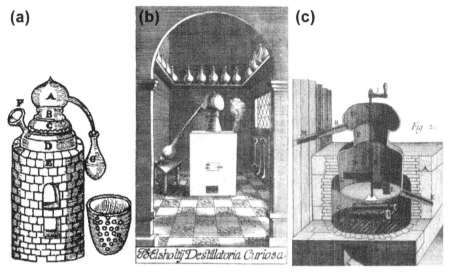

FIGURE 1.9 Standardized Distillation Still with Air-Cooled Alembic

From (a) John French (1651) [24]. (b) Johann Elsholtz (1674) [25]. (c) Anonymous from eighteenth century [26].

(1561—1620) for the first time described scientific methods for experimentation and presented related theoretical explanations. Robert Boyle (1626—1692), often called the father of modern chemistry, conducted systematic distillation experiments. He used fractionating distillation for analytical purposes. He distilled alcohol with fragments of acetic acid over several days and found that at the end of the process the liquid had a higher acid content than did the starting material. Boyle concluded from his experiments that acetic acid is evaporated at higher temperatures and can be separated from the low-boiling alcohol. He was therefore the first to recognize fractional distillation as a separation method for different boiling liquids. He worked under vacuum and elevated pressure conditions.

Johann Kunckel von Löwenstern, in his book *Laboratorium Chymicum* (1716) [29], describes the then-current knowledge of lab chemistry, but mixed with many alchemical elements such as the transmutation of metals. von Löwenstern was a glassmaker, apothecary, and chemist, and worked experimentally. Based on his intensive knowledge of different materials and processes, he preferred glass equipment in the laboratory. Knowledge of glass blowing and handling was important for chemists at that time [29].

During the early industrialization era, sulfuric acid played an important role as a bleaching agent. The number of mechanically fabricated linen or cotton sheets increased dramatically and had to be prepared for dyeing. This led to a bottleneck of meadows, as sunlight was the traditional method. The company of Roebuck (1718—1794) and Garbett (1717—1805) produced sulfuric acid in larger amounts. The price of the product was so high that stills made from platinum were used

[30]. This enormous development led Justus Liebig to make the following statement in 1843: "The economic development status of a country can be well described by its consumption of sulfuric acid (vitriol oil)."

Alcohol distillation gained more importance, especially with sugar cane fermentation and rum distillation in the Caribbean and South and Central America. Alcohol distillation and the increased alcohol content made it easier to transport and distribute the liquor. The Society of Rectifying Distillers [31] played a major role in technology development and distribution in the British colonies.

Figure 1.10 presents a typical chemical lab in the eighteenth century with a multitude of different distillation apparatuses. It shows the majority of chemical equipment known at that time, since not only separation but also chemical transformation occurred in retorts and stills. Antoine Beaumé (1728–1804) had already introduced connectors with plugs for filling, emptying, and connecting different devices. The standard glass grinding connection was introduced around 1900. Before that time, almost every laboratory or manufacturer had its own connection system.

Around 1800 Benjamin Thompson, Count Rumford, described the benefits of steam-heated distillation. In this type of distillation, the stills are rapidly heated up, and the bottoms of the glass bottles are less worn and not destroyed as rapidly. The gentle heating avoids decomposition of the distilled material. Beginning in the nineteenth century, a theory of heat was developed. Count Rumford set up a relationship between heat and friction and could show the transformation between different energy forms. The concept of latent heat was introduced by Joseph Black around 1750 and involves the heat of vaporization and condensation in comparison with sensible heat, which is accompanied by temperature change. A patent for steam heating of stills was given to Charles Wyatt in 1802 for distilling coffee. Birch

FIGURE 1.10 Typical Lab Arrangement in the Seventeenth and Eighteenth Centuries

Please note the triple alembic on the left side and the four-step receiver on the right side. Colored copper engraving from D. Diderot, J. d'Alembert [32].

was used in 1818 for the first time in a steam-heated jacket for heating technical equipment [11]. The book by the pharmacist Zeiser (1826) in Altona (now Hamburg, Germany) introduced steam as a safe and easy to use heating medium in chemical laboratories [33].

1.6 Scientific impact and industrialization (1800—1900)

In his first book on *Technical Chemistry* [34] in 1936, Friedrich Henglein stated that France was the leader in pure chemistry around 1800 but that England was leading in commercial-technical chemistry. In the chemical world exhibition of 1862, England was indisputably the world leader in the chemical industry. But in the 1893 world exhibition held in Chicago, Germany took the lead from England, in part because Germany had attained unification in 1870 and had managed to bundle its forces. It also became easier for Germany to engage in the exchange of goods and information. Another reason for Germany's new leadership role was the strength of its universities in chemistry research and teaching as well as the high standards of its engineering schools for science-based innovation.

The increasing amounts of mainly alcohol and mineral acids in industrial distillation led to a widespread growth of distillation equipment. In France many developments were initiated, based on alcohol distillery. Napoleon set a prize on sugar beet production and fermentation. Robert Forbes [1] described a series of patents issued from 1801 to 1818. Jean-Édouard Adam developed a discontinuous apparatus for fractionating distillation, which was further developed by Isaac Bérard with partial condensation. The work of Adam and Bérard led to the formulation of the two following principles [1]: (1) enrichment of a low-boiling component in the rising vapor by good contact with the downcoming liquid and (2) enrichment of the vapor by partial condensation and reflux into the still. Both principles combined led to the continuously working distillation patented by Jean-Baptiste Cellier-Blumenthal (1768—1840) in 1813. This distillation column is illustrated in Figure 1.11. The upper part of the column contains bubble-cap trays, while the lower part is structured by conical metal caps that also serve for contacting vapor and liquid. This setup was the basic one used for developments in France in the following 60 years, and it influenced constructions in Germany and England.

In 1817, Charles Derosne (1780—1846) also built a continuously working distillation column and brought it to industrial maturity [1]. His economic success relied on further industrial products such as components in sugar refining plants, locomotives, and other railway equipment. Anthony Perrier got a patent on "baffles" as tray construction in a whiskey distillery in 1822 to enhance the contact between vapor and liquid phase. Baffles look similar to current bubble-tray caps and inserts. A little later, in 1830, Aeneas Coffey developed perforated trays as sieve structures for vapor—liquid contact. Coffey's sieve plate columns had a distance of 6 in (15 cm) and more [36]. The sieve plate was primarily developed for higher viscous liquids, but it was not successful. Today, sieve trays are used for nonfoaming, low-viscous

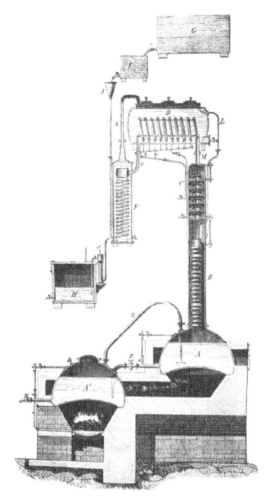

FIGURE 1.11 Continuously Working Distillation Column for Wine by Cellier-Blumenthal, Patented 1813 [11]

The arrangement consists of two vessels, the stripping region with conical metal sheets, the enrichment section, the partial condenser (dephlegmator), and the water-cooled condenser [35].

liquids such as liquefied air-separation columns. Higher viscous liquids could successfully be treated by the column, patented in 1854 by Henri Champonnois with bubble-cap trays, similar to current technology.

At nearly the same time as Cellier-Blumenthal, Heinrich Pistorius (1777–1858) invented his distillation plant in 1817 for alcohol from fermented potato mash (see Figure 1.12). Potato cultivation was promoted by the Prussian government started by Friedrich II in 1745. Brandy from fermented potatoes became very popular in

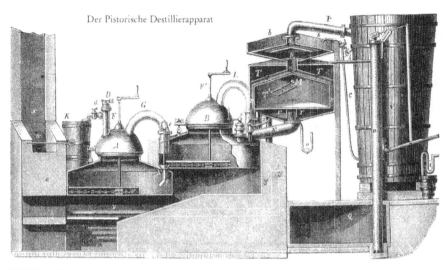

Der Pistorische Destillierapparat

FIGURE 1.12

Two-vessel still by Heinrich Pistorius—Prussian patent of 1817 [26], often used in small alcohol distilleries.

Germany in the beginning of the nineteenth century. Hence the Pistorius still was widely used in Germany until 1870 and produced liquors with an alcohol content of 60–80%.

Pierre Savalle (1791–1864) was an early co-worker of Cellier-Blumenthal and conducted his first experiments in a sugar refinery. Together with his son Francois Désiré Savalle (born 1838), Pierre Savalle further developed the equipment and performed many detailed improvements in cooling and heating [37]. The throughput was seven to eight times higher than in the Cellier column [1]. Savalle's simple, continuously working apparatus was especially used for alcohol made from cane sugar.

In his book written in 1873 [38], Savalle points out how more than 16,000 distilleries are working in Germany, but only 700 in France. He describes different kinds of equipment for alcohol distillation from mashes of molasses, sugar beets, malted grain, and potatoes to cane sugar and wine. From wine, 100 l of alcohol are produced with a 40 kg demand of coal. The Savalle columns typically contain 30 trays and reflux two-thirds of the total distillate. The columns produce 96% alcohol with a daily capacity of 500–20,000 l. Savalle also described the rectification of methyl alcohol and the fractionation of crude benzene. Two columns of crude benzene were built at Ludwigshafen for BASF [1] at that time.

Another pupil of Cellier-Blumenthal was Auguste-Pierre Dubrunfaut, who also improved the Cellier column. He designed and built columns with a diameter of 80–100 cm, with a daily throughput of 50,000–120,000 l and a yield of 200–4800 l alcohol (92–94%). He also wrote a book on distillation [39], which is one of the best sources for study of distillation in that period [1]. His book was translated into English and largely influenced British developments, especially for whiskey and rum distillation. One of his pupils was Champonnois. Dubrunfaut's columns were used as models for German companies such as Heckmann [1]. The senior engineer and designer of Heckmann Company was Eugen Hausbrand, whose activities are described in the next section.

The second product besides ethanol, by which the innovation in distillation is guided in the nineteenth century, was waste from the coal industry. Long before, in 1658, Johann Glauber had already described oil formation (black oleum) during coal retorting [40]. In the late eighteenth century, dry distilled coal (coke) was used for the first time for iron production and melting. Solid, liquid, and gaseous by-products were primarily dumped into the environment. Coke gases were found to burn with a bright flame and were introduced as private and public lighting. In 1826 gas lighting was introduced in Berlin; it had been introduced in London in 1807 and in Paris in 1822. Coal tar, the highly viscous waste, was produced in the gas factories for lighting and coke plants for steel furnaces, as well as railway fuel and heating. In 1822, the first industrial tar distillation plants were built in Britain [40]. In 1823, Friedlieb Runge discovered phenol and aniline in the coal tar and established the foundation of tar chemistry for colors and pharmacy. These products have to be separated and purified by distillation. The environmental impact of the gas and coke plants was recognized quite early, and the off gases had to be collected and processed. Ammonia, benzene, phenol, and other aromatic compounds in the coke gas were especially useful for further applications and were separated by washing, absorption, and distillation. All these processes, which are called the white side of coke plants, need gas—liquid contactors. The first absorption and distillation equipment was made from wood with stacked wooden internals [41]. Other materials were dumped, and there were unstructured rocks, coal, or coke. Ammonia distillation towers looked similar to alcohol plants (see Figure 1.13). The trays are similar to those in the Cellier column (Figure 1.11). Benzene washers were similar to the ammonia distillation columns [42].

In 1842, the first tar distillation in Germany was started in Offenbach near Frankfurt. Julius Rütgers built in 1860 a plant in Erkner near Berlin [40], followed by other plants in Dresden, Katowice, Vienna, Munich, and Rauxel. The tar was primarily used to soak railway sleepers, telegraph poles, and other wooden construction material. The high-boiling fraction or residue was used as pitch, as tar for street pavement, and as further sealing between stones. The typical composition of coal tar is water, 4—5%; light oil (443—453 K), 2—4%; middle fraction (513—518 K), 10—12%; heavy oil (543—548 K), 8—10%; anthracene oil (573—613 K), 18—25%; and finally pitch, 50—60% [43].

A comprehensive book on coal tar distillation was written by Georg Lunge in 1867 [44]. Lunge was born in 1839 in Breslau (Wroclaw), and he studied chemistry there and in Heidelberg. From 1864 to 1876 he worked in England, starting as chemist in a coal tar distillery, and then progressing to manager of a large soda factory in the Tyne district.

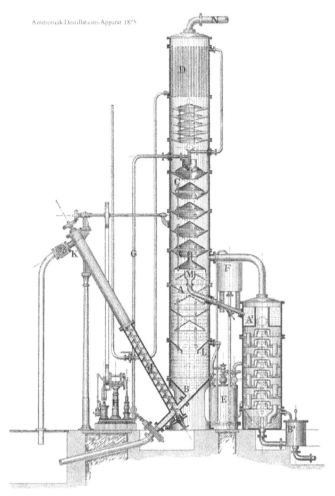

Ammoniak-Destillations-Apparat 1878

FIGURE 1.13

Ammonia distillation tower with a combination of dephlegmator and bubble-tray column, tall column with conical contactors, and right column with bubble-cap trays [26].

In 1876 Lunge was appointed head of technical chemistry at the Polytechnikum in Zürich, now ETH. He wrote several books on the tar and soda chemical industry, on analytical methods, and on the history of the Swiss chemical industry. He is considered the father of technical chemistry, bringing practical knowledge into academic circuits. He wrote on quantitative methods to design distillation plants and mentioned cost calculations. Lunge's contribution opened British technology to continental Europe and inspired new ideas. The young petroleum industry benefited from these developments, since the first petroleum stills looked like those shown in Figure 1.14(b). Phenol chemistry requires that components have high

(a)   **(b)**

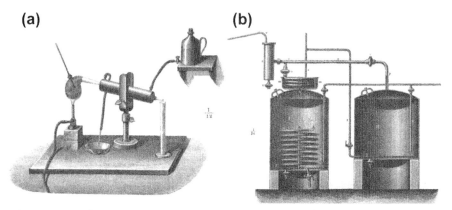

FIGURE 1.14 Equipment for Coal Tar Analysis and Distillation (1867) [44]

(a) Analytical setup for tar components, glass retort heated by a Bunsen burner and controlled by a thermometer; the outlet is cooled by water in countercurrent flow and collected in a graduated cylinder. (b) Distillation setup for fine purification.

purity, and so tall separation columns up to 60 m in height were constructed. This development demanded better construction material, enhanced gas–liquid contacting internals, and improved calculation procedures.

1.7 Engineering science (1900–1950)

In the beginning of the twentieth century, coal tar distillation was established for production of chemical raw materials [43]. Both discontinuous and continuous processes were developed, although the continuous processes are quite complex due to the high viscosity of tar and pitch. Frederic Lennard's invention, for which he had obtained German patents in 1889 and 1891, was path breaking. His still consisted of a long heated tube with gas generation and a typical throughput of 100–150 tons per day. The process was introduced in Westphalia in 1905. In 1916, the company Opitz & Klotz erected a distillation plan in Leipzig, Germany, with a daily capacity of 10–15 t. The plant consisted of four vessels running at 453, 473, 523, and 573 K, fired in a counterflow direction with the tar flow. The pitch at the outlet heated the incoming tar in a countercurrent flow. In 1914, a Raschig plant started in Linz/Rhein at Walter Feld GmbH consisted of three tube bundle heat exchangers and two columns. For the medium fraction, the column was equipped with bubble-cap trays, while the vacuum column had a Raschig ring filling for light oil and naphtha separation [43]. Phenol and pyridine bases were separated and purified under vacuum distillation, while naphtha was crystallized. A combination of different unit operations was employed to separate complex mixtures. (The term *unit operation* was coined by Arthur D. Little at MIT in 1915.)

Eugen Hausbrand (1845–1922), one of the first process engineers [45], published quantitative calculations for rectification and distillation [46]. In his famous monograph [47], he discussed heat and mass flow diagrams in rectification columns similar to the Sankey diagrams used today. Remarkably, his English-translated monograph remains an important source [36]. Hausbrand started as senior engineer at the Heckmann Company in Berlin, which was famous for its selective bubble-cap tray columns. After the death of its founder, Hausbrand served as director of the company for the next 40 years. He practiced the method of discussion to evaluate new technical systems, similar to the HAZOP method (hazard and operability studies) used for current safety evaluations. Hausbrand's other publications deal with heat and mass transfer of process equipment, later continued by Wilhelm Nußelt.

The first calculation for simple batch distillation was performed by Rayleigh in 1902 [36] using material balance and simplification of the relative volatility. Rayleigh also applied Henry and Raoult's law and compared his results with test data [48]. Hausbrand was a brilliant pioneer of distillation engineering [49], inspiring further work on distillation—notably, Robinson in 1922 [50], Lewis in 1923 [45], von Rechenberg in 1923 [51], Thiele and McCabe in 1925 [52], and Mariller in 1925 [53]. In addition, there was Kurt Thormann, who described tray columns and dumped packings, binary and ternary mixtures, as well as some properties of binary mixtures [49]. The last chapter of his book deals with the equipment of alcohol distillation, the best understood system at that time. Other systems are probably still being kept secret (air separation, purification of aromatics, hydrocarbons, etc.) or use equivalent equipment.

With increasing steel and coal production, oxygen demand was also increasing, parallel to the rising demand for nitrogen in chemistry. After the liquefaction of air by Carl von Linde in 1895, it became important to separate oxygen and nitrogen. Raoul Pictet developed his oxygen apparatus in 1899 with a special rectifying arrangement [54] (see Figure 1.15). Through his apparatus, the compressed air is cooled down in a countercurrent heat exchanger. Next the air is further cooled down and partly liquefied in the rectification zone. The heat is used to evaporate mainly nitrogen on the tray levels, which exit at the top of the column. The air is led to the top of the column and expanded in the upper two chambers. The fluid cools down through the Joule-Thompson effect, and precipitated CO_2 crystals are filtered there in order to avoid blockage of the rectification column. The oxygen concentration gradually increases in the downward-flowing liquid, and nearly pure oxygen is taken from the bottom of the column (tube S). Argon-rich air is taken from P and led over the heat exchanger to precool the incoming air.

The Linde oxygen column from 1902 used compressed and precooled air, which is split and condensed in two heat exchangers. Both liquefied air streams are expanded and fed at the top of the glass-bead-filled column. In the dumped packing, nitrogen-rich vapor streams up, while the oxygen-rich liquid trickles down to the column reboiler (upper heat exchanger). The liquid oxygen is partially evaporated and streams over the siphon T to the lower heat exchanger, where the entire oxygen

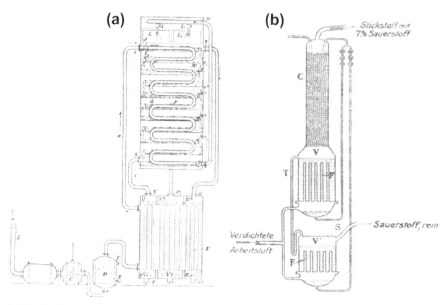

FIGURE 1.15

(a) Pictet oxygen plant [54]. (b) Linde oxygen apparatus with glass bead column (1902) [55].

is evaporated as product stream. Distillation columns with dumped glass beads 4 mm in diameter were tested by Walther Hempel 1881 for lab applications, and in 1890 Robert Ilges used porcelain balls that were 1−2 in (25−50 mm) in diameter in plant columns [36]. He also developed a temperature-controlled apparatus, which was then very popular in small distilleries.

Shortly thereafter, in 1903, Georges Claude (founder of the French company l'Air Liquide) presented an oxygen column with a quite similar setup, but only one heat exchanger. The column internals are bubble-cap trays with downcomers for the liquid on top of the heat exchanger section [54]. In 1907 Linde further improved his oxygen column (see Figure 1.16(a)). Compressed air is fed into the medium-pressure column (a), where first nitrogen enrichment is performed.

The bottom oxygen-rich liquid is expanded and fed into the middle of the low-pressure column. The top nitrogen-rich gas is expanded and fed to the top of the low-pressure column. There, too, the nitrogen exits the column, while oxygen is evaporated at the bottom of the column. No detailed description of the trays is given by Ludwig Kolbe [54], but probably already sieve trays were used. The double column in Figure 1.16(b) is similar to modern air separation units and demonstrates the relatively short path from first application to complete solution.

Two later modifications are noteworthy. In the middle of the low-pressure column, an argon-rich region ("argon belly") forms. This either has to be purged or is led to the adjacent argon purification part. This arrangement can already be identified as thermally coupled or dividing wall columns. Failure to purge this

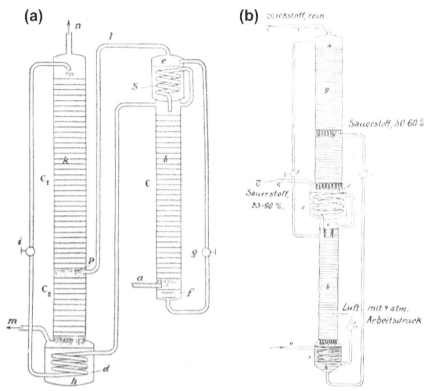

FIGURE 1.16 Linde Development of O_2/N_2 Separation for Pure Components, from Two Separate Columns to a Combined Column

(a) 1907. (b) 1910.

From Ref. [54].

enrichment section can block the entire column and separation process. Another issue is the enrichment of high-boiling compounds in the bottom part of the low-pressure column, mainly hydrocarbons such as acetylene or ethylene. This can be very dangerous and has led to severe explosions in the past. A continuous purge stream again prohibits a dangerous enrichment.

Together with improvements of the peripheral equipment, such as compressors, expanders, and integrated heat exchangers for higher energy efficiency, the energy demand to produce oxygen dropped from 1.5 kW/kg to the current 0.4 kW/kg. At the same time, plant capacity was increased from 1.3 t/h oxygen (98–99% purity) in 1920, to over 5.2 t/h oxygen (99%) in 1950, to 65 t/h oxygen (99.5%) in 2000. More information on the current development of air separation units can be found in Chapter 6 (Air Distillation) of [56].

With increasing industrial chemical production of various products, the number of special applications grew at the beginning of the twentieth century. Acetic acid and alcohol were recovered by direct spray condensation and distillation in a trickling over twisted wood chips [57]. Fatty acids were produced in batch distillation [58]. Glycerol was distilled in a dephlegmator with only three trays [59]. Perfumes and various odorous substances were distilled with water [60] assisted by oil extraction.

Petroleum distillation, which is probably today's largest distillation application, started from batch processes due to its small product amounts at the beginning. The main product was lamp oil until fuel for motor vehicles took over. The first plants looked similar to the still in Figure 1.14, because tar distillation was comparable due to the product consistency, viscosity, and components. With the eruption of the First World War, the demand for oil and fuel increased dramatically. Unfortunately, however, only 20—25% of the crude oil can directly be converted into gasoline or petroleum. The rest is low boiling and needs further treatment such as thermal or catalytic cracking. These processes gained increasing importance and also require many distillation steps and columns [61]. In the 1910s, thermal cracking units were installed to produce lighter oils from the long-chain hydrocarbons. In 1916, a Kubierschky apparatus was used for steam-assisted distillation of petroleum [62]. During the 1920s, gasoline production switched from batch to continuous distillation. The columns were mainly equipped with bubble-cap trays or similar gas—liquid contactors.

Measurement and prediction of mixture properties, especially vapor—liquid equilibrium, are very important to making accurate predictions of separation characteristics. Hausbrand, in his third edition (1916) [63], had already presented graphical data on the vapor—liquid equilibrium properties of eight nonaqueous mixtures and six aqueous mixtures. Thormann included ternary mixtures and a generalized graphical treatment of binary mixtures with different partial pressure and solubility characteristics [49]. Other authors included more and more data, collected in the Dechema Chemistry Data Series [64]. In 1969 Ulfert Onken proposed good test mixtures for characterizing column internals, especially packings under vacuum conditions [65]. During this period, computer programs were able to predict many properties of binary mixtures. The historical description of this development could well form a separate chapter on chemical thermodynamics; more information can be found in this handbook in the chapter on vapor—liquid equilibrium and physical properties [66].

Although Hausbrand [63] had earlier presented the mixture properties in vapor—liquid diagrams, it was left to McCabe and Thiele in 1925 [52] to put the operation of a distillation column into a diagram. The McCabe—Thiele diagram allowed a simple graphical display and design of a distillation column dealing with the complex correlation of vapor—liquid equilibrium and mass balances. Only 2 years later the method was mentioned in Thormann's textbook [49]. This reflects the power of journal publications to accelerate the spread of scientific results. In the 1920s, there were three widespread types of rectifying columns or towers with bubble-cap and sieve plates as well as towers with dumped or larger packed internals. Bubble-cap trays have a typical spacing of 6 in to 3 ft (0.15—1 m) in towers with up to 32 ft in diameter (approximately 10 m). In order to compare the different column types, in 1935 Thomas Chilton and Allan Colburn proposed a method of the height equivalent to a theoretical plate (HETP) reaching vapor–liquid equilibrium for packed columns in distillation and absorption [67]. Later, Colburn defined Murphree tray efficiency, which indicates the relative vapor concentration compared to the equilibrium and is a good measure of the efficacy of mass transfer over a tray. In 1942, distillation and rectification methods were standardized in Germany in DIN 7052 [68].

Unstructured packings were used in both absorption and distillation and consist of coke, rocks, or glass or porcelain beads. In 1913, Fritz Raschig patented in Germany rings that were 25 mm in height and diameter made from glass, porcelain, copper, iron, or other resistive material [36]. They exhibited a lower pressure loss and better mass transfer characteristics and could be set up in an arranged packing [34].

Mixtures with a close boiling point or separation of isotopes demand many separation steps, hence resulting in a tall column with intensive mass transfer. Raschig rings were already a step in this direction, but two other lines can be recognized: wire gauze or mesh and stretched metal sheets. In the 1930s, the Atomic Energy Research establishment in Harwell in the UK developed a special column internal called Spraypak filling for heavy water concentration [27]. The packing is made from stretched or expanded metal (see Figure 1.17). It showed good characteristics for high-duty loading for the gas and liquid phase and could be operated with two- to threefold gas velocity compared to bubble-cap trays. Metal mesh could reach approximately two theoretical stages per meter. Liquid loadings are of about 50 l/m^2 to 400 m^3/m^2, approximately two to three times that of conventional columns at that time.

The second development line of structured packing was the Stedman Column in 1936 from Knolls Atomic Power Laboratory, General Electric Company, for isotope

(a)   **(b)**

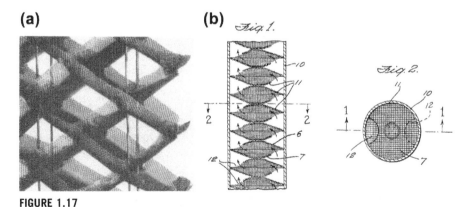

FIGURE 1.17

(a) Spraypak filling of an absorption/distillation column [27]. (b) Stedman column packing made of wire mesh from 1939 [69].

separation [69]. It was developed for low column height, weight, low-pressure drop, and holdup. The packing is made of wire cloth, which is punched, embossed, and welded to form a series of cells. The single-cell conical type consists of a series of stacked conical disks with changing order made of Monel wire cloth with 60 by 40 meshes/in and 0.009 in (0.23 mm) wire diameter. The cones are made in three different sizes from 3/8 in to 50 mm diameter similar to the ones shown in Figure 1.17 [69].

Another type of structured metal sheet packing, the multiple-cell triangular pyramid type of packing, is structured like pyramids, which are located on 3/8-in equilateral triangular centers. The sheets are perforated with 3/16-in-diameter holes located between the pyramids. The setup is hardly visible in Ref. [36], but it appears to be similar to a U.S. Patent of D. F. Stedman from 1935 (see Figure 1.18). The triangular packing type exhibits low-pressure drop and holdup and performs with HETP values of 1.5 in (38−50 mm) and 2.5−3.3 in (63−84 mm) for column diameters of 1.5−6 in (38−152 mm) and 12 in to 11 ft (300−3400 mm) diameter, respectively. Applications are low-boiling paraffin hydrocarbons (C1−C7 fraction), which are separated by low-temperature equipment. The column internals are consisting of a wire mesh guided by structured sheet metal following a proposal of Walter Podbielniak in 1931 [71].

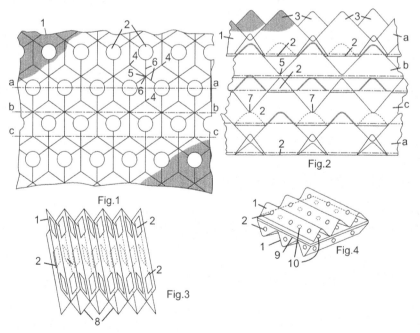

Fig.1

Fig.2

Fig.3

Fig.4

FIGURE 1.18 U.S. Patent for Perforated Metal Sheet, Structured Packing by D. F. Stedman, Toronto, Canada, 1935 [70]

Kenneth Hickman [36] was one of the pioneers in the development of molecular and short-path distillation from 1910 to 1925. Starting with isotope separation, the first application was probably made by Brönsted and Hevesy (1920) to separate the isotopes of mercury. Early molecular stills for labile products were designed by H. I. Waterman, while the "truly molecular still" was first applied to organic substances by B. C. Burch for separation of small carbon molecules [72]. The process was performed under high-vacuum conditions and frequently with a lab-style apparatus. Industrial molecular stills are often combined with a heated centrifuge in the center. The vacuum system is the most complex part with several pressure steps. A typical pump train for a large still consists of three pump stages of steam, two oil boosters, and a condensation system with a capacity of 1000–5000 l/s (see Figure 1.19).

Typical applications in the lab vary from vitamin E determination in food to purifying small samples of drugs, dyes, sterols, or hormones. Industrial applications are mainly combined with centrifugal units, 1–5 ft (0.3–1.5 m) in diameter, which are grouped in blocks of three to seven for multiple redistillation. The main applications are vitamin A esters from fish liver oils, stripping of vitamin E from vegetable oils, and the complete distillation of high-boiling synthetics, such as plasticizers, fatty acid dimers, and similar materials. In 1947 more than 5 M lb (2500 t) in sensible material were treated [36].

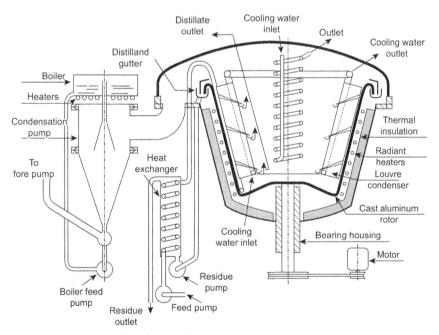

FIGURE 1.19 Molecular Still with Rotating Heat Exchanger and Vacuum System [73]

1.8 Improvements and integration (1950—1990)

Based on the pioneering work of Emil Kirschbaum in the 1930s [68], Reinhard Billet in Bochum, in Germany, worked on optimization studies in distillation, especially in vacuum columns [74]. He worked with several companies to optimize structured packings [75] and fluid distributors [76]. Ernst-Ulrich Schlünder in Karlsruhe combined extraction and adsorption with distillation [77] and worked on the heat and mass transfer and integration in columns. His pupil, Jerzy Maćkowiak, wrote the textbook on structured packings [78] and their transport characteristics. In 1960, the publication of Byron Bird, Warren Stewart, and Edwin Lightfoot's textbook on transport phenomena [79] was a landmark event, shifting the focus from unit operations to more fundamental mass and heat transfer processes [80]. Unfortunately, the book provided no examples of distillation columns; only an example of an absorption column can be found.

Supported by increasing computer performance, Gerhard Schembecker and co-workers developed the conceptual design strategy of process synthesis with detailed numerical simulation combined with heuristic methods and their combinations on different levels [81]. The method was applied for various combinations of reactive separation processes, especially catalytic distillation and its optimal control [82]. Together with the unit operation concept, Hannsjörg Freund and Kai Sundmacher developed the concept of elementary process functions [83] with a view on fundamental transport processes [84] and their combinations on different levels [85]. These novel methodological approaches will allow for better engineering of advanced and complex separation devices and processes.

The development of structured packings, such as sheet metal, was led by smaller companies such as Montz, Koch, Glitsch, Kühni, and Sulzer [86]. They took up the demand of the larger industrial companies and developed specialized solutions in gas—liquid contacting. This development is also reflected in the structure of Perry's Handbook (5th ed.) in 1973 [87]. The distillation chapter by B. D. Smith and Block contains only 60 pages on the thermodynamic derivation of material properties and activity coefficients as well as flash and binary, multicomponent, and azeotropic distillation. Column internals are now integrated in a separate chapter on mass transfer equipment [88]. John Fair and his co-authors give a detailed discussion of gas—liquid contacting trays, bubble-cap and sieve trays, and unstructured and structured packings, such as Raschig rings, Lessing rings, Berl saddle, Intalox saddle, Tellerette, and Pall rings. Stacked rings have higher loading capacity before flooding than dumped. The support and distribution systems for the gas and liquid phase play an important role. Interestingly, in the 1870s information to develop distillation in tar processing came from the experience in absorption; in 1973, both were described together for gas—liquid contacting [88]. And so another ring was closed. Only 10 years later [89], the gas—liquid contacting equipment featured the new arranged-type packings of Sulzer Mellapak or Koch Flexipac. Other packings of this general type include Goodloe, Hyperfil, Kloss, neo-Kloss, Glitsch-grid, and Spraypak.

A comprehensive cost analysis on the cost basis of 1979 [90] indicated that Raschig rings of porcelain are most cost efficient, while Propak and Hyperfil (knit metal), both from 316SS, are up to 100 times more expensive. Pall rings with 1 in (25 mm) diameter in stainless steel are slightly more expensive than Pall rings. Arranged packings are generally more expensive than dumped packings on an equivalent-tower-volume basis [89]. However, they can exhibit a lower pressure drop with great importance in high-vacuum distillations.

In extractive distillation, early work was done from 1945 to 1960 [87] to overcome azeotropic limits or to increase purity. The calculation is quite simple and easier than for azeotropic distillation, for example. An example is given in Perry's Handbook (1973) for the separation of toluene and methylcyclohexane. The chosen solvent is less volatile than the other components and can easily be recovered through a second distillation step from the low-boiling component. Selection of the additional solvent is the crucial part during the process design. A list of 32 candidates is given in Ref. [87] together with their activity coefficients. After a broad screening by functional group or chemical family, the individual candidates are identified by the specific criteria of the application. Salt effects can help promote extraction efficiency [91].

As a result of increasing computer power in the 1970s and 1980s, new cybernetic models and numerical tools were developed to design and operate distillation columns. Together with new vapor–liquid equilibrium property tools, these models made the prediction of distillation processes and column operation easier. The design of more complex systems is also assisted by cybernetic models, for example, degree-of-freedom analysis for control unit operations or homotopy-continuation methods, which lead to better convergence of numerical methods. Many of the methods presented are implemented in software packages of computer-aided distillation-design. Later in the 1980s, Junior Seader and others developed further graphical representations of distillation processes with distillation region diagrams (DRD), residue curve maps (RCM), and distillation region maps for multicomponent and azeotropic distillation [91]. These maps are a powerful tool for understanding all types of batch and continuous distillation operations; for system visualization, process synthesis, and modeling; as well as for analysis of lab data or process troubleshooting. These methods are also applicable to extractive and reactive distillation [92]. More details of the development and a description of the current status can be found in [93].

The first application of reactive distillation was published in 1948 by the Othmer group [94] and concentrated on the esterification of dibutyl phthalate from butanol and phthalic acid [91]. Further prominent examples are methyl-tert-butylether from isobutene and methanol, or nylon 6,6 prepolymer from adipic acid and hexamethylenediamine. Other reaction types include transesterifications, amidation, hydrolysis etherifications, alkylations and transalkylations, and nitration. A well-worked example published in 1984 is the integrated reactive-extractive distillation of methyl acetate production from methanol and acetic acid, which overcomes the chemical equilibrium as well as azeotropes [95]. Conceptually, the column can be treated as four heat-integrated distillation columns (one with the reaction) stacked on top

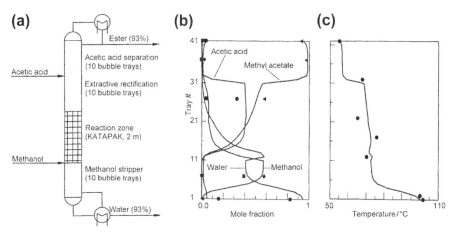

FIGURE 1.20 Integrated Reactive-Extractive Distillation Column to Produce Methyl Acetate

(a) Column with feed positions. (b) Concentration profile. (c) Temperature profile.

Adapted from Ref. [96].

of each other. The primary reaction zone consists of a series of countercurrent flashing stages in the middle of the column [91] (see Figure 1.20). The extractive distillation section above the reactive section is important to achieve high methyl acetate purity. The top rectification stages remove acetic acid from the methyl acetate product, while the stripping section removes any methanol and methyl acetate from water. Temperature and composition profiles for this reactive-extractive distillation column are shown in Figure 1.20(c). The reactive distillation scheme was widely discussed from 1995 to 2003, including in Refs [97] or [98]. Catalytic active packings such as Katapak were developed after 1999 and play an important role in process development.

In 1952, Fractionation Research Inc. (FRI) [99] was founded as a nonprofit research consortium by international petroleum, chemical, and engineering companies. It was established to perform research and investigations that were too expensive for a single company. The experimental site was first in Alhambra, California, with a 4-ft (1.2-m) diameter column. In 1989, the equipment was moved to Oklahoma State University, Stillwater. An historical overview is given in an AIChE paper [100]. In 1984 the Separations Research Program was founded at the University of Austin, in Texas [101]. This institute serves as a kind of national standard laboratory for distillation investigations. It allows for packing testing and distillation investigation in laboratory equipment and pilot plant columns.

With the energy consumption in focus, significant interest has arisen in thermally coupled systems and dividing wall columns for ternary or more mixtures. Two columns are thermally coupled if a vapor or liquid stream is sent from the first column to the second column, and then a return liquid or vapor stream is implemented between the same locations to provide partial reflux or boil-up to this column.

A fractionator with vapor-side stream and side-cut rectifier, also known as a Petlyuk tower, was first discussed by Stupin and Lockhart in 1972 [102]. This arrangement is particularly useful for reducing energy requirements for close boiling components. The dividing wall column is topologically equivalent to the fully thermally coupled system that was first patented in 1933 to produce three pure products from a single column [103]. One of the first industrial applications was the side rectifier configuration for air separation and argon production. By combining the two, we obtain the fully thermally coupled system of Petlyuk from 1965 [104]. Meanwhile, various design [105], simulation [106], and optimization methods [107] have been developed for coupled distillation systems. The fully thermally coupled system uses less energy than any other ternary column configuration [108], which can range from 30% to 50%, depending on the feed composition and volatilities of the components. In 1997, Fidkowski and Krolikowski [109] analytically solved the optimization problem for minimum steam consumption in the reboiler of the main column for the fully thermally coupled system.

Although invented long ago, dividing wall columns and fully thermally coupled distillation systems were not implemented in practice until the late 1980s [110]. The major objections concerned controllability and operability; however, several papers have shown that control of these systems is possible [111]. Computer-aided tools such as the disjunctive programming approach developed by Caballero and Grossmann in 2001 [112] allowed selection of the best thermally coupled column configuration. All the multicomponent thermally coupled configurations have a corresponding dividing wall column equivalent. In 1987, Gerd Kaibel [113] showed several examples of columns with multiple dividing walls, separating three, four, and six components (see a four-component example in Figure 1.21). In 2005, there

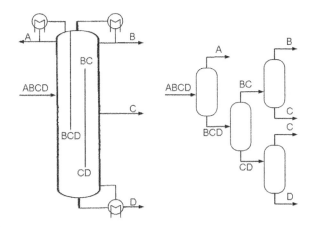

FIGURE 1.21

One possible dividing wall column for separation of a quaternary mixture [110]. The dividing wall column is not implemented in commercial CAPE programs and has to be translated to a similar topological system [114].

were about 60 dividing wall columns in operation; 42 of them were owned by BASF [115], which set up the first column in 1985 in Ludwigshafen, Germany. Nevertheless, the energy-saving issue and increasing robustness will lead to wider application as well as multipurpose use [114].

1.9 What will be the next innovation cycle (1990–2020 and beyond)?

Because of the long innovation cycle in process industries, the outlook starts quite deep in the history. Historical treatment needs distance for a clear overview. At the same time, the development of structured or arranged packing shows the increasingly shorter innovation cycles. The overall goal is increasing separation power per unit volume with better understanding of the physics of mass transfer processes, while decreasing equipment cost [116]. Optimization of the gas and liquid flows in future packing geometry will be of great importance. Distributors will develop more streamlined forms to increase capacity and robustness as well as to reduce plugging and maldistribution. Computational fluid dynamics is an important tool for flow visualization, pressure loss indication, and alignment with measured data. As an example, Luo et al. [117] studied the effect of channel entrance angle on the performance of structured packings in 2008. The inlet pressure drop was minimized, and further improvement of mass transfer could be measured. For gas–liquid contacting, a new random Intalox/metal packing type for high-void fraction was developed [118] with a more robust operation, wider operational window, and good partial load behavior.

Major cost reduction can be expected from improved manufacturing methods of structured packing and distribution manifolds. Standardized parts, modular setup, and a large number of equipment and applications will also drastically reduce cost. Heat integration with heat pumps and divided wall column are accepted mainly in chemical and petrochemical applications.

Dividing wall and heat integration will become more important for complex mixtures and higher throughput with better energy integration and efficiency [119]. Reducing energy consumption by dividing wall columns, which is now well established as a packed column [120], is realized by slow expansion into the applications dominated by tray columns.

A general trend is toward multifunctional packings and their application in combined systems, such as catalytic distillation. Still, it is necessary to have a better understanding of catalytic performance and reaction together with the distillation process [121]. Combined processes are often performed with separate equipment, such as hybrid separation processes made from rectification and melt crystallization to purify close boiling mixtures [122]. Membrane separation was integrated in a distillation column [123]. Another interesting combination is Li's vacuum distillation with biocatalytic active packing in 2012 [124]. Microbial biofilms are grown on the structure surface and employed for ethylene glycol conversion to

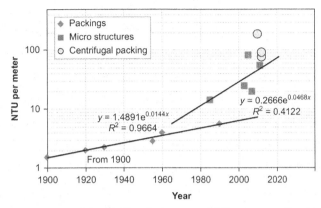

FIGURE 1.22 Development of the Packing Performance (NTU per meter, Number of Transfer Units) over the Last 110 Years

Data adapted from Ref. [127].

aldehydes. In a model study, the design and operation of a catalytic biofilm process were studied in a distillation trickle bed under vacuum conditions for optimum temperature conditions. The description of the development and the current status can be found in Ref. [125]. Another trend is toward use of alternative solvents such as ionic liquids [126].

Figure 1.22 displays the separation performance of dumped and structured packing from 1900 with rocks and Raschig rings over Berl and Intalox saddles to more complex structured Nutter rings and Sulzer structured packing in 1990 [1]. In earlier times, efficiency doubling needed nearly 50 years; see also Refs [128] and [129]. With miniaturized internal structures, an increase in separation performance is described with square symbols. Today only 15 years are necessary to double separation performance. Most of the data were gained with very tiny equipment in the millimeter range [127], from which the results are hard to scale to pilot or even production scale. The combination of microchannel flow and a centrifugal field leads to the highest performance as shown in Figure 1.22, displayed with circular points. In 2010, Jordan MacInnes and co-workers demonstrated a rotating spiral microchannel distillation [129], represented by the topmost circle. The combination of enhanced mass transfer in small channels and high acceleration by rotating equipment leads to this very good result. The other two circles represent experimental data from Wang et al., in 2011 [129,130]; see also Figure 1.23. The performance values were calculated in comparison to conventional structured packing given in their work. For rotating equipment and microchannel devices, the equipment height is no longer relevant. Hence, other performance criteria have to be defined.

Rotating internals in absorption and distillation columns for better mass transfer were already used for coke tar processing [35]. The concept of HiGee in which

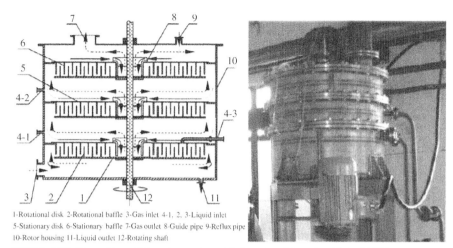

1-Rotational disk 2-Rotational baffle 3-Gas inlet 4-1, 2, 3-Liquid inlet
5-Stationary disk 6-Stationary baffle 7-Gas outlet 8-Guide pipe 9-Reflux pipe
10-Rotor housing 11-Liquid outlet 12-Rotating shaft

FIGURE 1.23 Three-Stage HiGee-Apparatus with Structured Rotating Disks [130]

centrifugal forces are used was introduced in the early 1980s by Colin Ramshaw [131]; Chinese activities started in the late 1980s. Currently, approximately 200 units are operated in laboratories and are in production, probably most of them in China.

With such multiple possibilities, the conceptual design of distillation systems will become more important [132] and serve as a guide through topographic methods selecting optimum configurations. To be successful, innovation needs to combine, compose, and synthesize new apparatuses and parts of them. Industry with both small and large companies and academia will lead equipment development through technology platforms and applications with close collaboration and open innovation forms.

1.10 Summary

Distillation has been important to humankind from the earliest civilizations; distilled products have determined and influenced cultural developments throughout the world. Important examples of such products are ethyl alcohol and petrol. Many brilliant minds have worked on the development of processes and apparatuses even until the present day.

Invented by unknown priests or craftsmen in Mesopotamia 5500 years ago, the first flowering was in Alexandria with the School of Alchemists, who systematically collected the knowledge and investigated new applications. In China, Japan, India, or Mongolia, distillation was known in ancient times and was developed to a certain art. In the Early Middle Ages the knowledge spread from the Arabs over Southern European towns to all of Europe and was used for drinking alcohol, medical products, and perfumes. The invention of book printing gave new momentum for knowledge collection and dissemination and was the base for systematic scientific work in laboratories. The Industrial Revolution initiated a gigantic market pull for mass products, such as mineral acids or drinking alcohol, as well as a distinctive technology push of new materials, fabrication methods, and applications.

The inventions of Cellier-Blumenthal and Pistorius ingeniously combined existing equipment and methods 200 years ago and led to efficient and robust distillation plants. Numerous followers improved the technology and applied it to products from coal tar or mineral oil. More than 100 years ago, Hausbrand developed engineering tools to design distillation columns and equipment, which were improved by McCabe, Kirschbaum, and Onken, just to name a few. New products and applications, complex mixtures or isotope separation, as well as increasing computer performance led to new equipment, column internals, and novel applications. Innovation sprouts from a combination of new and old concepts, while novel products and applications, increased purity, lower energy demand, and changing cost situation are the driving factors.

This chapter shows how ideas come and survive as innovation and novel products. Process intensification and integration of various process functions are the current guideline for science-based development and economic-driven innovations for higher efficiency. Large companies as users, and small companies as flexible suppliers, work closely together with academia and research institutes to push equipment development on technology platforms and its applications.

References

[1] R.J. Forbes, A Short History of the Art of Distillation, Brill, Leiden, 1948 reprint from 1970.

[2] L. Deibele, Die Entwicklung der Destillationstechnik im 19. Jahrhundert, Dissertation TU München, 1992.

[3] A.J.V. Underwood, The historical development of distilling plant, Trans. Inst. Chem. Eng. XIII (1935) 34−65.

[4] F. Ullmann (Ed.), Enzyklopädie der Technischen Chemie, Urban & Schwarzenberg, Berlin, 1914−1922.

[5] Perry's Chemical Engineer's Handbook, third−eighth ed., McGraw-Hill, New York, 1950−2008.

[6] H. Schelenz, Zur Geschichte der pharmazeutischen Destilliergeräte, Miltitz, Leipzig, 1911.

[7] E.O. von Lippmann, Entstehung und Ausbreitung der Alchemie, Springer, Berlin, 1919.

[8] E. Blass, T. Liebl, M. Häberl, Extraktion−ein historischer Rückblick, Chem. Ing. Tech. 69 (1997) 431−437.

[9] M. Levey, Chemistry and Chemical Technology in Ancient Mesopotamia, Elsevier, Amsterdam, 1959.

[10] J. Needham, H. Ping-Yü, L. Gwei-Djen, N. Sivin, Science and Civilisation in China, in: Chemistry and Chemical Technology Pt. 4, Spagyrical, Discovery and Invention: Apparatus, Theories and Gifts, vol. 5, Cambridge University Press, 1980.

[11] E. Krell, Zur Geschichte der Labordestillation, in: E.H.W. Giebeler, K.A. Rosenbauer (Eds.), Historia Scientiae Naturalis, GIT-Verlag, Darmstadt, 1982.

[12] The Articella, Medieval Manuscripts, National Library of Medicine. www.nlm.nih. gov/hmd/medieval/articella.html, (accessed 28.02.14).

[13] M.P. von Schrick, Von den ausgebrannten Wassern, Augsburg, 1481. urn:nbn:de:bvb: 12-bsb00013339−9.

[14] H. Brunschwygk, Das Buch der rechten Kunst zu distillieren, Strassburg, 1500. urn: nbn:de:bvb:12-bsb00031146−3.

[15] P. Ulsted, Coelum Philosophorum, Strassburg, 1526. urn:nbn:de:bvb: 12-bsb00029475−2.

[16] W.H. Ryff, Das new groß Distillier-Buch, wolgegründter künstlicher Distillation, Frankfurt/Main, 1545. urn:nbn:de:bvb:12-bsb00029508−6.

[17] H. Will, Vergleich der Indikationen des 'Kleinen Destillierbuches' des Chirurgen Hieronymus Brunschwig (Strassburg 1500) mit den nach derzeitigem wissenschaftlichem Erkenntnisstand belegten Indikationen, Dissertation Universität Würzburg, 2006.

[18] G. Pfeiffer, Technologische Entwicklung von Destilliergeräten vom Spätmittelalter bis zur Neuzeit, Dissertation Universität Regensburg, 1986.

[19] C. Gesner, Thesaurus Euonymi Philiatri, Ein köstlicher Schatz usw, Zürich, 1555. urn: nbn:de:bvb:12-bsb00020905−4.

[20] G. Agricola, De Re Metallica, Libri XII, Basel, 1556, Marix Verlag, Wiesbaden, 2007.

[21] G. Buchheim, R. Sonnemann (Eds.), Geschichte der Technikwissenschaften, Birkhäuser, Basel, 1990.

[22] F. Ullmann (Ed.), Holzverkohlung, Enzyklopädie der technischen Chemie, vol. 6, Urban & Schwarzenberg, Berlin, 1915, p. 454.

[23] A. Libavius, Praxis alchimiae, Frankfurt, 1604, Alchymia, Frankfurt, 1606 urn:nbn:de: bvb:12-bsb10073209−2.

[24] J. French, Art of Distillation, London, 1651.

[25] J.S. Elsholtz, Destillatoria Curiosa, Berlin, 1674. urn:nbn:de:bvb:12-bsb10072415−6.

[26] P. Berdelle-Hilge, Es ist alles viel älter − Die Verfahrenstechniken im 19. Jahrhundert, Fa. Philipp Hilge, Bodenheim, 1987.

[27] F.A. Henglein, Grundriß der Chemischen Technik, eleventh Aufl., Verlag Chemie, Weinheim, 1963.

[28] F. Bacon, Novum Organum, 1620.

[29] J. Kunckel von Löwenstern, Laboratorium Chymicum, Heyl, 1716 urn:nbn:de:bvb: 12-bsb10073115−6.

[30] A. Clow, N.L. Clow, Die Schwefelsäure in der industriellen Revolution, in: A.E. Musson (Ed.), Wissenschaft und Technik im 18. Jahrhundert, Suhrkamp Verlag, Frankfurt, 1977.

[31] F.W. Gibbs, Brain Higgins und sein Kreis, in: A.E. Musson (Ed.), Wissenschaft und Technik im 18. Jahrhundert, Suhrkamp Verlag, Frankfurt, 1977.

[32] D. Diderot, J. d'Alembert. Encyclopédie ou dictionnaire raisonné des sciences, des arts et des métiers, Chimie, planche 1, vol. 38, 1779. Geneve, p. 12, urn:nbn:de:bvb: 12-bsb10351472-1.

[33] P. Zeiser, Anleitung zur vorteilhaften und sicheren Benutzung der Wasserdämpfe von einfacher und mehrfacher Spannung zumeist zum pharmazeutischen Gebrauch, 1826. Altona.

[34] F.A. Henglein, Grundriß der Chemischen Technik, second Aufl., Verlag Chemie, Weinheim, 1939.

[35] F.M. Ress, Geschichte der Kokereitechnik, Verlag Glückauf, Essen, 1957.

[36] R.H. Perry, Perry's Chemical Engineer's Handbook, third ed., McGraw-Hill, New York, 1950. Chapter Distillation, p. 655.

[37] D. Savalle, Appareils et procédés nouveaux de distillation, Masson, Paris, 1876.

[38] D. Savalle, Progrès récents de la distillation, Masson, Paris, 1873.

[39] A.P. Dubrunfaut, Traité complet de l'art de la distillation, second ed., 1825. Bruxelles.

[40] G.-P. Blümer, G. Collin, H. Höke, Tar and pitch, in: Ullmann's Encyclopedia of Industrial Chemistry, Wiley-VCH, 2011.

[41] F. Ullmann (Ed.), Kokerei, Enzyklopädie der technischen Chemie, vol. 7, Urban & Schwarzenberg, Berlin, 1919, p. 89.

[42] F. Ullmann (Ed.), Benzol, Enzyklopädie der technischen Chemie, vol. 2, Urban & Schwarzenberg, Berlin, 1915, p. 360.

[43] F. Ullmann (Ed.), Steinkohlenteer, Enzyklopädie der technischen Chemie, vol. 10, Urban & Schwarzenberg, Berlin, 1922, p. 655.

[44] G. Lunge, Die Destillation des Steinkohlentheers und die Verarbeitung der damit zusammenhängenden Produkte, Vieweg, Braunschweig, 1867, urn:nbn:de:bvb: 12-bsb10305222−0.

[45] W.H. Walker, W.K. Lewis, W.H. McAdams, Principles of Chemical Engineering, McGraw-Hill, New York, 1923.

[46] G. Buchheim, R. Sonnemann, Lebensbilder von Ingenieurwissenschaftlern, VEB Fachbuchverlag Leipzig, 1989.

[47] E. Hausbrand, Die Wirkungsweise der Rektificir- und Destillir-Apparate mit Hülfe einfacher mathematischer Betrachtungen, first ed., Springer, Berlin, 1893.

[48] C.S. Robinson, Elements of Fractional Distillation, McGraw-Hill, New York, 1930.

[49] K. Thormann, Destillieren und Rektifizieren, Spamer, Leipzig, 1928.

[50] C.S. Robinson, The Elements of Fractional Distillation, New York, 1922.

[51] C. von Rechenberg, Einfache und fraktionierte Destillation in Theorie und Praxis, Leipzig, 1923.

[52] W.L. McCabe, E.W. Thiele, Graphical design of fractionating columns, Ind. Eng. Chem. 17 (1925) 605−611.

[53] C. Mariller, Distillation et rectification des liquids industriels, Paris, 1925.

[54] L. Kolbe, Flüssige Luft, Barth, Leipzig, 1920.

[55] F. Ullmann (Ed.), Sauerstoff, Enzyklopädie der technischen Chemie, vol. 10, Urban & Schwarzenberg, Berlin, 1922, p. 10.

[56] A. Moll, Air distillation, in: A. Gorak, et al. (Eds.), Distillation: Fundamentals and Principles, vol. 1, Elsevier, 2014.

[57] F. Ullmann (Ed.), Essig, Enzyklopädie der technischen Chemie, vol. 4, Urban & Schwarzenberg, Berlin, 1916, p. 730.

[58] F. Ullmann (Ed.), Fettsäuren, Enzyklopädie der technischen Chemie, vol. 5, Urban & Schwarzenberg, Berlin, 1914, p. 449.

[59] F. Ullmann (Ed.), Glycerin, Enzyklopädie der technischen Chemie, vol. 6, Urban & Schwarzenberg, Berlin, 1915, p. 280.

[60] F. Ullmann (Ed.), Riechstoffe, Enzyklopädie der technischen Chemie, vol. 9, Urban & Schwarzenberg, Berlin, 1921, p. 505.

[61] H. Ruf, Kleine Technologie des Erdöls, second Aufl., Birkhäuser Verlag, Basel, 1963.

[62] F. Ullmann (Ed.), Erdöl, Enzyklopädie der technischen Chemie, vol. 4, Urban & Schwarzenberg, Berlin, 1916, p. 672.

[63] E. Hausbrand, Die Wirkungsweise der Rektifizier- und Destillier-Apparate mit Hilfe einfacher mathematischer Betrachtungen, third ed., Springer, Berlin, 1916.

[64] J. Gmehling, U. Onken, Vapor liquid equilibrium data collection, in: Dechema Chemistry Data Series, Frankfurt, 1982.

[65] U. Onken, W. Arlt, Recommended Test Mixtures for Distillation Columns, second ed. The Institution of Chemical Engineers, 1990, first ed., in 1969.

[66] J. Gmehling, M. Kleiber, Vapor-liquid equilibrium and physical properties for distillation, in: A. Gorak, et al. (Eds.), Distillation: Fundamentals and Principles, vol. 1Elsevier, 2014.

[67] T.H. Chilton, A.P. Colburn, Distillation and absorption in packed columns, a convenient design and correlation method, Ind. Eng. Chem. 27 (1935) 255−260.

[68] E. Kirschbaum, Destillier- und Rektifiziertechnik, third ed., Springer, Berlin, 1969.

[69] D.F. Stedman, Liquid and Vapor Contacting Device, U.S. Patent 2227164, 1939.

[70] D.F. Stedman, Packing for Fractionating Columns and the Like, U.S. Patent 2,047,444, 1935.

[71] W. Podbielniak, Apparatus and methods for precise fractional-distillation analysis, Ind. Eng. Chem. Anal. 3 (1931) 177−188.

[72] S.B. Detwiler, K.S. Markely, Bibliography on molecular or short-path distillation, Oil & Soap 16 (1939) 2−5.

[73] P.R. Watt, Molecular distillation, Vacuum 6 (1956) 113−160.

[74] R. Billet, Optimierung in der Rektifiziertechnik unter besonderer Berücksichtigung der Vakuumrektifikation, BI, Mannheim, 1967.

[75] J.F. Maćkowiak, Random packings, in: A. Gorak, et al. (Eds.), Distillation: Fundamentals and Principles, vol. 1, Elsevier, 2014.

[76] S. Gabele, F. Simon, 100 Jahre Julius montz GmbH—Eine Unternehmenschronik, Julius Montz GmbH, Hilden, 2011.

[77] E.U. Schlünder, F. Thurner, Destillation, Absorption, Extraktion, G. Thieme, Stuttgart, 1986.

[78] J. Maćkowiak, Fluiddynamik von Füllkörperkolonnen und Packungen, second ed., Springer, Berlin, 2003.

[79] R.B. Bird, W.E. Stewart, E.N. Lightfoot, Transport Phenomena, John Wiley, New York, 1960.

[80] R. Taylor, H.A. Kooijman, Mass transfer, in: A. Gorak, et al. (Eds.), Distillation: Fundamentals and Principles, vol. 1, Elsevier, 2014.

[81] G. Schembecker, State-of-the-art and Future Development of Computer Aided Process Synthesis, Series DECHEMA Monographien, Wiley-VCH, Weinheim, 1998.

[82] H. Schmidt-Traub, A. Gorak (Eds.), Integrated Reaction and Separation Operations, Springer, Berlin, 2006.

[83] H. Freund, K. Sundmacher, Towards a methodology for the systematic analysis and design of efficient chemical processes: part 1. From unit operations to elementary process functions, Chem. Eng. Process. 47 (2008) 2051−2060.

[84] H. Freund, K. Sundmacher, Process intensification, 1. fundamentals and molecular level, in: Ullmann's Encyclopedia of Industrial Chemistry, Wiley-VCH, 2011.

[85] H. Freund, K. Sundmacher, Process intensification, 2. Phase level, in: Ullmann's Encyclopedia of Industrial Chemistry, Wiley-VCH, 2011.

[86] L. Spiegel, M. Duss, Structured packings, in: A. Gorak, et al. (Eds.), Distillation: Fundamentals and Principles, vol. 1, Elsevier, 2014.

[87] B.D. Smith, B. Block, K.C.D. Hickman, Distillation, chapter 13, in: R.H. Perry, C.H. Chilton (Eds.), Perry's Chemical Engineer's Handbook, fifth ed., McGraw-Hill, 1973.

[88] J.R. Fair, D.E. Steinmeyer, W.R. Penney, J.A. Brink, Liquid—gas systems, chapter 18, in: R.H. Perry, C.H. Chilton (Eds.), Perry's Chemical Engineer's Handbook, fifth ed., McGraw-Hill, 1973.

[89] J.R. Fair, D.E. Steinmeyer, W.R. Penney, B.B. Crocker, Liquid—gas systems, chapter 18, in: R.H. Perry, D.W. Green, J.O. Maloney (Eds.), Perry's Chemical Engineer's Handbook, sixth ed., McGraw-Hill, 1984.

[90] J.D. Seader, Z.M. Kurtyka, Distillation, chapter 13, in: R.H. Perry, D.W. Green, J.O. Maloney (Eds.), Perry's Chemical Engineer's Handbook, sixth ed., McGraw-Hill, 1984.

[91] J.D. Seader, J.J. Siirola, S.D. Barnicki, Distillation, chapter 13, in: R.H. Perry, D.W. Green (Eds.), Perry's Chemical Engineer's Handbook, seventh ed., McGraw-Hill, 1997.

[92] M.F. Doherty, J.D. Perkins, On the dynamics of distillation processes I. The simple distillation of multicomponent non-reacting, homogeneous liquid mixtures, Chem. Eng. Sci. 33 (1978) 281.

[93] E.Y. Kenig, S. Blagov, Modeling of distillation processes, in: A. Gorak, et al. (Eds.), Distillation: Fundamentals and Principles, vol. 1, Elsevier, 2014.

[94] S. Berman, A.A. Melnychuk, D.F. Othmer, Dibutyl phthalate — reaction rate of catalytic esterification, Ind. Eng. Chem. 40 (1948) 2139.

[95] V.H. Agreda, L.R. Partin, Reactive Distillation Process for the Production of Methylacetate, U.S. Patent 4,435,595, 1984; V.H. Agreda, L.R. Partin, W.H. Heise, High purity methyl acetate via reactive distillation, Chem. Eng. Prog. 86 (2) (1990) 40.

[96] J. Krafczyk, J. Gmehling, Einsatz von Katalysatorpackungen für die Herstellung von Methylacetat durch reaktive Rektifikation, Chem. Ing. Tech. 66 (1994) 1372—1375.

[97] B. Bessling, J.-M. Löning, A. Ohligschläger, G. Schembecker, K. Sundmacher, Investigations on the synthesis of methyl acetate in a heterogeneous reactive distillation process, Chem. Eng. Tech. 21 (1998) 393—400.

[98] R.S. Huss, F. Chen, M.F. Malone, M.F. Doherty, Reactive distillation for methyl acetate production, Comput. Chem. Eng. 27 (1855) 2003.

[99] Fractionating Institute, FRI. www.fri.org (accessed 28.02.14).

[100] M. Resetarits, D. King, Fifty years of FRI contributions to separation science, AIChE Annu. Meet. (2008).

[101] Separations Research Program, University of Texas at Austin. uts.cc.utexas.edu/~utsrp/index.html (accessed 28.02.14).

[102] W.J. Stupin, F.J. Lockhart, Thermally coupled distillation: a case history, Chem. Eng. Progr. 88 (1972) 71—72.

[103] D.A. Monro, Fractionating Apparatus and Fractionating Method, U.S. Patent 2,134,882, 1933.

[104] F.B. Petlyuk, V.M. Platonov, D.M. Slavinskii, Thermodynamically optimal method for separating multicomponent mixtures, Int. Chem. Eng. 5 (1965) 555—561.

[105] R.O. Wright, Fractionating Apparatus, U.S. Patent 2,471,134, 1949.

[106] R. Agrawal, Multicomponent distillation columns with partitions and multiple reboilers and condensers, Ind. Eng. Chem. Res. 40 (2001) 4258—4266.

[107] I.P. Nikolaides, M.F. Malone, Approximate design and optimization of a thermally coupled distillation with prefractionation, Ind. Eng. Chem. Res. 27 (1988) 811.

[108] R. Agrawal, Z.T. Fidkowski, Operable and Efficient Distillation Schemes for Multi-component Separations, U.S. Patent 6,106,674, 1998; R. Agrawal, Z.T. Fidkowski, New thermally coupled schemes for ternary distillation, AIChE J., 45 (1999) 485–496.

[109] Z.T. Fidkowski, L. Krolikowski, Minimum energy requirements of thermally coupled distillation systems, AIChE J. 33 (1987) 643–653.

[110] M.F. Doherty, Z.T. Fidkowski, M.F. Malone, R. Taylor, Distillation, chapter 13, in: J.O. Maloney (Ed.), Perry's Chemical Engineer's Handbook, eighth ed., McGraw-Hill, 2008.

[111] R. Agrawal, Z.T. Fidkowski, More operable arrangements of fully thermally coupled distillation columns, AIChE J. 44 (1998) 2565–2568.

[112] J.A. Caballero, I.E. Grossmann, Generalized disjunctive programming model for the optimal synthesis of thermally linked distillation columns, Ind. Eng. Chem. Res. 40 (2001) 2260.

[113] G. Kaibel, Distillation columns with vertical partitions, Chem. Eng. Technol. 10 (1987) 92–98.

[114] T. Grützner, D. Staak, B. Schwegler, D. Roederer, Dividing Wall Column for Industrial Multi-purpose Use, CHISA, PRES Conference, Prague (Czech Republic), 26 August 2012–29 August 2012.

[115] G. Parkinson, CEP (July 2005) 10.

[116] L. Spiegel, W. Meier, Distillation columns with structured packings in the next decade, Chem. Eng. R&D 81 (2003) 39–47.

[117] S.J. Luo, W.Y. Fei, X.Y. Song, H.Z. Li, Effect of channel opening angle on the performance of structured packings, Chem. Eng. J. 144 (2008) 227–234.

[118] J.R. Fair, D.E. Steinmeyer, W.R. Penney, B.B. Crocker, Gas absorption and gas-liquid system design, chapter 14, in: R.H. Perry, D.W. Green (Eds.), Perry's Chemical Engineer's Handbook, seventh ed., McGraw-Hill, New York, 1997.

[119] N. Asprion, G. Kaibel, Dividing wall columns: fundamentals and recent advances, Chem. Eng. Process. 49 (2010) 139–146.

[120] Z. Olujić, M. Jödecke, A. Shilkin, G. Schuch, B. Kaibel, Equipment improvement trends in distillation, Chem. Eng. Process. 48 (2009) 1089–1104.

[121] J. Richter, A. Gorak, E.Y. Kenig, Catalytic distillation, chapter 3, in: H. Schmidt-Traub, A. Gorak (Eds.), Integrated Reaction and Separation Operations, Springer, Berlin, 2006.

[122] J. Micovic, T. Beierling, P. Lutze, G. Sadowski, A. Górak, Entwurfsmethode zur Auslegung hybrider Trennverfahren aus Rektifikation und Schmelzekristallisation für die Aufreinigung engsiedender Gemische, Chem. Ing. Tech. 84 (2012) 2035–2047.

[123] Z. Olujić, M. Behrens, L. Sun, J. de Graauw, Augmenting distillation by using membrane based vapor–liquid contactors: an engineering view from Delft, J. Membr. Sci. 350 (2010) 19–31.

[124] X.Z. Li, Application of microbial biofilms for the production of chemicals (Ph.D. thesis), University of New South Wales, Australia. unsworks.unsw.edu.au, 2012 (accessed 28.02.14).

[125] P. Lutze, A. Gorak, Distillation in biotechnology, in: A. Gorak, et al. (Eds.), Distillation: Fundamentals and Principles, vol. 1, Elsevier, 2014.

[126] W. Arlt, New separating agents for Distillation, in: A. Gorak, et al. (Eds.), Distillation: Fundamentals and Principles, Vol. 1, Elsevier, 2014.

[127] A. Ziogas, G. Kolb, H.-J. Kost, V. Hessel, Entwicklung einer leistungsstarken Mikrorektifikationsapparatur für analytische und präparative Anwendungen, Chem. Ing. Tech. 83 (2011) 465−478.

[128] H.Z. Kister, P.M. Mathias, D.E. Steinmeyer, W.R. Penney, B.B. Crocker, J.R. Fair, Equipment for distillation, gas absorption, phase dispersion, and phase separation, chapter 14, in: J.O. Maloney (Ed.), Perry's Chemical Engineer's Handbook, eighth ed., McGraw-Hill, 2008.

[129] J.M. MacInnes, J. Ortiz-Osorio, P.J. Jordan, G.H. Priestman, R.W.K. Allen, Experimental demonstration of rotating spiral microchannel distillation, Chem. Eng. J. 159 (2010) 159−169.

[130] G.Q. Wang, Z.C. Xu, J.B. Ji, Progress on Higee distillation − introduction to a new device and its industrial applications, Chem. Eng. R&D 89 (2011) 1434−1442.

[131] C. Ramshaw, R.H. Mallinson, Mass Transfer Apparatus and Process, U.S. Patent 4400275, 1983.

[132] M.F. Doherty, M.F. Malone, Conceptual Design of Distillation Systems, McGraw-Hill, 2001.

Vapor–Liquid Equilibrium and Physical Properties for Distillation

2

Jürgen Gmehling[1], Michael Kleiber[2]

University of Oldenburg, Industrial Chemistry, Oldenburg, Germany[1], Hattersheim, Germany[2]

CHAPTER OUTLINE

Distillation: Fundamentals and Principles. http://dx.doi.org/10.1016/B978-0-12-386547-2.00002-8

2.1 Introduction

For solving the so-called MESH equations (see chapter 10), in addition to caloric properties, in particular a reliable knowledge of the vapor–liquid equilibrium (VLE) behavior is required. Therefore the typical question asked by the chemical engineer is: What is the pressure and the composition in the vapor phase (β), if the vapor phase at given temperature is in equilibrium with the liquid phase (α) of a given composition. Often multicomponent systems with nonpolar, polar, and sometimes also electrolytes and supercritical compounds have to be considered. For a four-component system containing an electrolyte and a supercritical compound, the task is shown in Figure 2.1. In the case of reactive distillation, additional information about the kinetic and the equilibrium conversions of the chemical reactions is needed.

Although VLE data for approximately 12,500 binary nonelectrolyte systems have been published, much fewer data are available for ternary systems and almost no data can be found for multicomponent systems [1]. Of course with the help of sophisticated experimental techniques the required VLE data can be measured. But measurements are very time consuming. As discussed by Novak et al. [2], the measurement of a 10-component system at atmospheric pressure in 10 mol% steps would require approximately 37 years [3]. Therefore it would be most desirable to have access to thermodynamic models, which allow the reliable calculation of the VLE behavior of multicomponent systems using only a limited amount of experimental data, e.g. binary data.

For VLE the number of phases is not limited to two. For example, in the case of separations by heteroazeotropic distillation, two liquid phases exist, e.g. for the systems n-butanol–water, ethanol–water–cyclohexane, etc. For these systems, the vapor–liquid–liquid equilibrium (VLLE) has to be known.

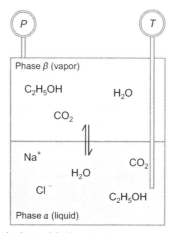

FIGURE 2.1 Typical Vapor–Liquid Equilibrium Problem

2.2 **Thermodynamic fundamentals**

VLEs or VLLEs exist if the different components i show identical fugacities in the liquid (L) and the vapor (V) phase [3]:

$$f_{L,i} = f_{V,i} \quad f'_{L,i} = f''_{L,i} = f_{V,i} \tag{2.1}$$

With the help of the auxiliary quantities fugacity coefficient ϕ_i and activity coefficient γ_i the following relations are obtained for VLE or VLLE [3]:

$$x_i \cdot \phi_{L,i} = y_i \cdot \phi_{V,i} \quad x'_i \cdot \phi'_{L,i} = x''_i \cdot \phi''_{L,i} = y_i \cdot \phi_{V,i} \tag{2.2}$$

$$x_i \cdot \gamma_i \cdot f_i^0 = y_i \cdot \phi_{V,i} \cdot p \quad x'_i \cdot \gamma'_i \cdot f_i^{0'} = x''_i \cdot \gamma''_i \cdot f_i^{0''} = y_i \cdot \phi_{V,i} \cdot p \tag{2.3}$$

In the so-called $\gamma-\phi$-approach (Eqn (2.3)), besides the activity coefficients γ_i, a value for the standard fugacity f_i^0 is required. Usually the fugacity of the pure liquid at system temperature T and system pressure p is used as standard fugacity f_i^0. This leads to the following expression for VLE [3]:

$$x_i \cdot \gamma_i \cdot \phi_i^s \cdot p_i^s \cdot \exp\left(\frac{v_{L,i} \cdot (p - p_i^s)}{R \cdot T}\right) = x_i \cdot \gamma_i \cdot \phi_i^s \cdot p_i^s \cdot Poy_i = y_i \cdot \phi_{V,i} \cdot p \tag{2.4}$$

Since at moderate pressures the ratio $\phi_i^s \cdot Poy_i / \phi_{V,i}$ shows values of nearly unity, often the following simplified relation is used to describe the VLE behavior for systems without strongly associating compounds, such as carboxylic acids [3]:

$$x_i \cdot \gamma_i \cdot p_i^s \approx y_i \cdot p \tag{2.5}$$

For supercritical components, Eqn (2.3) must be rewritten as:

$$x_i \cdot H_{i,j} = y_i \cdot \phi_{V,i} \cdot p \tag{2.6}$$

with $H_{i,j}$ as the temperature-dependent Henry constant of component i in the solvent mixture [3].

Using the different approaches, the following relations are obtained for the calculation of the required K-factors K_i and relative volatilities (separation factors) α_{ij}:

$\phi-\phi$ approach:

$$K_i = \frac{y_i}{x_i} = \frac{\phi_{L,i}}{\phi_{V,i}} \quad \alpha_{ij} = \frac{K_i}{K_j} = \frac{y_i/x_i}{y_j/x_j} = \frac{\phi_{L,i} \cdot \phi_{V,j}}{\phi_{V,i} \cdot \phi_{L,j}} \tag{2.7}$$

$\gamma-\phi$ approach:

$$K_i = \frac{y_i}{x_i} \approx \frac{\gamma_i \cdot p_i^s}{p} \quad \alpha_{ij} = \frac{K_i}{K_j} = \frac{y_i/x_i}{y_j/x_j} \approx \frac{\gamma_i \cdot p_i^s}{\gamma_j \cdot p_j^s} \tag{2.8}$$

The $\phi-\phi$ (equation of state) approach shows various advantages over the $\gamma-\phi$ approach, e.g. no problems occur with supercritical compounds. At the same time densities, enthalpies (including heats of vaporization), heat capacities, etc. as function of temperature, pressure, and composition can be calculated for both phases, which are required as additional information in the $\gamma-\phi$ approach. On the other hand, the strength of the $\gamma-\phi$ approach is its relative simplicity and the opportunity to use independent correlations for each other quantity, which can be fitted as accurately as possible.

Depending on the activity coefficients and the vapor pressures, very different kinds of VLE behavior are observed. In Figure 2.2 the observed VLE behavior for different values of the activity coefficients and vapor pressures is shown in the form of four typical binary VLE diagrams. In the figure, the $y-x$ diagram, the activity coefficients, the pressure in the case of isothermal data, and the temperature in the case of isobaric data as a function of the mole fraction are shown. In the two diagrams on the right-hand side the pressure and the temperature are not only given as a function of the liquid phase (continuous boiling point line) but also as a function of the vapor phase mole fraction (dashed dew-point line).

The first system, benzene–toluene, shows nearly ideal behavior ($\gamma_i \approx 1$) as assumed in Raoult's law. The activity coefficients for the next three systems steadily increase (positive deviations from Raoult's law). The influence of the activity coefficients can particularly be recognized from the pressure as a function of the liquid phase mole fraction x_1 at a given temperature. While a straight line for the pressure is obtained using Raoult's law, higher pressures are observed for the methanol–water system ($\gamma_i > 1$). With increasing activity coefficients as in the case of the 1-propanol–water system ($\gamma_i \gg 1$), the pressure even shows a maximum. At the same time a minimum of the temperature occurs in the isobaric case. At the pressure maximum (temperature minimum) the boiling-point line and the dew-point line meet. This means that the compositions in the liquid and the vapor phase become identical and in the $y-x$ diagram an intersection of the 45° line is observed. These points are called azeotropic points.

When the values of the activity coefficients further increase, two liquid phases can occur (see Section 2.5), as in the case of the 1-butanol–water system. When the two liquid phase region (shown by the horizontal line) intersects the 45° line in the $y-x$ diagram, a so-called heterogeneous azeotropic point occurs.[a] In the case of heterogeneous azeotropic points the condensation of the vapor leads to the formation of two liquid phases, where in the 1-butanol–water system a butanol- and a water-rich phase is formed. The pressure (temperature) and the vapor phase composition show constant values for the binary system in the whole heterogeneous region.

---

[a]Heterogeneous azeotropes are not formed in all systems with a miscibility gap. For example, in the heterogeneous systems 2-butanone–water, 2-butanol–water, and ethanol–water–xylene (o,m,p) a homogeneous azeotrope is observed.

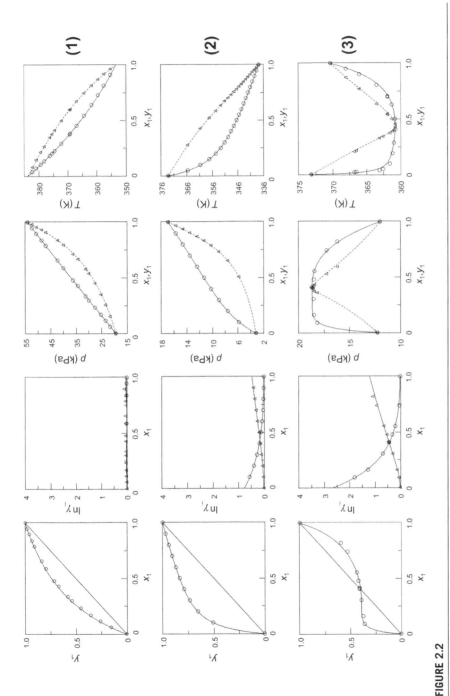

FIGURE 2.2

Different types of vapor–liquid equilibrium diagrams for the following binary systems: (1) benzene (1)–toluene (2) at 334.15 K and 101 kPa; (2) methanol (1)–water (2) at 298.15 K and 101 kPa; (3) 1-propanol (1)–water (2) at 323.07 K and 101 kPa; (4) 1-butanol (1)–water (2) at 298.15 K and 101 kPa; (5) ethyl acetate (1)–dichloromethane (2) at 348.15 K and 101 kPa; and (6) acetone (1)–chloroform (2) at 308.32 K and 101 kPa.

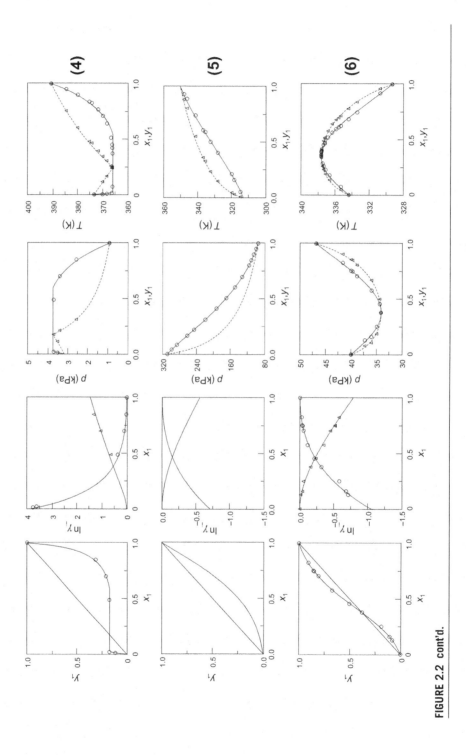

FIGURE 2.2 cont'd.

Besides the large number of systems with positive deviation from Raoult's law ($\gamma_i > 1$), sometimes systems with negative deviation from Raoult's law ($\gamma_i < 1$) are observed. In Figure 2.2 the ethyl acetate—dichloromethane and acetone—chloroform systems were chosen as examples. Because of the strong hydrogen bonding effects between the considered compounds, associates with low volatility are formed in these systems. This results in the fact that the equilibrium pressure of the mixture is lower than the pressure calculated using Raoult's law. Depending on the vapor pressures also systems with negative deviations from Raoult's law can show azeotropic behavior, as e.g. the acetone—chloroform system. However, in contrast to systems with positive deviations from Raoult's law, in these systems azeotropic points with a pressure minimum (temperature maximum) are formed. The occurrence and disappearance of binary azeotropes are discussed in more detail in Section 2.7. For negative deviation from Raoult's law, miscibility gaps are not possible.

2.3 Calculation of VLE using g^E models

For the calculation of VLEs using the $\gamma-\phi$ approach, in addition to the vapor pressure an activity coefficient model is required, which allows the calculation of the VLE behavior using only binary experimental data. An analytical expression for the activity coefficients can be derived from Eqn (2.9) starting from an expression for the excess Gibbs energy [3].

$$g^E = \sum_i x_i \cdot \overline{g}_i^E = \sum_i x_i \left(\frac{\partial G^E}{\partial n_i} \right)_{n_j \neq n_i, T, p} = RT \sum_i x_i \ln \gamma_i \qquad (2.9)$$

The excess Gibbs energy depends on temperature and pressure. Although the pressure influence on the activity coefficient can usually be neglected in the case of VLE, the consideration of the temperature dependence is recommended.

With the help of the Gibbs—Helmholtz equation (Eqn (2.10)), a relation for the temperature dependence of the activity coefficients can directly be derived [3]:

$$\left(\frac{\partial \left(g^E / T \right)}{\partial (1/T)} \right)_{p,x} = h^E \qquad \left(\frac{\partial \ln \gamma_i}{\partial (1/T)} \right)_{p,x} = \frac{\overline{h}_i^E}{R} \qquad (2.10)$$

The common approaches, e.g. the Wilson [4], NRTL [5], and UNIQUAC equations [6], allow an improved description of the real behavior of multicomponent systems from the information of the binary systems. All these expressions are based on the local composition concept introduced by Wilson [4]. This concept assumes that the local composition because of the interacting forces is different from the overall composition.

The different local composition g^E models are discussed in detail in Ref. [3]. There also the analytical expressions for the binary and multicomponent systems for the different models can be found. For the NRTL equation the required expressions are exemplarily given in Table 2.1.

Table 2.1 Analytical Expressions for the NRTL Equation [5]

Parameters	Expressions for the Activity Coefficients
Δg_{12} Δg_{21} α_{12}	$\ln \gamma_1 = x_2^2 \left[\tau_{21} \left(\dfrac{G_{21}}{x_1 + x_2 G_{21}} \right)^2 + \dfrac{\tau_{12} G_{12}}{(x_2 + x_1 G_{12})^2} \right]$ $\ln \gamma_2 = x_1^2 \left[\tau_{12} \left(\dfrac{G_{12}}{x_2 + x_1 G_{12}} \right)^2 + \dfrac{\tau_{21} G_{21}}{(x_1 + x_2 G_{21})^2} \right]$
$\Delta g_{ij}, \alpha_{ij}$	$\ln \gamma_i = \dfrac{\sum_i \tau_{ji} G_{ji} x_j}{\sum_k G_{ki} x_k} + \sum_j \dfrac{x_j G_{ij}}{\sum_k G_{kj} x_k} \left(\tau_{ij} - \dfrac{\sum_n x_n \tau_{nj} G_{nj}}{\sum_k G_{kj} x_k} \right)$
$\tau_{ij} = \Delta g_{ij}/T, \quad \tau_{ii} = 0$	
$G_{ij} = \exp(-\alpha_{ij}\tau_{ij}), \; G_{ii} = 1$	

For the interaction parameters Δg_{ij} usually the unit K is used. But often for published interaction parameters, e.g. given in Ref. [7], the unit of a molar energy can be found. That is the case when in the denominator of the exponential (G_{ij}) term RT instead of T is used. Then the unit depends on the choice of the unit for the general gas constant R (J/mol K, cal/mol K, etc.). In process simulation programs, a standardized notation is used to avoid this confusion [3].

2.3.1 Fitting of g^E model parameters

The quality of the design of a distillation column by solving the MESH equations mainly depends on the accuracy of the K-factors (separation factors). Using, e.g. the NRTL equation given in Table 2.1, these values can be calculated for the multi-component system if the binary parameters are available.

The K-factors and separation factors mainly depend on composition but also on temperature. The correct composition dependence is described with the help of activity coefficients. Following the Clausius–Clapeyron equation (see Section 2.9), the temperature dependence is mainly influenced by the slope of the vapor pressure curves (enthalpy of vaporization) of the compounds involved. But also the activity coefficients are temperature-dependent following the Gibbs–Helmholtz equation (Eqn (2.10)). This means that besides a correct description of the composition dependence also a reliable description of the temperature dependence of the activity coefficients is required. When a large temperature range is covered, temperature-dependent parameters have to be used to describe the temperature dependence of the activity coefficients with the required accuracy, following the Gibbs–Helmholtz relation (Eqn (2.10)) for the partial molar excess enthalpies in the temperature range covered.

The following temperature dependence of the binary interaction parameters can be used[b]:

$$\Delta g_{ij}(T) = a_{ij} + b_{ij}T + c_{ij}T^2 \tag{2.11}$$

To take into account the real vapor phase behavior, e.g. the virial equation or cubic equations of state (e.g. the Soave—Redlich—Kwong (SRK) or Peng—Robinson (PR) equations of state) can be used. In the case of strongly associating systems, such as carboxylic acids or HF, an association model (chemical theory) [3] has to be applied.

A prerequisite for the correct description of the VLE behavior of multicomponent systems is a reliable description of the binary subsystems with the fitted binary g^E model parameters. For fitting the binary interaction parameters nonlinear regression methods are applied. These methods allow adjusting the parameters in such a way that a minimum deviation of an arbitrary chosen objective function f_{obj} is obtained. For this job for example the Simplex—Nelder—Mead method [8] can be applied successfully. In contrast to many other methods [9], the Simplex—Nelder—Mead method is a simple search routine, which does not need the first and the second derivate of the objective function with respect to the different variables. This has the great advantage that computational problems, such as "underflow" or "overflow" with the arbitrarily chosen initial parameters, can be avoided.

For fitting the required g^E model parameters different types of objective functions for the experimental or derived properties X can be used, e.g. vapor phase mole fraction, pressure, temperature, K-factor K_i, separation factor α_{12}, etc., where either the relative or the absolute deviation of the experimental and correlated values (pressure, temperature, vapor phase composition, etc.) can be minimized. In the case of complete VLE data, this means where p, T, x_i, y_i is given, also the deviation between the experimental and predicted activity coefficients or excess Gibbs energies can be used in the objective function. Furthermore the parameters can be determined by a simultaneous fit to different properties to cover properly the composition and temperature dependence of the activity coefficients. For example, the deviation of the derived activity coefficients can be minimized together with the deviations in activity coefficients at infinite dilution, excess enthalpies, etc. Reliable activity coefficients at infinite dilution measured with special experimental techniques are of special importance, since they deliver the only reliable information about the real behavior in the dilute range [10],[c] e.g. at the top or the bottom of a distillation column. Excess enthalpies [11] measured,

[b]In process simulators often different temperature dependencies are used.
[c]With the required care it is possible to measure reliable activity coefficients at infinite dilution of a low-boiling substance in a high-boiling compound, e.g. with the help of the dilutor technique, gas—liquid chromatography, ebulliometry, or Rayleigh distillation. Unfortunately it is much more difficult to measure these values for high-boiling components in low-boiling compounds, e.g. for water in ethylene oxide, for NMP in benzene, etc.

e.g. with the help of isothermal flow calorimetry are important too, since they provide reliable information about the temperature dependence of the activity coefficients (see Eqn (2.10)), and thus for the separation and K-factors.

The impact of inaccurate g^E model parameters can be very serious. The parameters have a major influence on the investment and operating costs (number of stages, reflux ratio). The influence of the g^E model parameters on the results is especially large, if the separation factor is close to unity. Poor parameters can either lead to the calculation of nonexisting azeotropes in zeotropic systems or the calculation of zeotropic behavior in azeotropic systems. Poor parameters can also lead to a miscibility gap which does not exist.[d] In the case of positive deviation from Raoult's law the separation problem often exists at the top of the column, where the high boiler has to be removed, since at the top of a distillation column the most unfortunate separation factors are obtained.

Starting from Eqn (2.8), the following separation factors are observed at the top and the bottom of a distillation column for a binary system (low boiler: component 1):

top of the column $(x_1 \rightarrow 1)$:   bottom of the column $(x_2 \rightarrow 1)$:

$$\alpha_{12} = \frac{p_1^s}{\gamma_2^\infty \cdot p_2^s} \qquad\qquad \alpha_{12} = \frac{\gamma_1^\infty \cdot p_1^s}{p_2^s}$$

While for example for the acetone–water system at atmospheric pressure, separation factors a little above unity are obtained at the top of the column, separation factors greater than 40 are observed in the bottom of the column. In the case of negative deviations from Raoult's law the separation problem usually occurs in the bottom of the column.

For fitting reliable g^E model parameters accurate experimental VLE data should be used. Different types of VLE data have been published. An overview about the different types of VLE data published together with the proportion of such data from all published VLE data is given in Table 2.2.

In most cases the measurements are performed at isothermal or isobaric conditions. Occasionally measurements are also performed at constant composition. Sometimes none of the properties is kept constant. Only in approximately 45% of the VLE data all values (x_i, y_i, T, p) are measured. Complete VLE data allow the verification of the quality of the VLE data using thermodynamic consistency tests [3], since three of the four values (x_i, y_i, T, p) are sufficient to derive the fourth quantity. Because of the greater experimental effort required, dew-point data (T, p, y_i) are

---

[d]Process simulators often contain extensive databanks with pure component and mixture parameters, e.g. default g^E model parameters. This allows generating the required input very fast. But the user should use these data and parameters with care, as mentioned by the simulator companies. Prior to process simulation, the company expert for phase equilibrium thermodynamics should check the pure component data and mixture parameters carefully. In Ref. [3] it is shown what can happen when directly the default values stored in the process simulator are used.

Table 2.2 Percentage of VLE Published for Different Types of VLE Data [1]

Type of VLE Data	Measured Values				Percentage of the Published VLE Data in the Dortmund Data Bank (2012)
	x_i	y_i	p	T	
Isothermal complete	√	√	√	Constant	16.70
Isobaric complete	√	√	Constant	√	26.06
Isothermal px data	√	–	√	Constant	29.87
Isobaric Tx data	√	–	Constant	√	10.59
Isothermal xy data	√	√	–	Constant	3.31
Isobaric xy data	√	√	Constant	–	1.94
Isothermal yp data	–	√	√	Constant	0.58
Isobaric yT data	–	√	Constant	√	0.32
Isoplethic pT data	Constant	–	√	√	8.41
Other data	√	√	√	√	2.22

seldom published. But these data are of special importance to determine reliable separation factors for high boiling compounds in low boiling compounds (e.g. water in ethylene oxide) at the top of the column.

As can be seen from Table 2.2, the measurement of complete isobaric data is very popular. The reason is that a great number of chemical engineers like isobaric data, since distillation columns run at nearly isobaric conditions. But the measurement of isobaric data shows several disadvantages compared to isothermal data. This was already discussed in detail by Van Ness [12]:

> In the early unsophisticated days of chemical engineering VLE data were taken at constant pressure for direct application in the design of distillation columns, which were treated as though they operated at uniform pressure. There is no longer excuse for taking isobaric data, but regrettably the practice persists. Rigorous thermodynamic treatment of isobaric data presents problems that do not arise with isothermal data. Their origin is the need to take into account not only the composition dependence of the excess Gibbs energy but also its temperature dependence.

In Figure 2.3, an example is given where the use of isobaric data leads to serious errors in the determination of activity coefficients.

Since the measurement of temperature and pressure is more accurate than concentration measurements, Van Ness recommends the measurement of px data at isothermal conditions. Indeed, today mainly isothermal px data are measured by static equipments. The precise liquid composition is usually achieved by injection of the degassed liquids with the help of precise piston pumps. The change of the feed composition by evaporation can easily be taken into account, when the volume

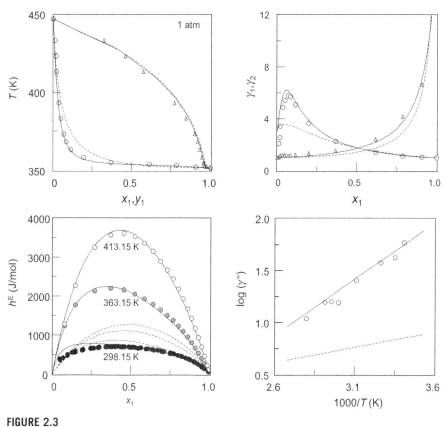

FIGURE 2.3

Results of the fit of temperature-independent and temperature-dependent Wilson parameters to consistent isobaric VLE data at $1.013 \cdot 10^5$ Pa. VLE, excess enthalpies and activity coefficients at infinite dilution of the ethanol (1)–n-decane (2) system [3]. ● Δ, Experimental [1]; - - -, temperature-independent parameters; –––, temperature-dependent parameters.

of the cell and the pressure is known. Depending on the vapor volume the change of the feed composition is usually smaller than 0.1 mol% at moderate pressures. This means by this method a much more precise determination of the liquid composition is achieved than by analytical measurements.

In contrast to complete VLE data, isothermal px data cannot be checked using thermodynamic consistency tests. But if the experimental px data can be described accurately with the help of a g^E model, these VLE data can be considered as thermodynamically consistent. The same is true for the other types of incomplete VLE data listed in Table 2.2.

It is the objective that with the help of the g^E model only the correct deviation from Raoult's law is described. This can only be achieved if the exact values of

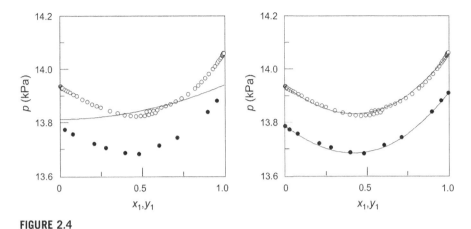

FIGURE 2.4

Influence of the pure component vapor pressures on the calculated VLE results for the 2-propanol (1)–*tert*-butanol (2) system at 313 K. – – –, Using default Antoine constants (left-hand side); – – –, using the pure component vapor pressures reported by the authors (right-hand side).

the pure component vapor pressures are used in the fitting procedure. This is demonstrated in Figure 2.4 for the nearly ideal but nevertheless azeotropic 2-propanol–*tert*-butanol system, where the interaction parameters are simultaneously fitted to two isothermal *px* datasets both measured at 313 K. Obviously, the two datasets show a systematic small pressure difference. A correct description of the *px* data and the azeotropic VLE behavior can only be obtained if the vapor pressure data of the authors are used for fitting the parameters. Total disagreement (nonazeotropic behavior) is observed if the VLE calculation is performed using the vapor pressures calculated using, e.g. default Antoine constants.

In practice mainly VLE data are used to fit the temperature-independent parameters. For a large number of binary systems the required binary g^E model parameters for the Wilson, NRTL, and UNIQUAC equations and the results of the consistency tests can be found in the VLE Data Collection of the DECHEMA Chemistry Data Series published by Gmehling, Onken et al. [7]. In Figure 2.5 one page of this series is shown. It shows the VLE data for the ethanol and water system at 343 K published by Mertl [13]. On every page of this data compilation the reader will find the system, the reference, the Antoine constants with the range of validity, the experimental VLE data, the results of two thermodynamic consistency tests, and the parameters of different g^E models, such as the Wilson, NRTL, and UNIQUAC equations. Additionally, the parameters of the Margules [14] and van Laar [15] equations are listed.[e] Furthermore the calculated results for the different models are given. For the model

---

[e]Both models (Margules and van Laar) are hardly used for process simulation today.

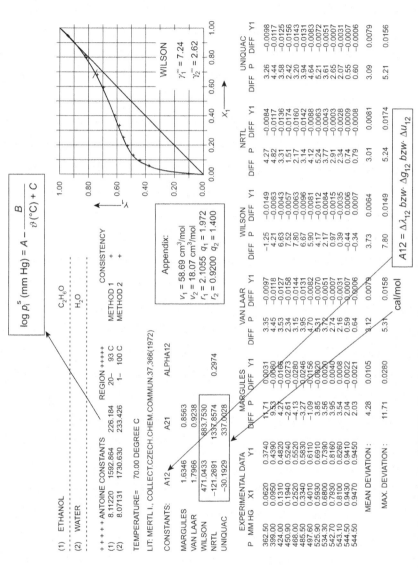

FIGURE 2.5 Example Page from the DECHEMA Chemistry Data Series [7]

that shows the lowest mean deviation in vapor phase mole fraction, the results are additionally shown in graphical form together with the experimental data and the calculated activity coefficients at infinite dilution. In the appendix of the data compilation the reader will find the additionally required pure component data, such as the molar volumes for the Wilson equation, the relative van der Waals properties for the UNIQUAC equation, and the parameters of the dimerization constants for carboxylic acids. Usually, the Antoine parameter A is adjusted to A' to start from the vapor pressure data measured by the authors. This causes the g^E model parameters to only describe the deviation from Raoult's law.[f] Since in this data compilation only VLE data up to 5000 mm Hg ($6.7 \cdot 10^5$ Pa) are presented, ideal vapor phase behavior is assumed when fitting the parameters. For systems with carboxylic acids the chemical theory [3] is used to describe the deviation from ideal vapor phase behavior.

Nowadays, phase equilibrium data are usually retrieved from databanks (Dortmund Data Bank (DDB) [1], DETHERM [16]). Modern regression programs can load the particular datasets directly into the database, and in contrast to the beginning of the computer age it is no longer a problem to regress many parameters simultaneously. However, many engineers have no access to regression programs, and in these cases the DECHEMA data series still provide valuable information.

As already discussed by Van Ness [12], the use of temperature-independent binary parameters can cause problems in the case of isobaric data, in particular if the boiling points of the two compounds are very different, as for the binary system ethanol−n-decane shown in Figure 2.3. In the case of isobaric data the temperature changes with composition. For the ethanol−n-decane system, the temperature change is nearly 100 K at low ethanol concentrations ($x_1 = 0−0.1$). Since the ethanol (1)−n-decane (2) system shows strong endothermic h^E behavior (see Figure 2.3), increasing temperature leads to a decrease of the activity coefficient of ethanol following the Gibbs−Helmholtz equation. This causes a maximum value of γ_1 at low ethanol mole fraction, since the temperature increases from nearly 360 to 450 K.

The results of the Wilson equation after fitting binary temperature-independent parameters to reliable consistent isobaric data for the ethanol−n-decane system at $1.013 \cdot 10^5$ Pa are shown in Figure 2.3 by the dashed line. As objective function the sum of the relative deviations of the activity coefficients was used. It can be seen that the curvature of the activity coefficients cannot be described using temperature-independent parameters, so that already for VLE poor results are obtained. In particular large deviations are obtained at low ethanol concentrations. Poor results are not only obtained for the Txy behavior but also for the activity coefficients, although the activity coefficients were used to fit the Wilson parameters. Also the excess enthalpies and activity coefficients at infinite dilution are not described reliably.

---

[f]Unfortunately, Mertl [13] has not given the pure component vapor pressures. But with the Antoine constants given, very reliable vapor pressures for both components are obtained for the given dataset.

To obtain the correct values at infinite dilution and the correct temperature dependence, in addition to VLE data further reliable thermodynamic information should be taken into account for fitting temperature-dependent parameters. Temperature-dependent Wilson parameters fitted simultaneously to VLE, excess enthalpies, and activity coefficients at infinite dilution of the ethanol—n-decane system are given in Ref. [3]. The results for VLE, activity coefficients as function of composition, activity coefficients at infinite dilution, and excess enthalpies as function of temperature using these parameters are shown in Figure 2.3. It can be seen that using temperature-dependent Wilson parameters not only the VLE behavior, but also the activity coefficients, the excess enthalpies as a function of composition and temperature, and the activity coefficients at infinite dilution are described correctly.

For the final design of a distillation column it would be desirable that all the binary parameters applied show a similar quality to describe the VLE behavior of the multicomponent system reliably. A high quality of the binary parameters is of special importance for the key components.

Nevertheless, it is not necessary to fit all the pairs of binary parameters. If both components only occur in small concentrations in each process steps, the binary parameters hardly have an influence on the results. The binary parameters of small concentration components with the key components can often be estimated. Predictive methods described in Section 2.8 are usually reliable enough to decide whether these components end up at the top or the bottom of the column. As long as they are not involved in the specification of the product streams, this information is fully sufficient.

2.4 Calculation of VLE using equations of state

As mentioned in Section 2.2, the ϕ—ϕ approach compared to the γ—ϕ approach has the great advantage that supercritical compounds, e.g. often inert compounds, can be taken into account easily and that besides the VLE behavior various other important thermodynamic properties like densities, enthalpies including enthalpies of vaporization, heat capacities, etc. can directly be calculated via residual functions for the pure compounds and mixtures at the given conditions (see Section 2.9). In principle, for the calculation of VLE any equation of state can be used which is able to describe the pvT behavior of the vapor and the liquid phase, e.g. cubic equations of state, further developments of the virial equation, or Helmholtz equations of state [3].

Most popular in chemical industry are further developments of the cubic van der Waals equation of state [17]. The van der Waals equation of state consists of a repulsive and an attractive part:

$$p = p_{\text{rep}} + p_{\text{att}}$$

$$p = \frac{R \cdot T}{v - b} - \frac{a}{v^2} \tag{2.12}$$

With only two parameters a and b, the van der Waals equation of state for the first time made it possible to describe the different observed phenomena, such as condensation, evaporation, the two-phase region, and the critical behavior. Therefore after the development of the van der Waals equation of state in 1873 an enormous number of different cubic equations of state have been suggested. Great improvements were obtained by a modification of the attractive part, by introducing a temperature dependence of the attractive parameter with the help of a so-called α-function and the development of improved mixing rules, the so-called g^E mixing rules. These developments allowed the application of equations of state also for asymmetric systems and systems with polar compounds. In this chapter only the most important cubic equations of state, such as the SRK [18], PR, [19] and volume translated [20] Peng−Robinson (VTPR) equation of state [21a−21c,22a,22b] are discussed. In Table 2.3 the analytical expressions for the repulsive and attractive part, the generalized α-functions together with the equations for the calculation of the required parameters a and b are given. In the VTPR equation of state, an additional c-parameter is introduced to obtain better results for liquid densities, and the α-function by Twu et al. is used [23]. For all mentioned equations of state the calculated vapor pressures can still be improved when instead of the generalized α-function the parameters of the α-function are directly fitted to experimental vapor pressure data. Typical results for vapor pressures and enthalpies of vaporization are shown in the figures below. In Figure 2.6 the experimental and calculated vapor pressures for five solvents are shown, where the VTPR equation of state with fitted parameters of the Twu α-function was used. It can be seen that nearly perfect agreement is obtained in the wide temperature range covered (see also Section 2.9.1). Even the slopes are described reliably. From the slopes it can be concluded that the enthalpies of vaporization increase from benzene to the alcohols, and then to water. Following the Clausius−Clapeyron equation (Eqn (2.28)) this leads to the fact that in binary systems, e.g. acetone−methanol, ethanol−benzene, etc., the low boiler at low temperature can become the high boiler at higher temperatures. At the so-called Bancroft point (intersections in Figure 2.6) the vapor pressures of the two components are identical.

In Figure 2.7 the experimental and calculated enthalpies of vaporization using the SRK and the VTPR equations of state for 11 different compounds in a wide temperature range up to the critical temperature are shown. It can be seen that both equations of state provide excellent agreement with the experimental findings. That is not surprising, since with the reliably calculated vapor pressures and volumes of the vapor phase, good enthalpies of vaporization should be expected following the Clausius−Clapeyron equation (see Eqn (2.28)).

2.4.1 Fitting of binary parameters of cubic equations of state

To be able to calculate also the behavior (phase equilibria, enthalpies, densities, heat capacities, Joule-Thomson coefficients, etc.) of mixtures using equations of state,

Table 2.3 Analytical Expressions for p_{rep}, p_{att}, the Determination of the Parameters a, b, and c, and the Generalized α-Functions of the SRK, PR, and VTPR Equation of State

	Soave–Redlich–Kwong (SRK)	Peng–Robinson (PR)	VTPR
p_{rep}	$\dfrac{RT}{v-b}$	$\dfrac{RT}{v-b}$	$\dfrac{RT}{v+c-b}$
p_{att}	$-\dfrac{a(T)}{v(v+b)}$	$-\dfrac{a(T)}{v(v+b)+b(v-b)}$	$-\dfrac{a(T)}{(v+c)(v+c+b)+b(v+c-b)}$
$a(T)$	$0.42748\dfrac{R^2 T_c^{2.5}}{p_c}\alpha(T)$	$0.45724\dfrac{R^2 T_c^2}{p_c}\alpha(T)$	$0.45724\dfrac{R^2 T_c^2}{p_c}\alpha(T)$
b	$0.08664\dfrac{RT_c}{p_c}$	$0.0778\dfrac{RT_c}{p_c}$	$0.0778\dfrac{RT_c}{p_c}$
$\alpha(T)$	$[1+(0.48+1.574\omega-0.176\omega^2)(1-T_r^{0.5})]^2$	$[1+(0.37464+1.54226\omega-0.26992\omega^2)(1-T_r^{0.5})]^2$	$T_r^{N(M-1)}\exp[L(1-T_r^{NM})]$
c			$v_{PR}-v_{exp}$ at $T_r=0.7$

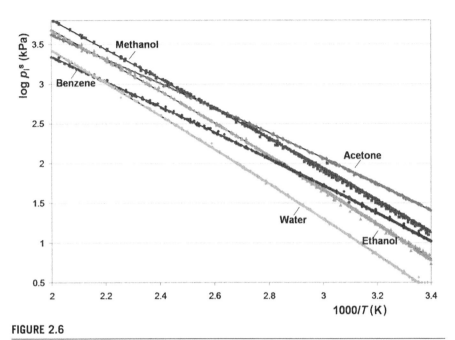

FIGURE 2.6

Experimental [1] and calculated vapor pressures for selected solvents using the
Peng–Robinson equation of state and the Twu α-function.

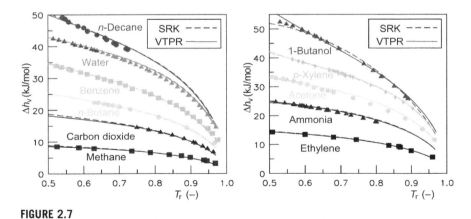

FIGURE 2.7

Experimental [1] and calculated enthalpies of vaporization using the
Soave–Redlich–Kwong and the volume-translated Peng–Robinson equation of state.

binary parameters are required. For cubic equations of state, for a long time empirical mixing rules for the parameters a and b were used. For the attractive parameter a, often the quadratic mixing rule is used:

$$a = \sum_i \sum_j z_i z_j a_{ij} \qquad (2.13)$$

For the calculation of the required cross-coefficient a_{ij}, the geometric mean of the pure component parameters is corrected using a binary parameter k_{ij} (combination rule):

$$a_{ij} = \left(a_{ii} a_{jj}\right)^{0.5} \left(1 - k_{ij}\right) \qquad (2.14)$$

Usually, the binary symmetric parameters k_{ij} (i.e. $k_{ij} = k_{ji}$) show small values. Nevertheless, they cannot be neglected, since they have a large influence on the results of phase equilibrium calculations.

For the calculation of the van der Waals co-volume b of a mixture often a simple linear mixing rule is used in most cases,

$$b = \sum_i z_i b_i \qquad (2.15)$$

These mixing and combination rules (Eqns (2.13)–(2.15)) are applied for the calculation of a and b in both the vapor and the liquid phase. Therefore instead of x_i or y_i for the mole fraction z_i is used in the equations above. For fitting the binary parameter usually VLE data are used. With the help of all the required binary parameters k_{ij} (in the case of a ternary system: k_{12}, k_{13}, k_{23}), a ternary or multicomponent system can be calculated using fugacity coefficients (see Table 2.4).

In Figure 2.8 for the binary system n-butane–CO_2 the experimental results are shown together with the calculated results using $k_{12} = 0$ and the fitted binary parameter $k_{12} = 0.1392$. It can be seen that the agreement is highly improved when going from $k_{12} = 0$ to $k_{12} = 0.1392$. Furthermore, it is remarkable that k_{12} seems to be temperature-independent over a wide temperature range.

For a long time these empirical mixing rules were successfully used in the gas-processing and petroleum industries. However poor results were obtained for systems with polar compounds. This was one of the reasons why equations of state were seldom used in the chemical industry. For the acetone–water system, the

Table 2.4 Analytical Expressions for the Residual Functions of the Gibbs Energy and Enthalpy [3]

$$(g - g^{id})_{T,p} = RT \ln \phi_i = \int_\infty^v \left(\left(\frac{\partial p}{\partial n_i} \right)_{T,v,n_j \neq n_i} - \frac{RT}{v} \right) dv - RT \ln z$$

$$(h - h^{id})_{T,p} = \int_\infty^v \left(T \left(\frac{\partial p}{\partial T_i} \right)_v - p \right) dv + pv - RT$$

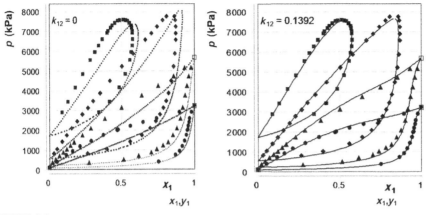

FIGURE 2.8

VLE results for the n-butane (1)–CO_2 (2) system using the binary parameter $k_{12} = 0$ and an adjusted binary parameter. ●, 270 K; ▲, 292.6 K; ◆, 325.01 K; ■, 377.6 K [1].

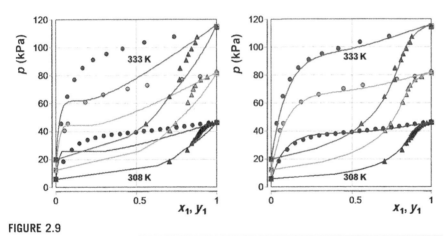

FIGURE 2.9

Experimental and calculated VLE data for the acetone (1)–water (2) system using the Soave–Redlich–Kwong equation of state with classical mixing rules (left-hand side) and with g^E mixing rules (right-hand side).

poor results obtained by fitting the binary parameter k_{12} to the experimental data are exemplarily shown on the left-hand side of Figure 2.9.

This situation was improved by Huron and Vidal [24]. They tried to combine the advantages of g^E models and equations of state and developed so-called g^E mixing rules. The principle of g^E mixing rules is that the excess Gibbs energy calculated by an activity coefficient model and by a cubic equation of state is identical:

$$g^{E,EOS} = g^{E,\gamma} = R \cdot T \sum_i x_i \cdot \ln \gamma_i \quad p = p^{ref} \tag{2.16}$$

Starting from Eqn (2.16) the following relation for the determination of the attractive parameter $a(T)$ at infinite pressure was derived by Huron and Vidal [24].

$$\frac{a(T)}{b} = \sum_i x_i \frac{a_{ii}(T)}{b_i} + \frac{g^E}{-0.6931} \quad p^{ref} = \infty \quad (2.17)$$

Later the g^E mixing rules have been further optimized with the idea to use a low reference pressure. This was achieved by Michelsen [25,26] with the so-called MHV1 and MHV2 mixing rules [3]. By analyzing the ratio of the liquid volumes/co-volumes for a large number of compounds, Holderbaum and Gmehling derived the following mixing rule for PSRK [27a,27b]:

$$\frac{a(T)}{b} = \sum_i x_i \frac{a_{ii}(T)}{b_i} + \frac{g^E + R \cdot T \sum_i x_i \cdot \ln\left(\frac{b}{b_i}\right)}{-0.64663} \quad p^{ref} = 1.013 \cdot 10^5 \text{ Pa} \quad (2.18)$$

Later on, further improved mixing rules were derived by Chen for the attractive parameter and the co-volume for the VTPR equation of state [28]:

$$\frac{a(T)}{b} = \sum_i x_i \frac{a_{ii}(T)}{b_i} + \frac{g^{E,R}}{-0.53087} \quad p^{ref} = 1.013 \cdot 10^5 \text{ Pa} \quad (2.19)$$

$$b_{ij}^{3/4} = \left(b_{ii}^{3/4} + b_{jj}^{3/4}\right)\Big/2 \quad b = \sum_i \sum_j z_i z_j b_{ij} \quad (2.20)$$

In g^E mixing rules, instead of the binary parameter k_{12}, parameters of a g^E model, e.g. of the Wilson, NRTL, or UNIQUAC models, are fitted to calculate the attractive parameter of the chosen cubic equation of state. The significant improvements that can be achieved using g^E mixing rules are shown on the right-hand side of Figure 2.9 for the acetone–water system, where also the SRK equation of state has been used. From the results shown it can be seen that the quadratic mixing rule with an adjusted k_{ij} interaction parameter fails completely, whereas the g^E mixing rule succeeds. The application of g^E mixing rules with adjusted parameters all of a sudden allowed applying equations of state for process simulation also for polar systems.

The direct adjustment of the interaction parameters of the g^E model (usually Wilson, NRTL, or UNIQUAC) to experimental data (VLE, h^E, γ^∞, ...) using a g^E mixing rule yields an accurate correlation of the data. Depending on the strength of the temperature dependence either constant or temperature-dependent g^E model parameters have to be fitted. For the acetone–water system, correlation results of the equation of state VTPR using temperature-dependent UNIQUAC parameters are shown in Figure 2.10. It can be seen that nearly perfect results are obtained for VLE, excess enthalpies, and the azeotropic composition. Even the temperature dependence of the excess enthalpies and the molar volumes is described with the required accuracy. Furthermore, it can be seen that also the VLE behavior at supercritical conditions can be calculated.

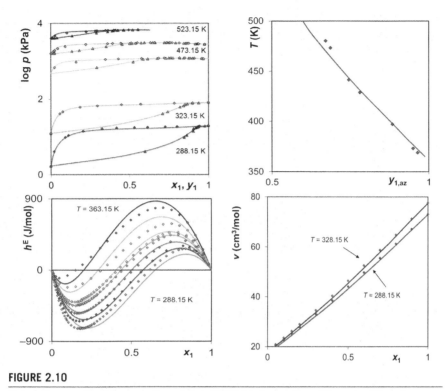

FIGURE 2.10

Correlation results of the equation of state VTPR for the acetone (1)–water (2) system using temperature-dependent UNIQUAC parameters.

2.5 Liquid–liquid equilibria

In Section 2.3.1 and Figure 2.2 it was shown that with increasing activity coefficients two liquid phases are formed. The concentration differences of the compounds in the different phases can be used, e.g. for the separation by extraction. In distillation processes, LLE occurs, e.g. in a decanter used to separate the condensate of the top product into two liquid phases. The knowledge of the VLLE behavior is of special importance for the separation of systems by heteroazeotropic distillation.

Many engineers have the opinion that the formation of an LLE does not take place in distillation columns. In fact, it does; however, in contrast to a phase equilibrium arrangement, the two phases do not separate due to turbulences (tray columns) or form thin layers that trickle down the internals of the column (packed columns). In both cases, it is useful to treat the two liquid phases as one homogeneous liquid for the determination of the overall liquid composition and the physical properties of the liquid phase, although there are attempts to take the LLE into account even for hydraulic calculations [29]. For the phase equilibrium calculations, there is no other way than to consider the liquid phase split to get the correct vapor composition.

As in the case of other phase equilibria, the fugacities in the different liquid phases are the same in the case of LLE (see Eqn (2.3)). Since the standard fugacities f_i^0 are identical for the two liquid phases, the following simple equation results from Eqn (2.3):

$$x_i' \cdot \gamma_i' = x_i'' \cdot \gamma_i'' \tag{2.21}$$

The product $x_i \cdot \gamma_i$ is called activity a_i. This means that the so-called iso-activity criterion has to be fulfilled in the case of LLE. Using fugacity coefficients, a similar relation results:

$$\left(x_i \cdot \phi_{L,i}\right)' = \left(x_i \cdot \phi_{L,i}\right)'' \tag{2.22}$$

At moderate pressures the LLE only depends on the temperature dependence of the activity coefficients respectively fugacity coefficients. For the calculation of LLE, again g^E models such as NRTL, UNIQUAC, or equations of state can be used. The formation of two liquid phases can result in the formation of binary and higher hetero-azeotropes, which can be used, e.g. for the separation of systems such as butanol—water or the production of anhydrous ethanol by heteroazeotropic distillation using e.g. cyclohexane as suitable solvent.

It is more difficult to calculate reliable LLEs of multicomponent systems using binary parameters than to describe VLEs or solid—liquid equilibria (SLEs). The reason is that in the case of LLE the activity coefficients have to describe not only the composition dependence but also the temperature dependence correctly, whereas in the case of other phase equilibria (VLE, SLE) the activity coefficients primarily have to account for the deviation from ideal behavior (Raoult's law or ideal solid solubility). The temperature dependence of VLE and SLE is mainly described by the standard fugacities (vapor pressure or melting temperature and heat of fusion).

This is the main reason why up to now no reliable LLE prediction (tie lines) using binary parameters is possible. Fortunately, it is quite easy to measure LLE data of ternary and higher systems at least up to atmospheric pressure.

In Figure 2.11 the calculated results (VLE, azeotropic data) using the UNIQUAC equation for the 1-butanol—water system are shown together with experimental VLE and LLE data. It can be seen that above 380 K instead of a heterogeneous azeotrope a homogeneous azeotrope is observed.

2.6 **Electrolyte systems**

The VLE behavior of electrolyte solutions is significantly influenced by long-range electrostatic interactions between charged ions (Coulomb—Coulomb interactions) and interactions of the charged ions and the dipoles of the solvent (Coulomb-dipole interactions). This has a considerable influence on the volatility of organic

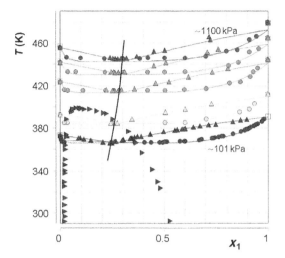

FIGURE 2.11

Calculated VLE and azeotropic data for the 1-butanol (1)—water (2) system using the UNIQUAC equation together with experimental VLE and LLE data [1]. ——, Azeotropic composition.

components or water in the presence of electrolytes [3]. If the volatility of a compound is reduced, it is called the "salting in" effect. An increase of the volatility is called the "salting out" effect. In Figure 2.12 the effects are shown for the ternary ethanol—water—salt system. It can be seen that primarily component 2 (water) forms a solvation shell with the ions, which leads to a lower partial pressure of component 2 (water) in the vapor phase (salting in effect). At the same time a salting out effect is observed for component 1 (ethanol).

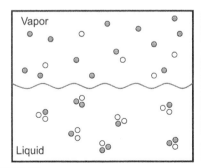 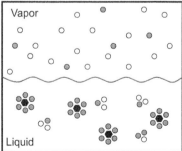

FIGURE 2.12

Influence of electrolytes, e.g. salts on the vapor—liquid equilibrium behavior component 1 (e.g. ethanol) ○; component 2 (e.g. water) ●; salt ●.

To be able to consider the effects of electrolytes in the thermodynamic models developed for nonelectrolyte systems, the long-range interactions have to be taken into account. For the development of activity coefficient models for electrolyte solutions, the theory of Debye and Hückel is usually applied as starting point. It can be regarded as an exact equation to describe the behavior of an electrolyte system at infinite dilution. But with increasing electrolyte concentration, the short-range interactions become more and more important. Therefore, in the activity coefficient models used in practice the Debye–Hückel term, which describes the long-range interactions, is extended by terms describing the short-range and middle-range interactions.

One of the most widely used activity coefficient model for electrolyte solutions has been proposed by Pitzer in 1973 [30,31]. In principle, it is a series expansion of the excess Gibbs energy, analogous to the virial equation of state. The terms take into account the short-range interactions, where binary and ternary parameters have to be adjusted as a function of the ionic strength.

The short-range interactions can also be described using local composition models, such as NRTL, UNIQUAC, or UNIFAC (see Section 2.8.1). The most widely used local composition model is the NRTL electrolyte model from Chen [32,33], which is implemented in most of the commercial process simulators.

Based on the UNIQUAC or UNIFAC model (see Section 2.8.1), electrolyte models have been developed by Li et al. [34] called LIQUAC, and by Yan et al. called LIFAC [35].

Extractive distillation with salts seems to be a possible technical application of the salting out effect. In Figure 2.13 the experimental and predicted VLE behavior using LIQUAC for the acetone–methanol–LiNO$_3$ system on the salt-free basis is shown. As can be seen, acetone–methanol forms an azeotrope at 313.15 K. By adding approximately 10 mol% lithium nitrate, the azeotrope vanishes due to the salting out effect of the acetone. From the diagram it can be concluded that there is a chance to separate the acetone–methanol system using LiNO$_3$ as entrainer.

The separation of azeotropic systems by extractive distillation using salts was applied long ago, e.g. for the production of anhydrous ethanol using a eutectic mixture of potassium and sodium acetate in the HIAG (Holzindustrie AG) process licensed by Degussa [36], where the molten salts are added to the reflux stream of the column. But this kind of process has not been applied for approximately 50 years, in particular because of the problems with solid processing and corrosion.

2.7 Conditions for the occurrence of azeotropic behavior

The greatest separation problem for distillation processes is the occurrence of binary, ternary, and quaternary azeotropic points [37]. At the azeotropic point, for homogeneous systems the mole fractions of all components in the liquid phase are identical with the mole fractions in the vapor phase. This leads to the fact that all K-factors (relative volatilities) show values of unity at the azeotropic point and that the system

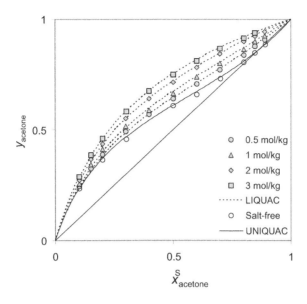

FIGURE 2.13

Experimental [1] and predicted vapor—liquid equilibria of the acetone—methanol system at 313.15 K in the presence of different amounts of $LiNO_3$ using the LIQUAC model.

cannot be separated by ordinary distillation. Therefore a reliable knowledge of all azeotropic points for the system to be separated is of great importance for the synthesis and design of the separation step.

For a binary system, the following relations can be derived for homogeneous systems at the azeotropic point using the simplified Eqn (2.8) of the $\gamma-\phi$ approach:

$$\alpha_{12} = \frac{K_1}{K_2} = \frac{y_1/x_1}{y_2/x_2} = \frac{\gamma_1 \cdot p_1^s}{\gamma_2 \cdot p_2^s} = 1 \quad \rightarrow \quad \frac{\gamma_2}{\gamma_1} = \frac{p_1^s}{p_2^s} \qquad (2.23)$$

Using the $\phi-\phi$ approach the following relation is obtained for the azeotropic point starting from Eqn (2.7):

$$\alpha_{12} = \frac{K_1}{K_2} = \frac{y_1/x_1}{y_2/x_2} = \frac{\phi_{L,1} \cdot \phi_{V,2}}{\phi_{V,1} \cdot \phi_{L,2}} = 1 \quad \rightarrow \quad \frac{\phi_{L,1}}{\phi_{V,1}} = \frac{\phi_{L,2}}{\phi_{V,2}} \qquad (2.24)$$

It can be seen that starting from Eqn (2.23) azeotropic behavior occurs if for a given composition the ratio of the activity coefficients γ_2/γ_1 is identical to the ratio of the pure component vapor pressures p_1^s/p_2^s.

The typical curvature of the γ_2/γ_1-ratio is shown in Figure 2.14 in logarithmic form for an azeotropic system with a positive (left-hand side) or negative (right-hand side) deviation from Raoult's law at constant temperature. The azeotropic composition can directly be obtained from the intersection of the straight line for the vapor pressure ratio p_1^s/p_2^s and the curve for the ratio of the activity coefficients γ_2/γ_1.

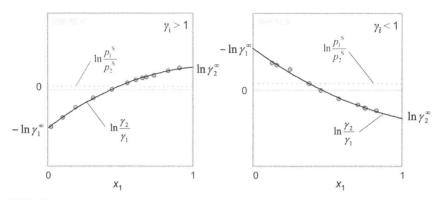

FIGURE 2.14

Examination of the azeotropic behavior of the homogeneous binary systems 1-propanol (1)–water (2) (left-hand side) and acetone (1)–chloroform (2) (right-hand side) at constant temperature (component 1 = low-boiling compound)[g].

Since the ratio of the vapor pressures $\ln p_1^s / p_2^s$ (component 1 low boiler) is always greater than $-\ln \gamma_1^\infty$ in the case of positive deviation from Raoult's law and always greater than $\ln \gamma_2^\infty$ in the case of negative deviation, azeotropic behavior always occurs when:

$$\ln \gamma_2^\infty > \ln \frac{p_1^s}{p_2^s} \text{ for } \gamma_i > 1 \quad -\ln \gamma_1^\infty > \ln \frac{p_1^s}{p_2^s} \text{ for } \gamma_i < 1 \quad (2.25)$$

From Eqn (2.23) it can be concluded that binary azeotropes easily occur if the vapor pressures of the two components are very similar, since in this case already very small deviations from Raoult's law are sufficient to fulfill Eqn (2.25) and to produce an azeotropic point either with positive or negative deviations from Raoult's law. At the Bancroft point, where the vapor pressure curves intersect, even ideal systems such as water–heavy water show azeotropic behavior [3].

From Figure 2.14 and Eqn (2.21), it can be concluded that the occurrence of azeotropic points as a function of temperature can be calculated if in addition to the activity coefficients at infinite dilution the ratio of the vapor pressures is known for the temperature range covered. The vapor pressures and the activity coefficients at infinite dilution depend on temperature following the Clausius–Clapeyron or Gibbs–Helmholtz equations [3]. The result can be that azeotropic behavior occurs or disappears with temperature (pressure). For the ethanol–1,4-dioxane and acetone–water systems, these values are shown in Figure 2.15 together with the calculated results using the NRTL equation. While for the ethanol–1,4-dioxane system the azeotropic behavior disappears at higher temperature; the opposite is true for the acetone–water system, which

---

[g]Two azeotropes can also rarely occur, if the activity coefficients show an unusual composition dependence, as in the case of the benzene–hexafluorobenzene system.

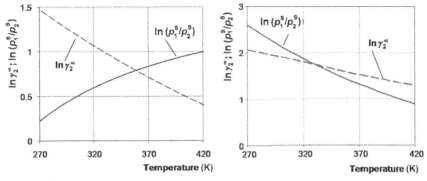

FIGURE 2.15

Temperature dependence of ln γ_2^∞ (- - -) calculated with the help of the NRTL equation and the ratio of the vapor pressures ln p_1^s/p_2^s (———) for the ethanol (1)–1,4-dioxane (2) system (left-hand side) and acetone (1)–water (2) (right-hand side).

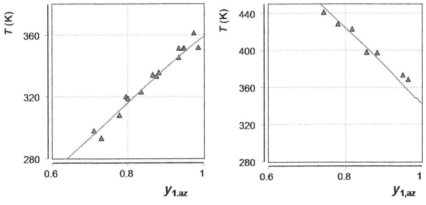

FIGURE 2.16

Experimental and calculated azeotropic compositions of the system ethanol (1)–1,4-dioxane (2) (left-hand side) and acetone (1)–water (2) (right-hand side). ▲, Experimental [1]; ———, calculated using NRTL.

shows azeotropic behavior at temperatures above 343 K. The experimental and calculated azeotropic points are shown in Figure 2.16. As can be seen, the occurrence and disappearance of the azeotropic behavior is described reliably with the NRTL model.

The procedure for the determination of ternary and quaternary homogeneous and heterogeneous azeotropes using g^E models or equations of state is explained in detail in Ref. [3].

Azeotropic behavior is not rare; approximately 45% of the published 55,500 azeotropic and zeotropic data for binary and higher systems stored in the DDB and published in a data compilation [37] show azeotropic behavior.

2.8 Predictive models

The g^E models and equations of state allow the calculation of multicomponent systems using binary information alone. However, often the required experimental binary data are missing. Assuming that 1000 compounds are of technical interest, experimental VLE information for about 500,000 binary systems are necessary to fit the required binary parameters, to be able to describe the VLE behavior of all possible binary and multicomponent systems. Although approximately 60,900 binary VLE datasets for nonelectrolyte systems have been published for the most important 1000 components [1], VLE data for only 8500 binary systems are available, since for some popular systems a large number of datasets were published, e.g. more than 300 datasets for ethanol–water. This means that parameters can be fitted for only 1.7% of the binary systems.

Since the assumption of ideal behavior can lead to very erroneous results and measurements are very time consuming, reliable predictive models with a large range of applicability would be desirable. Because of the importance of distillation processes, the objective at the beginning was the development of models for the prediction of VLE. The first predictive model with a wide range of applicability was the regular solution theory developed by Hildebrand and Scatchard [38a,38b]. But poor results were obtained for systems with polar compounds.

2.8.1 Group contribution methods (UNIFAC, modified UNIFAC)

Group contribution methods do not show the weaknesses for polar systems. In group contribution methods it is assumed that the mixture does not consist of molecules but of functional groups. It can be shown that the required activity coefficients can be calculated via group activity coefficients in the mixture Γ_k and the pure compounds $\Gamma_k^{(i)}$ using the solution of groups concept [3]:

$$\ln \gamma_i^R = \sum_k \nu_k^{(i)} \cdot \left(\ln \Gamma_k - \ln \Gamma_k^{(i)} \right) \tag{2.26}$$

when the group interaction parameters between the functional groups are known. For example, when the group interaction parameters between the alkane and the alcohol group are available, the activity coefficients (VLE behavior) for all other alkane–alcohol or alcohol–alcohol systems can be predicted.

In 1975, the UNIFAC (universal quasi-chemical theory functional group activity coefficients) group contribution method was published by Fredenslund et al. [39–41]. The UNIFAC method is based on the solution of groups concept (see Eqn (2.26)). In the UNIFAC method, the activity coefficients are calculated from a combinatorial and a residual part as in the UNIQUAC model:

$$\ln \gamma_i = \ln \gamma_i^C + \ln \gamma_i^R \tag{2.27}$$

While the temperature-independent combinatorial part takes into account the size and form of the molecules, i.e. the entropic contribution, the residual part considers the enthalpic interactions. In the UNIFAC method, for every main group combination two temperature-independent group interaction parameters (a_{nm}, a_{mn}) are required, which were fitted for a large number of group combinations to reliable experimental VLE data stored in the DDB [1]. The required expressions to calculate the combinatorial and residual part with the help of the group interaction parameters and the relative van der Waals properties for the UNIFAC method are given in Ref. [3]. Typical VLE results for alkane–ketone systems are shown in Figure 2.17 [3].

Because of the reliable results obtained for VLE, the method was directly integrated into the different process simulators. However, in spite of the reliable results obtained for VLE, UNIFAC also shows a few weaknesses [3]. Unsatisfying results are obtained:

- For activity coefficients at infinite dilution.
- For excess enthalpies, this means the temperature dependence of the activity coefficients following the Gibbs–Helmholtz relation.
- For asymmetric systems, i.e. compounds very different in size.

All these weaknesses are not surprising, since with the VLE data used to fit temperature-independent group interaction parameters no information about the temperature dependence (excess enthalpies), very asymmetric systems, and the very dilute region is used.

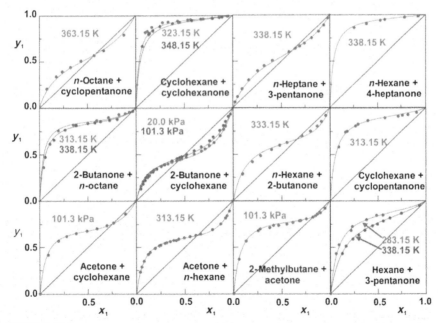

FIGURE 2.17

Experimental [1] and predicted VLE data for alkane–ketone systems using UNIFAC [3].

To reduce the weaknesses of UNIFAC, the modified UNIFAC method was developed [42]. The main differences compared to original UNIFAC are:

- An empirically modified combinatorial part was introduced to improve the results for asymmetric systems.
- Temperature-dependent group interaction parameters are used.
- Additional main groups, e.g. for cyclic alkanes, formic acid, etc. were defined.
- For fitting the temperature-dependent group interaction parameters of modified UNIFAC, in addition to VLE data, activity coefficients at infinite dilution, excess enthalpies, excess heat capacities, LLE data, azeotropic data, and SLE data of simple eutectic systems are used.

VLE (azeotropic data) provide the information about the activity coefficients for a wide composition range. Activity coefficients at infinite dilution deliver the required information for the dilute range. At the same time, γ^∞-values measured by gas—liquid chromatography supply reliable information about the real behavior of asymmetric systems. Excess enthalpies (excess heat capacities) deliver the required information about the temperature dependence. h^E-values at high temperature (>393 K) and SLE data of simple eutectic systems at low temperature are important supporting data for fitting reliable temperature-dependent group interaction parameters. To describe the temperature dependence, linear or quadratic temperature-dependent parameters are used.

Most important for the application of group contribution methods for the synthesis and design of separation processes is a comprehensive group interaction parameter matrix with reliable parameters. Since no government funds are available, the further extension, i.e. revision of the existing parameters, filling of gaps in the parameter matrix or the introduction of new main groups, has been carried out within the UNIFAC consortium [54] since 1996, with the help of systematically measured data and by using new experimental data stored in the DDB [1]. For the systematic further development of modified UNIFAC nearly 900 h^E datasets, in particular at high temperature, a large number of VLE data, SLEs of eutectic systems, and activity coefficients at infinite dilution were measured systematically in our laboratory. In the recent years new main groups were introduced for amides, isocyanates, epoxides, anhydrides, peroxides, carbonates, various sulfur compounds, N-formylmorpholine, ionic liquids [43], etc.

The present modified UNIFAC parameter matrix is shown in Figure 2.18. As can be seen today parameters for 94 main groups are available. A great part of the modified UNIFAC parameters was published by Gmehling et al. [44]. The revised and the new fitted parameters are only available for the members of the UNIFAC consortium.

Modified UNIFAC is an ideal thermodynamic model for process development. With the help of modified UNIFAC easily various process alternatives can be compared, suitable solvents for separation processes such as azeotropic distillation, extractive distillation, extraction can be selected, the influence of solvents on the chemical equilibrium conversion can be predicted, etc. Modified UNIFAC can of

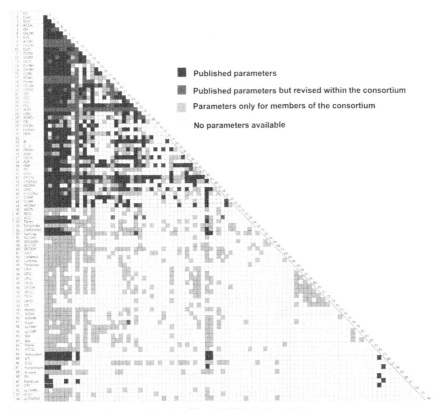

FIGURE 2.18 Present Status of the Modified UNIFAC Method [54]

course be applied to provide artificial data for fitting the missing binary parameters for a g^E model. But if the key components of a separation step are considered, for the final design an experimental examination of the results is recommended.

A comparison of the fitted results using the UNIQUAC method with the predicted results using UNIFAC and modified UNIFAC for 2200 consistent binary VLE data showed the progress when going from UNIFAC to modified UNIFAC. While using the UNIQUAC equation a mean absolute deviation of 0.0058 for the vapor phase mole fraction is obtained, a mean deviation of 0.0141 is obtained for UNIFAC and a mean deviation of 0.0088 using modified UNIFAC. But not only the results for VLE, but also for SLEs of eutectic systems, LLEs, excess enthalpies, excess heat capacities, activity coefficients at infinite dilution, and azeotropic data were distinctly improved when going from UNIFAC to modified UNIFAC.

Typical predicted results for VLEs, excess enthalpies, SLEs, azeotropic data, LLEs, and activity coefficients at infinite dilution, for systems of alkanes with ketones using modified UNIFAC, are shown in Figure 2.19 [3]. As can be seen in

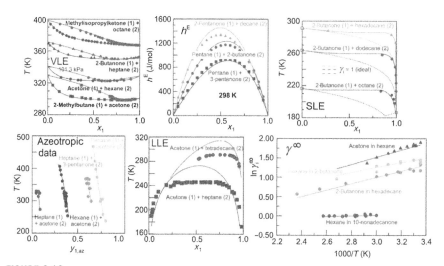

FIGURE 2.19

Experimental and predicted results for different phase equilibria for alkane–ketone systems. ——, Modified UNIFAC; ▲ ● ■, experimental [1].

all cases, good agreement is obtained for the different phase equilibria and excess properties, although a wide temperature range (170–450 K) is covered.

2.8.2 Weaknesses of the UNIFAC and modified UNIFAC group contribution methods

As shown before, the modified UNIFAC group contribution method is a powerful and reliable predictive g^E model. It is continuously further developed [54] to extend the range of applicability. But in spite of the great advantages compared to the original UNIFAC, it shows the typical weaknesses of a group contribution approach. So for example:

- Isomer effects cannot be predicted. This means the same activity coefficients are obtained, e.g. for *o-/m-/p*-xylene in the different solvents. But at least in the case of VLE or SLE calculations this is not a great problem, since the required standard fugacity, i.e. vapor pressure, melting point, and heat of fusion, are of much greater importance than small differences of the activity coefficients. By the way, similar problems with isomers are also observed for other predictive models, e.g. the quantum chemical approach.
- Unreliable results are obtained for group contribution methods in the case when a large number of functional groups have to be taken into account or when the molecule shows groups such as, e.g. $-C(Cl)(F)(Br)$. But also in these cases similar problems are observed for other approaches, e.g. the quantum chemical methods.

- Furthermore, poor results are obtained for the solubilities and activity coefficients at infinite dilution of alkanes or naphthenes in water. This was accepted by the developers of modified UNIFAC to achieve reliable VLE results, e.g. for alcohol—water systems. The problem was that starting from experimental γ^∞-values of approximately 250,000 for n-hexane in water at room temperature it was not possible to fit alcohol—water parameters that deliver γ^∞-values for hexanol in water of 800 and at the same time describe the VLE behavior of ethanol and higher alcohols with water, e.g. the azeotropic points as $f(T)$ reliably. To allow for a better prediction of hydrocarbon solubilities in water an empirical relation was developed [45,46] that allows the estimation of the solubilities of hydrocarbons in water and water in hydrocarbons [3].
- For the tertiary butanol—water system, a miscibility gap is predicted, although tertiary butanol in contrast to 1-butanol, 2-butanol, and i-butanol forms a homogeneous mixture with water.

2.8.3 Group contribution equations of state (PSRK, VTPR)

As can be recognized from the results shown before, modified UNIFAC is a powerful predictive model for the development and design of chemical processes, in particular distillation processes. However, modified UNIFAC is a g^E model. This means that it cannot handle supercritical compounds. For supercritical compounds either Henry constants or the $\phi-\phi$ approach have to be used. In practice often cubic equations of state like the SRK [18] or the PR equations [19] are applied.

The development of g^E mixing rules (see Section 2.4) enabled the combination of equations of state with group contribution methods. By combination of original UNIFAC with the SRK equation of state the predictive group contribution equation of state PSRK was developed [27a,27b]. The great advantage of this approach is that in the PSRK method, the already available UNIFAC parameters can be directly used. The possibility to handle systems at supercritical conditions all of a sudden allowed including gases such as CO_2, CH_4, H_2S, H_2, etc. as new functional groups in the parameter matrix. In total, 30 different gases were added as new main groups in PSRK.

Typical VLE results for different binary CO_2-alkane systems are shown in Figure 2.20. As can be seen, excellent results are obtained for all systems considered. This means that the group contribution concept can also applied for the new gases included in the PSRK matrix.

PSRK was even extended to systems with strong electrolytes replacing the original UNIFAC method by the electrolyte model LIFAC [35]. This allowed the prediction of salting in and salting out effects of strong electrolytes on VLEs and gas solubilities.

The PSRK model [27a,27b] provides reliable predictions of VLEs. Therefore PSRK was implemented in the different process simulators and is well accepted as a predictive thermodynamic model for the synthesis and design of the different

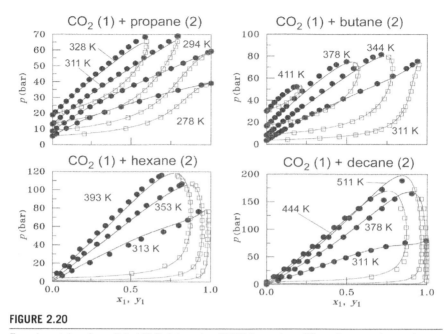

FIGURE 2.20

Experimental and predicted VLE data using PSRK for various CO_2–alkane systems at subcritical and supercritical conditions [3].

processes in chemical, gas-processing, and petroleum industries. But PSRK shows all the weaknesses of the models combined, i.e. of original UNIFAC and the SRK equation of state, i.e. poor results are obtained for liquid densities of the pure compounds and the mixtures. Furthermore, poor results are obtained for activity coefficients at infinite dilution, heats of mixing, and very asymmetric systems.

Ahlers and Gmehling [21a–21c] developed a group contribution equation of state called VTPR, where most of the above-mentioned weaknesses of PSRK were removed. A better description of liquid densities is achieved by using the VTPR (Peneloux [20]) instead of the SRK equation of state, which is used in the PSRK model. Based on the ideas of Chen et al. [28], improved g^E mixing rules are used (see Eqns (2.19) and (2.20)). Furthermore, temperature-dependent group interaction parameters are fitted. The improvements obtained when going from the group contribution equation of state PSRK to VTPR can be recognized from the predicted results using these models for symmetric and asymmetric alkane–alkane systems shown in Figure 2.21.

It has to be mentioned that for the prediction of alkane–alkane systems no interaction parameters are required for both models. This means that the results mainly depend on the mixing rules used. As can be seen from the pxy diagrams, much better results are predicted using VTPR in the case of the asymmetric ethane–tetradecane and ethane–octacosane systems, while nearly the same results are obtained for the symmetric ethane–propane and 2-methylpentane–n-heptane systems.

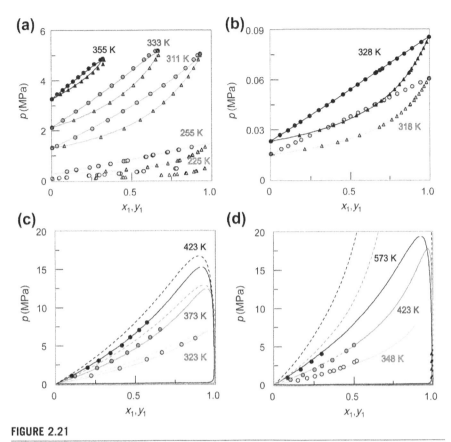

FIGURE 2.21

Experimental and predicted VLE data for symmetric and asymmetric alkane–alkane systems. (a) Ethane (1)—propane (2); (b) 2-methylpentane (1)—heptane (2); (c) ethane (1)—tetradecane (2); (d) ethane (1)—octacosane (2); - - - -,PSRK; ‾‾‾, VTPR.

The required temperature-dependent group interaction parameters of VTPR are fitted simultaneously to a comprehensive database. Besides VLE data for systems with sub- and supercritical compounds, gas solubilities, SLEs of simple eutectic systems, activity coefficients at infinite dilution, and excess enthalpies covering a large temperature and pressure range are used. The results obtained for the different pure component properties and various phase equilibria of the new group contribution equation of state are very promising [21a–21c].

In Figure 2.22 the results of the group contribution equation of state VTPR for alcohol—water systems are shown [22a,22b]. It can be seen that the predicted azeotropic data and VLEs are in very good agreement with the experimental data in the whole temperature range covered. Even the disappearance of the zeotropic point below 303 K and the VLE behavior at supercritical conditions of the ethanol—water system is described reliably. The main disadvantage of the VTPR group contribution

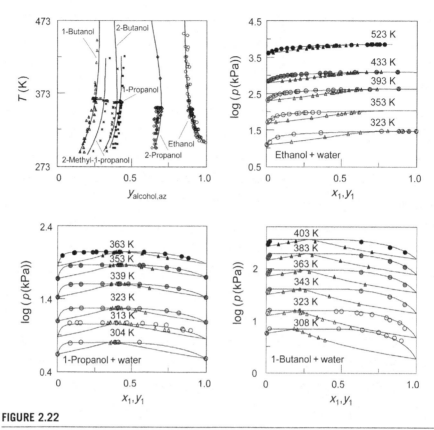

FIGURE 2.22

Experimental [1] and predicted azeotropic and VLE data using VTPR for different alcohol (1)–water (2) systems.

equation of state is that the available parameter matrix is still limited. But work is in progress to extend the existing group interaction parameter matrix [22a,22b]. Furthermore, the implementation in different process simulators is in progress.

A group contribution equation of state has great advantages compared to the usual equation of state approach in the case of multicomponent mixtures, if the components can be built up by the same functional groups as in the case of GTL (Fischer–Tropsch) processes. The reason is that the same parameters are used for all alkanes, alcohols, and alkenes, so that the number of required parameters becomes small in comparison to the typical equation of state approach. The experimental and predicted K-factors for a 12-component system consisting of nitrogen–methane–CO_2–alkanes using VTPR are shown in Figure 2.23 as a function of pressure. As can be seen, excellent results are obtained, although only group interaction parameters for six group combinations are used. Sixty-six binary k_{ij}-parameters would be required for the classical equation of state approach.

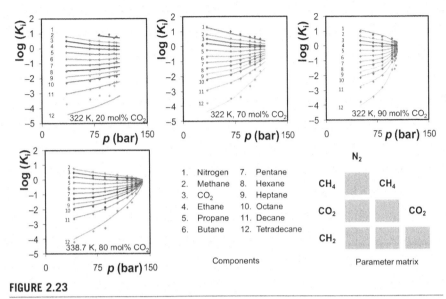

FIGURE 2.23

Experimental [55] and predicted K-factors for a 12 component system using VTPR.

An overview of the status of group contribution methods and group contribution equations of state for the prediction of phase equilibria and other thermophysical properties can be found in Refs [3,47].

2.9 Calculation of other important thermophysical properties

Besides the VLE, a number of other properties have an influence on distillation processes. The most important ones are the vapor pressure and the specific enthalpies. For hydraulic calculations in particular, the vapor density is important. The diffusion coefficients in both phases play a major role in rate-based calculations; the other transport properties and the liquid density should be known with reasonable accuracy.

2.9.1 Vapor pressure

The vapor pressure strongly influences the number of theoretical stages of distillation columns and the calculation of temperature profiles. It is an exponential function of temperature, starting at the triple point and ending at the critical point. It comprises several orders of magnitude. A logarithmic diagram allows the comparison between vapor pressures of different substances, e.g. deciding whether they intersect or not (see Figure 2.6).

Reliable vapor pressure data allow the estimation of other thermophysical properties, e.g. the enthalpy of vaporization via the Clausius–Clapeyron equation [3]:

$$\Delta h_v = T(v_V - v_L)\frac{dp^s}{dT} \tag{2.28}$$

Starting from the simplified Clausius–Clapeyron equation ($v_V \gg v_L$, ideal gas law) a simple vapor pressure equation can be derived [3]:

$$\ln p_i^s = A_i - \frac{B_i}{T} \tag{2.29}$$

A popular empirical extension of this vapor pressure equation is the Antoine equation:

$$\log p_i^s = A_i - \frac{B_i}{\vartheta + C_i} \tag{2.30}$$

The advantage of the Antoine equation is its simplicity. But it cannot be applied from the triple point to the critical point. In practice more capable vapor pressure correlations are needed. The Wagner equation allows very reliable correlations of vapor pressure data,

$$\ln \frac{p^s}{p_C} = \frac{1}{T_r} \cdot \left[A \cdot (1 - T_r) + B \cdot (1 - T_r)^{1.5} + C \cdot (1 - T_r)^3 + D \cdot (1 - T_r)^6 \right] \tag{2.31}$$

where $T_r = T/T_c$. Equation (2.30) is called the 3–6-form, where the numbers refer to the exponents of the last two terms. Some authors [48] prefer the 2.5–5-form. For the application of the Wagner equation, accurate critical data are required. As long as the experimental data points involved are far away from the critical point (e.g. only points below atmospheric pressure), estimated critical data are usually sufficient. As the critical point is automatically met due to the structure of the equation, the Wagner equation extrapolates reasonably to higher temperatures, even if the critical point is only estimated. However, like all other vapor pressure equations it does not extrapolate reliably to lower temperatures. Coefficients for the Wagner equation can be found in Ref. [49].

Similar to flexible α-functions used in equations of state, the above-mentioned relations can correlate the vapor pressures from the triple point to the critical point with excellent accuracy. In Figure 2.24 the relative vapor pressure deviations of propene for the 2.5–5-form of the Wagner equation and the Twu α-function of the VTPR equation of state are shown. It can be seen that for both approaches similarly good results are obtained. As expected for both models, larger relative deviations are obtained at lower temperatures (vapor pressures <100 Pa), which is caused by scattering of the data due to larger relative experimental uncertainties.

For a vapor pressure correlation, average deviations should be below 0.5%. Data points with correlation deviations larger than 1% should be rejected, as long as there are enough other data available. Exceptions can be made for vapor pressures below

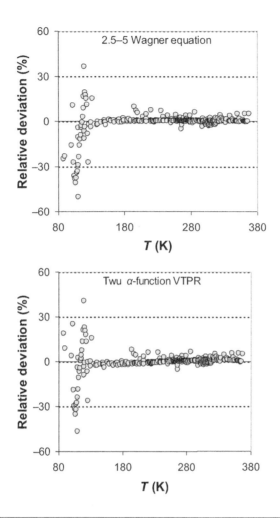

FIGURE 2.24

Relative deviation plots for the vapor pressure of propene ($T_c = 365$ K) using the Wagner 2.5–5 equation and the Twu α-function of the VTPR equation of state.

100 Pa, since the accuracy of the measurements is much lower in that range. The form of the deviations should always be carefully examined.

Despite this high accuracy demand for vapor pressures, there is also a need for good estimation methods, since often a lot of components are involved in a distillation process. But not all of these components are really important. However, one should know whether they end up at the top or at the bottom of the column. In many cases, vapor pressure measurements are not possible, as the effort for the isolation and purification of these components might be too large.

Estimation methods are mostly applied to medium and low pressures for molecules with a limited complexity. The estimation of vapor pressures is one of most difficult problems in thermodynamics. Due to the exponential relationship between vapor pressure and temperature, a high accuracy cannot be expected. Deviations in the range of 5—10% have to be tolerated, i.e. estimated vapor pressures should not be used for a main substance in a distillation column to evaluate the final design. However, they can be very useful to decide about the behavior of side components without additional measurements. In Ref. [3] different estimation methods are discussed.

Reliable vapor pressures play the most decisive role if isomers are separated by distillation. In this case, the separation factor (Eqn (2.7)) nearly only depends on the ratio of the vapor pressures, since the activity coefficients of isomers can often be set to 1 as an approximation.[h] Because separating isomers requires a large number of theoretical stages, it is strongly recommended to use vapor pressures of the isomers measured as accurately as possible with the same apparatus in the same laboratory.

Experimental vapor pressure data for more than 14,100 compounds have been published and are stored in the DDB [1]. Correlation parameters for a large number of compounds can be found in DDB [1], DIPPR [50], PPDS [51], and in the VDI Heat Atlas [52], so that there is usually no need to estimate the vapor pressure curves of key components in process simulation.

2.9.2 Specific enthalpy

The investment and operation costs of a distillation process are mainly determined by the size and the duty of the reboiler and condenser. For their evaluation, a correct description of the specific enthalpies is decisive. Furthermore, the enthalpy occurs in the heat balance of the MESH equations and is therefore necessary as well for the evaluation of the column profiles. As all components are more or less present in both the liquid and the vapor phase, the difficulty is that a continuous enthalpy description for both phases is necessary.

The extent to which particular qualities contribute to the enthalpy depends on the kind of distillation process. The most important quantities are the enthalpies of vaporization of the particular components, which are usually directly related to the reboiler duty. The standard enthalpies of formation become of course important if chemical reactions are involved. For strippers, most of the reboiler duty is often converted into sensible heat of the solvent. Therefore the energy balance is determined by the liquid heat capacities of the solvents. Except for a few well-known systems, the excess enthalpy is negligible in comparison to the enthalpy of vaporization; therefore it should be left out to avoid errors unless the binary parameters of the most important binary systems have been adjusted to excess enthalpy data as well.

---

[h]In fact, it often turns out that this is not a reasonable assumption. For example, a $\gamma^{\infty} = 0.98$, which still means that the mixture behaves nearly like an ideal mixture, would have an influence like a vapor pressure error of 2%. Therefore, it should also be considered to measure the VLE as well.

As mentioned in Section 2.4, modern cubic equations of state can directly be used to calculate the specific enthalpies and the enthalpy of vaporization via the expressions for the residual enthalpy (see Table 2.4).

The starting point for the enthalpy calculation is the definition of the standard state. Usually the ideal gas state at $T_o = 298.15$ K is used as standard state, where in process simulators the enthalpy $h_i^{id}(T_o)$ is often set to the value of the enthalpy of formation $\Delta h_{f,i}^o$ in the hypothetical ideal gas state.

For other temperatures T, the enthalpy of the different compounds in the ideal gas can be calculated using the molar heat capacities in the ideal gas state (c_p^{id}):

$$h_i^{id}(T) = \Delta h_{f,i}^o + \int_{T_o}^{T} c_{p_i}^{id}(T)dT \tag{2.32}$$

The enthalpy of the mixture in the ideal gas state $h^{id}(T)$ with the mole fractions z_i can then be obtained using the following expression:

$$h^{id}(T, z_i) = \sum_i z_i h_i^{id}(T) \tag{2.33}$$

The enthalpy for the vapor phase can then be calculated via the residual enthalpy for the vapor mixture:

$$h_V(T, p, z_i) = h^{id}(T, z_i) + \left(h_V - h^{id}\right)_{T,p,z_i} \tag{2.34}$$

For the enthalpy of the liquid mixture, the residual enthalpy for the liquid mixture has to be added to the enthalpy of the ideal gas mixture for the liquid composition.

$$h_L(T, p, z_i) = h^{id}(T, z_i) + \left(h_L - h^{id}\right)_{T,p,z_i} \tag{2.35}$$

For a pure fluid the procedure for the calculation of the specific enthalpies in the saturated phases (V, L) and the enthalpy of vaporization is shown in Figure 2.25.

The enthalpy of vaporization can also be obtained with the help of the Clausius–Clapeyron equation (see Eqn (2.28)) using the more accessible quantities of vapor pressure and molar liquid and vapor volume in the saturation state. But for associating substances, such as carboxylic acids, hydrogen fluoride, etc., chemical contributions have to be taken into account. These contributions can lead to a maximum in the enthalpy of vaporization as function of temperature.

In the case of the γ–ϕ approach, a different procedure has to be used to calculate the required enthalpies. The different procedures are described in detail in Refs [3] and [49].

2.9.3 **Density**

The density of a pure substance or a mixture is a fundamental quantity in any process calculation. For distillation processes, in particular the vapor density as a function of

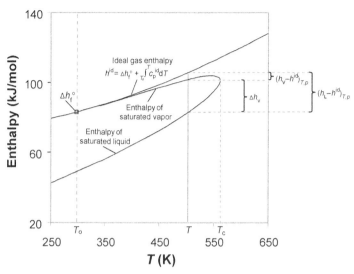

FIGURE 2.25

Calculation of the molar enthalpy (liquid, vapor) and the molar enthalpy of vaporization of a pure fluid using an equation of state.

temperature and pressure should be known, as it is decisive to determine the vapor volume, the most important quantity for hydraulic calculations. It is determined by the vapor model chosen in Eqn (2.4). The liquid density of pure components is treated only as a function of temperature. Appropriate correlations and their coefficients are given in Refs [3] and [49]. It is important to mention that for the estimation of the liquid density of mixtures, the reciprocal values of the density should be used:

$$\frac{1}{\rho_L} = \sum_i \frac{x_i}{\rho_i} \tag{2.36}$$

For the estimation of densities different methods are discussed in Ref. [3].

2.9.4 Viscosity, thermal conductivity, and surface tension

The viscosities of liquid and vapor, the thermal conductivities of liquid and vapor, and the surface tension play a subordinate role in comparison to phase equilibrium and enthalpies. Viscosities and surface tensions are used in hydraulic calculations for both tray and packed columns. High-precision correlations are not necessary, but large errors must of course be avoided. Wrong unit conversions in particular can cause erroneous orders of magnitude, and in this case even the surface tension can be responsible for a serious error in the column design. Viscosities and thermal conductivities are used in the design of the reboiler and condenser, where they play a decisive role. Viscosities and surface tension are used as well in rate-based models; the liquid conductivity is used if a heat transfer between the two phases is evaluated. Correlations, their coefficients, and estimation methods are given in Refs [3] and [49].

2.9.5 Diffusion coefficients

For all calculations where mass transfer is decisive (e.g. distillation processes when inert gases are present), the so-called rate-based approach is applied. In these cases, binary diffusion coefficients are required. Diffusion coefficients are rarely measured. But they can be estimated, and it is usually sufficient to get the correct order of magnitude. In the case of gases one can start from the kinetic gas theory. For liquids the situation is still more complicated. Estimation options for diffusion coefficients are described in Ref. [3].

2.9.6 Chemical equilibria

For reactive distillation processes, information about the enthalpy of reaction Δh_r, the equilibrium constant K and the reaction kinetics is also required. The thermodynamic properties can be calculated using tabulated standard thermodynamic properties, such as the standard enthalpies of formation $\Delta h_{f,i}^o$, standard Gibbs energies of formation $\Delta g_{f,i}^o$ and standard heat capacities $c_{p_i}^o$ using the Kirchhoff and van't Hoff equation from Ref. [3]:

$$\Delta h_r^o(T_o) = \sum_i \nu_i \Delta h_{f,i}^o(T_o) \quad \Delta h_r^o(T) = \Delta h_r^o(T_o) + \int_{T_o}^T \sum_i \nu_i c_{p_i}^o \, dT \tag{2.37}$$

$$\Delta g_r^o(T_o) = \sum_i \nu_i \Delta g_{f,i}^o(T_o) = -RT \ln K(T_o) \quad K = \frac{x_C^2}{x_A x_B} \frac{\gamma_C^2}{\gamma_A \gamma_B} = K_x K_\gamma \tag{2.38}$$

$$\frac{\partial \ln K}{\partial T} = \frac{\Delta h_r^o(T)}{RT^2} \tag{2.39}$$

To obtain correct results the real behavior, e.g. the activity coefficients γ_i, have to be taken into account. To be consistent, the real behavior has also to be used in the kinetic expressions. This means that activities instead of mole fractions should be used in the expressions for the reaction kinetics.

2.10 Application of thermodynamic models and factual databanks for the development and simulation of separation processes

The thermodynamic models introduced in Section 2.3−2.8 and factual databanks in connection with suitable software tools allow the processing of a large number of tasks during the development of distillation processes. Some of the important applications for distillation processes are:

- Calculation of VLEs, VLLEs, excess enthalpies, and activity coefficients at infinite dilution.
- Determination of separation problems (homogeneous and heterogeneous azeotropic data) in binary and higher systems.
- Construction of residue curves.
- Selection of the most suitable solvents for azeotropic or extractive distillation.
- Design of trayed or packed distillation columns.
- Consideration of the real behavior of the chemical equilibrium conversion and kinetic expression (e.g. for reactive distillation) or the driving force (rate-based model).

Using equations of state, in addition the required caloric properties (enthalpies, enthalpies of vaporization, heat capacities) e.g. required for the design in the MESH equations, the estimation of the energy consumption, etc. can be directly calculated.

Binary, ternary, and quaternary azeotropic points can lead to boundary distillation lines or surfaces that cannot be crossed by ordinary distillation. Azeotropic points can be taken directly from factual databanks or data compilations [1,36].

In multicomponent systems all the azeotropes of homogeneous systems can be determined with the help of thermodynamic models using the following objective function f_{obj}:

$$f_{obj} = \sum_j \sum_i |\alpha_{ij} - 1| = 0 \qquad (2.40)$$

For the determination of heterogeneous azeotropic points first the LLE behavior has to be calculated [2]. For two quaternary systems the results of this procedure are given in Table 2.5.

For a better understanding of the separation of ternary azeotropic systems by distillation, the construction of residue curves is quite helpful. Residue curves describe the alteration of the liquid composition of a mixture in a still during open evaporation. By introducing a nonlinear dimensionless timescale ξ one obtains the following relation (Rayleigh equation) [3]:

$$\frac{dx_i}{d\xi} = \frac{dx_i}{dL/L} = y_i - x_i \qquad (2.41)$$

where L is the molar amount of liquid in the still for the construction of distillation lines. This means the residue curves can directly be determined with the help of the thermodynamic models introduced in Section 2.3—2.8 [3]. End points of the residue curves are so-called singular[i] points. These are the pure components or the binary

---

[i]At a singular point (pure component, binary, ternary azeotrope) the bubble and dew-point line meet.

Table 2.5 Predicted Azeotropic Points for Two Quaternary Systems

| | Benzene (1)–Cyclohexane (2)–Acetone (3)–Ethanol (4) at 323 K | | | C₂H₆ (1)–H₂S (2)–Propane (3)–CO₂ (4) at 2000 kPa | | | |
| | Modified UNIFAC | | | VTPR | | | |
	Type of Azeotrope	Pressure (kPa)	$y_{1,az}$	$y_{2,az}$	Type of Azeotrope	T (K)	$y_{1,az}$	$y_{2,az}$
1–2	homPmax	40.35	0.5148		homPmax	265.55	0.9093	
1–3	–				–			
1–4	homPmax	50.63	0.6047		homPmax	248.13	0.3509	
2–3	homPmax	87.72		0.2279	homPmax	296.39		0.8643
2–4	homPmax	56.33		0.5910	–			
3–4	–				–			
1–2–3	–				–			
1–2–4	homPmax	57.12	0.1731	0.4485	–			
1–3–4	–				–			
2–3–4	–				–			
1–2–3–4	–				–			

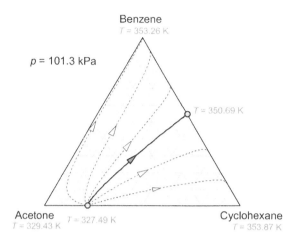

FIGURE 2.26

Predicted residual and boundary residual curves for the acetone—benzene—cyclohexane system at 101.325 kPa using modified UNIFAC.

and higher azeotropes. The topology depends on the number, type, and position of the singular points. While the residue curves are connections between the high-boiling compound or azeotrope (stable node) and the low-boiling compound or azeotrope (unstable node), the distillation boundaries are connecting lines between the saddle point and the stable or unstable node, respectively. In Figure 2.26 the calculated azeotropic points, residual curves, and boundary residual curve for the ternary benzene—cyclohexane—acetone system at 323.15 K are shown.

For the separation of azeotropic systems by distillation, different techniques can be applied. There is the possibility that the azeotropic behavior disappears at lower or higher pressure (see Section 2.7). This means that the separation can be performed in a column running at lower or higher pressure. Furthermore, a strong pressure dependence of the azeotropic composition can be used to separate the azeotropic system in two columns running at different pressure (pressure swing distillation). Heterogeneous azeotropic systems can be separated with the help of two columns and a decanter. Furthermore, hybrid processes, such as distillation and a membrane separation process or distillation and adsorption, can be applied to separate azeotropic systems, such as e.g. ethanol—water.

But in most cases special distillation processes like azeotropic or extractive distillation are applied to separate azeotropic systems. In the case of azeotropic distillation, a solvent is required that forms a lower-boiling azeotropic point, in most cases a heterogeneous azeotrope. In the case of extractive distillation, a high-boiling selective solvent is used, which alters the separation factor in a way that it becomes distinctly different from unity.

Although for azeotropic distillation the knowledge of the azeotropic points and of the miscibility gap is most important, for the selection of solvents (entrainers) for

Table 2.6 Suitable Solvents for the Separation of Benzene–Cyclohexane by Azeotropic and Extractive Distillation at Atmospheric Pressure

Azeotropic Distillation	Extractive Distillation
Acetone	Different ionic liquids
Ethanol	Sulfolane
Ethyl acetate	N-methylpyrrolidone (NMP)
Methyl acetate	N-formylmorpholine (NFM)

extractive distillation the knowledge of the influence of the entrainer on the separation factor is required.

$$\alpha_{12} \approx \frac{\gamma_1 \cdot p_1^s}{\gamma_2 \cdot p_2^s} \tag{2.42}$$

In most cases the selectivity at infinite dilution is used for selecting the most suitable solvent for extractive distillation. This means that only the knowledge about activity coefficients at infinite dilution is required.

$$S_{12}^\infty = \frac{\gamma_1^\infty}{\gamma_2^\infty} \gg 1 \ (\ll 1) \tag{2.43}$$

The required azeotropic points and activity coefficients at infinite dilution can be obtained either with the help of thermodynamic models or by access to the information stored in a factual databank, e.g. the DDB [1]. Sophisticated software packages for searching for suitable solvents using both possibilities were developed [53]. In Table 2.6 a few of the selected solvents for the separation of benzene from cyclohexane (representative for the separation of aliphatics from aromatics) are listed.

2.11 Summary

The development of sophisticated thermodynamic models (g^E models, equations of state, group contribution methods, group contribution equations of state, electrolyte models) in connection with sophisticated mathematical models and software tools allows the straightforward simulation of distillation columns or whole chemical plants. In this chapter the strengths and weaknesses of the thermodynamic models were introduced. In addition, the procedures for fitting reliable parameters and the application of these models to process design were discussed. Furthermore the calculation of additionally required thermophysical properties for distillation processes was described.

The different topics are discussed in more detail in a textbook [3], which can be recommended for further reading.

Acknowledgment

The authors are grateful to Prof. Dr U. Deiters and the diverse support received from DDBST GmbH, in particular from Dr Silke Nebig, Dr Jens Ahlers, Dr Bastian Schmid, and Wilfried Cordes during the preparation of this chapter.

References

[1] Dortmund Data Bank. www.ddbst.com.
[2] J.P. Novak, J. Matous, J. Pick, Liquid–Liquid Equilibria, Elsevier, Amsterdam, 1987.
[3] J. Gmehling, B. Kolbe, M. Kleiber, J. Rarey, Chemical Thermodynamics for Process Simulation, Wiley-VCH, Weinheim, 2012.
[4] G.M. Wilson, J. Am. Chem. Soc. 86 (1964) 127.
[5] H. Renon, J.M. Prausnitz, AIChE J. 14 (1968) 135.
[6] D. Abrams, J.M. Prausnitz, AIChE J. 21 (1975) 116.
[7] J. Gmehling, U. Onken, et al., Vapor–liquid equilibrium data collection, in: DECHEMA, Chemistry Data Series, 37 Parts, DECHEMA, Frankfurt Starting, 1977.
[8] J.A. Nelder, R. Mead, Comp. J. 7 (1965) 308.
[9] U. Hoffmann, H. Hofmann, Einführung in die Optimierung, Verlag Chemie, Weinheim, 1971.
[10] J. Gmehling, J. Menke, et al., Activity coefficients at infinite dilution, in: DECHEMA, Chemistry Data Series, Vol. IX, 6 Parts, Frankfurt Starting, 1986.
[11] J. Gmehling, T. Holderbaum, C. Christensen, et al., Heats of mixing, in: DECHEMA, Chemistry Data Series, DECHEMA, Frankfurt Starting, 1984.
[12] H.C. Van Ness, Pure Appl. Chem. 67 (1995) 859–872.
[13] L. Mertl, Collect. Czech. Chem. Commun. 37 (1972) 366.
[14] M. Margules, S.-B. Akad. Wiss. Wien. Math. Naturwiss. Kl. II 104 (1895) 1234.
[15] J.J. van Laar, Z. Phys. Chem. 72 (1910) 723.
[16] Detherm. www.dechema.de/Publikationen/Datenbanken/Detherm.html.
[17] J.D. van der Waals, (Ph.D. thesis), Leiden, 1873.
[18] G. Soave, Chem. Eng. Sci. 27 (1972) 1197.
[19] D.Y. Peng, D.B. Robinson, Ind. Eng. Chem. Fundam. 15 (1976) 59–64.
[20] A. Peneloux, E. Rauzy, R. Freze, Fluid Phase Equilib. 8 (1982) 7–23.
[21] [a] J. Ahlers, J. Gmehling, Fluid Phase Equilib. 191 (2001) 177–188.
     [b] J. Ahlers, J. Gmehling, Ind. Eng. Chem. Res. 41 (2002) 3489–3498.
     [c] J. Ahlers, J. Gmehling, Ind. Eng. Chem. Res. 41 (2002) 5890–5899.
[22] [a] B. Schmid, J. Gmehling, Fluid Phase Equilib. 317 (2012) 110–126.
     [b] B. Schmid, A. Schedemann, J. Gmehling, Ind. Eng. Chem. Res. 53 (2014) 3393–3405.
[23] C.H. Twu, D. Bluck, J.R. Cunningham, J.E. Coon, Fluid Phase Equilib. 8 (1991) 7–23.
[24] M.J. Huron, J. Vidal, Fluid Phase Equilib. 3 (1979) 255–271.
[25] M.L. Michelsen, Fluid Phase Equilib. 25 (1990) 323–327.
[26] S. Dahl, M.L. Michelsen, AIChE J. 36 (1990) 1829–1836.
[27] [a] T. Holderbaum, J. Gmehling, Fluid Phase Equilib. 70 (1991) 251–265.
     [b] S. Horstmann, K. Fischer, J. Gmehling, Fluid Phase Equilib. 167 (2000) 173–186.
[28] J. Chen, K. Fischer, J. Gmehling, Fluid Phase Equilib. 200 (2002) 411–429.

[29] M. Siegert, Dreiphasenrektifikation in Packungskolonnen, VDI-Verlag, Fortschritt-Berichte, Reihe 3: Verfahrenstechnik, Nr.586, 1999.

[30] K.S. Pitzer, J. Phys. Chem. 77 (1973) 268−277.

[31] K.S. Pitzer, G. Mayorga, J. Phys. Chem. 77 (1973) 2300−2308.

[32] C.C. Chen, H.I. Britt, J.F. Boston, L.B. Evans, AIChE J. 28 (1982) 588−596.

[33] C.C. Chen, L.B. Evans, AIChE J. 32 (1986) 444−454.

[34] J. Li, H.-M. Polka, J. Gmehling, Fluid Phase Equilib. 94 (1994) 89−114.

[35] W. Yan, M. Topphoff, C. Rose, J. Gmehling, Fluid Phase Equilib. 162 (1999) 97−113.

[36] HIAG Process, Int. Sugar J. 35 (1933) 266−268.

[37] J. Gmehling, J. Menke, J. Krafczyk, K. Fischer, Azeotropic Data, 3 Parts, Wiley-VCH, Weinheim, 2004.

[38] [a] G. Scatchard, Chem. Rev. 8 (1931) 321−333.
      [b] J. Hildebrand, S.E. Wood, J. Chem. Phys. 1 (1933) 817−822.

[39] Aa Fredenslund, R.L. Jones, J.M. Prausnitz, AIChE J. 21 (1975) 1086−1099.

[40] Aa Fredenslund, J. Gmehling, P. Rasmussen, Vapor−Liquid Equilibria Using UNIFAC, Elsevier, Amsterdam, 1977.

[41] H.K. Hansen, M. Schiller, Aa Fredenslund, J. Gmehling, P. Rasmussen, Ind. Eng. Chem. Res. 30 (1991) 2352−2355.

[42] U. Weidlich, J. Gmehling, Ind. Eng. Chem. Res. 26 (1987) 1372−1381.

[43] S. Nebig, J. Gmehling, Fluid Phase Equilib. 302 (2011) 220−225.

[44] J. Gmehling, J. Li, M. Schiller, Ind. Eng. Chem. Res. 32 (1993) 178−193.

[45] S. Batterjee, Environ. Sci. Technol. 19 (1985) 369−370.

[46] A. Jakob, H. Grensemann, J. Lohmann, J. Gmehling, Ind. Eng. Chem. 45 (2006) 7924−7933.

[47] J. Gmehling, J. Chem. Thermodyn. 41 (2009) 731−747.

[48] B.E. Poling, J.M. Prausnitz, J.P. O'Connell, The Properties of Gases and Liquids, McGraw-Hill, New York, 2001.

[49] M. Kleiber, Ind. Eng. Chem. Res. 42 (2003) 2007−2014.

[50] DIPPR Project 801, Design institute for physical property data, AIChE (2005).

[51] PPDS, TUV NEL Ltd., Glasgow. www.ppds.co.uk, 2007.

[52] M. Kleiber, R. Joh, VDI Heat Atlas, second ed., Springer, Berlin, 2010 (Chapter D3-1).

[53] J. Gmehling, C. Möllmann, Ind. Eng. Chem. Res. 37 (1998) 3112−3123.

[54] UNIFAC Consortium. http://www.unifac.org.

[55] E.A. Turek, R.S. Metcalfe, L. Yarborough, R.L. Robinson Jr, Soc. Petrol. Eng. J. 6 (1984) 308−324.

Mass Transfer in Distillation

Ross Taylor, Hendrik A. Kooijman
Clarkson University; Potsdam, NY, USA

CHAPTER OUTLINE

Distillation: Fundamentals and Principles. http://dx.doi.org/10.1016/B978-0-12-386547-2.00003-X

3.1 Introduction

The size of a distillation or absorption column depends on two things: the quantity of material to be processed and the rate at which material can move from one phase to another. This chapter focuses on the latter.

A highly idealized illustration of a section of a column is provided in Figure 3.1. An entire column is taken to consist of an array of such model sections; at the most fundamental level, the same equations are used to model both of the most common types of equipment: columns with trays and those filled with some sort of packing. In the segment shown in Figure 3.1, gas or vapor from the stage below is brought into contact with liquid from the stage above and is allowed to exchange mass and energy across the common interface represented in the diagram by the wavy line. (We limit this discussion to cases involving a gas/vapor phase and a single liquid-phase.) In modeling interphase mass transfer, it is usual to assume that these two phases are in equilibrium at the interface.

A model of this segment of column starts with component material balances for the gas/vapor and liquid phases:

$$V_l y_{i,l} = V_e y_{i,e} + \mathcal{N}_{V,i} \quad L_l x_{i,l} = L_e x_{i,e} - \mathcal{N}_{L,i} \tag{3.1}$$

where V denotes the upwards vapor molar flow rate, with the subscripts e and l meaning *entering* and *leaving*, respectively; y is the mole fraction of

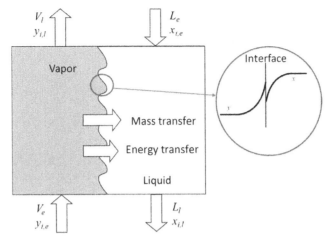

FIGURE 3.1 Schematic Diagram of a Section of a Distillation Column

The box represents some part of a distillation column, perhaps a tray in a trayed column or a slice of packing in a packed column. Inset shows representative composition profiles on either side of the phase interface; the latter usually is assumed to be at equilibrium.

species i; L denotes the downwards liquid molar flow rate; and x is the mole fraction of species i in the liquid.

The last terms on the right-hand side of Eqn (3.1) represent the compound molar flow rates out of the vapor phase and into the liquid phase.

3.2 Fluxes and conservation equations

To proceed, we define the mass transfer rates as follows:

$$\mathcal{N}_{V,i} = \int N_{V,i}\, \mathrm{d}a \quad \mathcal{N}_{L,i} = \int N_{L,i}\, \mathrm{d}a \tag{3.2}$$

where $N_{V,i}$ and $N_{L,i}$ are the molar fluxes of component i normal to the interface at a particular point in the two-phase dispersion and $\mathrm{d}a$ is the elemental interfacial area through which that flux passes.

The molar flux of species i is defined by

$$\mathbf{N}_i = c_i \mathbf{u}_i \tag{3.3}$$

which has the unit mol/m²s (or equivalent); c_i is the molar density of species i and $\mathbf{u}_i$ denotes the velocity of component i (with respect to a stationary coordinate reference frame).

The total molar flux is the sum of these quantities:

$$\mathbf{N}_t = \sum_{i=1}^{c} \mathbf{N}_i = c_t \mathbf{u} \tag{3.4}$$

We have defined the molar average velocity $\mathbf{u}$ as follows:

$$\mathbf{u} = \sum_{i=1}^{c} x_i \mathbf{u}_i \tag{3.5}$$

We may also define the molar diffusion flux by

$$\mathbf{J}_i = c_i(\mathbf{u}_i - \mathbf{u}) \tag{3.6}$$

These diffusion fluxes have the property

$$\sum_{i=1}^{c} \mathbf{J}_i = 0 \tag{3.7}$$

The molar flux N_i is related to the molar diffusion flux by

$$\mathbf{N}_i = \mathbf{J}_i + c_i\mathbf{u} = \mathbf{J}_i + x_i\mathbf{N}_t \tag{3.8}$$

These are the fluxes and velocities that are most useful for modeling mass transfer in distillation and absorption because the molar fluxes $\mathbf{N}_i$ appear in engineering design models. Other sets of fluxes could be defined; we could, for example, define a mass flux and a mass diffusion flux relative to a mass average velocity.

During interphase mass transfer, concentration gradients will be created in the fluid phase adjacent to a phase boundary. The concentration variations in each bulk phase is described by the differential balance relation for continuity of mass of species i:

$$c_t\left(\frac{\partial x_i}{\partial t} + \mathbf{u}\,\mathbf{\nabla}\cdot x_i\right) = -\mathbf{\nabla}\cdot\mathbf{J}_i \tag{3.9}$$

Equation (3.9) can be expressed in terms of the component molar fluxes N_i as the following (note that we do not consider chemical reactions in this chapter):

$$\frac{\partial c_i}{\partial t} + \mathbf{\nabla}\cdot\mathbf{N}_i = 0 \quad \frac{\partial c_t}{\partial t} + \mathbf{\nabla}\cdot\mathbf{N}_t = 0 \tag{3.10}$$

Readers are referred to standard texts on transport phenomena (e.g. Ref. [1]) for derivations. Reference [2] is also useful for additional discussion of the material in this chapter.

3.3 Constitutive relations

3.3.1 The Maxwell–Stefan equations for ideal gas mixtures

Diffusion in an ideal gas mixture with any number of different compounds is described by the Maxwell–Stefan equations:

$$\mathbf{d}_i = -\sum_{j=1}^{n} \frac{x_i x_j\left(\mathbf{u}_i - \mathbf{u}_j\right)}{Ð_{i,j}} \tag{3.11}$$

where $\mathbf{d}_i \equiv (1/p)\nabla p_i$ may be considered to be the driving force for diffusion of species i in an ideal gas mixture at constant temperature and pressure. (At non-isothermal and nonisobaric conditions, the driving force contains additional terms, but they do not impact mass transfer in distillation and are omitted here. See Ref. [2] for more details and other references.) $Ð_{i,j}$ is the Maxwell–Stefan diffusion coefficient, with the physical significance of an inverse drag coefficient.

These equations are named after the Scottish physicist James Clerk Maxwell and the Austrian scientist Josef Stefan [3,4]. These equations appeared, in more or less the form of Eqn (3.11), in an early edition of the *Encyclopedia Britannica* (incomplete forms had been published earlier) in a general article on diffusion by Maxwell [3]. Maxwell derived these equations based on his own development of the kinetic theory of gases. Stefan, who was aware of Maxwell's work, used a very different approach based on the principle of conservation of momentum. A complete derivation using Stefan's approach is available [5]. An abbreviated version of that treatment is given by Taylor and Krishna [2] and more rigorous (and complicated) developments are given in Refs [6–8].

Equation (3.11) is not in the form that is most useful to applications in chemical engineering; the species velocities may be eliminated using the definition of the molar fluxes, $N_i = c_i \mathbf{u}_i$, to get

$$\mathbf{d}_i = \sum_{j=1}^{n} \frac{x_i \mathbf{N}_j - x_j \mathbf{N}_i}{c_t Ð_{i,j}} = \sum_{j=1}^{n} \frac{x_i \mathbf{J}_j - x_j \mathbf{J}_i}{c_t Ð_{i,j}} \tag{3.12}$$

Only $n-1$ in Eqns (3.12) are independent because the mole fractions x_i sum to unity and the molar diffusion fluxes J_i sum to zero.

It is often useful to rewrite Eqn (3.12) in $n-1$ dimensional matrix form as:

$$(\mathbf{J}) = -c_t [B]^{-1} (\mathbf{d}) \tag{3.13}$$

where $[B]$ is a square matrix of order $n-1$ with elements given by

$$B_{i,i} = \frac{x_i}{Ð_{i,n}} + \sum_{\substack{k=1 \\ k \neq i}}^{n} \frac{x_k}{Ð_{i,k}} \quad B_{i,j} = -x_i \left(\frac{1}{Ð_{i,j}} - \frac{1}{Ð_{i,n}} \right) \tag{3.14}$$

and where $(\mathbf{J})$ and $(\mathbf{d})$ are column matrices of order $n-1$. For a two-component system, Eqn (3.13) becomes

$$\mathbf{J}_1 = -c_t B^{-1} d_1 = -c_t Ð_{1,2} \mathbf{d}_1 \tag{3.15}$$

where B is obtained from the first part of Eqn (3.14) as

$$B = \frac{x_1 + x_2}{Ð_{1,2}} = \frac{1}{Ð_{1,2}} \tag{3.16}$$

For two-component systems, it is straightforward to show that the Maxwell–Stefan binary diffusion coefficients are symmetric: $Ð_{1,2} = Ð_{2,1}$. For a multicomponent ideal gas mixture, a more detailed analysis [8,9] can show that $Ð_{i,j} = Ð_{j,i}$.

3.3.2 **Diffusion in nonideal fluids**

The much higher density of liquids and dense gases means that it is difficult to develop an analysis of liquid-phase diffusion in complete parallel to that which results in Eqn (3.12) for gases. However, the physical interpretation of Eqn (3.12) applies equally to gases and liquids. If (and only if) the constituents (i and j) are in motion relative to one another can we expect composition gradients to be set up in the system as a result of the frictional drag of one set of molecules moving through the other. The force acting on species i per unit volume of mixture tending to move the molecules of species i is $c_t R T \mathbf{d}_i$ where $\mathbf{d}_i$ is related to the relative velocities ($\mathbf{u}_i - \mathbf{u}_j$):

$$\mathbf{d}_i \equiv \frac{x_i}{RT} \nabla_{T,p} \mu_i = -\sum_{j=1}^{n} \frac{x_i x_j (\mathbf{u}_i - \mathbf{u}_j)}{D_{i,j}} = \sum_{j=1}^{n} \frac{x_i \mathbf{N}_j - x_j \mathbf{N}_i}{c_t D_{i,j}} \tag{3.17}$$

where the physical significance of the Maxwell–Stefan diffusivity remains that of an inverse drag coefficient. For nonideal fluids, the driving force, $\mathbf{d}_i$, involves the chemical potential gradient. These gradients arise naturally in the thermodynamics of irreversible processes as the fundamentally correct driving forces for diffusion. The subscripts T, p emphasize that the gradient in Eqn (3.17) is to be calculated under constant temperature, constant pressure conditions. (Pressure gradients and external forces also contribute to $\mathbf{d}_i$ but we ignore their influence here.) The driving force $\mathbf{d}_i$ reduces to $(1/p)\nabla p_i$ for ideal gases. Only $n-1$ driving forces are independent due to the Gibbs–Duhem restriction (e.g. [10]).

The driving force may be expressed in terms of the mole fraction gradients as follows:

$$\mathbf{d}_i = \sum_{j=1}^{n-1} \Gamma_{i,j} \nabla x_j \quad \text{where} \quad \Gamma_{i,j} = \delta_{i,j} + x_i \left. \frac{\partial ln\gamma_i}{\partial x_j} \right|_{T,p,\Sigma} \tag{3.18}$$

where $\delta_{i,k}$ is the Kronecker delta and γ_i is the activity coefficient of species i in the mixture. The symbol Σ is used to indicate that the differentiation of $ln\,\gamma_i$ with respect to mole fraction x_j is to be carried out while keeping constant the mole fractions of all other species except the n-th. The mole fraction of species n must be eliminated using the fact that the mole fractions sum to unity. More conventionally,

$$\left. \frac{\partial ln\gamma_i}{\partial x_j} \right|_{T,p,\Sigma} = \left. \frac{\partial ln\gamma_i}{\partial x_j} \right|_{T,p,x_k, k \neq j=1\ldots n-1} \tag{3.19}$$

The evaluation of $\Gamma_{i,j}$ for liquid mixtures from activity coefficient models is discussed at length by by Taylor and Kooijman [11] (see also, Appendix D of Ref. [2]).

For dense gas mixtures exhibiting deviations from ideal gas behavior, the above equation can be used with the activity coefficient γ_i replaced by the fugacity coefficient, ϕ_i:

$$\Gamma_{i,j} = \delta_{i,j} + x_i \left. \frac{\partial ln\phi_i}{\partial x_j} \right|_{T,p,\Sigma} \tag{3.20}$$

An equation of state needs to be used for the calculation of the molar density c_t and the derivatives of the fugacity coefficients for the evaluation of the thermodynamic factors from Eqn (3.20).

Equation (3.17) can be rewritten in $n - 1$ dimensional matrix form as

$$(\mathbf{J}) = -c_t[B]^{-1}[\Gamma](\nabla x) \tag{3.21}$$

where the square matrix $[B]$ and the column matrices $(\mathbf{d})$ and $(\mathbf{J})$ are as before and $[\Gamma]$ is a square matrix with elements given by Eqn (3.18) or Eqn (3.20).

3.3.3 Fick's law

Diffusion in binary mixtures most often is described by Fick's law [12,13], which takes the form:

$$\mathbf{J}_1 = c_1(\mathbf{u}_1 - \mathbf{u}) = -c_t D_{1,2} \nabla x_1 \tag{3.22}$$

Equation (3.22) defines the Fick diffusion coefficient $D_{1,2}$. A similar equation can be written for species 2, from which we see that $D_{1,2} = D_{2,1} = D$; that is, there is only one diffusion coefficient describing the molecular diffusion process in a binary mixture. There is also only one independent driving force ∇x_1 and only one independent flux $\mathbf{J}_1$.

For binary systems, Eqn (3.22) represents a linear relationship between the independent flux $\mathbf{J}_1$ and driving force ∇x_1. For n-component systems, there are $n - 1$ independent diffusion fluxes and composition gradients, and we simply continue to add terms and equations. Thus,

$$\mathbf{J}_i = -c_t \sum_{k=1}^{n-1} D_{i,k} \nabla x_k \tag{3.23}$$

Equation (3.23) represents the form of the generalization of Fick's law for multicomponent systems that is used most widely in applications in chemical engineering. It is important to recognize that many other forms exist: variations being due to the possible choices of reference velocity, units (mole or mass), and the selection of the driving force (concentration, mass or mole fraction, chemical potential etc.) [2].

The set of $n - 1$ equations (see Eqn (3.23)) is more useful when written in $n - 1$ dimensional matrix notation:

$$(\mathbf{J}) = -c_t[D](\nabla x) \tag{3.24}$$

where $(\mathbf{J})$ represents a column matrix of molar diffusion fluxes and (∇x) represents a column matrix of composition gradients with $n - 1$ elements.

The matrix $[D]$ of Fick diffusion coefficients is a square matrix of order $n - 1$. It is important to note that for multi-(n-)component diffusion, the nondiagonal or off-diagonal elements or cross-coefficients $D_{i,j}(i \neq j = 1, 2, ..., n - 1)$ are, in general, not zero.

Experimental evidence as well as theoretical work suggests that the matrix of multicomponent diffusion coefficients can be a complicated function of the composition of the mixture. The cross-coefficients $D_{i,k}$, $i \neq k$ can be of either sign; indeed, it is possible to alter the sign of these cross-coefficients by changing the order in which different compounds are numbered.

There are circumstances where the matrix $[D]$ is diagonal; the diffusion flux of species i does not, therefore, depend on the composition gradients of the other species. For a mixture made up of chemically similar species, the matrix of diffusion coefficients degenerates to a scalar times the identity matrix, that is,

$$[D] = D[I] \quad \text{(special)} \tag{3.25}$$

As the concentration of species i approaches zero, the off-diagonal elements $D_{i,k}$, $i \neq k$ also approach zero. Thus, for $n - 1$ components infinitely diluted in the n-th ($x_i \to 0$, $i = 1 \ldots n - 1$), the cross-coefficients, $D_{i,k}$, $i \neq k$, vanish. In this case, however, the diagonal elements $D_{i,i}$ are not necessarily equal. Dilute solutions occur quite often, and this special case is of some practical importance.

3.3.4 Effective diffusivity

The oldest, simplest, and still widely used method of modeling multicomponent diffusion in engineering applications employs the concept of an effective diffusion coefficient. In this class of model, the diffusion flux of species i is given by

$$\mathbf{J}_i = -c_t D_{i,\text{eff}} \nabla x_i \tag{3.26}$$

Using a model of this type greatly simplifies the mathematics involved in solving multicomponent diffusion problems. The increased simplicity comes at the cost of significantly reduced physical rigor. Equation (3.26) can be used with confidence only for systems where the binary $D_{i,k}$ displays little or no variation or in mixtures where one component is present in very large excess. These circumstances occur sufficiently often that the effective diffusivity retains some value, even though it is no longer strictly necessary.

3.4 Diffusion coefficients

Diffusion coefficients in binary liquid mixtures are of the order 10^{-9} m^2/s. Unlike the diffusion coefficients in ideal gas mixtures, those for liquid mixtures can be strong functions of concentration. The measurement of binary and multicomponent diffusion coefficients is beyond the scope of this chapter (see Refs [14–16] for descriptions of techniques and summaries of experimental results to their respective publication dates). A comprehensive collection of data on gaseous diffusion coefficients is contained in a review [17]. The most comprehensive summary of multicomponent Fick diffusion coefficients in liquids has been assembled by Mutoru and Firoozabadi [18], who explore the structural properties of the Fick matrix in some detail.

For process engineering work, it is important to be able to predict the diffusion coefficients. The experimental values of D or $Đ$, even if available in the literature, are unlikely to cover the entire range of temperature, pressure, and concentration that is of interest. We cannot here provide a comprehensive review of prediction methods (see, e.g. Refs [19–21]). Thus, in this section, we present only a modest selection of methods that have been found to be useful in engineering work.

3.4.1 Diffusion coefficients in binary mixtures

A straightforward comparison of the Maxwell–Stefan equation for diffusion in a two-component system can be made with Fick's law. Equation (3.22) leads to the following equivalence between the two diffusion coefficients for binary systems:

$$D = B^{-1}\Gamma = Đ\,\Gamma \tag{3.27}$$

where Γ is given by Eqn (3.18).

We see that the Fick D incorporates: (1) the significance of an inverse drag ($Đ$) and (2) thermodynamic nonideality (Γ). For systems that may be considered to be ideal, Γ is unity and the Fick D and the Maxwell–Stefan $Đ$ are identical.

$$D = B^{-1} = Đ \quad \text{(ideal)} \tag{3.28}$$

3.4.1.1 Estimation of diffusion coefficients in gas mixtures

The rigorous kinetic theory provides an explicit relation for the binary diffusion coefficient:

$$D = CT^{3/2}\frac{\sqrt{(M_1 + M_2)/M_1 M_2}}{p\sigma_{1,2}^2 \Omega_D} \tag{3.29}$$

where D is the diffusion coefficient (m²/s), C is 1.883×10^{-2}, T is the absolute temperature (K), p is the pressure (Pa), σ is a characteristic length (Å), and M_i is the molar mass of component i (g/mol). The parameter Ω_D, the diffusion collision integral, is a function of $k_B T/\varepsilon$ where k_B is the Boltzmann constant and ε is a molecular energy parameter. In practice, Ω_D may be calculated using a simple correlation that is suitable for computer calculations (see also Refs [19,21]). Values of σ and ε/k_B (which has units of Kelvin) can be found in the literature for a few species or estimated from critical properties [19]. The mixture σ is the arithmetic average of the pure component values. The mixture ε is the geometric average of the pure component values.

The most widely used method for gas mixtures is the correlation of Fuller et al. [22]:

$$D = CT^{1.75}\frac{\sqrt{(M_1 + M_2)/M_1 M_2}}{p(\sqrt[3]{V_1} + \sqrt[3]{V_2})^2} \tag{3.30}$$

T is expressed in Kelvin (K), P in Pascal (Pa), M_1 and M_2 in grams per mole (g/mol), $C = 1.013 \times 10^{-2}$, and D is square meters per second (m^2/s). The terms V_1 and V_2 are molecular diffusion volumes and are calculated from a molecular group contribution model. Further discussion can be found in Refs [19,21].

3.4.1.2 Diffusion coefficients in binary liquid mixtures

Collections of liquid diffusivity data can be found in other reviews [20,23,24]. The book by Tyrell and Harris [16] is a good place to begin a search for experimental measurements of D.

The binary Fick diffusion coefficient can be inferred from binary diffusion data; the Maxwell—Stefan $Đ$ may then be obtained from a rearrangement of Eqn (3.27)

$$Đ = D/\Gamma \tag{3.31}$$

The largest contribution to the composition dependence of D is due to the thermodynamic factor Γ. The Maxwell—Stefan $Đ$ calculated from Eqn (3.31) can be quite sensitive to the model used to compute Γ, an observation first made by Dullien [25]. One of the reasons for this sensitivity is that Γ involves the first derivative of the activity coefficient with respect to composition. The Maxwell—Stefan diffusion coefficients calculated from Eqn (3.31) may also be sensitive to the parameters used in the calculation of Γ. Illustrations of these points are provided elsewhere [2,11].

3.4.1.3 Estimation of diffusion coefficients in dilute liquid mixtures

As the mole fraction of either component in a binary mixture approaches unity, the thermodynamic factor Γ approaches unity and the Fick D and the Maxwell—Stefan $Đ$ are equal. The diffusion coefficients obtained under these conditions are the infinite dilution diffusion coefficients and given the symbol $Đ$.

Some of the models and correlations that have been developed are listed in Table 3.1. No correlation is best for all systems, and often alternative correlations for specific classes of system are preferred. Correlations for aqueous mixtures are most common. Comparative assessments of some correlations [21,26] recommend the Hayduk—Minhas and Tyn—Calus correlations over the Wilke-Chang and other methods not included here. Of the correlations for nonaqueous mixtures, those of Hayduk and Minhas [27], Tyn and Calus [28], and Siddiqi and Lucas [29] reportedly have about the same average error. The Modified Wilke-Chang method [30] is significantly better than the original when water is the solute.

Most of the correlations in Table 3.1 provide estimates of the infinite dilution diffusion coefficients at low pressure. Corrections for high pressure should be applied. Alternatively, a unified correlation [31] relates the diffusion coefficient in liquids at any temperature and pressure to that in dilute low-pressure gases (see Table 3.1). Average errors for this method are reported to be relatively low.

Table 3.1 Liquid Phase Diffusion Coefficients at Infinite Dilution

Stokes–Einstein

$$\mathcal{D}_{1,2}^{\circ} = \frac{k_B T}{6\pi\eta_2 r_1}$$

Wilke and Chang (1955)

$$\mathcal{D}_{1,2}^{\circ} = 7.4\cdot 10^{-8}\frac{\sqrt{\phi_2 M_2}\,T}{\eta_2 V_1^{0.6}}$$

ϕ_2 = Association factor for the solvent (2.26 for water, 1.9 for methanol, 1.5 for ethanol, and 1.0 for unassociated solvents)

Tyn–Calus [28]*

$$\mathcal{D}_{1,2}^{\circ} = 8.93\cdot 10^{-8} V_1^{1/6} V_2^{-1/3} \eta_2^{-1}\left(\frac{p_2}{p_1}\right)^{0.6} T$$

Hayduk–Minhas [27]*

$$\mathcal{D}_{1,2}^{\circ} = 1.55\cdot 10^{-8} V_2^{-0.23}\eta_2^{-0.92} p_2^{0.5} p_1^{-0.42} T^{1.29}$$

$$\mathcal{D}_{1,2}^{\circ} = 6.915\cdot 10^{-10}\eta_2^{-0.19} R_1^{0.2} R_2^{-0.4} T^{1.7}$$

Siddiqi–Lucas [29]

$$\mathcal{D}_{1,2}^{\circ} = 9.89\cdot 10^{-8}\eta_2^{-0.907} V_1^{-0.45} V_2^{0.265} T$$

He and Yu [84]

$$\mathcal{D}_{1,2}^{\circ} = A\cdot 10^{-5}\sqrt{\frac{T}{M_1}}\exp\left(-\frac{0.3887 V_{c,2}}{V_2 - 0.23 V_{c,2}}\right)$$

Where $A = 14.882 + 5.9081 k + 2.082 k^2$ $k = T_{c,1} V_{c,1}/1000\, M_1$

Modified Wilke-Chang [30]

$$\mathcal{D}_{1,2}^{\circ} = 7.4\cdot 10^{-8}\frac{\sqrt{\phi_2 M_2}\,T}{\eta_2 (\theta_1 V_1)^{0.6}} \text{ with } \theta_1 = 1;\ \theta_{H_2O} = 4.5$$

Kooijman [30]

$$\mathcal{D}_{1,2}^{0} = 1.58(1 - \phi_1)(1 - \phi_2)^{1/3}\theta_{1,2}\mathcal{D}_{1,2}^{MSE}$$

$\mathcal{D}_{1,2}^{MSE}$ = Modified Stokes–Einstein diffusion coefficient given by equation at the top of this table with $r_1 = 3.18 \times 10^{-10} R_1^{1/3}$.

$\phi_i = |1 - R_i/1.249 Q_i|$ = Normalized nonroundness of molecule i,

$\theta_{i,j} = 1 + (R_j/R_i)^{1/3}$ with R and Q = UNIQUAC parameters

Leahy-Dios and Firoozabadi [31]

$$\frac{\mathcal{D}_{1,2}^{\circ}}{D_{1,2}^{FSG}} = A_0\left(\frac{T_{r,2} p_{r,1}}{T_{r,1} p_{r,2}}\right)^{A_1}\left(\frac{\eta}{\eta^0}\right)^{A_2 + A_3}$$

$D_{1,2}^{FSG}$ = Fuller et al. diffusion coefficient

η^0 = Viscosity of gas mixture at low pressure

$A_0 = e^{-0.0472}$ $A_1 = 0.103$

$A_2 = -0.0147(1 + 10\omega_2 - \omega_1 + 10\omega_1\omega_2)$

$A_3 = -0.00053\left(P_{r,2}^{-2.663} - 6P_{r,1}^{-0.337} + 6T_{r,2}^{-1.852}\right)$

$$-0.1914 T_{r,1}^{0.1852} + 0.0103\left(\frac{T_{r,2} P_{r,1}}{T_{r,1} P_{r,2}}\right)$$

*   Water should be treated as a dimer; that is, parachor and molar volumes should be doubled. Organic acid solutes should be treated as dimers except when water, methanol, or butanol is the solvent. For nonassociating solutes in monohydroxy alcohols, the solvent parachor and molar volume should be multiplied by $8\eta_2$.

Continued

Table 3.1 Liquid Phase Diffusion Coefficients at Infinite Dilution—Cont'd	
Notation (units)	$Ð_{1,2}^o$ = Diffusion coefficient of species 1 infinitely diluted in species 2 (cm^2/s)
	M_2 = Molar mass of the solvent (g/mol), T = temperature (K)
	η_2 = Dynamic viscosity of the solvent (mPa s = cP)
	V_1 = Molar volume of solute 1 at its normal boiling point (cm^3/mol)
	$V_{c,1}$ = Critical molar volume of solvent 2 (cm^3/mol)

3.4.1.4 Estimation of diffusion coefficients in concentrated liquid mixtures

Most methods for predicting $Ð$ in concentrated solutions attempt to combine the infinite dilution coefficients $Ð_{1,2}^o$ and $Ð_{2,1}^o$ in a simple function of composition. The simplest expression

$$Ð_{1,2} = x_2 Ð_{1,2}^o + x_1 Ð_{2,1}^o \qquad (3.32)$$

proposed by Caldwell and Babb [32] has been recommended by Danner and Daubert [19]. A logarithmic average has been proposed by Vignes [33]:

$$Ð_{1,2} = \left(Ð_{1,2}^o\right)^{x_2} + \left(Ð_{2,1}^o\right)^{x_1} \qquad (3.33)$$

Equation (3.33) is not always the better way to estimate $Ð$, as had been shown earlier by Vignes himself. However, the Maxwell–Stefan diffusion coefficient can be quite sensitive to the correlation used to calculate the activity coefficients [25]. The Vignes equation is less successful for mixtures containing an associating component (e.g. an alcohol).

The sensitivity of the predicted Fick D to the thermodynamic model used to calculate Γ, as well as to the model parameters, is such that any recommendation is subject to some uncertainty. That said, the present authors make most frequent use of the Vignes method in Eqn (3.33). Other, sometimes related, methods are discussed by Poling et al. [21], who recommend the Vignes equation.

3.4.2 Estimation of multicomponent diffusion coefficients

The rules of matrix algebra do not allow us to assert that $[D]$ and $[B]^{-1}[\Gamma]$ are equal. The equality of these two matrices is an assumption, but it is the only reasonable way to relate the Fick diffusion coefficients $D_{i,j}$ to the Maxwell–Stefan diffusion coefficients $Ð_{i,j}$. Thus,

$$[D] = [B]^{-1}[\Gamma] \qquad (3.34)$$

is used throughout the literature on multicomponent mass transfer and allows us to estimate the Fick matrix $[D]$ from information on the binary Maxwell–Stefan diffusivities $Ð_{i,j}$ and activity (or fugacity) coefficients.

3.4.2.1 Estimation of multicomponent diffusion coefficients for gas mixtures

For ideal gases, the thermodynamic matrix $[\Gamma]$ reduces to the identity matrix. Equation (3.34) becomes

$$[D] = [B]^{-1} \tag{3.35}$$

For gases, the diffusion coefficients of each binary pair $Ð_{i,j}$ can be estimated from the kinetic theory of gases or from an appropriate correlation to a reasonable degree of accuracy, particularly for nonpolar molecules. The matrix of diffusion coefficients may therefore be calculated using Eqn (3.35). If all of the binary diffusion coefficients $Ð_{i,j}$ are equal, then the matrix of diffusion coefficients becomes a scalar times the identity matrix:

$$[D] = Ð[I] \quad \text{(special)} \tag{3.36}$$

For $n - 1$, components infinitely diluted in another, the Fick matrix simplifies to

$$D_{i,i} = Ð_{i,n} \quad D_{i,j} = 0 \tag{3.37}$$

3.4.2.2 Estimation of Maxwell–Stefan diffusion coefficients for multicomponent liquid mixtures

In order to use Eqn (3.34) to predict $[D]$, we need to be able to estimate the Maxwell–Stefan diffusivities of each binary pair in the multicomponent mixture. There are relatively few experimental values of Maxwell–Stefan diffusivities in multicomponent liquids. Most methods that have been suggested are based on extensions of the techniques proposed for binary systems discussed above. The Vignes equation, for example, may be generalized as follows [34,35]:

$$Ð_{i,j} = \prod_{k=1}^{n} Ð_{i,j,x_k \to 1}^{x_k} \tag{3.38}$$

where $Ð_{i,j,x_k \to 1}$ are the limiting values of the Maxwell–Stefan diffusivities in a mixture where component k is present in a very large excess.

Equation (3.38) must reduce to the binary Vignes equation when $x_i + x_j \to 1$ and $x_k \to 0 (k \neq i, j)$. Thus:

$$Ð_{i,j,x_j \to 1} = Ð_{i,j}^{\circ} \quad Ð_{i,j,x_i \to 1} = Ð_{j,i}^{\circ} \tag{3.39}$$

which leads to

$$Ð_{i,i} = \left(Ð_{i,j}^{\circ}\right)^{x_j} \left(Ð_{j,i}^{\circ}\right)^{x_i} \prod_{k \neq i,j=1}^{n} Ð_{i,j,x_k \to 1}^{x_k} \tag{3.40}$$

Several (mostly) ad-hoc models for the limiting diffusivities $Ð_{i,j,x_k \to 1}$ have been put forward. Kooijman and Taylor [34]; proposed the following model for the limiting diffusivities:

$$Ð_{i,j,x_k \to 1} = \left(Ð_{i,k}^{\circ} Ð_{k,i}^{\circ}\right)^{1/2} \tag{3.41}$$

A more complicated model was proposed by Rehfeldt and Stichlmair [36].

Using molecular simulation [37], the limiting MS diffusion coefficients may be obtained from molecular simulations; this led to the proposal of the model below:

$$D_{i,j,x_k \to 1} = \frac{Ð_{i,\text{self}} Ð_{j,\text{self}}}{Ð_{k,\text{self}}} = \frac{Ð_{i,k,x_k \to 1} Ð_{j,k,x_k \to 1}}{Ð_{k,\text{self}}} \tag{3.42}$$

which includes Eqn (3.41) as a (very) special case. If a reliable method of estimating the self-diffusion coefficients needed by Eqn (3.42) is available, then this method is the only one that has any theoretical basis. Further work is needed to develop this approach; absent a validated method, we suggest Eqn (3.41) as, arguably, among the least unsatisfactory.

3.5 Mass transfer coefficients

Consider the two-phase system shown in Figure 3.2, where representative composition profiles are shown. The bulk phase mole fractions are denoted by $x_{i,b}$ and $y_{i,b}$ for the two phases. The interface compositions are $x_{i,I}$ on the x side of the interface and $y_{i,I}$ on the y side of the interface. The interface itself is assumed to be a surface that offers no resistance to mass transfer. The mole fractions $y_{i,I}$ and $x_{i,I}$ are assumed to be in equilibrium. The starting point for any analysis of the interphase mass transfer process is a mass balance at the interface. For most practical purposes, this balance takes the form

$$N_i^x = N_i^y = N_i \quad N_t^x = N_t^y = N_t \tag{3.43}$$

where N_i^x is the normal component of $\mathbf{N}_i$ in the x phase and N_i^y is the corresponding flux in the y phase.

3.5.1 Binary mass transfer coefficients

The mass transfer coefficient k in phase x for a binary system may be defined in the way suggested by Bird et al. (1960, p. 639)

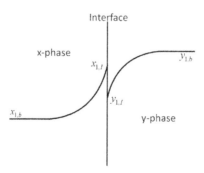

FIGURE 3.2 Representative Composition Gradients Adjacent to Phase Interface

$$k_b = \lim_{N_i \to 0} \frac{N_{1,b} - x_{1,b}N_{t,b}}{c_t(x_{1,b} - x_{1,I})} = \frac{J_{1,b}}{c_t \Delta x_1} \qquad (3.44)$$

where the driving force for mass transfer is $\Delta x_1 = x_{1,b} - x_{1,I}$. It is the bulk phase diffusion fluxes that appear in Eqn (3.44) and the bulk phase mass transfer coefficients that are obtained from Eqn (3.44). We may obtain the corresponding quantities at the interface by using the interface compositions when calculating the convective term $x_1 N_t$ in Eqn (3.44). Equation (3.44) can be applied to any phase.

The units of k_b as defined above are those of velocity. In fact, k_b is the maximum velocity (relative to the velocity of the mixture) at which a component can be transferred in the binary system [2].

Let us now discuss why the limit $N_i \to 0$ appears in Eqn (3.44). During the actual mass transfer process itself, the composition (and velocity) profiles are distorted by the flow (diffusion) of 1 and 2 across the interface. The mass transfer coefficient defined in Eqn (3.44) corresponds to conditions of vanishingly small mass transfer rates, when such distortions are absent. It is these low-flux or zero-flux coefficients that typically are available from empirical correlations of mass transfer data. These correlations usually are obtained under conditions where the mass transfer rates are low. For the actual situation under conditions of finite transfer, we have

$$k_b^\bullet = \frac{N_{1,b} - x_{1,b}N_t}{c_t \Delta x_1} = \frac{J_{1,b}}{c_t \Delta x_1} \qquad (3.45)$$

The superscript $\bullet$ indicates that the transfer coefficients $k_b^\bullet$ correspond to conditions of finite mass transfer rates. For further discussion of this point, see Bird et al. (1960, Chapter 21). The finite flux coefficient $k_b^\bullet$ usually can be related to the zero-flux coefficient by a relation of the form

$$k_b^\bullet = k_b \Xi_b \qquad (3.46)$$

where Ξ_b is a correction factor that accounts for the effect of finite fluxes on k_b.

3.5.2 Multicomponent mass transfer coefficients

For multicomponent mixtures, we define a matrix of finite flux mass transfer coefficients $[k_b^\bullet]$ by

$$(J_b) = (N) - (x_b)N_t = c_t[k_b^\bullet](x_b - x_I) = c_t[k_b^\bullet](\Delta x) \qquad (3.47)$$

The finite flux coefficients are related to the zero-flux or low-flux coefficients by a matrix equation of the form

$$[k_b^\bullet] = [k_b][\Xi_b] \qquad (3.48)$$

where $[\Xi_b]$ is a matrix of correction factors.

For a ternary system, Eqn (3.47) expands as follows:

$$\begin{aligned} J_1 &= c_t k_{1,1}^\bullet \Delta x_1 + c_t k_{1,2}^\bullet \Delta x_2 \\ J_2 &= c_t k_{2,1}^\bullet \Delta x_1 + c_t k_{2,2}^\bullet \Delta x_2 \end{aligned} \qquad (3.49)$$

Now, $k_{1,2}^{\bullet}$, $k_{2,1}^{\bullet}$, Δx_1, and Δx_2 can take either positive or negative signs. It is therefore possible for the term $k_{1,2}^{\bullet}\Delta x_2$ to overshadow $k_{1,1}^{\bullet}\Delta x_1$; if the terms have opposite signs, then $J_1/\Delta x_1 < 0$ and component 1 is said to experience *reverse diffusion* [38]. Two related phenomena are:

1. *Osmotic* diffusion, when there exists a nonvanishing flux of component 1, $J_1 \neq 0$, even when its constituent driving force is zero $\Delta x_1 = 0$.
2. A so-called *diffusion barrier* exists for component 1, $J_1 = 0$ despite a nonzero driving force, $\Delta x_1 \neq 0$.

None of these phenomena can take place in a two-component system.

3.5.3 **The bootstrap problem**

Section 3.6 is devoted to methods of estimating the low flux mass transfer coefficients k and $[k]$ and of calculating the high flux coefficients $k_b^{\bullet}$ and $[k_b^{\bullet}]$. In practical applications, we will need these coefficients to calculate the molar fluxes N_i. N_i values are needed because it is these fluxes that appear in the material balance equations for distillation and absorption processes. Thus, even if we know (or have an estimate of) the diffusion fluxes J_i, we cannot immediately calculate the molar fluxes N_i because all n of these fluxes are independent, whereas only $n - 1$ of the J_i are independent. We need one other piece of information if we are to calculate N_i. The problem of determining N_i knowing J_i has been called the *bootstrap problem* (see Ref. [2] for an extended discussion). In many cases, we can employ a generalized relationship between the fluxes that has the following form:

$$\sum_{i=1}^{n} v_i N_i = 0 \tag{3.50}$$

N_i can then be related to J_i by

$$N_i = -\sum_{k=1}^{n-1} \beta_{i,k} J_k \tag{3.51}$$

where $\beta_{i,k}$ is defined by the following equation (where $\delta_{i,k}$ is the Kronecker delta):

$$\beta_{i,k} = \delta_{i,k} - x_i \Lambda_k \quad \text{where} \quad \Lambda_k = (v_k - v_n) \bigg/ \sum_{j=1}^{n} v_j x_j \tag{3.52}$$

For equimolar counterdiffusion (often assumed to be true in distillation), the v_i must be equal.

$$v_i = v_n \quad \beta_{i,k} = \delta_{i,k} \quad (N_t = 0) \tag{3.53}$$

For nonequimolar distillation, the v_i may be set equal to the molar latent heats of vaporization:

$$v_i = \Delta H_{\text{vap},i} \tag{3.54}$$

where the $\nu_i = \Delta H_{vap,i}$ are the molar latent heats of vaporization. It can be seen that if the molar latent heats are equal, the total flux N_t vanishes. Equation (3.53) is, in fact, a special case of a more general relationship between the fluxes [2].

For Stefan diffusion, all of the ν_i are set to zero, except one that must have a nonzero value:

$$\nu_i = 0 \quad \nu_n = 1 \quad (N_n = 0) \tag{3.55}$$

This situation is encountered during condensation in the presence of a noncondensing gas as well as in processes where one or more species are removed from a gas stream by absorption into a liquid.

Equation (3.51) will be needed in $n - 1$ dimensional matrix form:

$$(N) = [\beta](J) \tag{3.56}$$

$[\beta]$ is known as the bootstrap matrix and has elements given by Eqn (3.52) [39]. For equimolar counterdiffusion, $[\beta] = [I]$. Departures from the identity matrix signify the increasing importance of the convective term $x_i N_t$.

3.5.4 Interphase mass transfer

Consider, once again, mass transfer across the phase boundary in Figure 3.2. Equation (3.43) states that we must have continuity of the fluxes across the interface. These fluxes can be written in terms of the driving forces for mass transfer on either side of the interface as

$$
\begin{aligned}
(N) &= c_t^L [k_L^{\bullet}] (x^L - x^I) + N_t (x^L) = c_t^L [\beta^L] [k_L^{\bullet}] (x^L - x^I) \\
(N) &= c_t^V [k_V^{\bullet}] (y^I - y^V) + N_t (y^V) = c_t^V [\beta^V] [k_V^{\bullet}] (y^I - y^V)
\end{aligned}
\tag{3.57}
$$

As noted earlier, we assume that at the interface itself the two phases are in equilibrium with each other. The compositions on either side of the interface are, therefore, related by

$$y_i^I = K_i x_i^I \quad i = 1, \ldots, n \tag{3.58}$$

K_i are the equilibrium ratios or "K values."

It is useful to linearize the vapor–liquid equilibrium relationship for the interface over the range of compositions obtained in passing from the bulk to the interface conditions

$$(y^I) = [M](x^I) + (b) \text{ where } M_{i,j} = \partial y_i^* / \partial x_j \quad i,j = i, 2, \ldots, n - 1 \tag{3.59}$$

y_i^* is the mole fraction of a vapor in equilibrium with a liquid of composition x_i and (b) is a column matrix of "intercepts." If the vapor liquid equilibrium ratios (K values) are given by

$$K_i = \gamma_i \frac{p_i^s}{p} \tag{3.60}$$

then it can be shown that the matrix $[M]$ is given by

$$[M] = [K][\Gamma] \tag{3.61}$$

where $[K]$ is a diagonal matrix with elements that are the first $n - 1$ equilibrium ratios and $[\Gamma]$ is the matrix of thermodynamic factors defined by the second part of Eqn (3.18).

3.5.5 Overall mass transfer coefficients

Summing the resistance in both phases in order to obtain a single expression for computing the fluxes without knowing the interface composition is widely discussed in the literature on binary mass transfer.

We define the composition of a vapor that would be in equilibrium with the bulk liquid using Eqn (3.61) as

$$(y^*) = [M](x^L) + (b) \tag{3.62}$$

For binary systems, we may define an overall mass transfer coefficient, $K_{OV}^{\bullet}$, by

$$J_1^V = c_t^V K_{OV}^{\bullet}(y_1^V - y_1^*) \tag{3.63}$$

For binary systems, all matrices contain must one element and we can show that

$$\frac{1}{c_t^V K_{OV}^{\bullet}\beta^V} = \frac{1}{c_t^V k_V^{\bullet}\beta^V} + \frac{M}{c_t^L k_L^{\bullet}\beta^L} \tag{3.64}$$

where, $M = K_1\Gamma$, with Γ defined in Eqn (3.18).

For multicomponent systems, the matrix generalization of Eqn (3.63) is

$$(J^V) = c_t^V[K_{OV}^{\bullet}](y^V - y^*) \tag{3.65}$$

where the matrix of multicomponent overall mass transfer coefficients, $[K_{OV}^{\bullet}]$, is defined by a generalization of Eqn (3.64):

$$[K_{OV}^{\bullet}]^{-1}[\beta^V]^{-1} = [k_V^{\bullet}]^{-1}[\beta^V]^{-1} + \frac{c_t^V}{c_t^L}[M][k_L^{\bullet}]^{-1}[\beta^L]^{-1} \tag{3.66}$$

In this case, it is necessary to *define* $[K_{OV}^{\bullet}]$ by Eqn (3.66) because we have only $n - 1$ independent equations and the matrix $[K_{OV}^{\bullet}]$ contains $(n - 1)^2$ elements.

If the total flux is near zero (as it will often be in distillation), the bootstrap matrices $[\beta]$ reduce to the identity matrix. Moreover, in distillation, the finite flux mass transfer coefficients will often be well approximated by their low flux limits and we can use the still simpler expression:

$$[K_{OV}]^{-1} = [k_V]^{-1} + \frac{c_t^V}{c_t^L}[M][k_L]^{-1} \tag{3.67}$$

For binary systems, Eqn (3.67) simplifies to

$$\frac{1}{K_{OV}} = \frac{1}{k_V} + \frac{c_t^V}{c_t^L}\frac{M}{k_L}$$

(3.68)

Equations (3.65) and (3.67) are used in the development of expressions for modeling mass transfer in multicomponent distillation, a topic we consider in Sections 3.8 and 3.9. The addition of the resistances concept has seen use in distillation models by, among others, Krishna et al. [88] and Gorak and Vogelpohl [40].

3.6 Estimation of mass transfer coefficients in binary systems

In practice, mass transfer coefficients may be obtained from either one of the following sources:

- Experimental data
- Empirical correlations (of experimental data)
- Theoretical models

In this section, we will briefly describe two important models of mass transfer.

3.6.1 The film model

In the film model, we assume that all of the resistance to mass transfer is concentrated in a thin film that is adjacent to the phase boundary. Mass transfer within this film occurs only by steady-state molecular diffusion; outside the film, the level of fluid mixing is so high that all composition gradients are wiped out. Figure 3.3 shows the model. It is essential to emphasize that this is a conceptual model; a

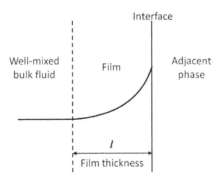

FIGURE 3.3 Film Model for Mass Transfer

Composition variations are restricted to a layer (film) of thickness adjacent to the phase interface. Model devised by Lewis and Whitman.

diffusion film does not actually exist, but if it did its thickness would be approximately 0.01–0.1 mm for liquid phases and in the range 0.1–1 mm in gases.

The diffusion process is fully described by

1. The one-dimensional steady-state of forms of Eqn (3.10)

$$\frac{dN_i}{dr} = 0 \quad \frac{dN_t}{dr} = 0 \tag{3.69}$$

which means that N_i and N_t are independent of position in the film.

2. The constitutive relations (see Eqn (3.17) or Eqn (3.23))
3. The determinacy condition (see Eqn (3.50))
4. The boundary conditions of the film model

$$\begin{aligned} r &= r_0 \; x_i = x_{i0} \\ r &= r_\delta \; x_i = x_{i\delta} \end{aligned} \tag{3.70}$$

For a two-component system, the generalized Maxwell–Stefan diffusion equations (see Eqn (3.17)) simplify to

$$\frac{dx_1}{dr} = \frac{x_1 N_2 - x_2 N_1}{c_t D} \tag{3.71}$$

where $D = Đ\,\Gamma$ is the Fick diffusion coefficient. It is usual to assume that the Fick D is constant; this is equivalent to assuming that the Maxwell–Stefan $Đ$ and the thermodynamic factor Γ are constant. The Fick D is, in fact, constant for ideal gas mixtures at constant temperature and pressure and the assumption of constant D is reasonable for modest concentration changes in nonideal systems.

The solution to Eqn (3.71) and the associated boundary conditions is:

$$\frac{x_1 - x_{10}}{x_{1\delta} - x_{10}} = \frac{\exp(\Phi\bar{z}) - 1}{\exp \Phi - 1} \tag{3.72}$$

where we have defined a dimensionless distance $\bar{z}$

$$\bar{z} = \frac{r - r_0}{r_\delta - r_0} = \frac{r - r_0}{l} \tag{3.73}$$

where l is the thickness of the diffusion layer or "film." We have also defined a mass transfer rate factor:

$$\Phi = \frac{N_1 + N_2}{c_t D / l} = \frac{N_t}{c_t D / l} \tag{3.74}$$

The diffusion flux at $\bar{z} = 0$, J_{10} is given by

$$J_{10} = -c_t \frac{D}{l} \frac{dx_1}{d\bar{z}} \Big|_{\bar{z}=0} = c_t \frac{D}{l} \frac{\Phi}{(\exp \Phi - 1)} (x_{10} - x_{1\delta}) \tag{3.75}$$

Similarly, the diffusion flux at $\bar{z} = 1$, $J_{1\delta}$ can be obtained:

$$J_{1\delta} = -c_t \frac{D}{l} \frac{dx_1}{d\bar{z}} \Big|_{\bar{z}=1} = c_t \frac{D}{l} \frac{\Phi \exp \Phi}{\exp \Phi - 1} (x_{10} - x_{1\delta}) \tag{3.76}$$

A comparison of Eqns (3.75) and (3.76) with the basic definition of the low flux mass transfer coefficient (see Eqn (3.44)) shows that

$$k_0 = k_\delta = k = D/l \tag{3.77}$$

with the correction factors given by

$$\Xi_0 = \frac{\Phi}{\exp \Phi - 1} \quad \Xi_\delta = \frac{\Phi \exp \Phi}{\exp \Phi - 1} \tag{3.78}$$

In the limit of N_t tending to zero, $\Phi = 0$ and the correction factors are $\Xi_0 = 1$ and $\Xi_\delta = 1$.

The flux N_1 can be calculated by multiplying the diffusion flux by the appropriate bootstrap coefficient:

$$N_1 = \beta_0 J_{10} = c_t \beta_0 k \frac{\Phi}{\exp \Phi - 1} (x_{10} - x_{1\delta})$$

$$= \beta_\delta J_{1\delta} = c_t \beta_\delta k \frac{\Phi \exp \Phi}{\exp \Phi - 1} (x_{10} - x_{1\delta}) \tag{3.79}$$

To compute the flux N_1 from Eqn (3.79) requires an iterative procedure because N_1 (and N_2) are involved in the rate factor Φ. Repeated substitution of the fluxes starting from an initial guess calculated with $\Xi = 1$ will usually converge in only a few iterations. It is possible, however, to derive an equation from which the flux N_t (and, hence, N_1 and N_2) may be calculated without iteration:

$$\Phi = N_t / c_t k = \ln(\beta_0 / \beta_\delta) \tag{3.80}$$

When the total molar flux is zero:

$$N_1 = -N_2 = c_t k(x_{10} - x_{1\delta}) \quad (N_t = 0) \tag{3.81}$$

For the case where $N_2 = 0$, Eqn (3.80) simplifies to

$$N_1 = c_t \, k \ln\left(\frac{1 - x_{1\delta}}{1 - x_{10}}\right) \quad (N_2 = 0) \tag{3.82}$$

The former situation can arise in, for example, a distillation column when the molar heats of vaporization of the two components are equal to each other; the latter situation arises very often during absorption or condensation operations.

3.6.2 Surface renewal models

In the so-called penetration or surface renewal models, fluid elements (or eddies) are pictured as arriving at the interface from the bulk fluid phase and residing at the interface for a period of time t_e (the exposure time). During the time t_e that the fluid element resides at the interface, mass exchange takes place with the adjoining phase by a process of unsteady-state diffusion. The fluid element is quiescent during this exposure period at the interface and the diffusion process is purely molecular. The element may, however, move in plug flow along the interface. After exposure and consequent mass transfer, the fluid elements return to the bulk fluid phase and are replaced by fresh eddies. A pictorial representation of this model, based on that in Scriven [41], is shown in Figure 3.4.

For one-dimensional, unsteady-state diffusion in a planar coordinate system Eqn (3.10) may be written as

$$c_t \frac{\partial x_i}{\partial t} + \frac{\partial N_i}{\partial z} = 0 \tag{3.83}$$

where z represents the direction coordinate for diffusion. For the mixture as a whole, we have:

$$\frac{\partial N_t}{\partial z} = 0 \tag{3.84}$$

If we substitute Eqn (3.8) for the molar fluxes N_i into Eqn (3.83), we obtain

$$c_t \frac{\partial x_i}{\partial t} + N_t \frac{\partial x_i}{\partial z} = \frac{\partial J_{i,z}}{\partial z} \tag{3.85}$$

The molar diffusion flux $J_{1,z}$ is given by Eqn (3.22) for a binary system and either Eqn (3.23) or Eqn (3.24) for a multicomponent system.

The assumptions of the model are incorporated into the initial and boundary conditions. During the diffusion process, the interface has the composition x_{i0}. This composition usually is assumed to be constant and may be expressed as

$$z = 0 \quad t > 0 \quad x_i = x_{i0} \tag{3.86}$$

Before the start of the diffusion process, the compositions are everywhere uniform in the phase under consideration, and equal to the bulk fluid composition, $x_{i\infty}$, and we have the initial condition

$$z \geq 0 \quad t = 0 \quad x_i = x_{i\infty} \tag{3.87}$$

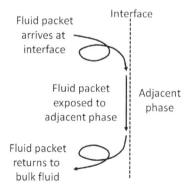

FIGURE 3.4 Surface Renewal Model

The final boundary condition that is valid for short contact times,

$$z = 0 \quad t > 0 \quad x_i = x_{i0} \tag{3.88}$$

which means that the diffusing component has not penetrated into the bulk fluid phase.

The penetration model [42] is based on the assumption that all the fluid elements reside at the interface for the same length of time. The surface age distribution for this model is

$$\psi(t) = 1/t_e \tag{3.89}$$

for all $t \leq t_e$ and $\psi(t) = 0$ for $t > t_e$. The surface renewal model [43] is based on the idea that the chance of an element of surface being replaced with fresh liquid from the bulk is independent of the length of time for which it has been exposed. The age distribution function assumed is

$$\psi(t) = s \exp(-st) \tag{3.90}$$

Here, s is the fraction of the area of surface that is replaced with fresh liquid in unit time.

3.6.3 Surface renewal model for binary systems

For a binary system with no convection perpendicular to the interface the solution to Eqn (3.85) is

$$\frac{x_1 - x_{10}}{x_{1\infty} - x_{10}} = \frac{1 - \mathrm{erf}\left((\zeta - \phi)/\sqrt{D}\right)}{1 + \mathrm{erf}\left(\phi/\sqrt{D}\right)} \tag{3.91}$$

where we have assumed c_t and D to be constant and where $\zeta = z/\sqrt{4t}$. The parameter ϕ is not a function of ζ (see Bird et al. (1960) for justification) and is defined by

$$\phi = (N_t/c_t)\sqrt{t} \tag{3.92}$$

The molar diffusion flux at the interface $z = 0$ is obtained from the one-dimensional form of Eqn (3.22):

$$
\begin{aligned}
J_{10} &= -c_t D \frac{dx_1}{dz}\bigg|_{z=0} \\
&= c_t \sqrt{D/\pi t} \, \frac{\exp\left(-\phi^2/D\right)}{\mathrm{erf}\left(\phi/\sqrt{D}\right)} (x_{10} - x_{1\infty})
\end{aligned}
\tag{3.93}
$$

In the limit that N_t goes to zero, Eqn (3.93) simplifies to

$$J_{10} = c_t \sqrt{D/\pi t}(x_{10} - x_{1\infty}) \tag{3.94}$$

The value of the low flux mass transfer coefficient k at any instant of time, defined by Eqn (3.44) (Bird et al., 1960) is

$$k(t) = \sqrt{D/\pi t} \tag{3.95}$$

The average mass transfer coefficient over the total exposure period t_e is given by

$$k = \int_0^{t_e} k(t)\psi(t)\mathrm{d}t \tag{3.96}$$

With the age distribution function for the classic Higbie model, Eqn (3.89), the average mass transfer coefficient is

$$k = 2\sqrt{D/\pi t_e} \tag{3.97}$$

The Danckwerts surface age distribution (see Eqn (3.90)) leads to an average value of the mass transfer coefficient given by

$$k = 2\sqrt{Ds} \tag{3.98}$$

The correction factor for finite mass transfer rates Ξ is given by (see Bird et al., 1960)

$$\Xi = \frac{\exp(-\Phi^2/\pi)}{1 + \mathrm{erf}(\Phi/\sqrt{\pi})} \quad \text{where } \Phi = N_t/c_t k \tag{3.99}$$

For a given mass transfer rate factor $\Phi = N_t/c_t k$, the film model and penetration model predictions of Ξ give very similar results.

The molar flux at the interface N_{10} can be calculated by multiplying the diffusion flux J_{10} by the appropriate bootstrap coefficient β_0 evaluated at the interface composition:

$$N_{10} = \beta_0 J_{10} = c_t \beta_0 k \Xi(x_{10} - x_{1\infty}) \tag{3.100}$$

It is necessary to use an iterative method to compute the flux N_{10} from Eqn (3.100). Repeated substitution of the fluxes, starting from an initial guess calculated with $\Xi = 1$, will usually converge in only a few iterations.

The calculation of k using Eqns (3.97) and (3.98) requires a priori estimation of the exposure time t_e or the surface renewal rate s. For example, for bubbles rising in a liquid, the exposure time is the time the bubble takes to rise its own diameter. During the flow of a liquid over column packing, t_e is the time for the liquid to flow over a packing element.

3.6.4 Mass transfer in bubbles, drops, and jets

In what follows, we present a brief selection of expressions for the mass transfer coefficients for diffusion in spherical and cylindrical geometries. The results presented here are useful in the modeling of mass transfer in, for example, gas bubbles in a liquid, liquid droplets in a gas, or gas jets in a liquid. Models of mass transfer on distillation trays sometimes are based on models of this kind (as will be discussed in Section 3.8.4).

For a binary system, under conditions of small mass transfer fluxes, the unsteady-state diffusion equations may be solved to give the fractional approach to equilibrium F defined by the following [44]:

$$F = \frac{x_{10} - \langle x_1 \rangle}{x_{10} - x_{1I}} \tag{3.101}$$

where x_{10} is the initial composition (at $t = 0$) within the particle, x_{1I} is the composition at the surface of the particle (held constant for the duration of the diffusion process), and $\langle x_1 \rangle$ is the cup-mixing composition of the dispersed phase. This average composition appears because the assumption that the diffusing component does not "see" the bulk fluid does not apply. For long contact times, the diffusing species will penetrate deep into the heart of the bubble or drop, and it is important to define the mass transfer coefficient in terms of the driving forces $\Delta x_i = x_{iI} - \langle x_i \rangle$.

The Sherwood number for a spherical particle $Sh = k \times 2r_0/D$, at time t, defined by taking the driving force to be $x_{1I} - \langle x_1 \rangle$, may be expressed in terms of F as follows [44]:

$$Sh = \frac{2}{3(1 - F)} \frac{\partial F}{\partial Fo} \tag{3.102}$$

where $Fo = Dt/r_0^2$ is the Fourier number r_0 is the radius of the particle. The time averaged Sherwood number is

$$\overline{Sh} = -2 \ln(1 - F)/3Fo \tag{3.103}$$

The time averaged mass transfer coefficient $\bar{k}$ is

$$\bar{k} = -\ln(1 - F)/a't \tag{3.104}$$

where a' is the surface area per unit volume of particle $a' = 3/r_0$ and t is the contact time.

For a rigid spherical particle (bubble or droplet), F is given by the following [44]:

$$F = 1 - \frac{6}{\pi^2} \sum_{m=1}^{\infty} \frac{1}{m^2} \exp(-m^2 \pi^2 Fo) \tag{3.105}$$

We see from Eqns (3.101) and (3.105) that when $t \rightarrow \infty$ equilibrium is attained, the average composition $\langle x_1 \rangle$ will equal the surface composition x_{1I}. The time-averaged Sherwood number and mass transfer coefficients for a rigid spherical particle may be obtained directly from Eqns (3.103) and (3.104) with F given by Eqn (3.105) above. The Sherwood number at time t is found to be

$$Sh = \frac{2}{3}\pi^2 \left(\sum_{m=1}^{\infty} \exp(-m^2 \pi^2 Fo) \middle/ \sum_{m=1}^{\infty} \frac{1}{m^2} \exp(-m^2 \pi^2 Fo) \right) \tag{3.106}$$

For large values of Fo, the zero-flux mass transfer coefficient is

$$k = \pi^2 D / 3r_0 \tag{3.107}$$

showing, as for the film model discussed above, a first-power dependence on the Fick coefficient D. At small values of Fo, the mass transfer coefficients has a square-root dependence on D.

Another situation that is of practical importance is radial diffusion inside a cylindrical jet of gas or liquid. The fractional approach to equilibrium is given by

$$F = 1 - 4 \sum_{m=1}^{\infty} \exp\left(-16 j_m^2 \text{Fo}\right) / j_m^2 \tag{3.108}$$

where the j_m are the roots of the zero-order Bessel function $J_0(j_m) = 0$. For this case, the time-averaged Sherwood number and mass transfer coefficient are given by

$$\overline{\text{Sh}} = -\ln(1 - F) / \text{Fo} \quad \overline{k} = -\ln(1 - F) / a't \tag{3.109}$$

where a' is the surface area per unit volume of particle $a' = 2/r_0$ and t is the contact time.

3.6.5 Estimation of binary mass transfer coefficients from empirical correlations

In many cases, it is impossible to estimate the film thickness or contact time, and we must resort to empirical methods of estimating the mass transfer coefficients. Binary mass transfer data usually are correlated in terms of dimensionless groups, such as the Sherwood number, $\text{Sh} = kd/D$, the Stanton number $\text{St} = k/u$, and the Chilton–Colburn j-factor $j_D = \text{St Sc}^{2/3}$, where d is some characteristic dimension of the mass transfer equipment, u is the mean velocity for flow, D is the Fick diffusion coefficient, and Sc is the Schmidt number ν/D. Empirical correlations often take the following form (where A, B, and C are numerical constants).

$$\text{Sh} = A\text{Re}^B \text{Sc}^C \tag{3.110}$$

Another often-used expression for estimating mass transfer coefficients is the Chilton–Colburn analogy:

$$j_D = f/2 \quad \text{or} \quad \text{St} = \frac{1}{2}\text{Sc}^{-2/3} \tag{3.111}$$

where f is the Fanning friction factor. Further discussion of the analogies between heat, mass, and momentum transfer can be found in the works of Bird et al. (1960), Sherwood et al. [45], and Churchill [86].

The film model appears to suggest that the mass transfer coefficient is directly proportional to the diffusion coefficient raised to the first power. This result is in conflict with most experimental data, as well as with more elaborate models of mass transfer such as surface renewal theory (Bird et al., 1960). However, if we insert the film theory expression for the mass transfer coefficient, Eqn (3.77), into the definition of the Sherwood number, we find $Sh = d/l$, which shows that the inverse Sherwood number Sh^{-1} may be considered to be a dimensionless film thickness. The film thickness obtained in this way will be a function of the flow conditions, system geometry, and physical properties such as viscosity and density. More importantly, l is proportional to the Fick diffusion coefficient D raised to a power that is less than 1 and removes the objection that film theory does not predict the correct dependence of k on D.

The binary mass transfer coefficients estimated from these correlations and analogies are the low flux coefficients and need to be corrected for the effects of finite transfer rates when used in design calculations. In some correlations, it is the binary mass transfer coefficient—interfacial area product that is correlated. In this case, then k should be considered to be this product and the molar fluxes so calculated from Eqn (3.79) et seq are the mass transfer rates themselves with units moles per second (mol/s or equivalent). Other correlations more applicable in distillation and absorption are discussed in Sections 3.8.3 and 3.9.1.

3.7 **Models for mass transfer in multicomponent mixtures**

Exact solutions of the Maxwell–Stefan equations for certain special cases involving diffusion in ternary ideal gas systems have been known for a long time. A general solution applicable to mixtures with any number of constituents and any relationship between the fluxes was obtained by Krishna and Standart [46], and the key results are summarized below. A complete derivation and a discussion of the relationships between the general solution and the many approximate solutions as well as exact solutions for special cases are explored by Taylor and Krishna [2].

Exact solutions of the governing equations for unsteady-state diffusion in multi-component systems are of limited use for computing the diffusion fluxes because they require an a priori knowledge of the composition profiles. A better approach to solving the unsteady-state diffusion equations is to make use of the Toor [47] and Stewart-Prober [48] approximations of constant $[D]$. Important results are summarized by Taylor and Krishna [2]. Multicomponent generalizations of the results in Section 3.6.4 for mass transfer in bubbles and drops are available in the same source.

3.7.1 **Exact method for mass transfer in multicomponent ideal gas mixtures**

For the film model described in Section 3.6.1, Krishna and Standart showed that the mass transfer coefficients and the associated correction factors are exact matrix analogs of the equations above for binary systems. That is, the zero flux mass transfer coefficients at the coordinate $\bar{z} = 0$ are given by

$$[k_0] = [D_0]/l = [B_0]^{-1}/l \tag{3.112}$$

The matrix of correction factors $[\Xi_0]$ is given by

$$[\Xi_0] = [\Phi][\exp[\Phi] - [I]]^{-1} \tag{3.113}$$

$[\Phi]$ is defined in Section 3.6.3 below. Alternatively, at $\bar{z} = 1$,

$$[k_\delta] = [D_\delta]/l = [B_\delta]^{-1}\big/l \tag{3.114}$$

with the corresponding matrix of correction factors

$$[\Xi_\delta] = [\Phi][\exp[\Phi]][\exp[\Phi] - [I]]^{-1} = [\Xi_0]\exp[\Phi] \tag{3.115}$$

By invoking the "bootstrap" solution, the fluxes (N) can be evaluated from one of two equivalent expressions:

$$\begin{aligned}(N) &= c_t[\beta_0][k_0][\Xi_0](x_0 - x_\delta) \\ &= c_t[\beta_\delta][k_\delta][\Xi_\delta](x_0 - x_\delta)\end{aligned} \tag{3.116}$$

The computation of N_i from Eqn (3.116) requires an iterative procedure; computational strategies are discussed in detail by Taylor and Krishna [2].

3.7.2 Multicomponent film model for nonideal fluids

The starting point for the analysis of mass transfer in nonideal fluid mixtures is the one-dimensional simplification of the set of generalized Maxwell—Stefan equation (Eqn (3.117)). Exact solutions to these equations may be obtained (see [2] for references) but they are of no practical value because they involve far too much computational effort, and any practical advantage over less rigorous methods has not been demonstrated. A more sensible approach is to employ an approximate solution developed by Krishna [49] based on the assumption that the coefficients $\Gamma_{i,k}$ and $Ð_{i,k}$ could be considered constant along the diffusion path. With these assumptions, the generalized Maxwell—Stefan equations represent a linear matrix differential equation, the solution of which can be written down in a manner exactly analogous to the ideal gas case. Thus, for nonideal fluids, the molar fluxes N_i can be calculated from Eqn (3.116), with the matrices of mass transfer coefficients given by:

$$[k_0] = [B_0]^{-1}[\Gamma_{av}]/l \quad [k_\delta] = [B_\delta]^{-1}[\Gamma_{av}]/l \tag{3.117}$$

The subscripts 0 and δ serve as reminders that the appropriate compositions, x_{i0} and x_{i0}, respectively, have to be used in the defining equations for $B_{i,j}$ (Eqn (3.14)). The subscript av on $[\Gamma]$ denotes that the matrix of thermodynamic factors is to be evaluated at the average composition. The correction factors $[\Xi_0]$ and $[\Xi_\delta]$ are given by Eqn (3.113) or Eqn (3.115) with the matrix $[\Theta] = [\Gamma_{av}]^{-1}[\Phi]$ replacing the matrix $[\Phi]$.

3.7.3 Estimation of multicomponent mass transfer coefficients from empirical correlations

Most published experimental works have centered on two-component systems and there are no correlations for the multicomponent $[k]$. How, then, can we estimate multicomponent mass transfer coefficients when all we have available are binary correlations?

In preparation for what follows, we define a matrix $[R]$ with elements

$$R_{i,i} = \frac{z_i}{\kappa_{i,n}} + \sum_{k \neq i = 1}^{n} \frac{z_k}{\kappa_{i,k}} \qquad R_{i,j} = -z_i \left(\frac{1}{\kappa_{i,j}} - \frac{1}{\kappa_{i,n}} \right) \tag{3.118}$$

where z_i are the mole fractions of the appropriate phase and $\kappa_{i,j}$ is a low flux mass transfer coefficient for the binary $i-j$ pair defined by

$$\kappa_{i,j} = Ð_{i,j}/l \tag{3.119}$$

We may also write the elements of the rate factor matrix $[\Phi]$ in terms of these binary mass transfer coefficients:

$$\Phi_{i,i} = \frac{N_i}{c_t \kappa_{i,n}} + \sum_{k \neq i = 1}^{n} \frac{N_k}{c_t \kappa_{i,k}} \qquad \Phi_{i,j} = -N_i \left(\frac{1}{c_t \kappa_{i,j}} - \frac{1}{c_t \kappa_{i,j}} \right) \tag{3.120}$$

Krishna and Standart suggested that, in situations where the film thickness is not known, the matrix of low flux mass transfer coefficients in (low-pressure) gas mixtures can be calculated directly from

$$[k] = [R]^{-1} \tag{3.121}$$

The binary $\kappa_{i,j}$ may be calculated as a function of the appropriate Maxwell−Stefan diffusion coefficient from a suitable correlation or physical model (e.g. the surface renewal models in Section 3.6.3). These binary $\kappa_{i,j}$ must also be used directly in the calculation of the matrix $[\Phi]$.

Comparison of Eqns (3.112) and (3.117) for the low flux mass transfer coefficients for ideal and nonideal systems, respectively, suggests that, in cases where the film thickness is not known, we may estimate the low flux mass transfer coefficient matrix for nonideal systems from

$$[k] = [R]^{-1}[\Gamma] \tag{3.122}$$

where the elements of the matrix $[R]^{-1}$ are given by Eqn (3.118) with the liquid-phase mole fractions x_i replacing the z_i.

Because most published correlations were developed with data obtained with nearly ideal or dilute systems where $\Gamma \approx 1$, this separation of diffusive and thermodynamic contributions is expected to work quite well. Krishna [50] formally defined the Maxwell−Stefan mass transfer coefficient $k_{i,j}$ as

$$\kappa_{i,j} = \lim_{N_i \to 0, \; \Gamma_{i,j} \to \delta_{i,j}} \frac{J_1}{c_t \Delta x_i} \tag{3.123}$$

The $\kappa_{i,j}$ may be estimated using an empirical correlation or alternative physical model (e.g. surface renewal theory) with the Maxwell–Stefan diffusivity of the appropriate i–j pair $Đ_{i,j}$ replacing the binary Fick D.

An alternative approach based on the work of Toor [47], Stewart, and Prober [48] was discussed by Krishna [2]. In our experience, the approach of Krishna and Standart and that of Toor almost always give very similar results. We therefore recommend the approach described above in view of the greater ease in computations.

3.7.4 Estimation of overall mass transfer coefficients: a simplified approach

We may use the results of Section 3.7.3 to develop a simple method for estimating overall mass transfer coefficients. The starting point for this development is Eqn (3.67) for $[K_{OV}]$ together with Eqn (3.121) for the vapor phase mass transfer coefficients, Eqn (3.122) for the liquid phase, and Eqn (3.61) for the linearized equilibrium matrix:

$$[K_{OV}]^{-1} = [R_{OV}] = [R^V] + \frac{c_t^V}{c_t^L}[K][R^L] \qquad (3.124)$$

For two-component systems, Eqn (3.124) simplifies to

$$\frac{1}{K_{OV}} = \frac{1}{\kappa_V} + \frac{c_t^V}{c_t^L}\frac{K_1}{\kappa_L} \qquad (3.125)$$

It is interesting to note that the thermodynamic factors cancel out of Eqns (3.124) and (3.125) and simplify the calculation of mass transfer rates in distillation, as discussed in the next section.

3.8 Mass transfer in tray columns

A schematic diagram of the two-phase dispersion on a distillation tray is shown in Figure 3.5. The composition of the vapor below the tray is y_{ie}; y_{il} is the composition of the vapor above the dispersion in the narrow slice of froth shown in the figure. The parameter h_f is the froth height.

Our interest is in being able to calculate the overall performance of a distillation tray. The equations that we use to calculate the mass transfer rates depend crucially on the way in which the liquid is assumed to flow across the tray and the way in which the gas/vapor phase is assumed to flow through (or past) the liquid. Several different models have been developed over the years, but it is beyond the scope of this chapter to consider all of them in the depth that this topic deserves. The simplest possible flow pattern is to assume that the liquid on a tray is well mixed both in the direction of liquid flow and vertically in the direction of vapor

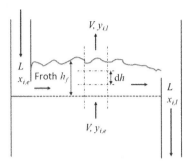

FIGURE 3.5

Schematic Illustration of the Froth on a Distillation Tray

flow. The latter may well be a good assumption for any size tray; the former may be a good approximation only for small-diameter trays, but not at all for large-diameter trays. It is also possible to assume that the vapor phase in the froth is well mixed in both flow directions; however, to assume the vapor is well mixed vertically is likely to be a poor approximation. The essential point of note here is that to assume that both the vapor phase and the liquid phase are well mixed in all directions, which will lead to the most conservative mass transfer model (lowest overall mass transfer rates). The opposite extreme is to assume that the vapor rises in plug flow through the liquid and that the liquid moves in plug flow across the tray. This model will lead to the higher overall mass transfer rates. The assumption that the vapor rises in plug flow is probably a good approximation, but it is more likely that the liquid flow is somewhere between completely mixed and plug flow. More realistic models of liquid flow then are the tanks in series model (in which the liquid is modeled as a series of well-mixed tanks) and dispersion models (in which superimposed on the basic plug flow pattern is a backmixing flow that is modeled by an eddy dispersion coefficient). Models of this kind are reviewed at length by Lockett [51]. The importance of proper flow modeling cannot be overemphasized; the relative increase in overall mass transfer rates can be as high as 20% for a liquid in plug flow as compared to a liquid that is well mixed on the tray [52].

We assume that the liquid phase in the vertical slice to be well mixed vertically. The vapor phase is assumed to rise through the liquid in plug flow.

The material balance for the vapor phase may be written as

$$\frac{dv_i}{dh} = -N_i a A_b \quad \frac{dV}{dh} = -N_t a A_b \tag{3.126}$$

where N_i is the molar flux of species/across the vapor–liquid interface, v_i represents the molar flow rate of component i in the vapor phase, $V = \sum v_i$ is the total vapor flow rate, a is the interfacial area per unit volume of froth, and A_b is the active bubbling area. Because $N_i = J_i^V + y_i N_t$ and $v_i = y_i V$, Eqn (3.126) can be written as:

$$V \frac{dy_i}{dh} = -J_i^V a A_b \tag{3.127}$$

which is valid even when $N_t \neq 0$—that is, when we have nonequimolar transfer.

Equation (3.127) must be integrated over the froth height to yield the composition profiles. The boundary conditions are

$$\begin{aligned} h &= 0 & y_i &= y_{i,e} \\ h &= h_f & y_i &= y_{i,l} \end{aligned} \tag{3.128}$$

where h_f is the froth height.

For the liquid flowing across the tray, the material balance equation is:

$$L \frac{dx_i}{d\zeta} = c_t^L D_e W h_L \frac{d^2 x_i}{d\zeta^2} - J_i^L a W h_f \tag{3.129}$$

where $\zeta = z/Z$ is the dimensionless distance along the flow path and D_e is the dispersion coefficient (and which is zero if the liquid is assumed to traverse the tray in plug flow). The boundary conditions are:

$$\begin{aligned} \zeta &= 0, & x_i &= x_{i,e} \\ \zeta &= 1, & x_i &= x_{i,l} \; dx_i/d\zeta = 0 \end{aligned} \tag{3.130}$$

The integration of Eqn (3.129) requires some additional assumptions regarding the diffusion flux term (the last term on the right hand side). Solutions for binary systems are available in the literature (see Ref. [51] for examples and literature sources). A solution for multicomponent systems was proposed by Kooijman [52].

There are more than 10 empirical correlations that can be used to estimate the dispersion coefficient. Convenient summaries, including equations, are given by Lockett [51] and Korchinsky [53]. Klemola and Ilme [54] provided comparisons of some of these methods. For further discussion, see Chapter 5 in Book 3.

3.8.1 Composition profiles in binary distillation

For a binary system, the material balance equation (see Eqn (3.127)) may be written for component 1 as

$$V \frac{dy_1}{dh} = -J_1^V a A_b \tag{3.131}$$

When the binary rate relation in Eqn (3.63) is combined with the material balance in Eqn (3.131), we obtain

$$V \frac{dy_1}{dh} = c_t^V K_{OV} (y_1^* - y_1) a A_b \tag{3.132}$$

Equation (3.132) may be solved, subject to the boundary conditions (see Eqn (3.128)) to give

$$\left(y_1^* - y_{1l}\right) = \exp(\text{NTU}_{OV})\left(y_1^* - y_{1e}\right) \tag{3.133}$$

where the overall number of transfer units for the vapor phase is defined by

$$\text{NTU}_{OV} = \int_0^{h_f} \left(\frac{c_t^V K_{OV} a A_b}{V}\right) dh \tag{3.134}$$

If we assume the integrand in Eqn (3.134) to be independent of froth height, we may complete the integration to give the overall number of transfer units NTU_{OV} as

$$\text{NTU}_{OV} = \frac{c_t^V K_{OV} a A_b}{V} h_f \tag{3.135}$$

and the vapor composition at any point in the dispersion may be obtained from

$$\left(y_1^* - y_1(\zeta)\right) = \exp(-\text{NTU}_{OV}\zeta)\left(y_1^* - y_{1e}\right) \tag{3.136}$$

where we have defined a dimensionless froth height by $\zeta = h/h_f$.

3.8.2 Numbers of transfer units in binary distillation

To predict performance at a point on a tray, we need to estimate the overall number of transfer units. A working relationship for NTU_{OV} may be obtained by combining Eqn (3.135) with Eqn (3.68) for K_{OV} to give

$$\frac{1}{\text{NTU}_{OV}} = \frac{1}{\text{NTU}_V} = \frac{S}{\text{NTU}_L} \tag{3.137}$$

where $S = MV/L$ is the stripping factor. NTU_V and NTU_L are the numbers of transfer units for the vapor and liquid phases defined by the following [51]:

$$\text{NTU}_V \equiv k^V a' t_V = \frac{k^V a h_f}{u_s} \quad \text{NTU}_L \equiv k^L \bar{a} t_L = \frac{k^V a h_f Z Q_L}{W} \tag{3.138}$$

where h_f is the froth height, Z is the liquid flow path length, W is the weir length, $Q_L = L/c_t^L$ is the volumetric liquid flow rate, $u_s = Q_V/A_b$ is the superficial vapor velocity based on the bubbling area of the tray, $Q_V = V/c_t^V$ is the volumetric vapor flow rate, a' is the interfacial area per unit volume of vapor, and $\bar{a}$ is the interfacial area per unit volume of liquid. The area terms a' and $\bar{a}$ are related to a, the interfacial area per unit volume of froth, by

$$a' = a/(1 - \alpha) \quad \bar{a} = a/\alpha \tag{3.139}$$

where $\alpha = h_L/h_f$ is the relative froth density. The parameters t_V and t_L are the residence times of the vapor and liquid phases, respectively:

$$t_V = (1 - \alpha)h_f/u_s \quad t_L = Z/u_L = h_L Z W/Q_L \tag{3.140}$$

where u_L is the horizontal liquid velocity.

Alternatively, if we combine Eqn (3.125) with Eqn (3.135), we may express the overall number of transfer units as

$$\frac{1}{\text{NTU}_{OV}} = \frac{1}{\mathcal{NTU}_V} + \frac{S}{\mathcal{NTU}_L} \tag{3.141}$$

where $\mathcal{NTU}_V$ and $\mathcal{NTU}_L$ are numbers of transfer units for the vapor and liquid phases defined as follows:

$$\mathcal{NTU}_V \equiv k^V a' t_V \quad \mathcal{NTU}_L \equiv k^L \bar{a} t_L \tag{3.142}$$

and where we have defined a modified stripping factor $S = K_1(V/L)$. Note the presence of the "ideal" mass transfer coefficients κ^V and κ^L in place of the conventional k^V and k^L. Taylor and Krishna [2] suggested that the numbers of transfer units NTU_V and NTU_L can be evaluated from the correlations presented below, with the Maxwell–Stefan diffusivities $Đ^V$ and $Đ^L$ replacing the Fick diffusivities.

3.8.3 Numbers of transfer units from empirical correlations

Three methods of estimating binary mass transfer coefficients in tray columns are summarized in Table 3.2. One of them comes from the *AIChE Bubble Tray Design Manual* [55], which represented the first comprehensive estimation procedure for numbers of transfer units. The method of Chan and Fair [56] was devised for sieve trays, but is widely used for valve trays as well. Table 3.2 includes the original version of the Chan and Fair method as well as a simplified version derived by

Table 3.2 Selected Correlations for Mass Transfer on Trays	
AIChE Method [55]	
Vapor phase	$\text{NTU}_V = (0.776 + 4.57 h_w - 0.238 F_s + 104.8 Q_L/W)/\sqrt{\text{Sc}}$
	$F_s = u_s \sqrt{\rho^V}$
Liquid phase	$\text{NTU}_L = 19{,}700\sqrt{(D^L)}(0.4 F_s + 0.17) t_L$
Chan and Fair Method [56]	
Vapor phase	Original: $\text{NTU}_V = \frac{(10{,}300 - 8670 F_f) F_f \sqrt{D^V} t_V}{\sqrt{h_L}}$ $\quad F_f = \frac{u_s}{u_{s,i}}$
	Simplified: $\text{NTU}_V = \frac{3060 \sqrt{D^V} t_V}{\sqrt{h_L}}$ based on maximum value (at $F_f = 0.6$)
Liquid phase	Same as AIChE method [55]
Residence time	$t_V = (1 - \alpha_e) h_L/(\alpha_e u_s)$ α_e from Eqn (3.144)
Zuiderweg Method [57]	
Vapor phase	$k^V = 0.13/\rho_t^V - 0.065/(\rho_t^V)^2 \quad (1 < \rho_t^V < 80 \text{ kg/m}^3)$
Liquid phase	$k^L = 2.6 \cdot 10^{-5} (\eta_L)^{-0.25}$ or $k^L = 0.024 (D^L)^{0.25}$

setting the fraction of flood to 0.6 (corresponding to the maximum in the quadratic function of F_f). The simplified version is more applicable to modern fixed valve designs that have a much higher turndown ratio. It is interesting to note that one of the methods in Table 3.2 is independent of the diffusion coefficient [57]. Other methods are summarized by by Lockett [51]. In a comparative assessment of these (and other) correlations [54], it is clear that the selection of the method used to estimate the numbers of transfer units is one of the keys to proper estimation of distillation tray performance.

The clear liquid height h_L and relative froth density α_e in the Chan and Fair method are to be calculated using the method of Bennett et al. [58] (see also Ref. [51]):

$$h_L = \alpha_e \left(h_w + (0.5 + 0.438 \exp(-137.8 h_w))(Q_L/(W\alpha_e))^{0.67} \right) \tag{3.143}$$

$$\alpha_e = \exp\left(-12.55(C_s)^{0.91} \right) \quad \text{where } C_s = u_s \sqrt{\frac{\rho_t^V}{\rho_t^L - \rho_t^V}} \tag{3.144}$$

C_s is the C-factor and has units of velocity (m/s) in Eqn (3.114)). This method can also be used for the same quantity in the AIChE method in Table 3.3. Zuiderweg [57] gave the correlation below for estimation of the clear liquid height for use with his own method in Table 3.3.

$$h_L = 0.6 h_w^{0.5} (pF_P/b)^{0.25} \tag{3.145}$$

where p is the hole pitch.

The parameter ah_f in the Zuiderweg method is dependent on the regime of operation. For the spray regime:

$$ah_f = \frac{40}{\phi^{0.3}} \left(\frac{u_s^2 \rho_t^V h_L F_{LV}}{\sigma} \right)^{0.37} \tag{3.146}$$

For the mixed froth-emulsion flow regime:

$$ah_f = \frac{43}{\phi^{0.3}} \left(\frac{u_s^2 \rho_t^V h_L F_{LV}}{\sigma} \right)^{0.53} \tag{3.147}$$

where $\phi = A_h/A_b$ is the fractional free area of the tray, A_h is the total area of the holes, and A_b is the bubbling area of the tray. The flow parameter (F_{LV}) is defined by

$$F_{LV} = \frac{M_L}{M_V} \sqrt{\frac{\rho_t^V}{\rho_t^L}} \tag{3.148}$$

where M_L and M_V are the mass flow rates of liquid and vapor phases. The transition from the spray regime to mixed froth-emulsion flow is described by $F_{LV} > 3bh_L$, where b is the weir length per unit bubbling area $b = W/A_b$.

Table 3.3 Selected Correlations of Mass Transfer Coefficients for Packed Columns

Onda's Correlations for Randomly Packed Columns

Vapor phase mass transfer coefficient	$\dfrac{k^V}{a_p D^V} = A \text{Re}_V^{0.7} \text{Sc}_V^{0.333} (a_p d_p)^{-2}$
Liquid phase mass transfer coefficient	$\dfrac{k^L}{(\rho_t^l / \eta_L g)^{0.333}} = 0.0051 (\text{Re}_L')^{0.667} \text{Sc}_L^{-0.5} (a_p d_p)^{0.4}$
Interfacial area density	$a' = a_p (1 - \exp(-1.45 (\sigma_c/\sigma)^{0.75} \text{Re}_L^{0.1} \text{Fr}_L^{-0.05} \text{We}_L^{0.2}))$
Reynolds numbers	$\text{Re}_V = \dfrac{\rho_t^V u_V}{\eta_V a_p} \quad \text{Re}_L = \dfrac{\rho_t^l u_L}{\eta_L a_p} \quad \text{Re}_L' = \dfrac{\rho_t^l u_L}{\eta_L a'}$
Schmidt numbers	$\text{Sc}_V = \dfrac{\eta_V}{\rho_t^V D^V} \quad \text{Sc}_L = \dfrac{\eta_L}{\rho_t^l D^l}$
Froude and Weber numbers	$\text{Fr}_L = a_p u_L^2 / g \quad \text{We}_L = \rho_t^l u_L^2 / (a_p \sigma)$
Other	d_p is the nominal packing size, and a_p is the specific surface area of the packing (m^2/m^3 of packing). The parameter A is a constant that takes the numerical value 2.0 if the nominal packing size, d_p, is less than 0.012 m and has the value 5.23 if the nominal packing size is greater than (or equal to) 0.012 m. σ_c is the critical surface tension of the packing. Values of the critical surface tension are tabulated by Onda et al. [59]

Bravo and Fair [60] Correlation for Randomly Packed Columns

Vapor and liquid phase mass transfer coefficients	As per method of Onda
Interfacial area density	$a' = 19.78 a_p (\text{Ca}_L \text{Re}_V)^{-0.392} \sigma^{0.5} / H^{0.4} \quad \text{Ca}_L = \dfrac{u_L^2 \eta_L}{\sigma}$

Rocha et al. [61] Method for Structured Packings

Vapor phase mass transfer coefficient	$\text{Sh}_V = 0.054 \text{Re}_V^{0.8} \text{Sc}_V^{0.33}$ where $\text{Sh}_V = k^V S / D^V$
Liquid phase mass transfer coefficient	$k^L = 2 (D^L u_{L,e} / \pi S)^{1/2}$
Interfacial area density (From Shi and Mersmann [90])	$\dfrac{a'}{a_p} = F_{SE} \dfrac{29.12 (\text{We}_L \text{Fr}_L)^{0.15} S^{0.359}}{\text{Re}_L^{0.2} \delta^{0.6} (1 - 0.93 \cos \gamma)(\sin \theta)^{0.3}}$
	$F_{SE} = $ Surface enhancement factor (see original source)
	$\cos \gamma = 0.9$ for $\sigma < 0.055$ N/m
	$\cos \gamma = 5.211 \times 10^{-16.835\sigma}$ for $\sigma > 0.055$ N/m
Reynolds number	$\text{Re}_V = \dfrac{\rho_t^V (u_{V,e} + u_{L,e}) S}{\eta_V}$
Velocities	$u_{V,e} = u_V / (\delta(1 - h_L) \sin \theta) \quad u_{L,e} = u_L / (\delta h_L \sin \theta)$
Other	ε is the void fraction of the packing
	θ is the angle of the channel with respect to the horizontal
	S is the corrugation spacing (channel side)
	h is the height of the triangle (crimp height)

Table 3.3 Selected Correlations of Mass Transfer Coefficients for Packed Columns—Cont'd	
Method of Billet and Schultes [4] for all packings	
Vapor phase mass transfer coefficient	$k^V a' = C_V (\grave{o} - h_L)^{-1/2} \dfrac{a_p^{3/2}}{d_h^{1/2}} D^V \left(\dfrac{u_V}{a_p \nu^V} \right)^{3/4} \left(\dfrac{\nu_V}{D^V} \right)^{1/3} \left(\dfrac{a'}{a_p} \right)^{3/4}$
Liquid phase mass transfer coefficient	$k^L = C_L 12^{1/6} u_L^{1/2} \left(\dfrac{D^L}{d_h} \right)^{1/2}$
Interfacial area density	$\dfrac{a'}{a_p} = 1.5 (a_p d_h)^{-1/2} \left(u_L \dfrac{d_h}{\nu^L} \right)^{-0.2} \left(u_L^2 \rho_t^l \dfrac{d_h}{\sigma} \right)^{0.75} \left(\dfrac{u_L^2}{g d_h} \right)^{-0.45}$
Other	C_V, C_L = packing specific parameters
	$d_h = 4\grave{o}/d_p \quad h_L = \left(12 \dfrac{\eta_L}{g\, \rho_t^l} u_L a_p^2 \right)^{1/3}$
Notes	The above applies up to the loading point. Equations to calculate the loading point and the flooding velocity are given in the original source, as are corrections to account for Marangoni effects

3.8.4 **Mechanistic models of tray performance**

The empirical model of Zuiderweg in Table 3.2 is motivated at least in part by the idea that there are two quite distinct flow regimes on a sieve tray: the froth regime and the spray regime. This picture of two different flow regimes has also served as motivation for some investigators to build detailed mechanistic models of mass transfer. The book by Lockett [51] describes in some detail a model of this class for mass transfer in binary systems. Other contributions are due to Lockett and Plaka [63], Kaltenbacher [64], Hofer [65], Prado and Fair [66], Garcia and Fair [67], Syeda et al. [68] and Vennavelli et al. [69].

The froth regime on a distillation tray really consists of three zones that are shown in Figure 3.6: Zone I, the jetting-bubble formation region; Zone II, the free bubbling zone, and Zone III, the splash zone.

In Zone I, the jetting-bubbling formation zone, the vapor issues through the perforations in the tray in the form of jets before breaking into bubbles. These jets can be modeled as a set of parallel cylindrical vapor jets. The model parameters are the diameter of the jet, the velocity of the vapor in the jet, and the height of the jetting zone. In the free bubbling zone, a distribution of bubble sizes is obtained. The model parameters for the bubbling zone are the bubble diameters, the bubble rise velocities, the height of the free bubbling zone, and the fraction of vapor that is in each bubble population.

Experimental observations suggest that the assumption of a bimodal bubble size distribution with fast-rising "large" bubbles and slow rising "small" bubbles is a

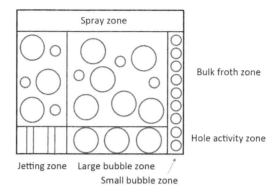

FIGURE 3.6 Schematic Diagram of Bubble-Jet Model of Froth on a Sieve Tray [66]

reasonable approximation. It is thought that about 90% of the incoming vapor is transported by the large bubbles. The small bubbles, despite their large interfacial area, are not very effective in mass transfer, contributing only about 10% of the total transfer [70]. Syeda et al. [68] list sources supporting this claim and provide expressions to estimate the bubble sizes; some other models treat the bubble sizes as free parameters used to tune the model to empirical data.

The large bubbles rise through the froth virtually in plug flow. It is assumed that the small bubble population also rise in plug flow through the froth [66]. The splash zone above the free bubbling zone consists of entrained droplets. The contribution to the total mass transfer of the splash zone is generally small and is neglected in some models in this class.

The key to developing an expression for estimating the overall mass transfer performance of a single tray using a model of this class is to recognize that in each region of the froth (bubble formation zone and each bubble population) the composition change in the vapor is given by Eqn (3.136), where y_e and y_l refer to the mole fractions entering and leaving each region. The overall numbers of transfer units must therefore be found for each region and their contributions summed appropriately (see Taylor and Krishna [2] for illustrative calculations). An extension to multicomponent mixtures of a model in this class is developed in [70]. Springer et al. [71] used a model of this type to describe distillation in a four-component system.

3.8.5 Multicomponent distillation

An extension of the material in Section 3.8.3 to multicomponent systems leads to the equation below for the overall number of transfer units:

$$[\text{NTU}_{OV}] \equiv \frac{c_t^V[K_{OV}]ah_f A_b}{V} = c_t^V[K_{OV}]a't_V \qquad (3.149)$$

The starting point for the prediction of the matrix NTU_{OV} for multicomponent systems is Eqn (3.67) for $[K_{OV}]$ and Eqn (3.124) for $[R_{OV}]$

$$[K_{OV}]^{-1} = [R_{OV}] = [R^V] + \frac{c_t^V}{c_t^L}[K][R^L]^{-1} \qquad (3.150)$$

The matrices $[R^V]$ and $[R^L]$ have elements defined by Eqn (3.118).

By combining Eqn (3.150) for $[K_{OV}]$ with Eqn (3.149) for $[NTU_{OV}]$, we get

$$[NTU_{OV}]^{-1} = [NTU_V]^{-1} + (V/L)[K][NTU_L]^{-1} \qquad (3.151)$$

where $[K]$ is a diagonal matrix of the first $n-1$ equilibrium K values. The matrices $[NTU_V]$ and $[NTU_L]$ are matrices of numbers of transfer units for the vapor and liquid phases, respectively. The inverse matrices are defined by

$$[NTU_V]^{-1} \equiv [R^V]/a't_V \quad [NTU_L]^{-1} \equiv [R^L]/\bar{a}t_L \qquad (3.152)$$

When we carry out the multiplications required by Eqns (3.151) and (3.152), we obtain explicit expressions for the elements of the inverse matrices in terms of the numbers of mass transfer units of each binary pair as follows:

$$
\begin{aligned}
NTU_{i,i}^{-1} &= \frac{z_i}{\mathcal{NTU}_{i,n}} + \sum_{k \neq i=1}^{n} \frac{z_k}{\mathcal{NTU}_{i,k}} \\
NTU_{i,j}^{-1} &= -Z_i\left(\frac{1}{\mathcal{NTU}_{i,j}} - \frac{1}{\mathcal{NTU}_{i,n}}\right)
\end{aligned}
\qquad (3.153)
$$

where $\mathcal{NTU}_{i,j}$ are binary numbers of transfer units for the phase in question defined by Eqn (3.142). The superscript -1 on the elements $N_{i,j}^{-1}$ indicates that these quantities are the elements of the inverse matrices $[NTU]^{-1}$. Thus, calculating $[NTU_V]^{-1}$ and $[NTU_L]^{-1}$ requires nothing more complicated than the determination of the binary numbers of transfer units $\mathcal{NTU}_{i,j}$ from an appropriate correlation, theoretical model, or experimental data, and the use of Eqn (3.153).

The numbers of transfer units for each binary pair may be obtained as described in Section 3.8.2 or from experimental data and these binary numbers of transfer units used directly in the estimation of the matrices of numbers of transfer units for multi-component systems.

3.9 Mass transfer in packed columns

Vapor and liquid flows in a packed column truly flow in opposite directions (in contrast with the flow in a tray column, where the vapor—liquid contact is between the vapor as it rises up through a liquid that is flowing laterally across the column prior to flowing down to the tray below). Thus, the task of finding an appropriate flow model is considerably simpler. Models that include back-mixing do exist, but most often mass transfer correlations are based on a model that assumes plug flow in both phases.

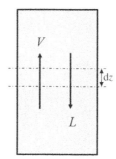

FIGURE 3.7 Differential Section of a Packed Column

For continuous contact equipment, the material and energy balances are written around a section of column of differential height as shown in Figure 3.7. For the vapor phase, the component and total material balances are

$$\frac{dv_i}{dz} = N_i^V a' A_c \quad \frac{dV}{dz} = N_t^V a' A_c \tag{3.154}$$

The differential material balances for the liquid phase are obtained in a similar way:

$$\frac{dl_i}{dz} = N_i^L a' A_c \quad \frac{dL}{dz} = N_t^L a' A_c \tag{3.155}$$

V and L are the total vapor and liquid flows in the column, respectively.

The terms on the right-hand sides of Eqns (3.154) and (3.155) are the rates of mass transfer of species i in the vapor and liquid phases, respectively; we assume that transfers from the vapor to the liquid phase are positive. Note that the liquid is countercurrent to the vapor phase; this explains why the sign on the right-hand side of the four preceding equations is positive.

By making use of the relationship between the molar and diffusion fluxes in Eqn (3.8), the differential material balances can now be rewritten in the following form:

$$V\frac{d(y)}{dz} = c_t^V [K_{OV}](y^* - y) a A_c \tag{3.156}$$

3.9.1 Transfer units for binary systems

It is traditional for chemical engineers to model packed columns through the concept of transfer units (in much the same way as we used transfer units in the treatment of mass transfer in the froth on a tray in Section 3.8). For two-component systems, Eqn (3.156) simplifies to:

$$\frac{dy_1}{d\zeta} = \text{NTU}_{OV}\left(y_1^* - y_1\right) \tag{3.157}$$

where $\zeta = z/H$ is a dimensionless coordinate with H being the total height of packing. The parameter NTU_{OV} is the overall number of transfer units for the vapor phase and may be defined by

$$\text{NTU}_{OV} \equiv \frac{c_t^V K_{OV} a' H A_C}{V} \tag{3.158}$$

There is an important difference between Eqn (3.158) and the corresponding expressions for binary mass transfer in the froth on a distillation tray in Eqn (3.134). In the model of mass transfer in the froth on a tray, the liquid is assumed to be well mixed vertically. Thus, y^* may be assumed to be constant. A similar assumption is not justified for a packed column, where both liquid and vapor flow in opposite directions and the equilibrium vapor composition y^* changes with position as the liquid composition changes. Analytical solutions of Eqn (3.157) for binary systems are available in the literature for certain special cases; in general, however, numerical or graphical techniques are used to solve Eqn (3.157) (see, e.g. [45]).

To evaluate the overall number of transfer units, we may proceed to combine Eqn (3.68) for K_{OV} with Eqn (3.158) for NTU_{OV} to give

$$\frac{1}{\text{NTU}_{OV}} = \frac{1}{\text{NTU}_V} + \frac{M(V/L)}{\text{NTU}_L} \tag{3.159}$$

where NTU_V and NTU_L are the numbers of transfer units for the vapor and liquid phases defined by

$$\text{NTU}_V \equiv \frac{k^V a' H}{u_V} \quad \text{NTU}_L \equiv \frac{k^L \overline{a} H}{u_L} \tag{3.160}$$

where $u_V = V/(c_t^V A_c)$ and $u_L = L/(c_t^L A_c)$ are the superficial vapor and liquid velocities.

Mass transfer coefficients for packed columns sometimes are expressed as the height of a transfer unit (HTU). Thus, for the vapor and liquid phases, respectively,

$$\text{HTU}_V \equiv \frac{H}{\text{NTU}_V} = \frac{u_V}{K_V a'} \quad \text{HTU}_L \equiv \frac{H}{\text{NTU}_L} = \frac{u_L}{K^L a'} \tag{3.161}$$

The overall height of a transfer unit for a two-component system is

$$\text{HTU}_{OV} \equiv \frac{H}{\text{NTU}_{OV}} = \frac{u_V}{K_{OV} a'} \tag{3.162}$$

If we use Eqn (3.68) for K_{OV}, we obtain the following relationship

$$\text{HTU}_{OV} = \text{HTU}_V + S \, \text{HTU}_L \tag{3.163}$$

where $S = MV/L$ is the stripping factor, with M being the slope of the equilibrium line.

The performance of a packed column is often expressed in terms of the height equivalent to a theoretical plate (HETP). The HETP is related to the height of packing by

$$\text{HETP} = H/N_{\text{eqm}} \tag{3.164}$$

where N_{eqm} is the number of equilibrium stages needed to accomplish the same separation possible in a real packed column of height H. The equilibrium-stage model of distillation and absorption is described in a number of textbooks (see, e.g. Seader and Henley, [89]). If the equilibrium line may be assumed straight, the HETP may be related to the HTU values by

$$\text{HETP} = \text{HTU}_{OV}\frac{\ln(S)}{S - 1} \tag{3.165}$$

3.9.2 Mass transfer coefficients for packed columns

More than 100 correlations for packed columns have been presented in the literature; wide-ranging reviews are byPonter and Au-Yeung [72] and by Wang et al. [73]. A summary of some of the more widely used correlations can be found in Chapter 5 of *Perry's Chemical Engineers Handbook* (Green and Perry [87]). A selection of methods for estimating binary mass transfer coefficients in packed columns is provided in Table 3.3. The method of Rocha et al. [61] for sheet metal packings in Table 3.3 is very similar to that of Brunazzi et al. [74] for gauze packings. Other important works include papers by Linek and coworkers (for sources see [75–77]); the latter is particularly useful as it provides a convenient and detailed summary of several other methods not included here.

3.9.3 Transfer units for multicomponent systems

Equations (3.156) may be written in dimensionless form as

$$\frac{\text{d}(y)}{\text{d}\zeta} = [\text{NTU}_{OV}](y^* - y) \tag{3.166}$$

where $\zeta = z/H$. The overall number of transfer units $[\text{NTU}_{OV}]$ is for the vapor phase and may be defined by

$$[\text{NTU}_{OV}] \equiv \frac{c_t^V[K_{OV}]a'HA_c}{V} = \frac{[K_{OV}]a'HA_c}{u_v} \tag{3.167}$$

The composition of the vapor along the length may be determined by integrating (numerically) Eqn (3.166). Each step of the integration requires the estimation of the matrix of overall number of transfer units.

To evaluate the overall number of transfer units, we may proceed to combine Eqn (3.67) for $[K_{OV}]$ with Eqn (3.167) for $[\text{NTU}_{OV}]$ to give the following (cf. Eqn (3.151)):

$$[\text{NTU}_{OV}]^{-1} = [\text{NTU}_V]^{-1} + V/L[K][\text{NTU}_L]^{-1} \tag{3.168}$$

where $[K]$ is a diagonal matrix of the first $n - 1$ equilibrium K values. The matrices $[\text{NTU}_V]$ and $[\text{NTU}_L]$ represent the numbers of transfer units for the vapor and liquid phases, respectively; they may be expressed in terms of the mass transfer coefficients for each phase as

$$[\text{NTU}_V] \equiv [K^V]/a'H/u_v \quad [\text{NTU}_L] \equiv [K^L]/a'H/u_L \tag{3.169}$$

If we use the simplified models discussed in Section 3.7.3 for the matrices of mass transfer coefficients, we may relate the inverse matrices of numbers of transfer units to the matrices of inverse binary mass transfer coefficients as follows (cf. Equations (3.152)):

$$[\text{NTU}_V]^{-1} \equiv [R^V]u_V/(a'H) \qquad [\text{NTU}_L]^{-1} \equiv [R^L]u_L/(a'H) \tag{3.170}$$

For multicomponent systems, it is possible to define matrices of HTU values:

$$[\text{HTU}_V]^{-1} \equiv (u_V/a')[k^V]^{-1} \qquad [\text{HTU}_L]^{-1} \equiv (u_L/a')[k^L]^{-1} \tag{3.171}$$

The multicomponent generalization of Eqn (3.163) is

$$[\text{HTU}_{OV}] = [\text{HTU}_V] = (V/L)[K][\text{HTU}_L] \tag{3.172}$$

To evaluate the matrices of heights or numbers of transfer units, we first use the empirical methods discussed in Section 3.8.3 to estimate the binary (Maxwell–Stefan) mass transfer coefficients as functions of the Maxwell–Stefan diffusion coefficients. The matrices of transfer units follow from Eqn (3.171) with the elements of the $[R]$ matrices computed with the aid of Eqn (3.118).

Experimental studies carried out with a view to testing these models are summarized by Taylor and Krishna [2].

3.10 Further reading

Taylor and Krishna [2] provide considerable additional discussion of mass transfer in multicomponent systems. A simplified treatment of the subject is given by Wesselingh and Krishna [35]. Krishna and Wesselingh [78] give a review of applications that goes well beyond this chapter. For analyses of the constitutive equations for diffusion see Refs [79–83].

References

[1] R.B. Bird, W.E. Stewart, E.N. Lightfoot, in: first ed.Transport Phenomena, Wiley, New York, 1960.

[2] R. Taylor, R. Krishna, Multicomponent Mass Transfer, Wiley, New York, 1993.

[3] J.C. Maxwell, in: W.D. Niven (Ed.), The Scientific Papers of James Clerk Maxwell, Dover, New York, 1952.

[4] J. Stefan, Uber das Gleichgewicht und die Bewegung, insbesondere die Diffusion von Gasmengen, Sitzungsber. Akad. Wiss. Wien 63 (1871) 63−124.

[5] R.D. Present, Kinetic Theory of Gases, McGraw-Hill, New York, 1958.

[6] S. Chapman, T.G. Cowling, The Mathematical Theory of Non-uniform Gases, third ed., Prepared in cooperation with D. Burnett, Cambridge University Press, Cambridge, England, 1970.

[7] R.E. Cunningham, R.J.J. Williams, Diffusion in Gases and Porous Media, Plenum Press, New York, 1980.

[8] J.O. Hirschfelder, C.F. Curtiss, R.B. Bird, Molecular Theory of Gases and Liquids, Second Corrected Printing, Wiley, New York, 1964.

[9] C. Muckenfuss, Stefan−Maxwell relations for multicomponent diffusion and the Chapman Enskog solution of the Boltzmann equations, Chem. Phys. 59 (1973) 1747−1752.

[10] J.M. Prausnitz, R.N. Lichtenthaler, E.G. de Azevedo, Molecular Thermodynamics of Fluid Phase Equilibria, Prentice-Hall, Englewood Cliffs, NJ, 1986.

[11] R. Taylor, H.A. Kooijman, Composition derivatives of activity coefficient models (for the estimation of thermodynamic factors in diffusion), Chem. Eng. Commun. 102 (1991) 87−106.

[12] A. Fick, On liquid diffusion, Phil. Mag. 10 (1855) 30−39.

[13] A. Fick, Uber diffusion, Poggendorff's Ann. 94 (1855) 59−86.

[14] E.L. Cussler, Multicomponent Diffusion, Elsevier, Amsterdam, The Netherlands, 1976.

[15] P.J. Dunlop, B.J. Steel, J.E. Lane, Experimental Methods for Studying Diffusion in Liquids, Gases and Solids, in: A. Weissberger, B.W. Rossiter (Eds.), Techniques of Chemistry, I, Physical Methods of Chemistry, Wiley, New York, 1972.

[16] H.J.V. Tyrell, K.R. Harris, Diffusion in Liquids, Butterworths, London, England, 1984.

[17] T.R. Marrero, E.A. Mason, Gaseous diffusion coefficients, J. Phys. Chem. Ref. Data 1 (1972) 3−118.

[18] J.W. Mutoru, A. Firoozabadi, Form of multicomponent Fickian diffusion coefficients matrix, J. Chem. Thermodyn. 43 (2011) 1192−1203.

[19] R.P. Danner, T.E. Daubert, Manual for Predicting Chemical Process Design Data, AIChE, New York, 1983.

[20] H. Ertl, R.K. Ghai, F.A.L. Dullien, Liquid diffusion of nonelectrolytes: part II, AIChE J. 20 (1974) 1−20.

[21] B.E. Poling, Prausnitz, J.P. O'Connell, The Properties of Gases and Liquids, fifth ed., McGraw-Hill, New York, 2001.

[22] E.N. Fuller, P.D. Schettler, J.C. Giddings, A new method for prediction of binary gas-phase diffusion coefficients, Ind. Eng. Chem. 58 (1966) 19−27.

[23] R.K. Ghai, H. Ertl, F.A.L. Dullien, Liquid diffusion of nonelectrolytes: part I, AIChE J. 19 (1973) 881−900.

[24] P.A. Johnson, A.L. Babb, Liquid diffusion of non-electrolytes, Chem. Rev. 56 (1956) 387−453.

[25] F.A.L. Dullien, Statistical test of Vignes' correlation of liquid phase diffusion coefficients, Ind. Eng. Chem. Fundam. 10 (1971) 41−49.

[26] W. Hayduk, Correlations for molecular diffusivities in liquids, in: N.P. Cheremisinoff (Ed.), Encyclopedia of Fluid Mechanics, vol. IGulf Publishing Corp., Houston, TX, 1986, pp. 48−72.

[27] W. Hayduk, B.S. Minhas, Correlations for prediction of molecular diffusivities in liquids, Can. J. Chem. Eng. 60 (1982) 295–299. Correction, 61, 132 (1983).

[28] M.T. Tyn, W.F. Calus, Temperature and concentration dependence of mutual diffusion coefficients of some binary liquid systems, J. Chem. Eng. Data 20 (1975) 310–316.

[29] M.A. Siddiqi, K. Lucas, Correlations for prediction of diffusion in liquids, Can. J. Chem. Eng. 64 (1986) 839–843.

[30] H.A. Kooijman, A modification of the Stokes–Einstein equation for diffusivities in dilute binary mixtures, Ind. Eng. Chem. Res. 41 (2002) 3326–3328.

[31] A. Leahy-Dios, A. Firoozabadi, Unified model for nonideal multicomponent molecular diffusion coefficients, AIChE J. 53 (2007) 2932–2939.

[32] C.S. Caldwell, A.L. Babb, Diffusion in ideal binary liquid mixtures, J. Phys. Chem. 60 (1956) 51–56.

[33] A. Vignes, Diffusion in binary solutions, Ind. Eng. Chem. Fundam. 5 (189–199) (1966).

[34] H.A. Kooijman, R. Taylor, Estimation of diffusion coefficients in multicomponent liquid systems, Ind. Eng. Chem. Res. 30 (1991) 1217–1222.

[35] J.A. Wesselingh, R. Krishna, Mass Transfer in Multicomponent Mixtures, Delft University Press, 2000.

[36] S. Rehfeldt, J. Stichlmair, Measurement and prediction of multicomponent diffusion coefficients in four ternary liquid systems, Fluid Phase Equil. 290 (2010) 1–14.

[37] X. Liu, A. Bardow, T.J.H. Vlugt, Multicomponent Maxwell–Stefan diffusivities at infinite dilution, Ind. Eng. Chem. Res. 50 (2011) 4776–4782.

[38] H.L. Toor, Diffusion in three component gas mixtures, AIChE J. 3 (1957) 198–207.

[39] R. Krishna, G.L. Standart, Mass and energy transfer in multicomponent systems, Chem. Eng. Commun. 3 (1979) 201–275.

[40] A. Gorak, A. Vogelpohl, Experimental study of ternary distillation in a packed column, Separ. Sci. Technol. 20 (1985) 33–61.

[41] L.E. Scriven, Flow and transfer at fluid interfaces, Chem. Eng. Educ. 150–155 (Fall 1968); 26–29 (Winter 1969); 94–98 (Spring 1969).

[42] R. Higbie, The rate of absorption of a pure gas into a still liquid during short periods of exposure, Trans. Am. Inst. Chem. Eng. 31 (1935) 365–383.

[43] P.V. Danckwerts, Significance of liquid film coefficients in gas absorption, Ind. Eng. Chem. 43 (1951) 1460–1467.

[44] R. Clift, J.R. Grace, M.E. Weber, Bubbles, Drops and Particles, Academic Press, London, England, 1978.

[45] T.K. Sherwood, R.L. Pigford, C.R. Wilke, Mass Transfer, McGraw-Hill, New York, 1975.

[46] R. Krishna, G.L. Standart, A multicomponent film model incorporating an exact matrix method of solution to the Maxwell–Stefan equations, AIChE J. 22 (1976) 383–389.

[47] H.L. Toor, Solution of the linearized equations of multicomponent mass transfer, AIChE J. 10 (1964) 448–455, 460-465.

[48] W.E. Stewart, R. Prober, Matrix calculation of multicomponent mass transfer in isothermal systems, Ind. Eng. Chem. Fundam. 3 (1964) 224–235.

[49] R. Krishna, A generalized film model for mass transfer in non-ideal fluid mixtures, Chem. Eng. Sci. 32 (1977) 659–667.

[50] R. Krishna, A simplified film model description of multicomponent interphase mass transfer, Chem. Eng. Commun. 3 (1979) 29–39.

[51] M.J. Lockett, Distillation Tray Fundamentals, Cambridge University Press, Cambridge, England, 1986.

[52] H.A. Kooijman, Dynamic Nonequilibrium Column Simulation (Ph.D. thesis), Clarkson University, 1995.

[53] W.R. Korchinsky, Liquid mixing in distillation trays : simultaneous measurement of the diffusion coefficient and point efficiency, Chem. Eng. Res. Des. 72 (1994) 472−478.

[54] K.T. Klemola, J.K. Ilme, Distillation efficiencies of an industrial-scale i-butane/ n-butane fractionator, Ind. Eng. Chem. Res. 35 (1996) 4579−4586.

[55] AIChE, Bubble Tray Design Manual: Prediction of Fractionation Efficiency, AIChE, New York, 1958.

[56] H. Chan, J.R. Fair, Prediction of point efficiencies on sieve trays. 1. Binary systems, Ind. Eng. Chem. Proc. Des. Deu. 23 (1984) 814−819.

[57] F.J. Zuiderweg, Sieve trays—a view of the state of the art, Chem. Eng. Sci. 37 (1982) 1441−1464.

[58] D.L. Bennett, R. Agrawal, P.J. Cook, New pressure drop correlation for sieve tray distillation columns, AIChE J. 29 (1983) 434−442.

[59] K. Onda, H. Takeuchi, Y. Okumoto, Mass transfer coefficients between gas and liquid phases in packed columns, Chem. Eng. Jpn. 1 (1968) 56−62.

[60] J.L. Bravo, J.R. Fair, Generalized correlation for mass transfer in packed distillation columns, Ind. Eng. Chem. Process. Des. Dev. 21 (1982) 162−170.

[61] J.A. Rocha, J.L. Bravo, J.R. Fair, Distillation columns containing structured packings: a comprehensive model for their performance. 2. Mass-transfer model, Ind. Eng. Chem. Res. 35 (1996) 1660−1667.

[62] R. Billet, M. Schultes, Prediction of mass transfer columns with dumped and arranged packings, Chem. Eng. Res. Des. 77 (1999) 498.

[63] M.J. Lockett, T. Plaka, Effect of non-uniform bubbles in the froth on the correlation and prediction of point efficiencies, Chem. Eng. Res. Des. 61 (1983) 119−124.

[64] E. Kaltenbacher, On the effect of the bubble size distribution and the gas phase diffusion on the selectivity of sieve trays, Chem. Eng. Fundam. 1 (1982) 47−68.

[65] H. Hofer, Influence of gas-phase dispersion on plate column efficiency, Ger. Chem. Eng. 6 (1983) 113−118.

[66] M. Prado, J.R. Fair, Fundamental model for the prediction of sieve tray efficiency, Ind. Eng. Chem. Res. 29 (1990) 1031−1042.

[67] J.A. Garcia, J.R. Fair, A fundamental model for the prediction of distillation sieve tray efficiency. 2. Model development and validation, Ind. Eng. Chem. Res. 39 (2000) 1818.

[68] S.R. Syeda, A. Afacan, K.T. Chuang, A fundamental model for prediction of sieve tray efficiency, Chem. Eng. Res. Des. 85 (2007) 269−277.

[69] A.N. Vennavelli, J.R. Whitely, M.R. Resetarits, New fraction jetting model for distillation sieve tray efficiency prediction, Ind. Eng. Chem. Res. 51 (2012) 11458−11462.

[70] R. Krishna, Model for prediction of point efficiencies for multicomponent distillation, Chem. Eng. Res. Des. 63 (1985) 312−322.

[71] P.A.M. Springer, R. Baur, R. Krishna, Composition trajectories for heterogeneous azeotropic distillation in a bubble-cap tray column: influence of mass transfer, Chem. Eng. Res. Des. 81 (2003) 413−426.

[72] A.B. Ponter, P.H. Au-Yeung, Estimating Liquid Film Mass Transfer Coefficients in Randomly Packed Columns, Chapter 20, in: N.P. Cheremisinoff (Ed.), Handbook of Heat and Mass Transfer, vol. II, Gulf Publishing Corp., Houston, TX, 1986, pp. 903−952.

[73] G.Q. Wang, X.G. Yuan, K.T. Yu, Review of mass-transfer correlations for packed columns, Ind. Eng. Chem. Res. 44 (2005) 8715−8729.

[74] E. Brunazzi, G. Nardini, A. Paglianti, L. Petarca, Interfacial area of mellapak packing: absorption of 1,1,1-trichloroethane by Genosorb 300, Chem. Eng. Technol. 18 (1995) 248−255.

[75] Ž. Olujic, A.B. Kamerbeek, J.A. de Graauw, Corrugation geometry based model for efficiency of structured distillation packing, Chem. Eng. Process. 38 (1999) 683.

[76] N. Razi, O. Bolland, H. Svendsen, Review of design correlations for CO_2 absorption into MEA using structured packings, Int. J. Greenhouse Gas Control 9 (2012) 193−219.

[77] L. Valenz, F.J. Rejl, J. Šíma, V. Linek, Absorption mass-transfer characteristics of Mellapak packings series, Ind. Eng. Chem. Res. 50 (2011) 12134−12142.

[78] R. Krishna, J.A. Wesselingh, The Maxwell−Stefan approach to mass transfer, Chem. Eng. Sci. 52 (1997) 861−911.

[79] C.F. Curtiss, R.B. Bird, Multicomponent diffusion, Ind. Eng. Chem. Res. 38 (1999) 2515−2522.

[80] P.J.A.M. Kerkhof, M. Geboers, Toward a unified theory of isotropic molecular transport phenomena, AIChE J. 51 (2005) 79−121.

[81] P.J.A.M. Kerkhof, M. Geboers, Analysis and extension of the theory of multicomponent fluid diffusion, Chem. Eng. Sci. 60 (2005) 3129−3167.

[82] A.F. Mills, The use of the diffusion velocity in conservation equations for multicomponent gas mixtures, Int. J. Heat Mass Tran. 41 (1998) 1955−1968.

[83] D. Matuszak, M.D. Donohue, Inversion of multicomponent diffusion equations, Chem. Eng. Sci. 60 (2005) 4359−4367.

[84] C.-H. He, Y.-S. Yu, New equation for infinite-dilution diffusion coefficients in supercritical and high-temperature liquid solvents, Ind. Eng. Chem. Res. 37 (1999) 3793−3798.

[85] R. Krishna, G.L. Standart, Addition of resistances for non-isothermal multicomponent mass transfer, Lett. Heat Mass Tran. 3 (1976) 41−48.

[86] S.W. Churchill, Critique of the classical algebraic analogies between heat, mass, and momentum transfer, Ind. Eng. Chem. Res. 36 (1997) 3866−3878.

[87] D. Green, R.H. Perry (Eds.), Perry's Chemical Engineers Handbook, McGraw-Hill, New York, 2008.

[88] R. Krishna, R.M. Salomo, M.A. Rahman, Ternary mass transfer in a wetted wall column.significance of diffusional interactions. Part II. equimolar diffusion, Trans. Inst. Chem. Engrs. 59 (1981) 44−53.

[89] J.D. Seader, E.J. Henley, Separation Process Principles, second ed., Wiley, New York, 2006.

[90] M.G. Shi, A. Mersmann, Effective interfacial area in packed columns, Ger. Chem. Eng. 8 (1985) 87−96.

Principles of Binary Distillation

4

Eva Sorensen

Department of Chemical Engineering, UCL, London, UK

CHAPTER OUTLINE

Distillation: Fundamentals and Principles. http://dx.doi.org/10.1016/B978-0-12-386547-2.00004-1

4.1 Introduction

Distillation is today the most widely used unit operation for separation of liquid mixtures in the chemical and petroleum industries. The method is based on boiling the mixture to create two phases, a vapor phase and a liquid phase, as most boiling mixtures will result in a vapor phase which is richer in the more volatile component than the liquid phase. By separating the two phases, separation of the components is achieved. The demand for purer products, coupled with the need for greater flexibility, has promoted continued research in distillation, despite the fact that the method has been used for millennia.

Distillation today accounts for 90−95% of all industrial separations. Approximately 40,000 distillation columns are in operation in the United States alone, requiring around 40% of the total energy consumption in the U.S. chemical process industries [1]. Distillation columns have a wider range in capacity than any other type of processing equipment, with single columns ranging from 0.3 to 12 m in diameter and 1 to 75 m in height.

The main challenges of distillation column design is to determine the right dimensions of the column, i.e. the column height and diameter, and the right operating conditions, generally determined by the heat input in the reboiler and the amount of reflux returned to the column at the top, in order to achieve the required purities of the products at minimum capital and operating costs, and often in the face of feed, and other, variations.

4.2 Vapor−liquid equilibrium

Calculations of distillation columns are based on vapor−liquid equilibrium data. Experimental determination, and in particular, estimation of such data, is constantly developing and will not be considered here, but the reader is referred to Chapter 2 of this book (Ghmeling and Kleiber, 2014), or to standard textbooks on mass transfer operations (e.g. [1−4] etc.). This chapter will give only a brief introduction to vapor−liquid equilibrium to illustrate which data is required and how this is used in distillation calculations.

At thermodynamic equilibrium, the components in a mixture will distribute themselves between the vapor phase and the liquid phase as illustrated in Figure 4.1. For binary mixtures, the equilibrium data is normally depicted in diagrams showing either temperature (T) as a function of vapor (y) and liquid (x) compositions, or as vapor composition as a function of liquid composition, both at a constant pressure (P) as shown in Figure 4.2. The data is normally shown with reference to the lightest,

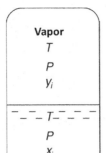

FIGURE 4.1 Vapor–Liquid Equilibrium

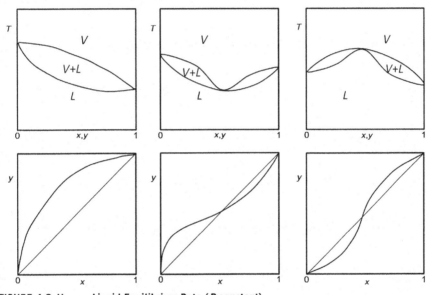

FIGURE 4.2 Vapor–Liquid Equilibrium Data (*P* constant)

Top: T-*xy* diagrams, Bottom: *x-y* diagrams. (Left: zeotropic mixture, Middle: minimum boiling azeotrope, Right: maximum boiling azeotrope).

i.e. lowest boiling or more volatile, component. The connection between the two diagrams is shown in Figure 4.3.

In the T-*xy* diagram (see Figure 4.2 top), the lower line is the *bubble point line* for the binary mixture and corresponds to the temperature T_{bp} to which a liquid mixture with composition x of the lightest component needs to be heated for the first bubble of vapor to form. The upper line is the *dew point line* and corresponds to the temperature T_{dp} to which a vapor mixture with composition y of the lightest component needs to be cooled for the first drop of condensed liquid to form. The region between

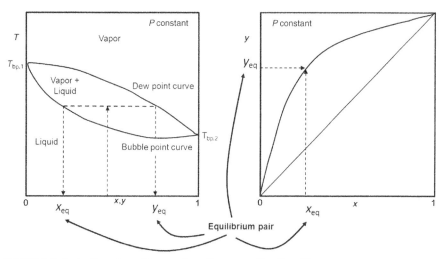

FIGURE 4.3 Derivation of x-y Diagram from Corresponding T-xy Diagram

Smith Chapter 4

the two temperatures is the two-phase region in which both a liquid phase and a vapor phase will exist in equilibrium. At temperature T, the liquid phase will have a composition x of the lightest component, and the vapor phase will have a corresponding composition y of the lightest component in this region. It can be seen that for any liquid composition x, the vapor with composition y formed when this mixture boils is always richer in the lightest component (see Figure 4.2).

The distribution of component i between the vapor phase, with composition y_i, and the liquid phase with composition x_i, is normally represented by the *distribution coefficient* K_i:

$$K_i = y_i/x_i \qquad (4.1)$$

The distribution coefficient, or *K-value*, is generally a function of temperature and pressure only and not of compositions. A component with a high K-value will be quite volatile, as the vapor phase will have a high concentration of that component. Similarly, a low K-value means the component is not particularly volatile.

For successful separation of a mixture, the main interest is the difference between the volatilities of the components, as a large difference will mean an easy separation. This difference is usually expressed in terms of the *relative volatility* α_{ij} where component i is more volatile than component j:

$$\alpha_{ij} = \frac{K_i}{K_j} = \frac{y_i/x_i}{y_j/x_j} \qquad (4.2)$$

The higher the relative volatility is, then the easier is the separation. Distillation will rarely be considered for mixtures with relative volatilities below 1.05 [5].

4.2.1 **Ideal homogeneous systems**

When the liquid phase behaves as an ideal solution, then [6]:

1. all molecules have the same size
2. all intermolecular forces are equal
3. the properties of the mixtures depend only on the properties of the pure components.

Mixtures of isomers, or of an adjacent number in a homologous series, give close to ideal liquid phase behavior.

In a homogeneous ideal mixture, the number of molecules of any one component per unit area of vapor–liquid interphase surface will be less than if that component exposed the same area of surface alone as a pure component. For this reason, the rate of vaporization of a component per unit area will be lower in a mixture than for a pure component. This can be expressed in terms of *Raoult's law*, which states that the partial pressure of a component in the vapor phase p_i is a function of the composition of that component in the liquid phase x_i and the vapor pressure p_i^{vap} of the component:

$$p_i = x_i \cdot p_i^{\text{vap}} \tag{4.3}$$

Experimental values for vapor pressures of many components can be found in the literature. In calculations, however, it is easier to have this expressed in the form of a mathematical equation. Many different equations are available, and the easiest of these is probably the *Antoine equation* for which values for the three coefficients A_i, B_i, and C_i can be found in the literature for a number of components i (e.g. Ref. [5]):

$$\ln p_i^{\text{vap}} = A_i - \frac{B_i}{T + C_i} \tag{4.4}$$

A word of warning, Antoine's equation is not dimensionless, hence when using this equation care must be taken to use the right units for temperature and vapor pressure as the values of the coefficients will depend on the units used.

The partial pressure p_i is by definition equal to the product of the vapor phase composition y_i and the total pressure P:

$$p_i \equiv y_i \cdot P \tag{4.5}$$

Dalton's law states that the total pressure of an ideal gas mixture, P, equals the sum of the partial pressures:

$$P = \sum p_i \tag{4.6}$$

Note that the different pressures are often mixed up. The *total pressure P* is the pressure of the system, the *partial pressure p_i* is the hypothetical pressure of pure component i if component i alone occupies the same volume as the mixture at the same temperature, and the *vapor pressure p_i^{vap}* is a thermodynamic quantity of component i and is the pressure exerted by a vapor phase in thermodynamic equilibrium with a liquid phase at a given temperature in a closed system. It should be noted that different textbooks use different notation for vapor pressure, and this can be represented with the symbols p_i^{vap}, p_i^{sat}, p_i, or p_i^0.

Equations (4.3) and (4.5) can be combined to:

$$y_i \cdot P = x_i \cdot p_i^{vap} \qquad (4.7)$$

The distribution coefficient K_i can then be expressed as (Eqn 4.1):

$$K_i = \frac{y_i}{x_i} = \frac{x_i p_i^{vap}/P}{x_i} = \frac{p_i^{vap}}{P} \qquad (4.8)$$

and the relative volatility α_{ij} as (Eqn 4.2):

$$\alpha_{ij} = \frac{K_i}{K_j} = \frac{p_i^{vap}/P}{p_j^{vap}/P} = \frac{p_i^{vap}}{p_j^{vap}} \qquad (4.9)$$

For a binary mixture of components A and B, the compositions of component B in the two phases are given by $y_B = 1 - y_A$ and $x_B = 1 - x_A$, respectively, and therefore:

$$P = p_A + p_B = x_A p_A^{vap} + x_B p_B^{vap} = x_A p_A^{vap} + (1 - x_A) p_B^{vap}$$
$$= x_A \left(p_A^{vap} - p_B^{vap} \right) + p_B^{vap} \qquad (4.10)$$

or

$$x_A = \frac{P - p_B^{vap}}{p_A^{vap} - p_B^{vap}} \qquad (4.11)$$

The vapor—liquid equilibrium characteristics for an ideal binary mixture can thus be calculated based only on information about their respective vapor pressures, as Eqn (4.11) can be used to find x_A, and Eqn (4.7) to find y_A, at selected temperatures between the boiling points of the two pure components at pressure P, respectively.

For a binary mixture of components A and B, the relative volatility α_{AB}, if known, can also be used to find the equilibrium conditions, as the vapor composition y_A can be found as a function of the liquid composition x_A and the relative volatility α_{AB} from:

$$\alpha_{AB} = \frac{y_A/x_A}{y_B/x_B} = \frac{y_A/x_A}{(1 - y_A)/(1 - x_A)} = \frac{y_A}{(1 - y_A)} \frac{(1 - x_A)}{x_A} \qquad (4.12)$$

or

$$y_A = \frac{\alpha_{AB} x_A}{1 + (\alpha_{AB} - 1)x_A} \qquad (4.13)$$

Figure 4.4 shows the vapor composition y_A for different values of the relative volatility α_{AB} for a binary liquid mixture as a function of the composition x_A of

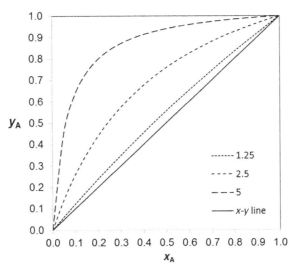

FIGURE 4.4 Vapor–Liquid Equilibrium Conditions for Different Relative Volatilities

the lightest component A, i.e. the x-y or equilibrium diagram, calculated based on Eqn (4.13). It can clearly be seen that, as the relative volatility increases, the composition of component A in the vapor phase becomes progressively higher for a given liquid composition x_A.

At a given total pressure, vapor pressures are a function of temperature, and the relative volatility is therefore also a function of temperature, however, less so than for the distribution coefficients because it is proportional to the ratios of the vapor pressures. In general, the vapor pressure of the more volatile component tends to increase at a slower rate with increasing temperature than the less volatile component, and the relative volatility therefore generally decreases with increasing temperature [7]. In many instances though, the relative volatility can be considered constant, which greatly simplifies distillation calculations.

4.2.2 Nonideal homogeneous mixtures

In an ideal binary mixture, the intermolecular forces between two molecules of component A are the same as the forces between two molecules of component B, and between one molecule of component A and one molecule of component B. This is, however, not the case for most real liquid mixtures[1]. Deviations from Raoult's law are due to changes in the intermolecular forces, of which the

---

[1]The vapor phase may also exhibit nonideality, however, normally only at quite high pressures, e.g. above $7 \cdot 10^5 - 10 \cdot 10^5$ Pa, and assuming ideal vapor phase behavior is therefore a common assumption in distillation calculations.

hydrogen bond plays an important role. A real liquid mixture may exhibit either positive or negative deviations from Raoult's law, i.e. can either have a higher or a lower vapor pressure than the corresponding ideal mixture. Positive deviations from ideality are more common when the molecules are dissimilar and exhibit repulsive forces. Negative deviations occur when there are attractive forces between molecules of different components that do not occur for molecules of either component alone.

If the deviations from ideality are large, the mixture may have either a maximum or a minimum vapor pressure when boiling at a constant temperature, or correspondingly, have a minimum or a maximum temperature when boiling at a constant pressure (see Figure 4.2, middle and right). At this point, the composition of the vapor phase equals that of the liquid phase and is called a minimum boiling or maximum boiling *azeotrope*[2]. The azeotropic point is a function of total pressure and can for some mixtures be shifted, or even disappear, by changing the total pressure. Azeotropic mixtures cannot be separated by standard distillation, but there are methods that may still be used to separate such mixtures, e.g. extractive or azeotropic distillation (see Chapters 6 and 7 in [8]).

Deviations from ideal behavior in the liquid phase are taken into account by modifying Raoult's law by introducing a liquid phase *activity coefficient* γ_i, and Eqn (4.3) then becomes:

$$p_i = x_i \gamma_i p_i^{\text{vap}} \tag{4.14}$$

Correspondingly, Eqns (4.7)–(4.9) become:

$$y_i = x_i \gamma_i p_i^{\text{vap}} / p \tag{4.15}$$

$$K_i = \frac{\gamma_i p_i^{\text{vap}}}{p} \tag{4.16}$$

$$\alpha_{ij} = \frac{\gamma_i p_i^{\text{vap}}}{\gamma_j p_j^{\text{vap}}} \tag{4.17}$$

The activity coefficient γ_i is a function of temperature T and composition of component i in the liquid phase x_i. The value of the activity coefficient γ_i approaches unity as the liquid concentration x_i approaches unity, and the highest value of γ_i occurs as the concentration approaches zero [2].

Many equations have been suggested for calculation of the activity coefficient, and the simplest versions are the van Laar, Margules, or Wilson equations. The van Laar equations for a binary mixture of component A (most volatile) and component B (least volatile) are given by the following equations, where the constants A_{AB}

---

[2]Minimum boiling azeotropes are far more common.

and A_{BA} are different and can be found estimated from experimental values in reference literature (e.g. [5]):

$$\ln \gamma_A = \frac{A_{AB} x_B^2}{\left[x_A \left(\frac{A_{AB}}{A_{BA}} \right) + x_B \right]^2} \tag{4.18}$$

$$\ln \gamma_B = \frac{A_{BA} x_A^2}{\left[x_B \left(\frac{A_{BA}}{A_{AB}} \right) + x_A \right]^2} \tag{4.19}$$

4.2.3 Heterogeneous mixtures

A mixture can also consist of *two* liquid phases, both of which are at equilibrium with the vapor phase. Condensation of a component can take place only at the restricted area where the vapor molecules of a component are in contact with its own molecules in the liquid phase. Thus, the vapor pressure of one of the liquid phases is unaffected by the presence of the other liquid phase. The total pressure is the sum of the vapor pressures of the two liquid phases. A mixture with two liquid phases can also exhibit an azeotrope, which is then referred to as a *heterogeneous azeotrope*.

4.3 Differential distillation

Differential distillation, also often referred to as simple or *Rayleigh distillation*, is the most elementary example of batch distillation and is illustrated in Figure 4.5.

The unit consists of a heated still or reboiler, a condenser, and normally just one receiving tank. No trays or packing are used, and there is no reflux of condensed material back to the still. At the start, feed material is charged into the still and brought to the boil. The resulting vapor is condensed in the condenser and collected in the receiving tank. As the vapor is richer in the more volatile component compared to

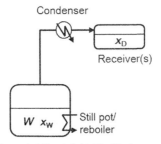

FIGURE 4.5 Schematic of Differential (Rayleigh) Distillation

the liquid in the still, the liquid remaining in the still will become progressively weaker in the more volatile component and therefore richer in the less volatile component. The process is continued until the still is either depleted of the more volatile component, a product specification is reached for either the still or the receiver composition, or the batch time has come to an end.

The still may be assumed to be the only *equilibrium stage* as there are no trays above the still. Letting W_0 represent the moles of initial feed to the still, x_{W_0} the initial mole fraction of light component in that feed, W the number of moles remaining in the still at any time with x_W the corresponding mole fraction of light component, and x_D the average mole fraction of light component in the vapor, a material balance for the light component over the unit at any point in time yields:

$$x_D dW = d(W x_W) = x_W dW + W dx_W \qquad (4.20)$$

Here dW represents the amount of material leaving the still, i.e. the amount of vapor produced during an infinitesimal time interval dt. Rearranging the variables, and integrating from the initial charge conditions of W_0 and x_{W_0} to the final conditions of W and x_W, gives the well-known Rayleigh equation:

$$\int_{W_0}^{W} \frac{dW}{W} = \int_{x_{W_0}}^{x_W} \frac{dx_W}{x_D - x_W} \qquad (4.21)$$

From Rayleigh's equation it is possible to calculate the amount of liquid that must be distilled $(W_0 - W)$ in order to obtain a liquid of specified composition (x_W) in the still. The integral may be solved numerically or graphically, however, in some cases, the integral can be carried out analytically. If equilibrium is assumed between liquid and vapor, the right-hand side of the equation may be evaluated by plotting $1/(x_D - x_W)$ versus x_W and measuring the area under the curve between the limits x_{W_0} and x_W. If the mixture is a binary system for which a relative volatility α is constant, or if an average value that will serve for the range considered can be found, Eqn (4.13) can be substituted into Rayleigh's equation, and a direct integration can be made resulting in [3]:

$$\ln\left(\frac{W_0}{W}\right) = \frac{1}{\alpha - 1}\left[\ln\left(\frac{x_{W_0}}{x_W}\right) + \alpha \ln\left(\frac{1 - x_W}{1 - x_{W_0}}\right)\right] \qquad (4.22)$$

It should be noted that a complete separation is impossible based on Rayleigh distillation unless the relative volatility of the mixture is infinite. Its application is therefore limited to wide-boiling mixtures that are easy to separate or to laboratory-scale separations where high purities are not required. The process is nevertheless also used in the beverage industry for the production of whisky, for instance, where the initial distillate is cooled and distilled a second time to improve the separation.

4.4 Flash distillation

The simplest continuous distillation process is the single-stage flash process shown in Figure 4.6 in which a feed is partially vaporized, or condensed, to give a vapor richer in the more volatile component than the remaining liquid. The two-phase mixture is obtained by either [3]:

1. heating a pressurized liquid mixture and flashing the mixture adiabatically across a valve to a lower pressure, resulting in the creation of a vapor phase,
2. heating and partially vaporizing a low-pressure liquid in a heat exchanger, resulting in the creation of a vapor phase, or
3. cooling and partially condensing a vapor mixture in a heat exchanger, resulting in the creation of a liquid phase.

The vaporized fraction of the feed, $f = V/F$, depends on the amount of heat added to, or removed from, the system. The drum provides disengagement space to allow the vapor to separate from the liquid. An entrainment eliminator or a demister is often employed to prevent liquid droplets from being carried over by the vapor. Unless the relative volatility is very large, the degree of separation achievable between

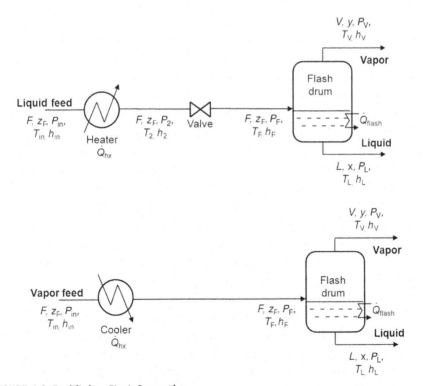

FIGURE 4.6 Equilibrium Flash Separation

two components in a flash process is generally quite poor. Flashing is therefore normally used as an auxiliary process to prepare streams for further purification. An exception is seawater desalination, in which pure drinking water can be obtained from seawater through a multistage flash (MSF) operation.

Flash calculations are of fundamental importance, not only to understand the operation of the flash unit in Figure 4.6 but also to determine the phase conditions of any stream given its pressure, temperature, and compositions. A total material balance over the flash drum is given by:

$$F = V + L \tag{4.23}$$

where F is the feed molar flow rate, V is the vapor product molar flow rate and L is the liquid molar flow rate and the corresponding component balance for component i is:

$$F z_{F,i} = V y_i + L x_i \tag{4.24}$$

The energy balance is given by:

$$F h_F + \dot{Q}_{flash} = V h_V + L h_L \tag{4.25}$$

where $\dot{Q}_{flash}$ is the energy added or removed in the flash drum. The variables h_F, h_L, and h_V are the enthalpies of the feed, liquid and vapor streams, respectively, which are all functions of pressure, temperature, and compositions:

$$h_F = h_F(P_F, T_F, z_F) \tag{4.26}$$

$$h_L = h_L(P_L, T_L, x) \tag{4.27}$$

$$h_V = h_V(P_V, T_V, y) \tag{4.28}$$

The unit is generally assumed to be under mechanical equilibrium ($P_F = P_V = P_L = P$), thermal equilibrium ($T_F = T_V = T_L = T$) and phase equilibrium ($y_i = K_i y_i$) where the distribution coefficient K_i is generally a function of pressure and temperature.

Seader et al. (2013) [3] provide detailed descriptions of the calculation procedures for different flash conditions such as isothermal flash (T_V and P_V specified), adiabatic flash ($\dot{Q}_{flash} = 0$ and P_V specified) etc. Multicomponent flash calculations can be very tedious because of their iterative nature, which makes manual calculations unsuitable mainly due to the complexity of the expression for the thermodynamic quantities, i.e. the enthalpies and the distribution coefficients. Luckily, robust calculation methods are incorporated into most simulation software such as ChemCAD, HYSYS, Aspen, and gPROMS.

4.4.1 Binary flash calculations

In a simple binary flash, the two material balances are:

$$F = V + L \tag{4.29}$$

$$F z_F = V y + L x \tag{4.30}$$

where x and y are the liquid and vapor compositions, respectively, of the more volatile component. By introducing the vaporized fraction $f = V/F$ and combining the equations, the following expression is obtained[3]:

$$Fz_F = Vy + (F - V)x = fFy + (F - fF)x$$

or

$$z_F = fy + (1 - f)x$$

which can be rearrange to

$$y = \frac{z_F}{f} - \frac{(1-f)}{f}x \qquad (4.31)$$

Assuming the total drum pressure P, as well as the feed composition z_F and the fraction of vapor f, are known, then this equation has two unknowns, x and y. However, as x and y are coordinates of a point on the equilibrium curve at pressure P (see Figure 4.2 and Figure 4.3), if one of the compositions is known then the other is given by the equilibrium conditions. Eqn (4.31) is thus the equation of a straight line, the *operating line*, with slope $-(1 - f)/f$ and y-intercept z_F/f, in the x-y diagram as shown in Figure 4.7. It is, however, easier not to draw the line from the y-intercept but rather from the interception between the operating line and the $x = y$ line, as substituting $x = y$ into Eqn (4.31) gives $x = y = z_F$, i.e. the line should start from (z_F, z_F)[4].

Once the operating line has been drawn, then the interception between the operating line and the equilibrium line give the composition in the vapor and liquid streams, respectively. From the compositions the corresponding temperature T can be found from the equilibrium data as shown in Figure 4.3.

With the compositions, temperature, and pressure known, the enthalpies of the outlet streams, h_L and h_V, can be found from Eqns (4.27) and (4.28)[5]. The feed enthalpy, h_F, can be found from the enthalpy balance (see Eqn (4.25)), and from Eqn (4.26) then also the feed temperature T_F. The amount of energy required in the heater, $\dot{Q}_{hx}$, can be determined from an energy balance around the heater:

$$\dot{Q}_{hx} + Fh_{in}(P, T_{in}, z_F) = Fh_F(P, T_F, z_F) \qquad (4.32)$$

The amount of energy required will then in turn determine the size of the heat exchanger required. Equally, the flash drum can be sized once the compositions and flow rates have been determined (see example by Wankat, 2012 [1]; page 48) for both vertical and horizontal drums.

---

[3]Note that the fraction f is not fixed directly but is a result of the enthalpy of the hot incoming feed as well as the enthalpies of the vapor and liquid streams leaving the flash drum.

[4]It is important to understand that the line $x = y$ has no practical significance and is only used to make the graphical representation easier.

[5]For ideal mixtures the enthalpies can be calculated from heat capacities (for h_F and h_L) and from heat capacities and latent heats (for h_V).

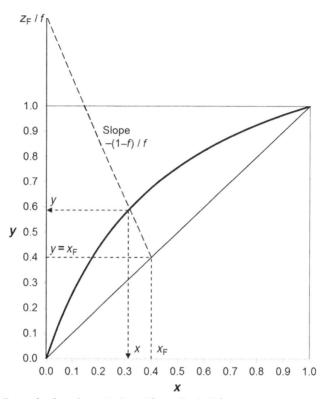

FIGURE 4.7 Determination of *x* and *y* for a Binary Flash Unit

4.4.2 **Multistage flash processes**

As the separation performance in a single flash process is low, several flash units can be combined in a cascade to gradually improve the separation. In the example in Figure 4.8 (top), the vapor stream from the first flash is directed to the second flash operated at a higher pressure, and then that of the second to the third, etc. Similarly, the liquid from the first flash can become the feed of a second flash at a lower pressure, etc. (see Figure 4.8 bottom), or both the vapor and the liquid streams can be directed to additional flash units. As previously mentioned, seawater desalination can be realized in MSF units, and this process currently accounts for over half of the desalinated water produced in the world.

4.5 **Continuous distillation with rectification**

The achievable separation in a flash unit, and even in a train of flash units, is generally fairly low. An improvement in the separation performance is achieved by condensing the vapor and recycling some of this condensate as liquid. Continuous

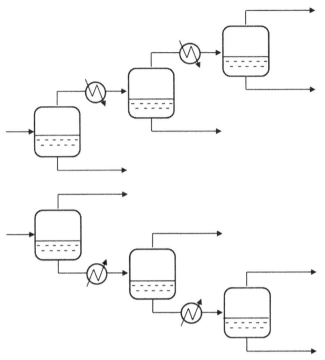

FIGURE 4.8 Cascades of Flash Units

rectification, or *fractionation*, is a multistage countercurrent distillation operation. Continuous distillation with rectification of a type used even today was invented by Cellier Blumenthal in 1808[6].

A schematic of continuous rectification is shown in Figure 4.9 (left). The trays in the column are here numbered from the top and down, i.e. tray 1 is the top tray, and tray N is the bottom tray, in the column section. It should be noted that some textbooks or authors may number the trays from the bottom and up (e.g. [5]) so care should be taken when using any equations. (The same goes for simulation software, i.e. always check in which direction the trays are numbered.)

In this process, vapor from the reboiler bubbles through the liquid on the bottom tray in the column section (tray N) and is partially condensed. The heat liberated by condensing the vapor from the reboiler to liquid on the bottom tray will in turn revaporize some of the liquid on the tray, resulting in a vapor of a higher composition. The vapor from the bottom tray travels upwards to the second lowest tray (tray $N - 1$) and is there partially condensed, which will thus revaporize some of the

---

[6]France was cut off from supplies of cane sugar, and Cellier Blumenthal was attracted by a prize of 1 million francs offered by Napoleon for a good method for obtaining large quantities of white sugar from beet root. The process involved the use of alcohol that had to be recovered [9].

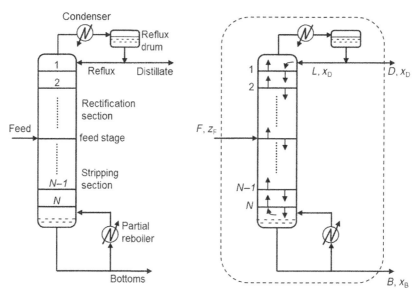

FIGURE 4.9 Continuous Rectification Distillation Column

(Left: schematic, Right: overall mass balance envelope).

liquid on that tray. This process of partial condensation and revaporization is repeated on each tray throughout the column. The vapor leaving the top of the column section (tray 1) is condensed in the condenser and directed to a reflux drum. Liquid from the reflux drum is either removed as product (distillate) or returned to the column as *reflux*. The liquid on the trays in the column is thus provided by this reflux stream, which will flow downwards in the column from tray to tray, meeting the vapor on its way up. The liquid leaving the bottom tray (tray N) is returned to the reboiler or removed as product (bottoms). The countercurrent operation of the unit ensures that every time material is vaporized, the more volatile component tends to concentrate in the vapor phase and the less volatile component in the liquid phase.

It is important to note that the vapor entering a tray from below is *not* in equilibrium with the reflux entering from the top. On each tray, however, the system tends toward equilibrium because [2]:

1. Some of the less volatile component condenses from the rising vapor into the liquid thus increasing the concentration of the more volatile component remaining in the vapor phase.
2. Some of the more volatile component is vaporized from the liquid on the tray thus decreasing the concentration of the more volatile component remaining in the liquid phase.

The number of molecules passing in each direction from vapor to liquid, and in reverse, can be approximately the same if the heat given by 1 mol of the vapor on condensing is approximately equal to the heat required to vaporize 1 mol of the liquid, i.e. equimolar counter diffusion. (This assumption is commonly referred to as *constant molar overflow* as will be explained later.)

In the arrangement in Figure 4.9, the feed is introduced continuously to the column and two product streams are continuously withdrawn, one from the top of the column as condensed vapor from the reflux drum and one from the bottom of the column or the reboiler. The top stream, called the *distillate*, is richer in the more volatile component, and the bottom stream, called the *bottoms* or *residue*, is richer in the less volatile component. The section above the feed is referred to as the *rectification* section and the section below the feed as the *stripping* section. The purity obtained from this process will depend on the relative volatility between the components, the amount of reflux returned relative to the vapor from the reboiler, and the number of trays in the distillation column.

4.5.1 Column balances

The first step in any distillation calculation is to establish the material and energy balances over the unit. A total material balance over the whole column unit at steady state can be described over the dashed area as (see Figure 4.9 right):

$$F = B + D \tag{4.33}$$

where F is the molar flow rate of the feed, D is the molar flow rate of the distillate and B is the molar flow rate of the bottoms. The corresponding component balance for a binary mixture as (the mole fractions are with reference to the most volatile component):

$$F z_F = B x_B + D x_D \tag{4.34}$$

Separate balances can also be set up over subsections of the column, e.g. over the top of the column (see Figure 4.10 left):

$$V_{n+1} = L_n + D \tag{4.35}$$

$$V_{n+1} y_{n+1} = L_n x_n + D x_D \tag{4.36}$$

where V_{n+1} is the molar flow rate of the vapor into the top section and L_n is the molar flow rate of liquid leaving the top section.

Equivalently, balances over the bottom of the column (see Figure 4.10 right):

$$V_m = L_{m-1} + B \tag{4.37}$$

$$V_m y_m = L_{m-1} x_{m-1} - B x_B \tag{4.38}$$

where L_{m-1} is the molar flow rate of the liquid into the bottom section and V_m is the molar flow rate of vapor leaving the bottom section.

Balances can also be established over each stage. For stage n, four streams are involved as shown in Figure 4.11 (left): the vapor stream entering stage n from the stage below $(n+1)$, the liquid stream entering stage n from the stage above

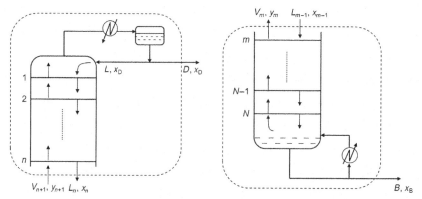

FIGURE 4.10 Mass Balance over Column Sections

(Left: top, Right: bottom).

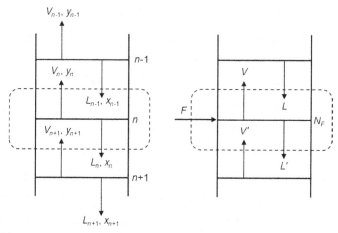

FIGURE 4.11

Balances over (left) distillation stage n and (right) over the feed stage.

$(n-1)$, and the vapor and liquid streams leaving stage n, respectively. The total and component material balances over stage n at steady state are thus given by:

$$V_{n+1} + L_{n-1} = L_n + V_n \tag{4.39}$$

$$V_{n+1}y_{n+1} + L_{n-1}x_{n-1} = L_n x_n + V_n y_n \tag{4.40}$$

4.5.2 Feed conditions

The feed conditions require some extra attention. The total material over the feed stage (see Figure 4.11 right) is similar to that of a general stage with the added stream from the feed, i.e.

$$V' + L + F = L' + V \tag{4.41}$$

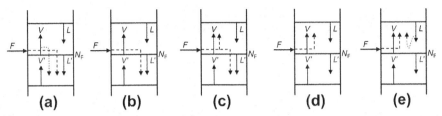

FIGURE 4.12 Possible Feed Conditions

(a) sub-cooled liquid, (b) liquid at bubble point, (c) partially vaporized, (d) vapor at dew point, and (e) superheated vapor.

where V' is the molar flow rate of the vapor entering the feed stage, V is the molar flow rate of vapor leaving the feed stage, L is the molar flow rate of liquid entering the feed stage and L' is the molar flow rate of liquid leaving the feed stage.

The feed to a distillation column is usually in liquid form or the feed might be partially vaporized and therefore in two-phase form, but pure vapor feeds are of course also possible. There are therefore five possible feed conditions:

1. Subcooled liquid, i.e. liquid below the bubble point
2. Liquid at the bubble point
3. Partially vaporized, i.e. two phase
4. Vapor at the dew point
5. Superheated vapor, i.e. vapor above the dew point

The relationship between the vapor streams above and below the feed point, V and V', respectively, and that between the corresponding liquid streams, L and L', respectively, depends on the conditions of the feed as shown in Figure 4.12. From Figure 4.12 it can be seen that for conditions 2−4 (recall that f is defined as the fraction of the feed that is vaporized):

$$V = V' + fF \quad \text{and} \quad L' = L + (1 - f)F$$

If the feed is liquid at the bubble point ($f = 0$), then this simplifies to:

$$V = V' \quad \text{and} \quad L' = L + F$$

and if it is vapor at the dew point, ($f = 1$) then:

$$V = V' + F \quad \text{and} \quad L' = L$$

For subcooled liquid and superheated vapor feeds, the flow rates will also depend on energy balances and cannot be described by simple material balances alone. The relationship is usually defined as [5]:

$$L' = L + qF \tag{4.42}$$

and from a material balance around the feed stage:

$$V = V' - (q - 1)F \tag{4.43}$$

Table 4.1 Possible Feed Conditions and Corresponding q Values

Feed Condition	f	q	q Calculation
Subcooled liquid	<0	>1	$q = [c_{P,L}(T_{bp} - T_F) + \Delta H_{vap}]/\Delta H_{vap}$
Liquid at bubble point	0	1	$q = \Delta H_{vap}/\Delta H_{vap} = 1$
Partially vaporized	$0 < f < 1$	$1 < q < 0$	$q = [(1 - f) \Delta H_{vap}]/\Delta H_{vap}$
Vapor at dew point	1	0	$q = 0$
Superheated vapor	$f > 1$	<0	$q = - [c_{P,V}(T_F - T_{dp})]/\Delta H_{vap}$

where

$$q = \frac{\text{energy to convert 1 mol of feed to saturated vapor at dew point}}{\text{molar heat of vaporization}}$$

For conditions 2–4, i.e. between the bubble point and the dew point, q is the fraction of feed that is liquid, i.e. $q = 1 - f$.

Values of q, and how to calculate these, are given in Table 4.1 for the five feed conditions. For the subcooled liquid, the energy required to convert 1 mol of feed to saturated vapor at the dew point involves first heating the feed to the boiling point ($c_{P,L}(T_{bp} - T_F)$) and then vaporizing the feed (ΔH_{vap}). As the liquid is cold as it enters the column, a fraction of the vapor that was moving up the column is instead condensed on the feed stage and will follow the liquid flow down to the stage below, i.e. $L' > L + F$ (see Figure 4.12(a)). For the superheated vapor, the feed must be cooled down to the dew point ($c_{P,V}(T_F - T_{dp})$). As the feed is superheated as it enters the column, it will vaporize some of the material on the feed stage, i.e. $V > V' + F$ (see Figure 4.12(e)).

Note that when calculating q, the heat capacities and heat of vaporization are for the feed *mixture*. In most cases, assuming a weighted average of the components in the feed should give a good estimate, e.g. for NC components:

$$c_{p,L} = \sum_{i=1}^{NC} z_{F,i} c_{p,L,i}$$

4.5.3 Reflux ratio

The *reflux ratio* R is defined as the ratio between the material returned as reflux, L, and the material removed as distillate, D:

$$R \equiv \frac{L}{D} \tag{4.44}$$

Although distillation operations can be specified by either the reflux ratio or the reboil ratio $R_B = V'/B$, by tradition R is usually used as the distillate is most often the most important product [3]. The reflux ratio in Eqn (4.44) is the *external reflux ratio*. The reflux ratio is also sometimes referred to as the *internal reflux ratio*, i.e. $R_{int} = L/V$. The advantage of the internal reflux ratio is that it will always be a number between 0 (no reflux) and 1 (total reflux), whilst the external reflux ratio can vary from 0 to any number, although will rarely be above $5-10$. A reduction in the reflux ratio reduces the heat and cooling requirements of the unit as less material is being returned to the column, but increases the number of stages needed in the column section to achieve the same separation.

The *minimum reflux ratio*, R_{min}, is defined as the reflux ratio that will require an infinite number of stages for the given separation, and this is the lower limit for the ratio. R_{min} is clearly infeasible as an operating point, and most existing columns are designed for reflux ratios between 1.1 and 1.5 times R_{min}.

The *minimum number of stages* N_{min} is found at total reflux, i.e. when $R = \infty$. In this case, no product is withdrawn and all the overhead vapor is condensed and returned to the column as reflux. Again, this is not a practical operating point but gives an indication of what the lowest value of number of stages N is.

The optimal reflux ratio R_{opt} is the reflux ratio that gives the lowest sum of capital costs and running costs for the column over the lifetime of the unit. This can only be found either (1) by trial and error by repeated design calculations, or (2) by rigorous optimization. The reflux ratio has a strong impact on operating costs as when the reflux ratio increases, the heat input in the reboiler must also increase to cope with the vaporization demand of the increased liquid flow[7]. The reflux ratio also has an impact on the capital costs of the column as can be seen in Figure 4.13.

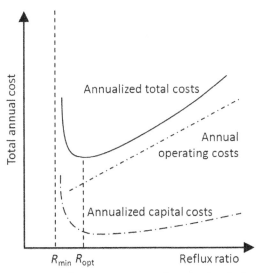

FIGURE 4.13 Operating and Running Costs as a Function of Reflux Ratio

---

[7]Note that, for a fixed column size, increasing the reflux ratio will eventually lead to column *flooding*, which must be avoided, i.e. there is a practical upper limit [10].

The lower limit is R_{min}, and as the reflux ratio is reduced toward this value, the capital costs will increase dramatically as the required height of the column is increased. The costs will approach R_{min} asymptotically as this reflux ratio corresponds to an infinitely large column. Alternatively, as the reflux ratio is increased, the capital costs will again start to increase as the increased liquid flow in the column will require a wider column to cope with the increased material, i.e. the column diameter is increased. The total costs, which determine R_{opt}, are the combined capital costs (normally annualized) and operating costs.

Finding the optimal reflux ratio from calculating total annualized costs is, however, rather time consuming, and detailed cost data is often unavailable at the early design stage. The reflux ratio is therefore in practice normally estimated from R_{min}. It is often assumed that a reasonable value for R is a simple multiple of R_{min}, e.g. $R = [1.1 - 1.5] R_{min}$. It should also be noted that most plants are operated at reflux ratios somewhat higher than the optimal as the total costs are not very sensitive to the reflux ratio in this range and better operating flexibility may be obtained if a reflux ratio above the optimal is used [11].

4.5.4 McCabe Thiele's method

The main design variable for a distillation column is the number of trays or stages in the column section. There are numerous ways of calculating this and computational software such as ChemCAD, HYSYS, Aspen, or gPROMS is normally used for this task, based on detailed material and energy balances coupled with detailed thermodynamic calculations of phase equilibrium, enthalpies, etc. One graphical method is still very popular from a teaching perspective, however, mainly due to its simplicity and its ability to explain the fundamentals of the fractionation. The *McCabe–Thiele method* was first published by Warren L. McCabe and Ernest W. Thiele in 1925 and is mainly used to calculate the total number of stages required in the column section N, the minimum reflux ratio R_{min}, and the minimum number of stages N_{min}.

The main assumptions of the method are:

1. Binary mixture
2. Steady state operation
3. Constant pressure
4. The feed stream is mixed with the feed stage fluids prior to separation
5. No heat losses
6. No heat of mixing
7. Equal, and constant, molar heats of vaporization for the two components
8. Equilibrium on each stage

The assumption of constant pressure (assumption 3) is usually good except if the column is operated under vacuum. For vacuum systems, the equilibrium curve must be adjusted for pressure variations [12]. The assumption of mixing on the feed stage (assumption 4) is good for single-phase feeds, but less so for partially vaporized feeds as a vaporized feed splits prior to mixing [12]. Assumptions 4–6 lead to constant molar vapor streams and constant molar liquid streams, in both the rectification and the stripping section of the column, and is often referred to as the *constant molar overflow* assumption[8], i.e. above the feed:

$$V_{n+1} = V_n = V \quad \text{and} \quad L_{n+1} = L_n = L \tag{4.45}$$

and equivalently below the feed:

$$V_{m-1} = V_m = V' \quad \text{and} \quad L_{m-1} = L_m = L' \tag{4.46}$$

The column component balances above the feed thus simplify to (see Eqn 4.36):

$$V y_{n+1} = L x_n + D x_D \tag{4.47}$$

or

$$y_{n+1} = \frac{L}{V} x_n + \frac{D}{V} x_D \tag{4.48}$$

and below the feed (see Eqn 4.38):

$$V' y_m = L' x_{m-1} - B x_B \tag{4.49}$$

or

$$y_m = \frac{L'}{V'} x_{m-1} - \frac{B}{V'} x_B \tag{4.50}$$

Equation 4.48 is often called the *operating line* for the rectification section of the column, or the *top operating line*[9], as it gives the relationship between the composition of the vapor entering a stage, y_{n+1}, and the composition of the liquid on the stage, x_n (see Fig. 4.10). The relationship is linear with a slope given by L/V, which is a constant value below 1 as V is less than or equal to L in the rectification section (as $V = L + D$). The operating line goes through the point (x_D, x_D) since when $x_n = x_D$

$$y_{n+1} = \frac{L}{V} x_n + \frac{D}{V} x_D = \frac{L}{V} x_D + \frac{D}{V} x_D = \frac{L+D}{V} x_D = \frac{V}{V} x_D = x_D$$

The y-intercept is given by $(x_D D)/V$ or $x_D/(R + 1)$. Equivalently, the operating line for the stripping section of the column (see Fig. 4.11), or the *bottom operating line*, is given by Eqn (4.50) and is also linear with slope V'/L', and passing through the point (x_B, x_B) since when $x_{m-1} = x_B$:

$$y_m = \frac{L'}{V'} x_{m-1} - \frac{B}{V'} x_B = \frac{L'}{V'} x_B - \frac{B}{V'} x_B = \frac{L'-B}{V'} x_B = \frac{V'}{V'} x_B = x_B$$

---

[8]The assumption holds well for e.g. benzene-toluene, isobutene-*n*-butane, and propane-*n*-butane, but is less satisfactory for e.g. acetone-water or methane-ethylene systems, and is poor for ammonia-water [12].

[9]Although the term *operating line* is normally used, it is a bad choice of words since it states little about the physical nature of these lines. The term *component balance lines* is more descriptive and would have been a far better term [12].

The intercept with $x = 1$ is given by $(L' - Bx_B)/V'$.

Note that the operating lines are both straight due to the assumption of constant molar overflow, i.e. because it has been assumed that the liquid and vapor streams are constant above and below the feed location[10]. A point on an operating line represents two passing streams in the column, and the operating line itself is the locus of all possible pairs of passing streams within the column section to which the operating line applies [5].

4.5.4.1 Constructing the McCabe–Thiele diagram

McCabe–Thiele's method involves plotting the two operating lines in the equilibrium, or x-y, diagram as shown in Figure 4.14. An operating line can be located on the diagram if either (1) two points are known (e.g. from substituting two values of x into the relevant equation, i.e. Eqns (4.48) or (4.50)), or (2) one point and the slope are known. The known points are usually the intersection with the diagonal $(x = y)$ and the intersection with another operating line, or with $x = 0$ or $x = 1$. In Figure 4.14, both lines are shown in the diagram from their origins at (x_D, x_D)

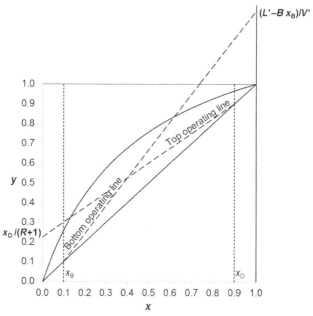

FIGURE 4.14 McCabe–Thiele's Method

---

[10]When the constant molar overflow assumption does not hold, then the operating lines become curves instead of straight lines.

and (x_B, x_B), respectively, until either the y-intercept ($x = 0$) or the intersection with the $x = 1$ line. Although the operating lines have been shown across the whole diagram to illustrate the intersections, it should be noted that they are only valid until the feed location, as this is the validity range for the respective material component balances that the operating lines represent (Eqns (4.48) and (4.50)).

As the rectification and the stripping sections intersect at the feed location, the same x-y point must satisfy both equations. Subtracting the bottom material balance from the top material balance (Eqn (4.50) from Eqn (4.48)), and noting that $y_{n+1} = y_m = y$ and $x_n = x_{m-1} = x$ at this point yields:

$$Vy - V'y = Lx + Dx_D - L'x + Bx_B$$

$$(V - V')y = (L - L')x + Dx_D + Bx_B$$

Recall the component material balance over the column Eqn (4.34) and the total material balance over the feed stage Eqns (4.42)–(4.43), then

$$y = \left(\frac{q}{q-1}\right)x - \left(\frac{z_F}{q-1}\right) \tag{4.51}$$

This equation is known as the *feed line* or the *q-line*, and is located in the McCabe–Thiele diagram by noting that when $x = z_F$, then $y = z_F$, which lies on the diagonal. The slope of the line is given by $q/(q-1)$, i.e. it depends on the conditions of the feed. The five different alternatives are illustrated in Figure 4.15. Note that the q-line intersects with the intersection between the top and the bottom operating lines (see Figure 4.14).

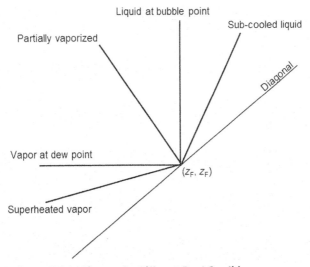

FIGURE 4.15 McCabe–Thiele Diagram for Different Feed Conditions

4.5.4.2 Determining the number of equilibrium stages

The McCabe–Thiele diagram is used to determine the number of theoretical stages in the column. An example is shown in Figure 4.16 where the feed is assumed to be liquid at the boiling point, i.e. the q-line is vertical. Starting from the top of the column, the mole fraction y_1 of the more volatile component leaving the column section from the top stage ($n = 1$) is equal to the mole fraction of the distillate, x_D. Assuming equilibrium on each stage (assumption 8), the liquid composition on the top stage x_1 is found from the equilibrium line as the liquid that is in equilibrium with vapor of composition y_1 (x_1, y_1). A material balance over the top stage is given by Eqn (4.48) ($n = 1$), and this will give the mole fraction of the vapor from the second stage, y_2. This can also be read from the diagram on the top operating line (x_1, y_2). The liquid composition on the second stage can be found from the equilibrium line as y_2 is known (x_2, y_2).

The procedure is continued stage by stage until the q-line is reached, i.e. until the composition on the feed stage is reached. If the feed is at its bubble point, then the composition at the feed stage equals the composition of the feed, z_F. The number of equilibrium stages needed between the feed stage and the top stage can thus be determined graphically from the diagram as shown, or alternatively by computation if the equilibrium data is available either as tabulated values or in the form of an equation, e.g. the constant relative volatility equation (see Eqn (4.13)). The number of stages in the stripping section is determined in a similar way (see Figure 4.16).

The stages can be stepped off from the top to the feed, and from the bottom to the feed, alternatively, from the top all the way to the bottom or vice versa. Normally the stages are counted from the top down to the bottom. Above the feed location, the top operating line is used, and below the feed location the bottom operating line is used, as this is where the corresponding material balances are valid. The number of stages

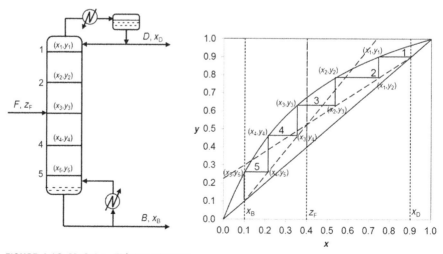

FIGURE 4.16 McCabe–Thiele's Method

(feed at the bubble point).

may also include a fraction of a stage, but note that the number of stages for a tray column must always be rounded up to the nearest integer to ensure that the purity specifications are met.

4.5.4.3 The effect of wrong feed location

In the example above, the feed stage was determined to be where the stepping procedure between the equilibrium line (corresponding to equilibrium conditions on each stage) and the operating line (corresponding to movement from one stage to the next) met the q-line. This will correspond to the optimal location of the feed stage. Figure 4.17 shows that more stages are required if the feed is not placed in the optimal location (see also [3] or [12]). The optimum feed location is on stage 4 with a total of 6.7 stages required (see Figure 4.17(a)). If the feed location is moved downwards to stage 6, then a total of 8.1 stages are needed (see Figure 4.17(b)), and if the location is too high on stage 3, then 7.1 stages are needed (see Figure 4.17(c)).

4.5.4.4 Determining the minimum reflux ratio R_{min}

As mentioned earlier, there is a lower limit for the reflux ratio corresponding to an infinitely tall column. This minimum reflux ratio, R_{min}, can be found using the

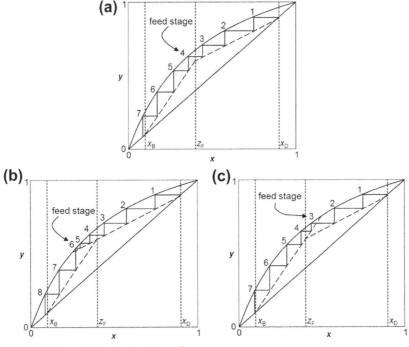

FIGURE 4.17 Determining the Feed Location

(a) optimal feed location, (b) feed stage above optimal, and (c) feed stage below optimal.

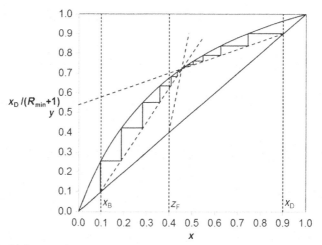

FIGURE 4.18 Minimum Reflux Ratio from McCabe–Thiele's Method

McCabe–Thiele diagram. The intersection of the top operating line with the y-axis depends on the reflux ratio ($x_D/(R+1)$). As the reflux ratio is reduced, the y-intercept will increase. The top operating line can be moved until its intersection with the q-line also intersects with the equilibrium line (but no further as this would violate the laws of thermodynamics). The y-intercept for this limiting line is $x_D/(R_{min}+1)$, and R_{min} is thus determined as illustrated in Figure 4.18[11]. Note that if the equilibrium curve is not smooth, a pinch point may occur, and if this is located above z_F then the intersection of the top operating line will not be with the equilibrium curve at the intersection with the q-line but rather with the equilibrium curve at the pinch point. Again this is because the operating line would otherwise cross the equilibrium curve, which would violate the laws of thermodynamics.

4.5.4.5 Determining the minimum number of stages N_{min}

The other limiting condition is the minimum number of stages, N_{min}, which corresponds to total reflux, in other words there is no feed ($F = 0$) and no product is produced ($D = B = 0$) (i.e. all the condensed vapor is returned to the column as reflux, and all the material reaching the reboiler is vaporized and returned to the column as vapor). This will give the best possible separation in the column but is of course impractical. When $R = \infty$, then the top and the bottom operating lines will both be equal to the diagonal, as $D = B = 0$ and $L = V$. The stepping procedure will then be between the equilibrium line and the diagonal (see Figure 4.19).

---

[11] Note that the true R_{min} may often be larger than that obtained from McCabe–Thiele's method due to the assumption of constant molar overflow not holding [11].

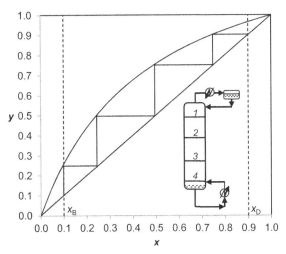

FIGURE 4.19 Minimum Number of Stages from McCabe–Thiele's Method

4.5.5 **Multiple feed streams**

Continuous distillation columns are often operated with multiple feed streams. This is for instance the case when material is being recycled back from downstream units. If the fresh feed and the recycled stream are sufficiently different, they will not be mixed but rather the recycle stream will be returned to the column at the most appropriate stage. This is to avoid mixing components that were already separated out in the recycle stream.

The most appropriate feed stage for each of the two feed streams will depend on the compositions. A feed should be introduced to the column at a location where the conditions within the column resemble most closely the composition of that feed. The state of the feed, i.e. whether it is liquid or vapor, is not important as both liquid and vapor can be introduced to the column at any stage and will indeed be present at any stage.

The introduction of the second feed does not change the material balances much, and the total and component balances become (see Figure 4.20 left):

$$F_1 + F_2 = B + D \tag{4.52}$$

$$F_1 z_{F_1} + F_2 z_{F_2} = B x_B + D x_D \tag{4.53}$$

The balances over the top and the bottom of the column are the same as before. Balances can also be set up between the top of the column and further down in the column:

$$V' y_{n+1} + F_1 z_{F_1} = L' x_n + D x_D \tag{4.54}$$

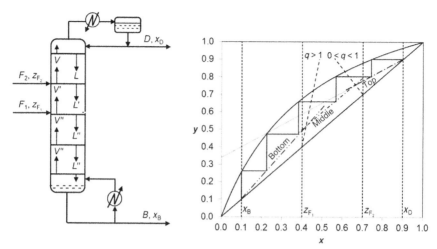

FIGURE 4.20 Distillation with Two Feed Streams

(Feed 1: subcooled liquid, Feed 2: partially vaporized) (left: schematic, right: McCabe–Thiele diagram).

or

$$y_{n+1} = \frac{L'}{V'}x_n + \frac{Dx_D - F_1 x_{F_1}}{V'}$$ (4.55)

Note that a balance over the lower part of the column can also be set up and will also be valid, however, the top half is usually used.

4.5.5.1 Multiple feeds in a McCabe–Thiele diagram
The McCabe–Thiele diagram can also be used for analysis of columns with multiple feeds. Each feed is considered separately, i.e. as if neither "knew" of the other's presence [4]. The top operating line and the bottom operating lines are located as outlined above. The material balance given by Eqn (4.55) becomes the *middle operating line*. The line has a slope L'/V' and is valid between the two feeds, i.e. between z_{F_1} and z_{F_2}. The top operating line and the middle operating line intersect at the q-line of the top feed F_1, and the bottom operating line and the middle operating line intersect at the q-line of the lower feed F_2 (see example in Figure 4.20, right). For the least number of stages at a given reflux ratio, the two feeds are each introduced at the stage indicated by the diagram.

4.5.6 Side streams
A distillation column can also have side streams. Side streams are normally only used for multicomponent mixtures but can in principle also be used for binary separations. A side stream may be withdrawn from the liquid phase or the vapor phase

from any location in the column section, however, a liquid side stream from a tray is easiest from a practical point of view. A total material balance over the side stream stage for a liquid withdrawal gives

$$L' = L - S \quad \text{and} \quad V' = V \tag{4.56}$$

and for a vapor withdrawal

$$L' = L \quad \text{and} \quad V' = V + S \tag{4.57}$$

The side stream can be placed above or below the feed; either way, the introduction of the side stream does not change the material balances much, and the total and component balances become (see Figure 4.21 left and Figure 4.22 left):

$$F = B + D + S \tag{4.58}$$

$$Fz_F = Bx_B + Dx_D + Sz_S \tag{4.59}$$

The balances over the top and the bottom of the column are the same as before ($Vy_{n+1} = Lx_n + Dx_D$ and $V''y_m = L''x_{m-1} - Bx_B$). The balance over the middle part of the column will depend on the location of the side stream relative to the feed location. If the side stream is located *above* the feed (see Figure 4.21 left), then the balance is:

$$V'y_{n+1} = L'x_n + Dx_D + Sz_S \tag{4.60}$$

or

$$y_{n+1} = \frac{L'}{V'}x_n + \frac{Dx_D + Sz_S}{V'} \tag{4.61}$$

If the side stream is located *below* the feed (see Figure 4.22 left), then the balance is:

$$V'y_{n+1} + Fz_F = L'x_n + Dx_D \tag{4.62}$$

or

$$y_{n+1} = \frac{L'}{V'}x_n + \frac{Dx_D - Fx_F}{V'} \tag{4.63}$$

4.5.6.1 Side streams in a McCabe–Thiele diagram

The McCabe–Thiele diagram can also be used for analysis of columns with side streams. The side stream is considered separately, i.e. independent of the feed. The top operating line and the bottom operating lines are located as outlined above.

The material balance given by Eqn (4.61) or Eqn (4.63) becomes the *middle operating line*. The line has a slope L'/V' and is valid between the side stream and the feed (side stream located above feed), i.e. between z_S and z_F (see Eqn (4.61)), or between the feed and the side stream (side stream located below the feed), i.e. between z_F and z_S (see Eqn (4.63)).

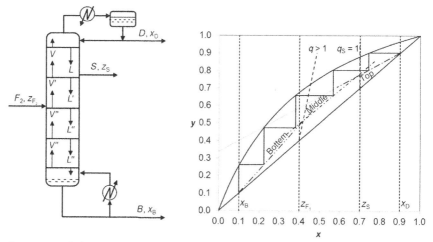

FIGURE 4.21 Distillation with Side Stream above Feed

(left: schematic, right: McCabe–Thiele diagram).

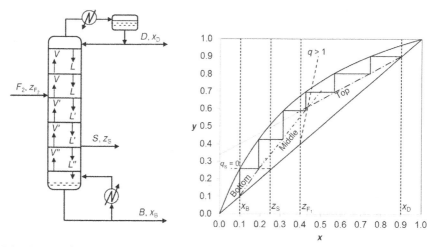

FIGURE 4.22 Distillation with Side Stream below Feed

(left: schematic, right: McCabe–Thiele diagram).

For a side stream located above the feed, the top operating line and the middle operating line intersect at the q-line of the side stream. For a side stream located below the feed, the middle operating line and the bottom operating line intersect at the q-line of the side stream. For a liquid phase withdrawal, $q = 1$ and the q-line is vertical, and for a gas phase withdrawal, $q = 0$ and the q-line is horizontal (see Figure 4.21 right and Figure 4.22 right) [5]. For the least number of stages at a given reflux ratio, the feed and the side stream are introduced at the stage indicated by the diagram.

4.5.7 **Steam distillation**

In some cases where the bottom product is water, and some where the distilled mixture is immiscible with water, steam is introduced directly into the still. The steam acts as an energy input and the reboiler is therefore not needed. Direct steam is also used in vacuum distillation of high-boiling organic materials that would decompose if they were distilled directly at atmospheric pressure. The steam then acts as an inert that reduces the partial pressure and thereby the temperature of the organic compounds. Direct steam is also used to obtain agitation in liquids with poor heat-transfer characteristics, e.g. distillation of fatty acids from tall oil.

The steam acts as a second feed below the bottom stage as shown in Figure 4.23 (left). The total and component balances become:

$$F + F_S = B + D \tag{4.64}$$

$$Fz_F = Bx_B + Dx_D \tag{4.65}$$

Note that the steam does not add to the component balance as $z_{FS} = 0$. The material balance over the top of the column is the same as before, however, the bottom material balance is different as $V' = F_S$ and $L' = B$:

$$F_S y_m = Bx_{m-1} - Bx_B \tag{4.66}$$

or

$$y_m = \frac{B}{F_S} x_{m-1} - \frac{B}{F_S} x_B \tag{4.67}$$

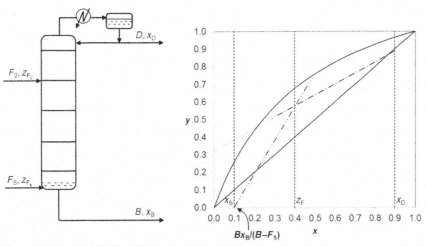

FIGURE 4.23 Steam Distillation

(left: column configuration, Right: McCabe–Thiele diagram).

4.5.7.1 *Steam distillation in a McCabe–Thiele diagram*

The McCabe–Thiele diagram can also be used for analysis of columns with direct steam injection. The bottom operating line is given by Eqn (4.67) and has slope B/F_S. Note that this equation does not start from the diagonal, as when $x = y$ then

$$y_m = x = \frac{B}{F_S}x - \frac{B}{F_S}x_B \quad \Rightarrow \quad x = \frac{B}{B-S}x_B$$

The steam distillation operating line is shown in the McCabe–Thiele diagram in Figure 4.23 (right).

4.5.8 Reboilers and condensers

So far, there has been no mention of reboilers or condensers. Whether or not they are included in the number of stages, in other words, whether or not they are considered equilibrium stages, will depend on their type. The criteria are [12]:

1. The stage is at steady state and has both a liquid outlet and a vapor outlet
2. All vapor and liquid entering the stage are intimately contacted and perfectly mixed
3. Total vapor leaving the stage is in equilibrium with total liquid leaving the stage

If these criteria hold, then the unit is an equilibrium stage, and will either be the top stage in the McCabe–Thiele diagram (for a condenser), or the bottom stage (for a reboiler).

For reboilers, different types are used to provide vapor to the stripping section of the column. For plant-size distillation columns, the reboiler is usually an external heat exchanger, e.g. a kettle reboiler or a thermosyphon reboiler. In a *kettle reboiler*, liquid leaving the bottom of the column section is partially vaporized in the kettle reboiler, normally by heat transfer with tubes of steam or other heating mediums[4]. The liquid leaving the reboiler is assumed to be in equilibrium with the vapor, which is returned to the column section, i.e. the reboiler is a partial reboiler and thus a distillation stage, and the lowest stage in the McCabe–Thiele diagram will correspond to the reboiler.

Thermosyphon reboilers are favored when the bottom product contains thermally sensitive components, only a small temperature difference is available for heat transfer, or heavy fouling occurs [7]. A thermosyphon reboiler can either be vertical (most common) or horizontal [3]. In the vertical thermosyphon, the liquid withdrawal can be either from the bottom sump of the column section (most common) or from the downcomer of the bottom stage. The former is not considered to correspond to an equilibrium stage as the liquid, which is a combination of liquid from the column section and liquid in the reboiler, is not in equilibrium with the vapor leaving the stage (thus violating the third criteria) [12].

---

[12]The kettle reboiler is sometimes located at the bottom of the column to avoid piping.

A condenser can either be a total condenser or a partial condenser. In a *total condenser*, all the vapor from the column section is condensed and directed to the reflux drum[13]. For a *partial condenser*, only a portion of the vapor is condensed and returned as reflux from the reflux drum whilst the rest is removed as a vapor distillate product from the reflux drum. As liquid and vapor are in equilibrium in the drum, then the top stage in the McCabe–Thiele diagram will correspond to the condenser.

The *heat duties* of the reboiler and condenser can be estimated from an energy balance assuming that kinetic and potential energy as well as work terms can be neglected:

$$Fh_F + \dot{Q}_R = Dh_D + Bh_B + \dot{Q}_C + \dot{Q}_{loss} \tag{4.68}$$

where $\dot{Q}_R$ is the energy added in the reboiler, $\dot{Q}_C$ is the energy removed in the condenser and $\dot{Q}_{loss}$ is the energy loss from the column (all energy/time).

The enthalpies h_F, h_D, and h_B can be found from heat capacities and latent heats of vaporization. Except for small and/or uninsulated distillation equipment, $\dot{Q}_{loss}$ is negligible and can be ignored [3].

If the column is well insulated with no heat loss so that the only heat transfer is from the reboiler and condenser ($\dot{Q}_{loss} = 0$), the condenser is a total condenser, the feed is at the bubble point, and the assumptions of the McCabe–Thiele method hold, then the total energy balance over the column is ([3], [7]):

$$\dot{Q}_R = \dot{Q}_C = D(R + 1)\Delta H_{vap} \tag{4.69}$$

In general, the cost of cooling can be neglected for an initial estimate, as the cost of heating, usually via steam, will be at least an order of magnitude higher.

4.5.9 Fenske-Underwood-Gilliland method

The McCabe–Thiele method is a convenient graphical method for preliminary column design but relies on the availability of an equilibrium diagram for a binary mixture. Approximate calculation methods are often used instead to give a first estimate of the design, also for multicomponent mixtures, and a popular method for this is the Fenske-Underwood-Gilliland (FUG) method.

4.5.9.1 Fenske's equation
The first step of the method is the calculation of the minimum number of theoretical stages N_{min} from *Fenske's equation* (a derivation of the equation for multicomponent mixtures can be found in [3]):

$$N_{min} = \frac{\log\left[\left(\frac{x_D}{(1-x_D)}\right)\left(\frac{1-x_B}{x_B}\right)\right]}{\log \alpha_{ave}} \tag{4.70}$$

---

[13]The condenser may also subcool the liquid (most common), i.e. the material leaving the condenser is below the bubble point. The McCabe–Thiele method does not account for subcooled condensers.

where α_{ave} is the average relative volatility of the light component relative to the heavy component. If the relative volatility varies over the column, an average can be taken over the column, i.e.

$$\alpha_{ave} = \sqrt{\alpha_B \alpha_D} \quad \text{or} \quad \alpha_{ave} = \sqrt[3]{\alpha_B \alpha_F \alpha_D}$$

Note that Fenske's equation is sometimes written as $N_{min} + 1$ rather than N_{min} as in the equation above. The "1" then relates to the reboiler, which in the equation above has been included, i.e. the first stage from the bottom is the reboiler (see Section 4.5.8).

Fenske's equation allows a rapid estimation of the minimum number of stages required. The equation is quite reliable except when the relative volatility varies appreciably over the column and/or when the mixture is nonideal.

4.5.9.2 Underwood's equation

The development of the *Underwood equation* is quite complex, and the reader is directed to the reference literature for a derivation. For a binary mixture, the minimum reflux ratio R_{min} can be estimated from Underwood's binary equation [2]:

$$R_{min} = \frac{1}{(\alpha - 1)} \left[\frac{x_D}{x_F} - \frac{\alpha(1 - x_D)}{(1 - x_F)} \right] \tag{4.71}$$

This version of Underwood's equation assumes that the feed is liquid at the bubble point [7]. Underwood's equation for multicomponent mixtures is given by most textbooks, e.g. [5].

4.5.9.3 Gilliland's correlation

A general shortcut method for determining the number of stages N and the corresponding reflux ratio R would be extremely useful, but unfortunately such a method has yet to be developed. However, *Gilliland's correlation* can be used to provide an estimate [1]. The correlation is normally given in the form of a graph with $(R - R_{min})/(R + 1)$ on the x-axis and $(N - N_{min})/(N + 1)$ on the y-axis (see Figure 4.24).

The correlation requires knowledge of N_{min} and R_{min}, and these parameters can be estimated from Fenske's and Underwood's equations, respectively, as explained above. The reflux ratio R is often estimated as a multiple of R_{min} as mentioned before, e.g. $R = [1.1 - 1.5]R_{min}$ (see Section 4.5.3). The number of stages N can then be found from Gilliland's correlation by reading off the y-axis term $(N - N_{min})/(N + 1)$ for the given x-axis term $(R - R_{min})/(R + 1)$.

It is inconvenient to use Gilliland's correlation in graphical form. A number of equations have been suggested that approximate the correlation, and two of these are given below:

$$\frac{N - N_{min}}{N + 1} = 0.545827 - 0.591422 \frac{R - R_{min}}{R + 1} + \frac{0.002743}{\frac{R - R_{min}}{R + 1}}$$

$$\text{when } 0.01 < \frac{R - R_{min}}{R + 1} < 0.90$$

([1], p. 255)

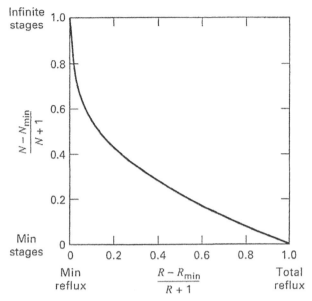

FIGURE 4.24 Gilliland's Correlation

(with permission from [3])

$$\frac{N - N_{min}}{N + 1} = 1 - \exp\left[\left(\frac{1 + 54.4 \frac{R - R_{min}}{R+1}}{11 + 117.2 \frac{R - R_{min}}{R+1}}\right)\left(\frac{\frac{R - R_{min}}{R+1} - 1}{\left(\frac{R - R_{min}}{R+1}\right)^{0.5}}\right)\right]$$

([3], p. 509)

Simple rules of thumb that avoid the correlation altogether are to assume $N = 2.5 N_{min}$ [1] or $N = 2 N_{min}$ [3].

Gilliland's correlation is very useful for exploration of preliminary designs, but rigorous tray-by-tray calculation will be required to obtain a final design, and in this day and age, computation software such as ChemCAD, HYSYS, Aspen or gPROMS will be used for this. Note that these programs may in fact use the FUG calculation method as outlined above for shortcut design.

4.5.9.4 Kirkbride's equation

Implicit in the application of Gilliland's correlation is the specification that the stages are distributed optimally between the rectification and stripping sections. An estimation of the distribution between the two sections is provided by Kirkbride's equation:

$$\frac{N_R}{N_S} = \left[\left(\frac{1 - x_F}{x_F}\right)\left(\frac{x_B}{1 - x_D}\right)^2 \frac{B}{D}\right]^{0.206} \tag{4.72}$$

where N_R is the number of stages in the rectification section and N_S is the number of stages in the stripping section, respectively.

The total number of stages in the column section, N', is given by:

$$N_R + N_S = N' \qquad (4.73)$$

Note that it is here assumed that N' does not include the reboiler and/or condenser.

4.5.10 Column efficiency

The equations and calculation methods considered so far have all assumed that equilibrium is established on each stage. This is, however, rarely the case. It would be extremely difficult to take the deviation from equilibrium into account in all the equations, and the engineering approach is therefore generally to design the columns assuming equilibrium, and then at the end to account for the fact that this is not the case. This is done by transferring the number of equilibrium stages found using McCabe–Thiele's method, the Fenske-Underwood-Gilliland method, software, etc., into an equivalent number of *actual* stages. In order to do this, the *efficiency* of the stages must be known.

Column efficiency can be defined in different ways. The *overall stage efficiency* E_O is defined as:

$$E_O = \frac{\text{number of theoretical stages}}{\text{number of actual stages}} \qquad (4.74)$$

A common estimate of column efficiency is *Murphree's efficiency* E_{MV}, which is defined as the ratio between the actual change in composition on the stage over the change in composition for an equilibrium stage:

$$E_{MV} = \frac{y_n - y_{n+1}}{y_n^* - y_{n+1}} \qquad (4.75)$$

where y_n^* is the vapor composition which would be in equilibrium with the liquid on stage n, x_n. A schematic showing the location of the compositions is shown in Figure 4.25. An analogous efficiency can be defined for the liquid phase. Note that it has been assumed that the vapor streams are completely mixed and uniform in composition.

A number of other efficiency definitions exist, for instance, the *Murphree point efficiency*, which takes into account variation along the length of a stage. Correlations also exist for estimation of column efficiencies for certain common systems, for instance, for hydrocarbon systems. The overall stage efficiency depends on a number of factors and is a complex factor of [3]:

1. Geometry and design of the contacting trays or packing
2. Flow rates and flow paths of vapor and liquid streams
3. Compositions and properties of vapor and liquid streams

Note that the concept of stage efficiency is only applicable to devices in which there are actual stages where material is contacted and then separated, i.e. in tray

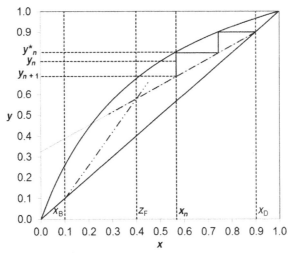

FIGURE 4.25 Murphree Stage Efficiency

columns. For columns with internals of random or structured packing, the concept of efficiency is imbedded in the HETP (height equivalent to a theoretical plate).

4.5.11 Summary

The methods presented herein are for the most common continuous distillation systems. Rectifying columns where the feed is introduced to the bottom stage, and stripping columns where the feed is added to the top stage, have not been considered, nor have systems with partial condensers [3]. The examples used have been for a simple zeotropic mixture, and systems exhibiting pinch point in the vapor–liquid equilibrium have not been discussed but do introduce further complexity to the analysis. Finally, other popular calculation methods, such as that of Ponchon-Savarit method, can be found in some mass transfer textbooks (e.g. [3,4]).

4.6 Concluding remarks

The equations and methods outlined herein are greatly simplified and rely on numerous assumptions, such as steady state, constant pressure, constant molar overflow, etc. Vapor-liquid equilibrium is clearly of key importance, and although the equations presented in this chapter will be applicable for many separations, they are nevertheless simplifications. A detailed account of vapor–liquid equilibrium as applicable to distillation can be found in Chapter 2. Of equal importance are mass transfer phenomena, which are covered in Chapter 3. The mass transfer characteristics, and thus the operation of the column, are also dependent on the design of the column internals, i.e. trays, random packing, or structured packing that is used. A detailed account of column internals can be found elsewhere in this series [8].

Most industrial separations are not binary but rather involve multiple components. Quantitative design techniques for multicomponent fractionation are therefore required. Whilst the principles established for binary solutions generally apply to multicomponent distillation, new problems of design are introduced that require special consideration. For a start, graphical methods are no longer applicable, and rigorous calculations are required. The reader is referred to other chapters or other textbooks for an analysis of design and operation of multicomponent distillation.

The presence of more than two components raises additional questions. A single distillation column can only separate two components or streams into pure products, and for multicomponent mixtures NC − 1 columns are generally required in order to separate a mixture with NC components. Which components should be separated into which columns, and the order of these columns, is of great importance as the cost implications of a nonoptimal sequence can be significant. The issue of *column sequencing* is still receiving considerable interest both in industry and in academia. Sequencing is clearly an optimization problem, and in an ideal world, the overall design problem should be posed as a large optimization problem encompassing all NC − 1 columns and their individual designs. This is, however, in most instances not practical due to the significant computational complexity involved. Much research has been devoted to obtaining good estimation methods, and an account of some of this work can be found in Chapter 7.

This chapter has mainly considered continuous operation of distillation. Distillation is also often carried out in its batch variant, which adds complexity to the operation due to the time dependence. Batch distillation does, however, add significant flexibility as a multicomponent mixture can be separated into NC components in a *single* batch column whilst NC continuous columns would have been required for continuous operation. A detailed account of batch distillation is given in the next chapter of this book, Chapter 5.

Many mixtures typically encountered in industry are nonideal, and the simple methods of McCabe−Thiele or Fenske-Underwood-Gilliland are not sufficiently accurate to deal with these, even for preliminary design, and more sophisticated methods are needed. Azeotropic mixtures are common and require special consideration as the separation of these mixtures is not possible by standard distillation. A detailed account of extractive distillation can be found in Ref. [13].

Finally, even though the unit operation of distillation has been used in various forms for millennia, the process is still very much under active research, in particular, in finding novel, more efficient column internals, but more importantly in reducing the very large energy requirement that is associated with column operation. An overview of the latter can be found in Chapter 6.

References

[1] P.C. Wankat, Separation Process Engineering, third ed., Pearson, 2012.

[2] J.H. Harker, J.M. Coulson, J.R. Backhurst, J.F. Richardson, Coulson and Richardson's Chemical Engineering, in: Particle Technology and Separation Processes, fifth ed., vol. 2, Butterworth-Heinemann Ltd, 2002.

[3] J.D. Seader, E.J. Henley, D.K. Roper, Separation Process Principles, third ed., Wiley and Sons, 2013.

[4] R.E. Treybal, Mass-Transfer Operations, third ed., McGraw-Hill, 1980.

[5] M.F. Doherty, Z.T. Fidkowski, M.F. Malone, R. Taylor, Distillation, in: D.W. Green, R.H. Perry (Eds.), Perry's Chemical Engineers' Handbook, eighth ed., McGraw-Hill, 2008.

[6] R. Smith, Chemical Process Design and Integration, John Wiley & Sons, 2005.

[7] A.B. de Haan, H. Bosch, Industrial Separation Processes, De Gruyer, 2013.

[8] A. Gorak, Z. Olujic (Eds.), Distillation — Equipment and Processes, Elsevier, 2014.

[9] A. Lyderson, Mass Transfer in Engineering Practice, Wiley & Sons, 1983.

[10] J.M. Coulson, J.F. Richardson, R.K. Sinnott, Coulson and Richardson's Chemical Engineering, in: Chemical Engineering Design, second ed., vol. 6, Butterworth-Heinemann Ltd, 1996.

[11] W.L. McCabe, J.C. Smith, P. Harriott, Unit Operations of Chemical Engineering, seventh ed., McGraw-Hill Higher Education, 2005.

[12] H.Z. Kister, Distillation Design, McGraw Hill, 1992.

[13] A. Gorak, H. Schoenmakers (Eds.), Distillation — Operation and Applications, Elsevier, 2014.

Design and Operation of Batch Distillation

Eva Sorensen

Department of Chemical Engineering, UCL, London, UK

CHAPTER OUTLINE

Distillation: Fundamentals and Principles. http://dx.doi.org/10.1016/B978-0-12-386547-2.00005-3

5.1 Introduction

Batch distillation refers to the use of distillation in steps, or in batches, and is used extensively in laboratory separations and in the production of fine and specialty chemicals, pharmaceuticals, polymers, and biochemical products either for purification purposes or for the recovery of valuable solvents. In these industries, batch distillation is more favorable than its continuous counterpart for several reasons:

1. Batch distillation provides a single unit solution, unlike continuous distillation, which requires at least $NC-1$ columns to separate a feed of NC components; a single properly designed batch distillation column can separate mixtures with any number of components into its pure constituents.
2. Seasonal or customer demand may require one column to be used for different feed mixtures and/or to produce different products.
3. The required capacity of the processing unit may be too small to warrant continuous operation at a practical scale.
4. If material to be separated is high in solid content, or contains tar or resins that would otherwise plug up or foul a continuous column, then the use of a batch unit can keep solids separated and permit convenient removal at the end of the batch.
5. Batch-wise operation enables batch identity, that is, product traceability, which is important in the production of pharmaceuticals and foodstuffs that have strict quality control requirements.

The greatest advantage of batch distillation is thus its ability to cope with different separation duties, for example, varying feed mixture, feed compositions, and product specifications, by simply changing the operating conditions of the column. However, this flexibility, and the inherent unsteady-state nature of batch distillation, poses additional design and operational challenges when compared with continuous distillation. The operation may also require higher energy consumption than the equivalent continuous counterpart and the operating time may be longer.

Batch distillation is a very old unit operation that has been run in more or less the same way for centuries. Although distillation is one of the most intensively studied and better understood processes in the chemical industry, the batch version still represents an interesting field for academic and industrial research for a number of reasons. The demands for efficiency and productivity are greater than ever, and the development of new batch column configurations, and operating and/or control policies, is therefore essential to meet these demands. Even for simple binary distillation, there are several possible alternative operations, and with complex trade-offs as a result of the many degrees of freedom available, there is ample scope for optimization. Since the process is intrinsically dynamic, its optimization leads to a challenging optimal control problem. Industrial batch columns are often run as if they are merely a larger version of laboratory batch columns, meaning that the operating strategies are simple. This is a general characteristic for smaller companies who are not able to support research or development in engineering or in the synthesis and production of their products. There is, therefore, still a need for reliable methods for optimization of batch distillation design and operation and for the development of practical control strategies.

This chapter first introduces the concepts of batch distillation and considers the various batch operating modes and batch column configurations that are currently available, as well as different strategies for batch column control. It then highlights the challenges of batch distillation of complex systems such as azeotropic, extractive, reactive, and pressure-swing operations before examining the key concerns involved in the modeling and optimization of batch distillation processes. Finally, some thoughts for the future of batch distillation are offered.

5.1.1 Differential (Rayleigh) distillation

Differential distillation, also often referred to as *simple* or *Rayleigh distillation*, is the most elementary example of batch distillation and is illustrated in Figure 5.1.

The unit consists of a reboiler or heated still, a condenser, and normally just one receiving tank. No trays or packing are provided, and there is no reflux of condensed material back to the reboiler. At the start, feed material is charged into the reboiler and brought to a boil. The resulting vapor is condensed in the condenser and collected in the receiving tank. Because the vapor is richer in the more volatile component compared to the liquid in the reboiler, the liquid remaining in the reboiler will contain progressively less of the more volatile component and therefore becomes richer in the less volatile component. The process is continued until the

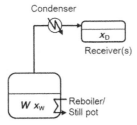

FIGURE 5.1 Schematic of Differential (Rayleigh) Distillation

reboiler is either depleted of the more volatile component, a product specification is reached for either the reboiler or the receiver composition, or the batch time has come to an end.

The reboiler may be assumed to be the only *equilibrium stage* because there are no trays above the reboiler. Letting W_0 represent the moles of initial feed to the reboiler, x_{W_0} the initial molar fraction of light component in that feed, W the number of moles remaining in the reboiler at any time, x_W the corresponding mole fraction of light component, and x_D the average mole fraction of light component in the vapor, a material balance for the light component over the unit at any point in time yields:

$$x_D dW = d(W x_W) = x_W dW + W dx_W \tag{5.1}$$

Here, dW represents the amount of material leaving the reboiler, that is, the amount of vapor produced during an infinitesimal time interval dt. Rearranging the variables and integrating the initial charge conditions of W_0 and x_{W_0} to the final conditions of W and x_W gives the well-known Rayleigh equation:

$$\int_{W_0}^{W} \frac{dW}{W} = \int_{x_{W_0}}^{x_W} \frac{dx_W}{x_D - x_W} \tag{5.2}$$

From Rayleigh's Eqn (5.2) it is possible to calculate the amount of liquid that must be distilled $(W_0 - W)$ to obtain a liquid of specified composition (x_W) in the reboiler. The integral may be solved numerically or graphically; in some cases, however, the integral can be determined analytically. If equilibrium is assumed between liquid and vapor, the right-hand side of the equation may be evaluated by plotting $1/(x_D - x_W)$ versus x_W and measuring the area under the curve between the limits x_{W_0} and x_W. If the mixture is a binary system for which a relative volatility α is constant, or if an average value that will serve for the range considered can be found, the relative volatility equation

$$\alpha = \frac{x_D/x_W}{(1 - x_D)/(1 - x_W)} \quad \text{or} \quad x_D = \frac{\alpha x_W}{1 + (\alpha - 1)x_W} \tag{5.3}$$

can be substituted into Rayleigh's equation and a direct integration can be made, resulting in Eqn (5.4) [1]:

$$\ln\left(\frac{W_0}{W}\right) = \frac{1}{\alpha - 1}\left[\ln\left(\frac{x_{W_0}}{x_W}\right) + \alpha \ln\left(\frac{1 - x_W}{1 - x_{W_0}}\right)\right] \tag{5.4}$$

It should be noted that a complete separation is impossible based on Rayleigh distillation unless the relative volatility of the mixture is infinite. Its application is therefore limited to wide-boiling mixtures that are easy to separate or to laboratory-scale separations where high purities are not required. The process is, however, also used in the beverage industry in the production of whisky, for instance, where the initial distillate is cooled and distilled a second time to improve the separation.

5.1.2 **Conventional batch rectification**

For a sharp-split separation, a tray or packed column section is added between the still and the condenser, and a reflux drum and a splitter are added after the condenser, allowing the reflux to be directed back to the column during operation, as shown in Figure 5.2. In addition, one or more product receivers or drums are provided to collect distillate cuts as either product cuts *Pi* or offcuts *Oi*. In multicomponent separation, the most volatile component is removed first, then the second most volatile component, and so on, leaving the heaviest component in the still at the end.

The basic difference between a batch rectification column and a continuous distillation column is therefore that in continuous distillation the feed is continuously entering the column, whereas in batch distillation the feed mixture is usually fed to the still or reboiler only at the start of the operation. Also, while the top and bottom products are both removed continuously in continuous distillation, there is no bottom product in conventional batch distillation. A continuous column is thus operated at a steady state: the amount of feed entering the column equals the total amount leaving the column as top and bottom products, respectively. In contrast, in batch distillation the contents in the reboiler are depleted over time as material is removed from the top—hence the process is inherently unsteady.

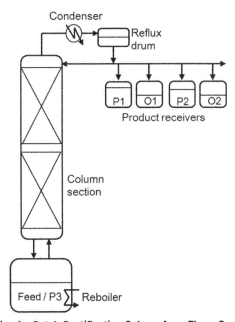

FIGURE 5.2 Schematic of a Batch Rectification Column for a Three-Component Mixture

O1, offcut between components 1 and 2; O2, offcut between components 2 and 3; P1, product cut of component 1; P2, product cut of component 2; P3, product cut of component 3.

5.2 **Batch column operation**

Operation of a conventional batch distillation column can be conveniently described in three parts: (1) the startup period, (2) the production period, and (3) the shutdown period. Before startup, the feed is normally charged to the reboiler and then heated. During the startup period, the system is operated under total reflux, that is, with no product removed from the top, until either a steady state is achieved or the distillation composition product reaches a desired purity. The production period is normally the most time consuming, but for difficult separations such as high purity or azeotropic separations, the startup time may also be significant.

During the production period, a fraction of the overhead condensate is continuously removed from the column in accordance with the chosen reflux policy (see Section 5.2.1, below). As time proceeds, the material being distilled becomes less rich in the more volatile components, and the distillation of a *cut* is stopped when the composition of the distillate accumulated in the product receiver reaches a desired value. For each component or product, cuts are made by switching to alternate product receivers, at which point the operating conditions may also be changed. Cuts may also be made between components to give *offcuts*, which are collected and then processed together in the same column at a later date or time (see Section 5.2.3). The example in Figure 5.2 and Figure 5.3 shows the distillate or top composition for the separation of a ternary mixture, where P1, P2 and P3 are the product cuts of components 1, 2 and 3, respectively, and O1 and O2 are the offcuts between components 1 and 2 (O1) and between components 2 and 3 (O2), respectively. The separation will be stopped after the second offcut O2 to allow component 3 to be collected in the still/reboiler as P3.

During the shutdown period, the column operation is stopped and the remaining material is removed from the reboiler and the column. This is normally done by

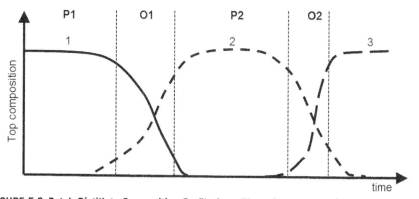

FIGURE 5.3 Batch Distillate Composition Profile for a Three-Component Mixture

O1, offcut between components 1 and 2; O2, offcut between components 2 and 3; P1, product cut of component 1; P2, product cut of component 2.

allowing the material on the trays or in the packing in the column section to run back into the reboiler.

5.2.1 **Reflux operating policies**

Two different batch column *operating policies*, or modes of operation, are cited most frequently in the literature because they are the easiest both to operate and to model (see Figure 5.4):

- *Constant reflux operation*: The reflux ratio is set at a predefined value, which is maintained for the duration of the given cut. The instantaneous distillate composition is therefore changing continuously. Distillation is continued until the average distillate has reached the desired value, and the cut is then changed.
- *Constant distillation composition operation*: When maintaining a constant distillation composition is desired, the amount of reflux returned to the column must be constantly increased throughout the cut, or run. As time progresses, the lightest component in the still will be continually depleted. The reflux ratio will eventually reach a very high value, and the cut is therefore changed or stopped.

Although the operations for each operating policy can give the same amount of product with the same average product purity, the total time required for each operation generally differs. In the case of a constant reflux operation, the product composition is allowed to vary but the reflux is kept constant. This means that the reflux ratio used may be higher than necessary at the start of the cut, and therefore the distillate product is at a purity well above the specification. On the other hand, toward the end of the cut, the reflux ratio is too low and the distillate product is then below the specification. (The average composition is, however, at the specification.) A similar argument can be made for the constant distillate composition policy, where the composition is fixed but the reflux ratio is allowed to vary. Both operating policies thus have their limitations. The constant distillate composition policy is generally found to yield a better performance than the constant reflux policy but is more complex to implement because a controller is needed.

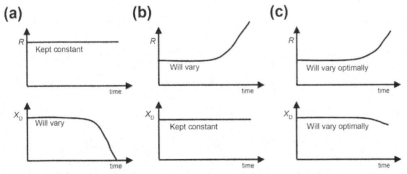

FIGURE 5.4 Reflux Ratio and Distillate Composition for Different Batch Operating Policies

(a) constant reflux policy. (b) constant distillate policy. (c) optimal reflux policy.

Other operating policies have been considered, in particular, optimal reflux ratio and cyclic operation.

Optimal reflux operation: The obvious improvement on the two standard operating policies is to let both the reflux ratio *and* the distillate composition vary during the cut or run (see the comparison in Figure 5.4). It is relatively easy to control the reflux ratio directly, so the ratio is normally used as the degree of freedom in an optimization where the optimal reflux ratio is found based on some optimal criterion, for example, minimum batch time, maximum amount of product, minimum cost, or maximum profit. The optimal reflux ratio is generally found to be one of increasing reflux as the batch progresses and is reported to lie between those from the constant reflux ratio and the constant distillate composition policies. Optimization of batch distillation is considered in Section 5.8.

Cyclic operation: The accurate measurement and control of small flow rates in laboratory columns can be quite difficult, which would favor operation without control or accurate flow setting, as in the policies mentioned earlier. A cyclic procedure can therefore be used instead which is characterized by three periods of operation: (1) filling up, (2) total reflux and (3) dumping or emptying (see Figure 5.5). During the filling up period, the condenser drum is filled up by condensed material and no, or only very little, reflux is returned to the column. During the second period, the column is run under total reflux, similar to during startup for conventional batch reflux operation. During this period, the light component accumulates in the condenser drum at the highest possible purity until the column reaches a steady state. Distillate is then taken off during the final period as total draw-off for a short period of time, after which the column is again filled up as another cycle is started. This three-step procedure is continued until the specifications are satisfied for either the top product in the product receiver or for the residue in the reboiler and column.

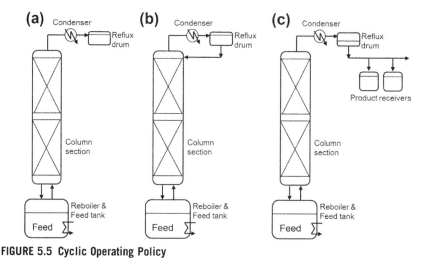

FIGURE 5.5 Cyclic Operating Policy

(a) filling up period. (b) total reflux period. (c) dumping period.

The main degrees of freedom in this policy are the condenser drum holdup, the number of cycles, as well as the duration of the total reflux period. The drum holdup and the duration of the total reflux period may vary between cycles. Cyclic operation of batch distillation can in some cases lead to significant savings in batch time when compared to reflux policies [2].

5.2.2 Vapor loading policies

In addition to the reflux ratio, the other main degree of freedom in batch distillation is the vapor loading to the reboiler. Several options have been considered in the literature [3], although having constant reboiler duty is by far the most common option used industrially because it is the easiest to implement.

1. *Constant reboiler duty*: In this policy, the reboiler heat input is typically set at or near its highest capacity without flooding the column and is held constant throughout the batch processing time. Because of the transient state of the composition and holdup in the reboiler, and the rate of vaporization of the components, the vapor rate out of the reboiler varies continuously.
2. *Constant boilup rate*: This policy requires the vapor rate out of the reboiler and into the column section to be kept at a constant level throughout by varying the reboiler duty accordingly.
3. *Constant condenser vapor load*: The rate at which vapor moves out of the column section and into the condenser is held at a constant value throughout the operation by varying the reboiler duty accordingly.
4. *Constant distillate or bottom rate*: The distillate (regular column) or bottom (inverted column) flow rate (see Section 5.4.2) is kept constant throughout by varying the reboiler duty accordingly.

The last three policies are more of academic interest and are assumptions that have been made in the past by some authors to simplify the modeling of batch columns.

5.2.3 Offcut handling

When a mixture is distilled in a batch column it is not necessary, or even desirable, that each feed charge be separated into a distillate and a residue that simultaneously meet the imposed product specifications. Any off-spec material, i.e. offcuts, may be reprocessed in a number of ways [4,5]:

1. The offcuts are mixed with fresh feed in the next batch (Alternative A).
2. A number of offcuts are collected and subsequently processed together (Alternative B).
3. Each offcut is stored and reprocessed separately, that is, all the first offcuts together, all the second offcuts together, and so on (Alternative C).
4. The offcuts are charged to the reboiler (or to the column section) at an appropriate time during the course of the next batch (Alternative D).

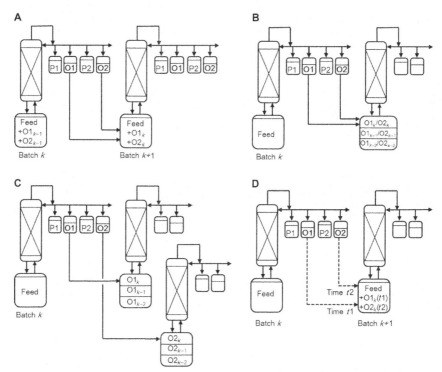

FIGURE 5.6 Different Alternatives for Recycling of Offcuts in Batch Distillation

O1, offcut between components 1 and 2; O2, offcut between components 2 and 3; P1, product cut of component 1; P2, product cut of component 2.

The different offcut handling options are illustrated in Figure 5.6 for a three-component mixture with two product cuts, two offcuts, and one residue product. The numbers refer to the products in order of decreasing volatility.

Recycling offcuts has been found to be advantageous for difficult separations with potentially significant time savings. Reprocessing the offcuts increases the recovery of the products but at the expense of a reduced processing capacity or a longer batch time, resulting in an interesting trade-off problem. The simplest option is to mix the offcuts with the next feed (Alternative A) because no extra storage is needed and complex charging procedures are avoided. In a binary separation, the composition of the offcut may not be much different from that of the feed, and mixing with fresh feed is therefore justified. But in the case of multiple components, this means mixing components that have already been separated, which is clearly not optimal. However, recycling offcuts at different times during the next batch (Alternative D) may be difficult to achieve in practice because of disturbances and changing conditions.

When the offcuts are recycled to the next batch, a cyclic operation across batches is established that is quasi-steady state, that is, subsequent batches follow the same trajectories and produce the same product cuts and offcuts. Luyben [5] indicated that in practice such a quasi-steady-state mode will be achieved after approximately three or four cycles.

It should be noted that there are two common policies regarding the size of the consecutive batches of fresh feed, keeping in mind that the size of the offcuts may vary slightly from batch to batch:

1. The reboiler is full for every cycle such that successive admissions of fresh feed are smaller as the size of the offcuts increase. The size of the total feed charge to be processed (fresh and recycled combined) remains the same.
2. The amount of fresh feed is constant such that the total feed charge (fresh and recycled combined) increases or decreases with the size of the offcuts.

5.2.4 **Further remarks**

The operating modes described above apply primarily to a regular column. For other configurations (see the next section), the operation modes are a variation of the sequences mentioned above. It is also possible to combine the various operating modes for a given separation. Apart from the reflux, withdrawal, and column vapor loading policies, there are other operating decision variables, especially for more flexible configurations such as the multivessel configuration; and the best choice of operating variables depends on the properties of the mixture being separated as well as on economic and practical considerations.

5.3 **Design of batch distillation**

The design of a batch distillation column is similar to the design of a continuous distillation column in that the following main design parameters must be determined:

1. Number of stages or plates in the column section (and thereby the column height)
2. Column diameter
3. Reboiler size (batch size and heat transfer area)
4. Condenser size (heat transfer area)
5. Reflux drum size

In addition to the design variables, the operation of the column must also be determined according to the chosen operating mode, that is, the correct reflux ratio must be found. In most practical applications, the boilup rate from the reboiler will be given by the reboiler heat duty and is therefore not a separate degree of freedom (see Section 5.2.2). The reboiler heat duty is usually kept near its maximum value while still avoiding flooding in the column section.

The design and operation of any distillation column—batch or continuous—are interdependent and should be considered simultaneously. There is a clear trade-off between design and operation: A column with many stages can be operated with a

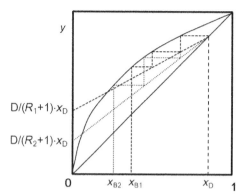

FIGURE 5.7 McCabe Diagram for Batch Distillation at Different Points during a Batch under Constant Distillate Composition Operation

lower reflux ratio, with a lower corresponding reboiler heat duty, and for a shorter batch time than a shorter column, which would require a higher reflux ratio, a higher heat duty, and a longer batch time.

Compared to continuous column design, the main challenge of batch column design is the nonstationary nature of the batch column. For continuous operation, the column is design to a fixed operating point (although normally it has some flexibility); for the batch alternative, an *operating trajectory* must be considered because the conditions in the reboiler and in the column section change with time as the lighter components are being removed, thus leaving the heavier components behind in the column and still. This can be illustrated as shown in Figure 5.7, which shows how the operating line and compositions for a simple binary mixture in a McCabe–Thiele diagram change depending on the time during the batch when the column is operated with a fixed constant distillate composition. As the distillate composition stays fixed, the reflux ratio continues to increase during the batch, and the operating line therefore changes. A similar diagram can be drawn for constant reflux ratio operation, in which case the distillate composition changes during the run whereas the reflux ratio remains fixed (not shown).

For a multicomponent separation, the column section must be designed according to the most difficult separation, that is, according to the two components that are the most difficult to separate, even if this means that the column is then overdesigned for all other component splits.

5.4 Batch distillation configurations

The *configuration* of a batch column can be classified primarily by the position of the feed charge, the number of column sections, and the points from which products are withdrawn. In most industrial cases, the batch column is traditionally used in a rectifying mode in what is often termed the *conventional*, or regular, batch column

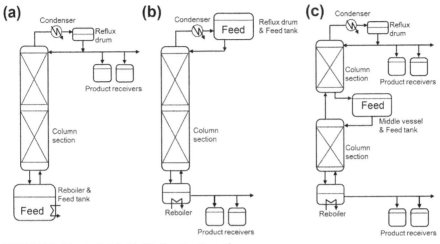

FIGURE 5.8 Simple Batch Distillation Configurations

(a) regular configuration. (b) inverted configuration. (c) middle-vessel configuration.

(or batch rectifier). *Unconventional* columns were first proposed by Robinson and Gilliland [6] in 1950. These novel unconventional column configurations are still mainly found only in theoretical research and pilot plant development stages and have yet to be implemented in general industrial usage. The different column configurations considered in the open literature are described below (see Figure 5.8 and Figure 5.9).

5.4.1 Regular batch column

As already explained, this traditional configuration consists of a bottom reboiler where the feed is charged, a rectifying column section with trays or packing, and a top, usually subcooled, condenser system. The products are withdrawn from the top of the reflux drum into a series of product accumulator tanks. The overhead composition varies during the operation, with the most volatile component being withdrawn first, then the second most volatile component, and so on. A number of cuts are usually made; these are either desired product cuts or intermediate fractions, or offcuts, that may be withdrawn to subsequent batches (see Section 5.2.3). The least volatile component may be recovered as a product in the reboiler at the final stage of the operation. This configuration is also referred to as a *batch rectifier* or *batch rectifying* column.

5.4.2 Inverted batch column

The inverted configuration was first proposed by Robinson and Gilliland [6] in 1950 (Figure 5.8(b)). In this stripping column, liquid feed is charged to the top reflux drum while the products and offcuts are withdrawn from the bottom reboiler. High boiling components are withdrawn first, followed by the more volatile components. This

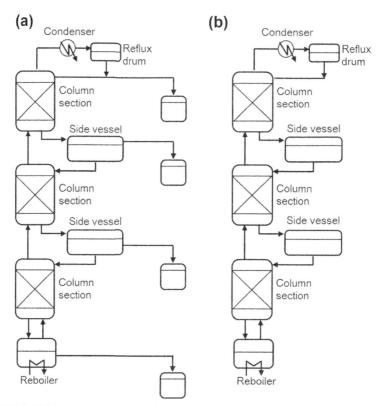

FIGURE 5.9 Multivessel Batch Distillation Column Configurations

(a) configuration with product removal. (b) total reflux configuration.

configuration is also referred to as a *batch stripper* or *batch stripping* column. It should be noted that the inverted column is not the true inverse of the regular column since the feed and product are still in liquid form and not in vapor form.

5.4.3 Middle vessel batch column

The middle vessel configuration was also originally proposed by Robinson and Gilliland [6] in 1950 and considered by Bortolini and Guarise [7] in 1970. Interest was rekindled in 1992 when it was reintroduced by Hasebe et al. [8] (Figure 5.8(c)). The feed is mostly charged into a vessel located between two column sections, splitting the column into rectifying and stripping sections. Products and offcuts can then be withdrawn simultaneously: the most volatile component first from the reflux drum at the top, and the least volatile component first from the reboiler at the bottom. There can be great variety within this configuration, which is defined by the way in which the liquid and vapor streams are arranged between the middle vessel and the rectifying and stripping sections. This configuration is sometimes referred to as a *complex* column.

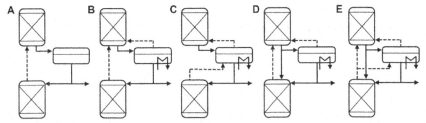

FIGURE 5.10 Stream Configurations around the Vessel in a Middle or Multivessel Column Configuration (A–E: Configuration alternatives)

Most studies of middle vessel columns involve only a fixed middle section stream configuration. However, there are several combinations of vapor and liquid streams possible between the column and the vessel, as shown in Figure 5.10. Studies of middle vessel columns have mainly concentrated on Configuration A because this alternative is clearly the easiest to handle. In this configuration, the whole liquid stream is diverted into the vessel as a result of weir overflow from the bottom tray in the upper column section, and the liquid stream from the vessel is fed back into the top tray of the lower column section. The vapor from the lower column section is directed to the bottom of the upper section. This configuration can operate without heat addition to the vessel and can be easily constructed by modifying an existing regular column. However, there is also an opportunity to introduce heating to the vessel, and an additional vapor stream originating from the vessel is then possible (see Configuration B in Figure 5.10). Furthermore, the liquid stream leaving the bottom tray of the upper column section and the vapor leaving the top tray of the lower column section can be directed in different ways (although at least one stream must be diverted to the vessel). These additional degrees of freedom lead to different vessel configurations, as shown in Configurations D–E in Figure 5.10. Low and Sorensen [9] considered optimizing Configuration E and found that by allowing both streams to vary in an optimal way, the profit increased by 45% compared to Configuration A and the batch processing time and energy consumption decreased by 19 and 14%, respectively. Warter and Stichlmair [10] compared the operation of extractive batch distillation in Configurations A and C.

5.4.4 Multivessel batch column

The multivessel configuration (see Figure 5.9) was proposed by Hasebe et al. [11] in 1995 and is an extension of the middle vessel concept. The configuration comprises three or more column sections with side vessels located between the columns. The feed and products can be charged to and withdrawn from these side vessels, respectively, as well as from the reflux drum and reboiler. The multivessel configuration has mainly been considered without product removal from the vessels (i.e. Configuration B in Figure 5.9) because this mode is much easier to operate.

This type of column gives the greatest flexibility and number of degrees of freedom and it can be converted into the regular, inverted, or middle vessel column by fixing some of the streams. This configuration is also referred to as a *multieffect* column.

5.4.5 Comparison between configurations

A number of authors have considered the relative merits of the different operating configurations, mainly in terms of optimal operation. Robinson and Gilliland [6], who originally proposed the use of inverted columns, stated that the main advantage of the configuration was that the highly pure light component would be collected in the condenser drum. For binary separations, the inverted column has generally been found to be better than the regular column in terms of less operating time when the light component is present in only small amounts and when a large amount of heavy component is removed from the column at the bottom (see, e.g. Ref. [12]). This is because when high purity of a light component is required, it is more time consuming to remove a small amount of light component overhead from a regular column using a very high reflux ratio than to remove a large amount of heavy component from the bottom of an inverted column using a low to moderate reboiler ratio. Also, in such cases the dynamic responses are slower in the regular column than in the inverted column.

For the middle and multivessel column configurations, the energy efficiency and production rate of the multivessel system are generally found to be greater than that of the regular column, and the benefit is more prominent when separating mixtures with more components. The annual profitability achievable by adopting a multivessel system can be more than twice that of the regular column (see, e.g. Ref. [13]).

As a further note, in terms of operation, the flexibility of the multivessel system allows additional degrees of freedom in terms of varying vessel holdups during the separation process. Several optimal control studies have indicated that the performance of the multivessel system can be improved by allowing the vessel holdups to be optimized: values of 11−43% in production rate, 12% in batch time, and 21% in mean energy consumption have been claimed by Hasebe et al., Noda et al., and Furlonge et al. [14−16], respectively. In practice, however, the implementation of an optimal holdup policy, although feasible, involved a much more complicated on-line control system (such as that proposed by Noda et al. [16]) than a simpler level controller for maintaining the holdup (see Section 5.5.4). This trade-off suggests that further comparative studies should be conducted by the design engineer to evaluate whether a more complicated operating policy, and its associated control system, are indeed worthwhile.

5.5 Control of batch distillation

It should be noted that many of the references to work on "*optimal control*" of batch distillation in the literature actually have nothing to do with control but rather refer to studies considering the *optimization* of batch distillation, which is covered in Section 5.8. (See the textbook by Bryson and Ho [17] and similar ones for a general description of optimal control.) These optimal control solutions only provide *open-loop* control strategies for a given column. In this section, "*control*" is taken in the traditional sense of the word, that is, it refers to a closed-loop system where a desired operating point is compared to an actual operating point and the difference between the two is used to drive the system toward the desired operating point.

5.5.1 Constant reflux control

Given the simplicity of a conventional regular batch column and its long history as a separation unit, the control strategies commonly used to operate the unit are also very basic. If operated at atmospheric pressure and according to a constant reflux policy, then only the condenser drum level needs to be controlled. For small units, such as laboratory or pilot plant columns, this may be done with a simple overflow arrangement, hence no actual controller is needed. For reflux control, reflux splitters are often used on small columns [18]. A reflux splitter is an on-off solenoid-operated device operated by a timer, which operates a slide gate that diverts all the liquid either to the column or to the product tank. The timer setting corresponds to a fixed reflux-to-distillate flow ratio and the setting is manipulated manually. The reflux splitter is easy to operate, avoids the problem of flow control in a small pipeline, and is therefore a low-cost alternative to more complex batch column control.

The constant reflux policy does have its limitations, however, as it is inherently an open-loop strategy. Unless proper feedback from the plant is considered, the average product composition is known only at the end of the batch. Therefore, if an off-specification product was to be obtained, the batch would have to be blended or rerun, with potential significant economic loss.

5.5.2 Constant distillate composition control

Constant distillate composition operation is more difficult than constant reflux operation because to keep the distillate composition at the desired value throughout the whole batch one needs to continuously adjust (increase) the reflux rate. On the other hand, the product composition is then controlled accurately during the batch; in other words, this mode is inherently a feedback operation. A simple feedback controller, for example, a proportional and integral (PI) controller, can be used for this purpose; however, the control can still be challenging because of the inherent nonlinear and nonstationary nature of the process, and the controller settings may need to be adapted to the conditions in the column during operation.

5.5.3 Advanced batch distillation control

As mentioned in Section 5.2, optimal reflux operation is an attractive alternative to the simple operating policies of constant reflux or constant distillate composition.

With this policy, the optimal reflux ratio trajectory is determined off-line by minimizing or maximizing a prespecified objective function (see Section 5.8). This operating policy is often simplified to a piecewise constant reflux ratio operation, which can be implemented in an open-loop fashion but then has drawbacks similar to those of constant reflux operation, as outlined above.

Because of the nonlinear, nonstationary, and finite time duration nature of the underlying dynamics, taking full advantage of the flexibility provided by a batch rectifier potentially gives rise to challenging control problems if the column is to be tightly controlled. As batch distillation is inherently an unsteady-state process, there exists no normal operating condition at which the conventional input/output linear model can be formulated and the employed controller can be tuned. Many academic researchers have, therefore, used the process as an example to demonstrate more advanced forms of control, such as inferential control using Kalman filters (e.g. [19,20]) and model-based control and/or neural networks (e.g. [21−27]).

5.5.4 Control of middle vessel or multivessel batch distillation

An interesting example of batch distillation control, that has received a fair bit of attention in the academic literature, is that of middle vessel and multivessel columns. These more complex column arrangements have more degrees of freedom, and some form of control of at least some of these is required. To recap Sections 5.4.3 and 5.4.4, three or more components are being purified simultaneously in these column configurations. For the most common mode of total operation, this involves controlling the compositions in each vessel to meet their respective specifications, and this is typically done by adjusting the holdup in each vessel, as well as in the reflux drum, to the required levels, leaving the reboiler drum to complete the mass balance. Alternative control strategies to achieve this are (see Figure 5.11):

1. Open-loop operation where the column is "controlled" by calculating a priori, using the mass balance from the feed composition, the required final holdup in each vessel and then using a level control system to keep the holdup in each vessel constant for multivessel operation [11,28] (see Strategy A in Figure 5.11). For cases where the feed composition is not known exactly, the open-loop operation can subsequently adjust the holdup in each vessel if composition measurements are available.
2. The actual compositions can be controlled directly by manipulating the reflux flows out of each vessel to obtain the required purities within the vessel and thus adjusting the holdups [29], as illustrated in Figure 5.11, Strategy B.
3. Composition measurements may sometimes be difficult to obtain and a simpler solution is used to indirectly control the composition based on temperature measurements. Skogestad et al. [30] and Wittgens et al. [31] developed a simple feedback control strategy whereby the temperature in each column section was controlled, thus indirectly controlling the holdup in the intermediate vessel

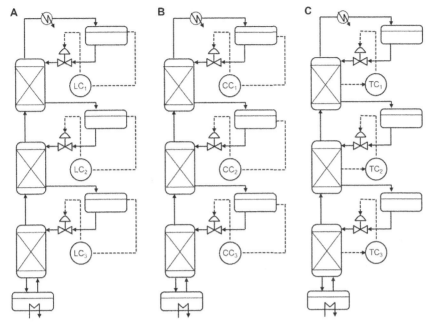

FIGURE 5.11 Control Strategies for Middle-Vessel Batch Distillation

(a) Level control (LC) strategy. (b) Composition control (CC) strategy. (c) Temperature control (TC) strategy.

above the section and thereby the composition (see Figure 5.11, Strategy C). The set point for each controller is set to the arithmetic mean value of the boiling temperatures of the two key components to be separated in that column section. Gruetzmann et al. [32] considered the same control strategy but also took into account startup of the column.

5.6 **Complex batch distillation**

Close boiling and low relative volatility mixtures are difficult and often uneconomical to distil and azeotropic mixtures are impossible to separate into pure components using ordinary distillation. Yet such mixtures are quite common, and many industrial processes depend on efficient methods for their separation. There are different alternatives for how to deal with such mixtures in a batch distillation framework, for instance:

1. Pressure-swing batch distillation
2. Extractive batch distillation
3. Hybrid batch distillation

5.6.1 **Pressure-swing batch distillation**

Pressure-swing distillation is an efficient method for the separation of pressure-sensitive azeotropic mixtures; many mixtures form an azeotrope, whose position can be shifted by changing the system pressure; that is, the azeotrope is pressure-sensitive. Examples of such mixtures are water and tetrahydrofuran, water and acetonitrile, methanol and acetone, ethanol and benzene, and even ethanol and water. The azeotropic composition can either decrease (e.g. ethanol-water) or increase (e.g. ethanol-benzene) as the pressure is increased [33]. For instance, the mole fraction of ethanol in the ethanol-water azeotrope increases from 0.8943 at 101.315 kPa to more than 0.9835 as the pressure is reduced to 11.955 kPa, and it even disappears at pressures below 9.321 kPa [1]. Knapp and Doherty [34] list 36 pressure-sensitive binary azeotropes.

When appreciable changes can be achieved over a moderate pressure range (>5 mol %), this effect can be exploited to separate azeotropic mixtures by so-called *pressure-swing distillation*. An advantage of pressure-swing distillation, when compared to alternative processes for separating azeotropic mixtures, is that there is no need for feeding and recycling additional substances (entrainer). A significant amount of literature has been published on pressure-swing distillation in a *continuous* system, although in recent years work also has focused on the separation of binary azeotropic mixtures by pressure-swing *batch* (e.g. [35]) and *semicontinuous* distillation (e.g. [36]) in a single column or in *double-column* systems [37], and a few studies have also investigated the separation of ternary mixtures (e.g. [38,39]).

In batch pressure-swing distillation of minimum boiling mixtures, only azeotropic mixtures can be withdrawn as distillate product (because they have the lowest boiling points), and the pure components remain as residue in the reboiler/still. If using only one batch column, two batches need to be run—one at low pressure and one at high pressure—to recover both pure components.

The minimum boiling azeotropic feed is initially charged into the reboiler as shown in Figure 5.12(c). Depending on the initial feed composition, the first batch step is either a high-pressure (HP) or a low-pressure (LP) distillation run. Consider the mixture shown in Figure 5.12(a and b) in which the feed concentration of the light component 1 is *lower* than the azeotropic point. (The feed can, of course, also lie between the two azeotropic compositions or above the highest azeotropic composition.) If the feed composition x_F is below the azeotropic composition, then the first step is an LP step and component 2 is accumulated in the reboiler. At the top, a mixture close or equal to the LP azeotropic composition Az_{LP} is withdrawn as a distillate product (D_1). This step runs until a certain concentration of component 2 is achieved in the reboiler.

Once the first step has been finished, component 2 is removed from the reboiler (R_1), and the process is switched over to the second step: the second feed, which was the azeotropic distillate of the LP run (D_1), is charged into the reboiler (as F_2) and the pressure is increased to HP. The second step is the HP distillation run, and component 1 is now accumulated in the reboiler (B_2) and a distillate with composition close

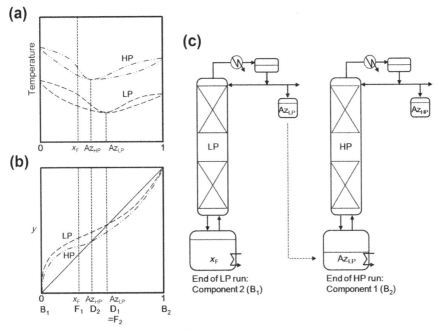

FIGURE 5.12 Pressure-Swing Distillation

(a) T-*x-y* diagram for a minimum boiling azeotrope at low pressure (LP) and high pressure (HP). (b) corresponding *x-y* diagram. (c) distillation sequence for the separation of pressure-sensitive minimum boiling azeotrope (Az) by regular batch distillation.

or equal to the HP azeotrope Az_{HP} is withdrawn at the top. The HP azeotrope product (D_2) can subsequently be recycled to the next LP batch. If the initial feed concentration is *higher* than the azeotropic point, then the first step is an HP step and component 1 is produced first, followed by an LP step during which component 2 is produced.

Pressure-swing distillation can, of course, also be operated in other configurations, for example, in inverted columns [35], and can be applied to the separation of the less-common maximum boiling binary azeotropes (see the continuous example provided by Luyben [40,41]).

5.6.2 Extractive batch distillation

Extractive distillation is an important process in chemical industries for the separation of azeotropes and close boiling mixtures, and *continuous* extractive distillation is a well-known and widespread technology. In contrast, extractive distillation in *batch* mode is a relatively new process in the literature, with the first studies published by Bernot et al. [42,43], and so far the process has had limited instances (or at least reports) of industrial implementation. As the process offers the advantages of both batch and extractive distillation, it has been extensively studied in the academic literature over the past decade. The choice of entrainer is clearly of great importance and should ideally be considered as part of the overall synthesis and design problem (although this is not considered here). The focus in the literature has been on two strands: either focusing on graphical methods with the primary objective of assessing the feasibility of different entrainers and column configurations for given mixtures, or considering the optimal design and/or operation for given entrainers and configurations.

The most common azeotropes encountered industrially are minimum boiling azeotropes, that is, the azeotrope is the lightest fraction, and these are normally separated using a low volatile entrainer, that is, the entrainer is the heaviest fraction. The easiest and most commonly used mode of operation to separate these mixtures in a regular batch still is to add the entrainer directly to the still at the start; this is also called *solvent enhanced batch distillation* [44], as illustrated in Alternative A in Figure 5.13. It is also possible to continuously add the entrainer to the still during the operation (Alternative B). In the academic literature, the process is commonly operated using a four-step operating policy for binary minimum azeotropic mixtures in a regular column based on Alternative C in Figure 5.13:

1. Total reflux operation *without* entrainer feeding. This establishes the azeotropic composition at the top.
2. Total reflux *with* entrainer feeding near the top of the column. This breaks the azeotrope and achieves a high purity of the light component at the top.
3. Finite reflux operation *with* entrainer feeding to withdraw the light component.

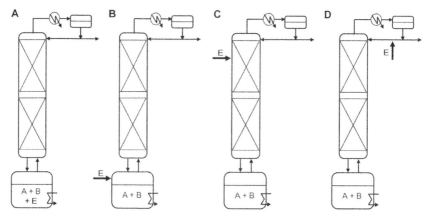

FIGURE 5.13 Extractive distillation of a minimum boiling azeotrope using an entrainer with low volatility in a regular column (A–D: different alternatives)

4. Finite reflux operation *without* entrainer feeding. The heavy component from the original feed mixture is withdrawn over the top and the highly pure entrainer is left in the reboiler.

Offcut fractions may also be taken off between each product. The entrainer is normally added either at the top of the column section, in which case the whole column section becomes an extractive zone, or to a point in the column section, separating the column into a rectifying section above the entrainer feed and an extractive section below the entrainer feed. Some authors have considered using intermediate boiling entrainers to separate minimum boiling azeotropes (e.g. [45]), or have studied the separation of maximum boiling azeotropes (e.g. [46]).

Extractive batch distillation in middle vessel configurations has also been considered, with different alternatives for addition of entrainer, as illustrated in Figure 5.14; that is, either feed to the middle vessel, to a location toward the top of the rectification section, or to a location toward the top of the stripping section (e.g. [9,47]). The feed can be added either to the middle vessel or the reboiler or can be distributed between the two. The intermediate component can be withdrawn from the middle vessel, although most authors have not considered this option, only allowing removal as distillate or as bottoms.

For the separation of binary minimum azeotropic mixtures in a middle vessel column using a low boiling entrainer, the following operations have been considered [48]:

1. Total reflux and total reboil operation *without* entrainer feeding.
2. Total reflux and reboil operation *with* entrainer feeding toward the top of the rectification section.
3. Entrainer feeding toward the top of the rectification section *with* distillate removal of the lightest component but *without* bottom removal, that is, total reboil operation.

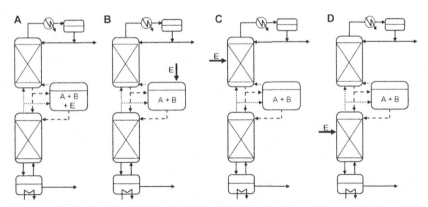

FIGURE 5.14 Extractive distillation of a minimum boiling azeotrope using an entrainer with low volatility in a middle-vessel column (A—D: different alternatives)

4. Entrainer feeding toward the top of the rectification section *with both* bottom and distillate removal (the bottom stream contains mainly the entrainer, which is therefore recycled back to the top of the column section).

5.6.3 **Hybrid batch distillation**

Many mixtures commonly encountered in the fine chemical and pharmaceutical industries are difficult or impossible to separate by normal distillation because of azeotropic behavior, tangent pinch, or low relative volatilities. Membrane separation has been hailed as an alternative to distillation for such mixtures; the separation mechanism is different, relying on differences in diffusivity and solubility between the components in the mixture and not vapor–liquid equilibrium, as in distillation. Membrane separations are still in general a more costly process than distillation; for some separations, however, the costs are comparable or in favor of membranes. Hybrid processes have recently been proposed where a distillation column unit and a membrane unit are integrated into one process [49]. The most commonly considered membrane process is pervaporation, but other hybrid alternatives are, of course, also possible, for example, distillation with vapor permeation or nanofiltration, as well as with crystallization and chromatography [50].

In a hybrid process, the shortcomings of one method are outweighted by the benefits of the other, allowing for significant savings in terms of energy consumption and cost. Considering hybrids between membranes and distillation, the advantages of distillation (robustness and high capacity) are combined with the advantages of membrane separation (high selectivity and a separation that exceeds distillation boundaries).

In a hybrid process combining batch distillation and membrane separation, the two units can be integrated in different ways; the membrane unit can be positioned before the column, after the column, or fully integrated (see Figure 5.15). In a pre-distillation hybrid system, the membrane unit is placed before the distillation column to pretreat the feed stream to the column. The permeate from the membrane can be fed to the column reboiler/still (Alternative A). Of course, it is also possible for the retentate to become the distillation feed (not shown). In a post-distillation hybrid system (Alternative B), the distillate from the distillation column enters the membrane, and the permeate and/or the retentate is collected as product(s). It can be argued that neither of these two systems are true hybrids but can instead be considered as two separate units in series. If the retentate (Alternative C) or the permeate (Alternative D) are returned to the column section, then the two units are integrated as the combined system relies on the separation characteristics of the combination of the two processes, and a change in one of the units has an immediate effect on the other and vice versa.

The combination of the two units adds complexity to the system but also more degrees of freedom, which, if chosen properly, can further increase the profitability of the system, particularly for a difficult separation such as that of azeotropic

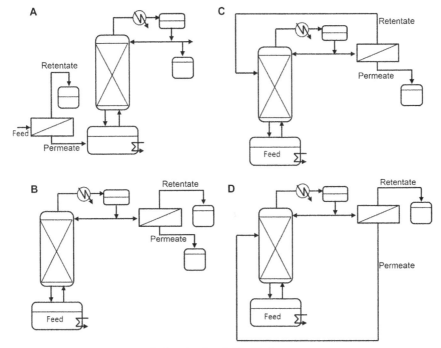

FIGURE 5.15 Examples of Hybrid Regular Batch Distillation

(a) predistillation hybrid; (b) postdistillation hybrid system; (c) and (d) integrated hybrid systems.

mixtures. For instance, the number of membrane stages and the number of modules within each stage, as well as the column stage to which the retentate/permeate is returned, must be determined. The conceptual design of continuous hybrid separations has been considered by Skiborowski et al. [51], and the batch variant by Barakat and Sorensen [52].

5.6.4 Reactive batch distillation

Continuous reactive distillation has recently found important industrial applications leading to higher reactant conversion, higher product selectivity, and significant energy savings (see also Chapter 8, Reactive distillation in Gorak and Olujic [53]). In this intensified process, a reactor and a distillation column are combined into one unit, generally by letting the reaction take place in the column section containing, for instance, catalytic packing. The batch variant is usually operated slightly differently, with the reaction normally taking place in the feed tank, that is, in the reboiler in a regular configuration (see Figure 5.16, Configuration A).

This process is appropriate if one of the products has a lower boiling point than the other product(s) and the reactants, as this product is removed continuously from

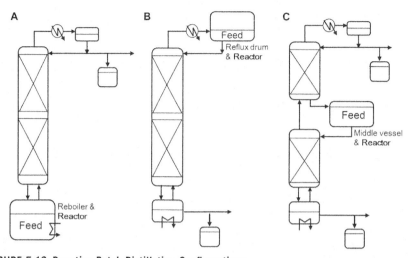

FIGURE 5.16 Reactive Batch Distillation Configurations

(a) regular configuration. (b) inverted configuration. (c) middle-vessel configuration.

the reaction mixture by distilling it over the top. In the case of an irreversible reaction, the continuous removal of this more volatile product increases the liquid temperature and hence the reaction rate in the reboiler/reactor.

With a reversible reaction, the elimination of one of the products will drive the forward reaction. In both cases, higher conversion of the reactants is expected than what occurs by reaction alone. An example of a reaction that can take place in a regular batch reactive column is acetic acid ($T_{bp} = 391.1$ K) and methanol ($T_{bp} = 337.8$ K) reacting to form methyl acetate ($T_{bp} = 330.1$ K) and water ($T_{bp} = 373.2$ K) [3]. Here, methyl acetate, which is the lightest component, will be removed over the top in the regular column.

The reaction products in a reaction scheme do not always have a lower boiling point than the reactants, and the use of a regular batch column would then result in the removal of reactants, which would clearly not be beneficial. If one or more of the products have a higher boiling point than the reactants, then the inverted configuration would be favored (Figure 5.16, Configuration B). An example of a reaction that can take place in an inverted batch reactive column is ethylene oxide ($T_{bp} = 283.5$ K) and water ($T_{bp} = 373.2$ K) reacting to form ethylene glycol ($T_{bp} = 470.4$ K) [3]. Here, ethylene glycol, which is the heaviest component, will be removed through the bottom in the inverted column.

Equally, if some of the products had a lower boiling point and some had a higher boiling point than the reactants, then a middle vessel configuration should be considered (Figure 5.16, Configuration C) [54]. Finally, there are cases where the distribution of boiling points between reactants and products is such that none of the configurations would be suitable; in that case an intensified process should not be considered.

5.7 **Modeling of batch distillation**

The theoretical analysis of elementary *simple distillation* presented by Rayleigh [55] marks the first theoretical work on batch distillation. The concept of reflux, and the use of plates and packing materials to increase the mass transfer, converts this simple still into a distillation column. The earliest work that considered the complete dynamics of batch distillation was presented by Huckaba and Danly in 1960 [56] and was restricted to ideal binary systems. Meadows [57] presented the first comprehensive model of batch distillation for the constant reflux mode of operation, and Distefano [58] presented the first detailed analysis of the characteristics of differential mass and energy balances associated with the complete dynamics of multicomponent batch distillation. The level of detail in these early works was naturally rather basic, but since then there has been a gradual relaxation of modeling assumptions, and more complex systems are being considered. Modeling complexity is thus relative; a model that was considered rigorous in the early 1980s would most likely be considered rather simple now as computers have become more powerful and numerical solvers have become more efficient, and this trend is likely to continue in the future. The level of modeling detail required is the most important decision when attempting to model any system because the modeling assumptions may in some cases have a significant impact on the results, and all simulation and optimization results must be considered carefully in the light of the assumptions made.

It should be noted that the system of equations for batch distillation is more complex than for continuous distillation; the differential algebraic equation (DAE) system is much stiffer because of the large differences in component volatilities and/or the much greater ratio of reboiler to tray holdup. Thus, the tray dynamics is significantly faster than the reboiler dynamics. In addition, batch distillation displays severe transients compared to continuous distillation, where the variations are relatively smaller.

In the following, the main modeling assumptions typically made when considering batch distillation are considered, and the range of models are split into categories depending on complexity. The range of model abstraction studied in the previous literature can be classified broadly according to the assumptions used (see Table 5.1):

- Short-cut models
- Simple models
- Rigorous models
- Rate-based models

Note that this classification is somewhat arbitrary, and different combinations may, of course, also be considered. The mathematical solution methods for how to solve the equation sets resulting from these models are not considered here as this goes beyond the scope of this chapter. Matlab, ProSim BatchColumn (e.g. [46]), ChemCAD (e.g. [46]), Aspen [59], and gPROMS (e.g. [35,60]) are commonly used simulations tools.

Table 5.1 Classification of Batch Distillation Models

Modeling Assumptions	Shortcut Models	Simple Models	Rigorous Models	Rate-based Models
Negligible tray liquid holdup	√			
Constant molar overflow	√	√		
Negligible vapor holdup	√	√		
Phase equilibrium	√	√	√	
Perfect mixing	√	√	√	
Adiabatic operation	√	√	√	√

5.7.1 Short-cut batch distillation models

The simplification of short-cut models involves tackling the stiffness during numerical integration caused by the large difference in time constant between the reboiler and the trays. The modeling solution to this problem is based on splitting the system into two levels: the reboiler, where the dynamics are slower, is represented by differential equations, and the column section, normally considered to be a tray section, is simplified in some way.

For example, the column section can be assumed to be in a quasi-steady state where the composition and enthalpy changes in the condenser and trays are assumed to be zero. Another example is to simplify the column section by developing a more direct relationship between the distillate and bottom reboiler compositions based on the Hengstebeck-Geddes' equilibrium relationship, Fenske and Underwood equations, and the Gilliland correlation. The approach assumes negligible tray liquid and vapor holdups and constant molar overflow, that is, negligible sensible heat effects and similar latent heats of components (see summary by Diwekar [61]).

5.7.2 Simple batch distillation models

Unlike the short-cut models, where the holdup in the column section is neglected, the simple models consider the stage-by-stage column dynamics in addition to the dynamics of the major holdups in the reboiler and condenser. The negligible tray liquid holdup assumption is relaxed, and the liquid stage holdup is responsible for the dynamics of each stage, with differential mass and energy balances being the governing equations.

5.7.3 Rigorous batch distillation models

An even greater level of detail can be included by relaxing the assumptions of negligible vapor holdup and/or constant liquid holdup on the trays. Variable liquid and vapor holdup can be taken into account by the inclusion of equations describing tray hydraulics such as liquid weir overflow, downcomer dynamics, and pressure drop—vapor flow rate relationships.

5.7.4 **Rate-based models**

The categories of models described above assume theoretical stages where the vapor and liquid streams leaving a stage are in equilibrium with each other. The common method of representing nonequilibrium or real trays is to use an efficiency factor of some kind, which is most likely to be either an overall efficiency or the Murphree efficiency, which can be easily incorporated into the equilibrium set of equations.

Rate-based, or nonequilibrium, models consider mass and energy transfer dynamics by including balance equations for both liquid and vapor phases within the tray and then link these by transport equations around the phase interface, using, for example, Maxwell–Stefan diffusion equations for mass transfer. To relax the assumption of perfect mixing, higher hierarchy models consider partial mixing, that is, non-uniform composition on a tray, which greatly increases the complexity of the model. For a more thorough overview of rate-based phenomena, see Chapter 3 (Mass Transfer Distillation) and Chapter 10 (Modeling of Distillation Processes).

5.8 **Optimization of batch distillation**

Research on the optimization of batch distillation has involved many aspects such as binary or multicomponent mixtures, multiproduct facilities, and recycling of offcuts. Rigorous optimization of batch distillation presents serious theoretical and computation difficulties due to:

1. Time dependence of the operation with discontinuities.
2. Modeling equations that involve DAEs whose dimensions strongly depend on the number of plates in the column section and the number of components in the feed mixture.
3. A mixture of discrete (e.g. sequence of product cuts and offcuts, recycling of offcuts) and continuous variables (e.g. reflux ratio, energy input).
4. The presence of several objectives, particularly when multicomponent, multifraction distillation is considered.

In the past, the column performance was generally described by very simple dynamic models because of limited computer resources; however, with the continuing increase in computational power, the complexity of models has increased, as discussed in Section 5.7. As a result, in the more recent literature the emphasis has been more on the methods used to solve the problem rather than on the actual results of the optimization.

Even for binary separations there is a large number of degrees of freedom—some in terms of design, but most in terms of operation. The number of product cuts, the number of offcuts, and the possible recycling of offcuts are typical *discrete* decision variables. The height of the column section is also a discrete decision if a tray column is used. Whether an offcut should be recycled is a yes/no decision, and so is the decision as to how it is to be recycled. Other decision variables vary *continuously* during the batch. Although the main operating variables are the reflux ratio and the energy input, the pressure and the condenser drum holdup may also vary. The complexity increases substantially with the number of components and even more so when multiple feeds are to be processed in the same batch facility, and the column will need to be sized according to the most difficult separation. Selecting the best design and operation thus presents a formidable challenge because the problem could, in principle, be posed as a large dynamic mixed-integer nonlinear optimization problem (dynamic MINLP).

5.8.1 Optimization of the reflux ratio

In most of the early literature on optimization of batch distillation, only the reflux ratio was considered as an optimization variable. The configuration and the operating strategy, that is, the number of product cuts and offcuts and the size of the column, were specified a priori. The optimal control problem was usually formulated in one of three ways:

1. *Minimum time problem.* The optimal policy is that which produces a required quantity of distillate of specified purity in the shortest possible time. The optimization can be stated as:

$$\min_{R(t)} t_f$$

$$x_A \geq x_A^{spec} \tag{5.5}$$

$$H_A \geq H_A^{spec}$$

   or, in words, find the optimal reflux ratio as a function of time $R(t)$, which minimizes the total operating time t_f subject to the constraints that the composition x_A and amount of distillate H_A must meet their specifications, that is, the concentration x_A^{spec} and the amount H_A^{spec} of the accumulated product, respectively.

2. *Maximum distillate problem.* The optimal policy is that which produces the maximum amount of distillate of specified purity in a fixed duration of time. The optimization can be stated as:

$$\max_{R(t)} H_A$$

$$x_A \geq x_A^{spec} \tag{5.6}$$

$$t_f = t_f^{spec}$$

   equivalent to finding the optimal reflux policy as a function of time $R(t)$, which maximizes the amount of distillate H_A subject to the constraint that the composition of distillate must meet its specification x_A^{spec} with the total operating time given a priori as $t_f = t_f^{\text{spec}}$.

3. *Maximum profit problem.* It also is possible to determine the optimal operating policy by optimizing some economic criterion. The optimization problem can then be stated as, for example:

$$\max_{R(t)} P = \max_{R(t)} \frac{C_A H_A - C_F H_F - \int_0^{t_f} \dot{Q}_{\text{total}}(t)\,dt}{t_f}$$

$$x_A \geq x_A^{\text{spec}}$$

$$H_A \geq H_A^{\text{spec}}$$

(5.7)

or finding the optimal reflux ratio as a function of time $R(t)$, which maximizes the hourly profitability P, where C_A and C_F are the prices of the distillate product and the feed, respectively, and the integral is the total energy consumption. Possible constraints are that the composition and amount of accumulated distillate, x_A and H_A, must meet their specifications, x_A^{spec} and H_A^{spec}, respectively.

Other types of constraints, such as recovery of a component and conversion of a reactant in reactive batch distillation, are also possible and have been considered in the literature.

The three control objectives differ in the way the optimal performance of the column is defined. In the first objective, it is considered to be optimal to use as little time as possible for the production of a fixed quantity, for example, a given amount of material of a set purity required in a downstream process. Also, storage may be limited, thereby restricting the amount produced. On the other hand, the time frame may be fixed, for example, the duration of one shift. The control objective, then, will be to produce as much as possible within this one shift—in other words, a maximum quantity of distillate. Normally, less production time will give a higher profit. The third objective, maximum profit, is rarely used because it requires additional cost parameters that can be difficult to obtain or estimate. A variation of the cost objective can be to consider only the energy contribution to the costs, that is, minimizing the energy consumption.

5.8.2 The general optimal control problem

Since batch distillation is an intrinsically dynamic process, the equality constraints for the optimization problem are ordinary DAEs. This type of problem is generally referred to as an *optimal control* problem (even though control in the usual sense of the word is not optimized) and is an element of the field of control theory, which has experienced significant development since the 1950s.

The optimal control problem can generally be stated as a minimization problem (although it can of course be reformulated to a maximization problem):

$$\min_{\dot{x}(\cdot),x(\cdot),y(\cdot),u(\cdot),v,x(t_0),t_f} f_{obj}\left(\dot{x}(t_f),x(t_f),y(t_f),u(t_f),v,t_f\right) \qquad (5.8)$$

subject to

$$g\left(\dot{x}(t),x(t),y(t),u(t),v,t_f,t\right) = 0 \qquad (5.9)$$

$$h\left(\dot{x}(t_f),x(t_f),y(t_f),u(t_f),v,t_f\right) \geq 0, \qquad (5.10)$$

where f_{obj} is the objective function depending on the conditions of the final time t_f, g is the equality constraints (DAEs), and h is the inequality constraints at the end point. Interior or path constraints may also exist. Upper and lower bounds may be defined for the vectors of control variables u, design variables v, initial states x_0, and final time t_f. The variable x denotes the vector of state variables and y the algebraic variables.

For a batch distillation problem, x may include the amount of each component on a tray and x_0 is the corresponding initial amounts; y may include the flow rates leaving and entering a tray (in addition to other variables), u is the time-varying reflux ratio, and v is a fixed reboiler heat input.

The optimal control problem is often referred to as an *infinite dimensional* problem because u is a function of time rather than a specific value at the optimum, as in continuous distillation. This type of problem can be solved using Pontryagin's maximization principle; however, this solution method is rather time consuming. It may therefore be more efficient to use either nonlinear programming (NLP) or a stochastic approach rather than the maximization principle to compute the optimal control.

5.8.3 Nonlinear programming approaches

The optimization problem can be solved by numerical techniques that convert the problem into an NLP problem. Two general approaches have been considered in the literature for batch distillation: control vector parameterization and full parameterization.

1. *Control vector parameterization*: By parameterizing the control vector, one can think of the true optimal control profile as being approximately represented by a simple basic function, for example, constant, linear, quadratic, or exponential. The basic function is defined for a control interval, and the number of control intervals is determined a priori. With this method, the control profile can be fully described by the number of control intervals, the control functions (constant value, linear, etc.), and the duration of the intervals. These parameters are then added to the optimization problem to form a finite set of decision variables.

   This approach is commonly referred to as a *feasible path approach*. The advantage is that the resulting optimization problems are small in terms of the number of variables. If the process is terminated before reaching an optimal solution, the termination point is still feasible and may be accepted as a practical, although suboptimal, solution to the problem. However, since the feasible path approach requires a complete solution of the DAEs for each trial value of the decision variables, this may be computationally expensive. Also, the simulation level may fail since a feasible region may not exist for certain values of the decision variables, for example, a heat input that may cause the reboiler to run dry before the end of the batch time.

2. *Full parameterization, that is, the collocation method*: This method discretizes the control functions, as well as the ordinary differential equations in the original DAE model, using collocation over finite elements. The result is a large system of algebraic equations that, together with the objective function and the inequality constraints, forms a sizable NLP problem. For the same problem, however, the degrees of freedom for this large NLP are the same as for the small-scale NLP in the control parameterization method. Since the DAEs and the optimization problem are solved simultaneously, this approach is called an *infeasible path approach*. The main advantage of the method is that the repeated integration of the DAEs is avoided and therefore the method should be faster.

5.8.4 Mixed integer dynamic optimization

When the optimization problem also contains *integer variables*, for example, number of stages in a column section or the number of product cuts, the problem translates into a complex mixed integer dynamic optimization problem. This type of problem is difficult to solve, and there is much ongoing research on developing practical solution methods, but these will not be covered here.

5.8.5 Stochastic approaches

A number of different stochastic approaches have been applied to solve batch column optimization problems, for example, genetic algorithm (GA) [52,60,62,63], simulated annealing [64], and modified simulated annealing [65]. The most commonly used stochastic method for the solution of batch distillation problems is probably GA, a class of evolutionary algorithms that is inspired by the natural genetic process. Thus candidate solutions are encoded as *genomes* that contain a set of *genes* representing the decision variables analogous to the DNA in a natural organism. In general, all decision variables (genes) are coded in the genome as direct real and integer values, for example, reflux ratio, heat input, and number of trays. The initial population of genomes, that is, different column design and operation alternatives, is created randomly. Solutions are assigned a fitness score based on, for instance, the annual profitability of each genome, and a penalty function procedure is applied when necessary to encourage the GA to drive the population toward feasibility.

5.8.6 Multi-objective optimization

In batch separation processes there is a trade-off between capital investment and energy consumption in terms of equipment sizing and performance. For instance, the design of high-performance equipment generally requires a large investment of capital but incurs low operational and energy costs. Alternatively, low-performance equipment, requiring a smaller investment of capital but higher operational and energy costs, can also be considered. Longer batch times required for high product recovery also lead to higher energy consumption than would low product recovery. Thus, the multicriteria balance between capital investment and energy consumption must be considered.

In general, the goal of *multiobjective optimization* is to determine a range of different process alternatives to explore the trade-offs between two or more conflicting design and/or operational criteria. All optimal process alternatives need to satisfy given constraints in meeting these criteria. To achieve this goal, it is important that all of these criteria be considered *simultaneously* through an effective multiobjective optimization procedure. Multiobjective optimization is thus the simultaneous consideration of two or more objective functions that are completely or partially in conflict with each other, for example, minimum cost and maximum amount of distillate. The optimization of such functions is largely defined through the Pareto optimality, which is based on the Pareto dominance criteria. Multiobjective optimization of batch distillation has been considered by only a few authors (e.g. [59,66]).

5.9 **The future of batch distillation**

Batch distillation is one of the oldest separations known to man and has for millennia been used for a multitude of purposes. Apart from the major improvement of the inclusion of a column section above the still and the associated rectification, that is, the return of reflux to the column section, the process is still operated in the same traditional manner. Regular batch columns dominate industry and are typically operated either with open-loop constant reflux ratio policies or with very simple control strategies.

Research over the past 2–3 decades has, however, clearly demonstrated that large capital and operational savings can be achieved by the application of novel column configurations and novel operating strategies. Equally, the application of batch distillation for the separation of challenging mixtures via extractive batch distillation, pressure-swing batch distillation, and hybrid batch separations have also demonstrated the possibilities offered by batch distillation.

The main difficulty in the design of batch processes is the consideration of the time-variant nature of the process, which means dynamic process models and more sophisticated simulation and optimization techniques are required compared to continuous processes. Given the improvements in computer power and numerical solution algorithms over the past decade, accurate simulation of batch distillation is now certainly possible, and modeling and optimization should no longer be the bottleneck in batch distillation design.

Batch distillation is currently a separation process widely used in the fine chemical and pharmaceutical processing industries, and it is expected to remain a process of significant interest as long as these industries continue to expand. This chapter has demonstrated that considerable progress has been made in the study of the column configuration, design, and operation of batch distillation processes, but there is nevertheless plenty of scope for future work in better exploring the degrees of freedom offered by novel process alternatives and in considering these factors—configuration, design, operation, and control—simultaneously to determine the true potential that operation in batch distillation can offer, either alone or in intensified combinations with reactions or other separation processes.

The future is bright.

References

[1]   J.D. Seader, E.J. Henley, D.K. Roper, Separation Process Principles, third ed., Wiley and Sons, 2013.

[2]   E. Sorensen, A cyclic operating policy for batch distillation — theory and practice, Comput. Chem. Eng. 23 (4−5) (1999) 533−542.

[3]   I.M. Mujtaba, Batch Distillation. Design and Operation, Imperial College Press, London, 2004.

[4]   F.M. Christensen, S.B. Jørgensen, Optimal control of binary batch distillation with recycled waste cut, Chem. Eng. J. 34 (2) (1987) 57−64.

[5]   W.L. Luyben, Multicomponent batch distillation. 1. Ternary systems with slop recycle, Ind. Eng. Chem. Res. 27 (4) (1988) 642−647.

[6]   C.S. Robinson, E.R. Gilliland, Elements of Fractional Distillation, fourth ed., McGraw-Hill Inc., New York, 1950.

[7]   P. Bortolini, G.B. Guarise, A new practice of batch distillation (in Italian), Quad. Ing. Chim. Ital. 6 (1970) 150−159.

[8]   S. Hasebe, B.B. Abdul Aziz, I. Hashimoto, T.Watanabe, Optimal design and operation of complex batch distillation column, Preprints of the IFAC Workshop on Interactions between Process Design and Process Control, (1992) 177−182.

[9]   K.H. Low, E. Sorensen, Optimal operation of extractive distillation in different batch configurations, AIChE J. 48 (5) (2002) 1034−1050.

[10] M. Warter, J. Stichlmair, Batchwise extractive distillation in a column with a middle vessel, Comput. Chem. Eng. 23 (Suppl. 1) (1999) S915−S918.

[11] S. Hasebe, T. Kurooka, I. Hashimoto, Comparison of the separation performances of a multi-effect batch distillation system and a continuous distillation system, Preprints DYCORD'95, Helsingor, Denmark, (1995) 249−254.

[12] E. Sorensen, S. Skogestad, Comparison of regular and inverted batch distillation, Chem. Eng. Sci. 51 (22) (1996) 4949−4962.

[13] K.H. Low, E. Sorensen, Simultaneous optimal design and operation of multivessel batch distillation, AIChE J. 49 (10) (2003) 2564−2576.

[14] H.I. Furlonge, C.C. Pantelides, E. Sorensen, Optimal operation of multivessel batch distillation columns, AIChE J. 45 (4) (1999) 781−801.

[15] S. Hasebe, M. Noda, I. Hashimoto, Optimal operation policy for multi-effect batch distillation system, Comput. Chem. Eng. 21S (1997) 523−532.

[16] M. Noda, T. Chida, S. Hasebe, I. Hashimoto, On-line optimization system of pilot scale multi-effect batch distillation system, Comput. Chem. Eng. 24 (2−7) (2000) 1577−1583.

[17] A.E. Bryson, Y.C. Ho, Applied Optimal Control: Optimization, Estimation and Control, Taylor and Francis, 1975.

[18] H.Z. Kister, Distillation Design, McGraw-Hill, 1989.

[19] R.M. Oisiovici, S.L. Cruz, State estimation of batch distillation columns using an extended Kalman filter, Chem. Eng. Sci. 55 (20) (2000) 4667−4680.

[20] R.M. Oisiovici, S.L. Cruz, Inferential control of high-purity multicomponent batch distillation columns using an extended kalman filter, Ind. Eng. Chem. Res. 40 (12) (2001) 2628−2639.

[21] A. Bahar, C. Özgen, State estimation and inferential control for a reactive batch distillation column, Eng. Appl. Artif. Intell. 23 (2) (2010) 262−270.

[22] A.K. Jana, P.V.R.K. Adari, Nonlinear state estimation and control of a batch reactive distillation, Chem. Eng. J. 150 (2−3) (2009) 516−526.

[23] K. Konakom, P. Kittisupakorn, I.M. Mujtaba, Neural network-based controller design of a batch reactive distillation column under uncertainty, Asia-Pac. J. Chem. Eng. 7 (3) (2012) 361−377.

[24] R. Monroy-Loperena, J. Alvarez-Ramirez, A note on the identification and control of batch distillation columns, Chem. Eng. Sci. 58 (20) (2003) 4729−4737.

[25] R. Monroy-Loperena, J. Alvarez-Ramirez, Backstepping-based cascade control scheme for batch distillation columns, AIChE J. 50 (9) (2004) 2113−2129.

[26] R. Monroy-Loperena, J. Alvarez-Ramírez, Dual composition control in continuous, middle-vessel distillation columns, with a draw stream in the middle vessel, Ind. Eng. Chem. Res. 51 (12) (2012) 4624−4631.

[27] W. Weerachaipichasgul, P. Kittisupakorn, A. Saengchan, K. Konakom, I.M. Mujtaba, Batch distillation control improvement by novel model predictive control, J. Ind. Eng. Chem. 16 (2) (2010) 305−313.

[28] S. Hasebe, T. Kurooka, B.B.A. Aziz, I. Hashimoto, T. Watanabe, Simultaneous separation of light and heavy impurities by a complex batch distillation column, J. Chem. Eng. Jpn. 29 (6) (1996) 1000−1006.

[29] M. Barolo, G.B. Guarise, S.A. Rienzi, A. Trotta, S. Macchietto, Running batch distillation in a column with a middle vessel, Ind. Eng. Chem. Res. 35 (12) (1996) 4612−4618.

[30] S. Skogestad, B. Wittgens, R. Litto, E. Sorensen, Multivessel batch distillation, AIChE J. 43 (4) (1997) 971−977.

[31] B. Wittgens, R. Litto, E. Sorensen, S. Skogestad, Total reflux operation of multivessel batch distillation, Comput. Chem. Eng. 20 (Suppl. 2) (1996) S1041−S1046.

[32] S. Gruetzmann, G. Fieg, S. Skogestad, Experimental and theoretical studies on the start-up operation of a multivessel batch distillation column, Ind. Eng. Chem. Res. 48 (11) (2009) 5336−5343.

[33] D.W. Green, R.H. Perry, Perry's Chemical Engineers' Handbook, eighth ed., McGraw-Hill Professional, 2007.

[34] J.P. Knapp, M.F. Doherty, A new pressure-swing distillation process for separating homogeneous azeotropic mixtures, Ind. Eng. Chem. Res. 31 (1) (1992) 346−357.

[35] J.U. Repke, A. Klein, D. Bogle, G. Wozny, Pressure-swing batch distillation for homogenous azeotropic separation, Chem. Eng. Res. Des. 85 (4) (2007) 492−501.

[36] J.R. Phimister, W.D. Seider, Semicontinuous, pressure-swing distillation, Ind. Eng. Chem. Res. 39 (1) (2000) 122−130.

[37] G. Modla, P. Lang, Feasibility of new pressure swing batch distillation methods, Chem. Eng. Sci. 63 (11) (2008) 2856−2874.

[38] G. Modla, P. Lang, F. Denes, Feasibility of separation of ternary mixtures by pressure swing batch distillation, Chem. Eng. Sci. 65 (2) (2010) 870−881.

[39] G. Modla, Separation of a chloroform-acetone-toluene mixture by pressure-swing batch distillation in different column configurations, Ind. Eng. Chem. Res. 50 (13) (2011) 8204−8215.

[40] W.L. Luyben, Pressure-swing distillation for minimum- and maximum-boiling homogeneous azeotropes, Ind. Eng. Chem. Res. 51 (33) (2012) 10881−10886.

[41] W.L. Luyben, Comparison of extractive distillation and pressure-swing distillation for acetone/chloroform separation, Comput. Chem. Eng. 50 (5) (2013) 1−7.

[42] C. Bernot, M.F. Doherty, M.F. Malone, Patterns of composition change in multicomponent batch distillation, Chem. Eng. Sci. 45 (5) (1990) 1207−1221.

[43] C. Bernot, M.F. Doherty, M.F. Malone, Feasibility and separation sequencing in multicomponent batch distillation, Chem. Eng. Sci. 46 (5−6) (1991) 1311−1326.

[44] I. Rodríguez-Donis, V. Vargaa, V. Gerbaud, Z. Lelkes, E. Rév, Z. Fonyó, X. Joulia, Feasibility Study of Heterogeneous Batch Extractive Distillation, ESCAPE-15, 1995.

[45] Z. Lelkes, E. Rev, C. Steger, Z. Fonyo, Batch extractive distillation of maximal azeotrope with middle boiling entrainer, AIChE J. 48 (11) (2002) 2524−2536.

[46] I. Rodriguez-Donis, V. Gerbaud, X. Joulia, Thermodynamic insights on the feasibility of homogeneous batch extractive distillation. 3. Azeotropic mixtures with light entrainer, Ind. Eng. Chem. Res. 51 (12) (2012) 4643−4660.

[47] K.J. Kim, U.M. Diwekar, K.G. Tomazi, Entrainer selection and solvent recycling in complex batch distillation, Chem. Eng. Commun. 191 (12) (2004) 1606−1633.

[48] B.T. Safrit, A.W. Westerberg, U. Diwekar, O.M. Wahnshafft, Extending continuous conventional and extractive feasibility insights to batch distillation, Ind. Eng. Chem. Res. 34 (1995) 3257−3264.

[49] F. Lipnizki, R.W. Field, P.K. Ten, Pervaporation-based hybrid process: a review of process design, applications and economics, J. Membr. Sci. 153 (2) (1999) 183−210.

[50] P. Lutze, A. Gorak, Reactive and membrane-assisted distillation: recent developments and perspective, Chem. Eng. Res. Des. 91 (10) (2013) 1978−1997.

[51] M. Skiborowski, A. Harwardt, W. Marquardt, Conceptual design of distillation-based hybrid separation processes, Annu. Rev. Chem. Biomol. Eng. 4 (2013) 45−68.

[52] T.M.M. Barakat, E. Sorensen, Simultaneous optimal synthesis, design and operation of batch and continuous hybrid separation processes, Chem. Eng. Res. Des. 86 (3) (2008) 279−298.

[53] A. Gorak, Z. Olujic (Eds.), Distillation − Equipment and Processes, Elsevier, 2014.

[54] H. Arellano-Garcia, I. Carmona, G. Wozny, A new operation mode for reactive batch distillation in middle-vessel columns: start-up and operation, Comput. Chem. Eng. 32 (1−2) (2008) 161−169.

[55] L. Rayleigh, On the distillation of binary mixtures, Philos. Mag. Ser. 6 4 (23) (1902) 521−537.

[56] C.E. Huckaba, D.E. Danly, Calculation procedures for binary batch rectification, AIChE J. 6 (2) (1960) 335−342.

[57] E.L. Meadows, Multicomponent batch distillation calculations on a digital computer, Chem. Eng. Prog. Symp. Ser. 59 (1963) 48−55.

[58] G.P. Distefano, Mathematical modelling and numerical integration of multicomponent batch distillation equations, AIChE J. 14 (1) (1968) 190−199.

[59] M. Leipold, S. Gruetzmann, G. Fieg, An evolutionary approach for multi-objective dynamic optimization applied to middle vessel batch distillation, Comput. Chem. Eng. 33 (4) (2009) 857–870.

[60] K.H. Low, E. Sorensen, Simultaneous optimal design and operation of multipurpose batch distillation columns, Chem. Eng. Process. 43 (3) (2004) 273–289.

[61] U. Diwekar, Batch Distillation. Simulation, Optimal Design and Control, second ed., CRC Press, Boca Raton, 2012.

[62] K.H. Low, E. Sorensen, Simultaneous optimal configuration, design and operation of batch distillation, AIChE J. 51 (6) (2005) 1700–1713.

[63] S. Mukherjee, R.K. Dahule, S.S. Tambe, D.D. Ravetkar, B.D. Kulkarni, Consider genetic algorithms to optimize batch distillation, Hydrocarbon Process. 80 (9) (2001) 59–66.

[64] M. Hanke, P. Li, Simulated annealing for the optimization of batch distillation processes, Comput. Chem. Eng. 24 (1) (2000) 1–8.

[65] M.M. Miladi, I.M. Mujtaba, Optimisation of design and operation policies of binary batch distillation with fixed product demand, Comput. Chem. Eng. 28 (11) (2004) 2377–2390.

[66] E. Sorensen, T. Barakat, E.S. Fraga, Multi-objective optimisation of batch separation processes, Chem. Eng. Process. 47 (12) (2008) 2303–2314.

Energy Considerations in Distillation

Megan Jobson

School of Chemical Engineering and Analytical Science,
The University of Manchester, Manchester, UK

CHAPTER OUTLINE

Distillation: Fundamentals and Principles. http://dx.doi.org/10.1016/B978-0-12-386547-2.00006-5

6.1 Introduction to energy efficiency

Distillation processes involve mass transfer between a liquid phase (or two liquid phases) and a vapor phase flowing in countercurrent fashion. The vapor and liquid phases are generated by vaporization of a liquid stream and condensing a vapor stream, which in turn requires heating and cooling. Distillation is thus a major user of energy in the process industries and globally. The sudden rise in crude oil prices in the 1970s, the general price increases since the start of the twenty-first century [1] and growing concerns about the environmental impact of using energy, in particular increasing emissions of carbon dioxide [2], drive efforts to increase the energy efficiency of this inherently energy-intensive process.

The underpinning physical and thermodynamic phenomena of distillation result in its energy intensity. Distillation design and operation need to be based on a clear understanding of these phenomena, "energy efficiency" and the process economics of distillation processes and the associated energy-supply processes.

This chapter first introduces the concept of energy efficiency in the context of distillation in the process industries. The influence of design and operation of individual, conventional, simple, continuous distillation columns on energy efficiency is explored and the role of heat recovery for enhancing energy efficiency is discussed. Advanced and complex distillation configurations that exploit opportunities to reduce the energy requirements or increase heat recovery are presented. A short concluding section provides a summary and notes areas of ongoing research.

6.1.1 Energy efficiency: technical issues

It is significant that energy may take various forms, including heat (or cooling) and work, in the form of mechanical or electrical power. This section provides an overview of energy supply technologies applied in the process industries; more detail may be found in various sources (e.g. [3,4]). The process economics of energy efficiency is discussed in Section 6.1.2.

6.1.1.1 Heating

In the process industries, process heating requirements may be met efficiently and relatively safely using steam, although electrical heating is also applied, especially at smaller scales where steam generation is impractical. Heat released by the steam on condensing (dominated by the latent heat of condensation of the steam, rather than its sensible heat), is transferred to the process stream being heated. On condensing, the steam needs to be hotter than the stream being heated, by the second law of thermodynamics. There is an upper limit to the temperature at which the steam can condense, since the critical temperature of steam is 374 °C. In practice, steam is used in the process industries at pressures of up to about 4000 kPa, corresponding to a condensing temperature of 250 °C [5].

Heat is typically generated on a process site at various temperatures in a site utility system, as illustrated in Figure 6.1, where fuels are combusted to generate steam at a very high pressure (e.g. 11,000 kPa) with a relatively low heat of vaporization. This very high pressure (VHP) steam then has its pressure reduced to two or more lower pressure levels. While pressure-reducing valves may be used to "let down" the steam pressure, steam turbines facilitate generation of mechanical and/or electrical power as a by-product. On smaller sites, steam may be generated at a high pressure (around 4000 kPa) rather than a very high pressure [3].

The steam is then distributed across the process site at various pressures (e.g. as high-pressure (HP) steam, medium-pressure (MP) steam at around 2000 kPa and low-pressure (LP) steam at around 400 kPa), with corresponding condensing temperatures and additional power generation [3]. While the condensing temperature decreases as the pressure decreases, its latent heat of vaporization increases, as does power generation. It is therefore beneficial to use steam for heating at the lowest pressure available that allows effective heat transfer—a minimum temperature difference between the process stream and condensed steam of 20−40 °C is typical.

At higher temperatures, it is common practice to use fired heaters, as shown in Figure 6.2, where a fuel (gas, liquid, or solid) is combusted and the flue gases of the

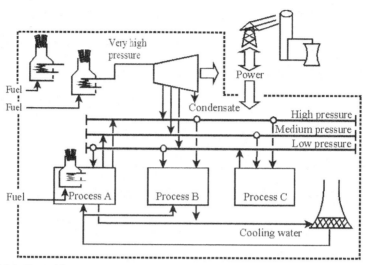

FIGURE 6.1 Site Utility System

Reprinted from Klemes et al., Targeting and design methodology for reduction of fuel, power and CO$_2$ on total sites, Applied Thermal Engineering 17 (8–10) 993–1003, Copyright 1997, with permission from Elsevier.

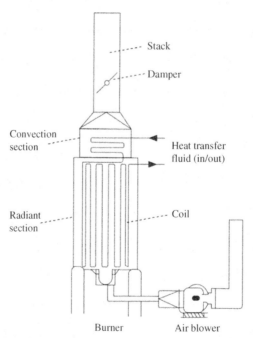

FIGURE 6.2 Fired Heater (or Furnace)

Reprinted from Wikipedia under the terms of the GNU Free Documentation License. http://en.wikipedia.org/wiki/Image:Furnace2.png.

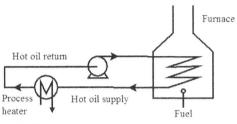

FIGURE 6.3 Hot Oil Circuit [3]

Reprinted from Chemical Process Design and Integration, Copyright 2005, with permission from John Wiley and Sons.

combustion process provide heat to the process by radiant and convective heat transfer. Typically, modern furnaces transfer 80–90% of the heat of combustion of the fuel to the process stream [5,6], with temperatures within the radiant section of the furnace of 700–900 °C [3]. The flue gas, which contains mainly nitrogen, carbon dioxide and water, undergoes sensible cooling to its dew point, typically 150–160 °C, where corrosive liquids form that can damage the chimney stack. In the case of natural gas, temperatures below 100 °C can be achieved, allowing latent heat of the water to be recovered [3]. The temperature to which the flue gas is cooled represents a limitation to the amount of heat that can be recovered from the stack gas.

Heat transfer fluids are also used to provide heat at higher temperatures, e.g. 350 °C; these fluids in turn require heating by a high-temperature heat source such as a fired heater, as shown in Figure 6.3. Using heat transfer fluids, pumped to a location some distance away from the fired heater, can bring safety or layout benefits. However, the need for intermediate heat exchange (i.e. flue gas–hot oil, hot oil–process stream) increases equipment costs and decreases the temperature to which the process stream may be heated.

At temperatures below the condensation temperature of low-pressure steam (130–150 °C), heat may be provided to a process by other utilities, such as hot water.

6.1.1.2 Cooling—above ambient temperatures

When cooling is required in a process plant above the ambient temperature, it is usually convenient to use ambient air or cooling water as the cold utility. The process stream may be cooled using an air-cooled heat exchanger, such as that shown in Figure 6.4, to a temperature about 20 °C hotter than the ambient air temperature; to achieve a smaller temperature difference than this would require significantly more heat transfer area. Fans or blowers increase the effectiveness of the coolers but consume power.

When available in an industrial site, cooling water is the most common cooling medium. The cooling water system, illustrated in Figure 6.5, uses cooling towers, in which evaporation of the water takes place, causing the temperature of the

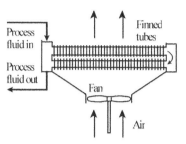

FIGURE 6.4 Forced-Draft Air-Cooled Heat Exchanger

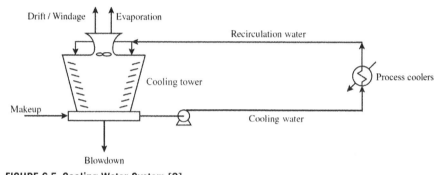

FIGURE 6.5 Cooling Water System [3]

Reprinted from Chemical Process Design and Integration, Copyright 2005, with permission from John Wiley and Sons.

remaining water to fall. The temperature of the cooling water is in practice higher than the wet-bulb temperature of the ambient air. The process stream can be cooled to around 10 °C warmer than the cold cooling water. The overall heat transfer coefficient of water cooling in a shell and tube heat exchanger is typically 500–900 W m^{-2} K^{-1}, which is significantly higher than that of air cooling (typically around 100 W m^{-2} K^{-1}) [5]; therefore, a lower temperature difference is achievable using cooling water, without requiring an excessively large heat exchanger.

6.1.1.3 Cooling—below ambient temperatures

Refrigeration is needed to cool a process stream to temperatures below ambient. In refrigeration systems, a refrigerant fluid acts as a heat sink (i.e. accepts heat) at a low temperature and rejects the heat at a higher temperature. Moving the heat from a lower to a higher temperature requires work to be done on the refrigerant.

The most common way of doing work on the refrigerant is to compress it when it is in the vapor or gas state, as illustrated in Figure 6.6. The refrigerant, a vapor–liquid mixture at a low pressure (point 1 on Figure 6.6), acts as a low-temperature heat sink. The temperature of the refrigerant at this point is lower than that of the

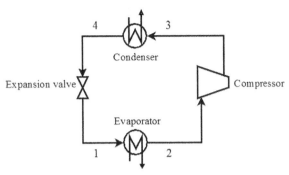

FIGURE 6.6 Refrigeration Cycle

process stream requiring cooling; the temperature difference is usually in the range 1–5 °C. On absorbing heat from the process requiring cooling, the fluid is evaporated (point 2). This vapor is at its saturation temperature at a low pressure; when compressed, its temperature rises (point 3). The high-pressure refrigerant can be condensed using a relatively warm coolant, e.g. ambient air or cooling water, forming a liquid (point 4). The pressure of the refrigerant is then reduced through an expansion valve or an expander, generating a cold, low-pressure, two-phase mixture (point 1) that can take up heat at a low temperature.

The main drawback of compression refrigeration is that raising the pressure of the fluid requires a compressor, which is a relatively expensive equipment item and mechanical or electrical power to drive the compressor. Low-molecular weight hydrocarbons, hydrofluorocarbons (HFCs), ammonia and nitrogen are commonly used as the refrigerant; since 1987, chlorofluorocarbons (CFCs) and hydrochlorofluorocarbons (HCFCs) are being phased out [7,8].

Single cycles, such as that shown in Figure 6.6, can cool process streams to around −40 °C. In so-called "cascaded refrigeration cycles", temperatures as low as −165 °C can be reached in a refrigeration cycle that rejects heat to one or more other refrigeration cycles.

An alternative to using compression to raise the pressure of the refrigerant fluid is to use an absorption–desorption process, as illustrated in Figure 6.7. As before, a two-phase refrigerant at a low pressure is evaporated by absorbing heat from the process stream requiring cooling. The vapor-phase refrigerant is then absorbed into a liquid absorbent, the pressure of which may be raised simply and inexpensively using a pump. The high-pressure mixture of the refrigerant and absorbent is separated in a desorption unit operating at a suitably high pressure. Desorption requires heat for vapor generation; cooling may also be required. The absorbent is then condensed, at the high pressure; once its pressure has been reduced, using an expansion valve, it can again act as a low-temperature heat sink.

While an absorption refrigeration cycle avoids using vapor compression, it requires heat as well as the absorption-desorption equipment. Absorption refrigeration is particularly attractive when low-grade heat (above 95 °C) is available and

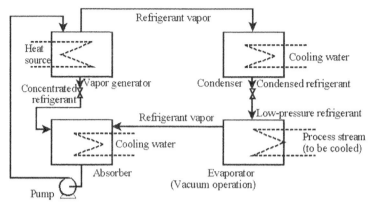

FIGURE 6.7 Absorption Refrigeration Cycle [8]

Adapted and reprinted from Applied Process Design for Chemical & Petrochemical Plants, Copyright 2001, with permission from Elsevier.

inexpensive [3]. The solutions applied in these schemes are usually water based (common refrigerant–absorbent pairs are ammonia–water and water–lithium bromide) [9]; the lowest temperature reached by the refrigerant in these cycles is limited to around −40 °C [3,8].

6.1.1.4 Mechanical or electrical power

It is important to consider energy in the form of work because of the need for refrigerated cooling in distillation processes as well as compression of distillation feeds and products. The steam system may provide power in the form of mechanical work generated by steam turbines, as illustrated in Figure 6.8, or generated using gas turbines. Alternatively, electricity may be imported from off-site generators.

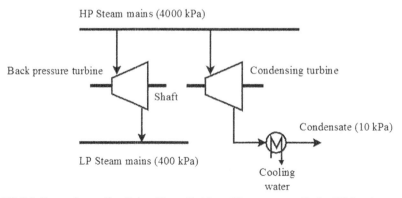

FIGURE 6.8 Power Generation Using Steam Turbines (Pressures are Typical Values)

6.1.1.5 Summary—technical aspects of energy efficiency

Energy may be employed in various forms. When considering energy efficiency in distillation it is important to account for the form of the energy being consumed and the quality of the heating and cooling required. Trade-offs exist between the energy sources, in terms of equipment requirements, cost and environmental impact.

6.1.2 Energy efficiency: process economics

Energy required for distillation processes—heat, cooling and power—require combustion of fuels, in fired heaters, for heat and power generation, as well as water or air for cooling. Energy-efficient distillation can reduce the demand for these utilities, bringing a range of benefits, in terms of the associated greenhouse gas emissions and use of water for process cooling, the size and cost of various equipment items and process operating costs. The main benefit is the reduction of operating costs, i.e. costs of fuel, power, electricity and water. A site may export electricity, thereby generating revenue.

This section considers the operating costs of various forms of energy for distillation. Investment and other costs will not be discussed, except briefly in Section 6.6. It is widely recognized that operating costs—in particular, costs of utilities for heating and cooling—dominate distillation process economics. There are also important trade-offs between operating and capital costs.

6.1.2.1 Heating

A site steam system can provide heat at several temperatures. Given that the "path" or process by which the steam is generated depends on the pressure at which it is provided to the process, its associated cost does too. A useful way of evaluating the relative cost of different steam levels is to use "marginal costing" [3], where the value of any power generated offsets the cost of generating the steam using a given fuel. The unit cost of steam is case specific as it depends on the steam conditions and flows at each level, the efficiency of the steam turbines, the type and price of fuel, the cost of importing electricity rather than generating it, etc. Figure 6.9 provides an illustration, Table 6.1 summarizes the energy flows and Table 6.2 shows that the higher the temperature at which heat is required, the higher the cost, where conventional steam tables have been used to determine steam properties such as enthalpies and entropy. (A drawback of this method for costing steam is that in some cases the value of power generated can be so high that the cost of the steam at the lowest pressure becomes negative, which could perversely suggest that using more steam is more energy efficient [3].)

Estimating the operating cost of fired heating is relatively straightforward. The type of fuel and its cost and energy content (heat of combustion), as well as the efficiency of the furnace, determine the cost per unit of heat provided to the process. For a fired heater burning natural gas costing £6.3 GJ^{-1} with 80% efficiency, the operating costs would be £7.9 GJ^{-1} of high-temperature heat.

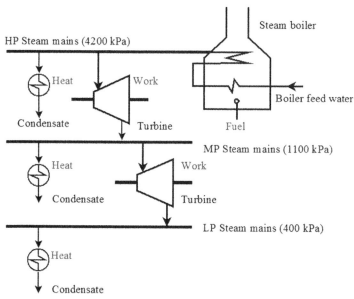

FIGURE 6.9 Marginal Steam Costing: Illustrative Example

Assumptions: Overall efficiency of boiler and steam distribution system: 80%; isentropic efficiency of steam turbines: 80%; cost of fuel (natural gas): £6.9 GJ^{-1}; value of power generated: £22.2 GJ^{-1}; steam condenses completely at constant pressure.

Table 6.1 Steam Properties for Illustrative Example of Figure 6.9

Stream	Temperature (°C)	Condensate Temp. (°C)	Pressure (kPa)	Enthalpy (kJ kg^{-1})	Entropy (kJ kg^{-1} K^{-1})	Enthalpy of Condensate (kJ kg^{-1})
Boiler feed water	100		101	419	1.303	–
HP steam	400	253	4200	3210	6.742	1101
MP steam	250	184	1100	2939	6.876	781
LP steam	159	144	400	2772	6.975	605

The operating cost of heat delivered by a heat transfer fluid is based on the cost of the heat from the furnace, taking into account heat losses of the heat distribution system. The operating cost of electrical heating, similarly, takes into account the cost or value of the electricity consumed and the efficiency of the heating system.

Table 6.2 Summary of Energy Flows (per kg of Steam) and Costs (£ per GJ of Heating) for the Illustrative Example of Figure 6.9

Steam Generated	Fuel Consumed (kJ kg^{-1})	Net Power Generated (kJ kg^{-1})	Heat Provided (kJ kg^{-1})	Net Cost of Heat (£ GJ^{-1})
HP steam	3488	–	2108	11.5
MP steam		271	2158	8.4
LP steam		167	2167	6.7

6.1.2.2 Cooling

When air is used as the cooling medium, the dominant operating cost is that of the power consumed by the fans causing a draft through the air-cooled heat exchanger. Typically, the fan power demand is 0.5–1.5% of the cooling duty [10]. The power requirements for a given cooling duty may be estimated [5]. The lower rates of heat transfer to air than to water mean that higher minimum temperature approaches are needed, around 20 °C [10].

The operating cost of cooling using cooling water depends strongly on the operating context. Rules of thumb may be used: the operating cost per unit of cooling is typically 1% of the cost of power per unit of energy [3], or 3–5% of the unit cost of heating using low-pressure steam. In the illustrative example presented in Figure 6.9 and Table 6.2, the unit cost of cooling using water would be around £0.3 GJ^{-1} cooling.

When subambient cooling is required, compression refrigeration is most commonly applied. Compressors are relatively expensive units and also use a power, a relatively expensive form of energy. Therefore, it is often preferable to avoid using refrigeration if another technical solution is feasible. Compressor costs can be estimated using standard approaches for estimating capital costs (e.g. Ref. [5,11]).

The unit cost of refrigeration depends strongly on both the temperature at which refrigerant is evaporated (point 2 in Figure 6.6) and the temperature of the liquid leaving the refrigerant condenser (point 4). A simple approach is to estimate the power demand of an ideal refrigeration cycle, i.e. a reverse Carnot engine and to apply a factor to consider the inefficiencies that apply to a real refrigeration cycle.

The "Carnot model" (Eqn (6.1)) relates the ideal compression power demand, W_{ideal}, to raise the refrigerant from its evaporation temperature, T_{evap}, to its condensing temperature, T_{cond}, to satisfy the process cooling duty, $\dot{Q}_{cool}$:

$$W_{ideal} = \dot{Q}_{cool}\frac{T_{cond} - T_{evap}}{T_{evap}} \tag{6.1}$$

A factor representing the ratio of ideal to actual power demand of 0.6 is typically used for moderate evaporation temperatures, e.g. down to −40 °C, to estimate the actual compression work [3,7]; this factor accounts for mechanical and thermodynamic inefficiencies in the refrigeration cycle. The condensing and evaporation

Table 6.3 Unit Operating Costs of Refrigeration—Illustrative Example[a]

Process Stream Temperature (°C)	Ideal Work (kJ kJ^{-1} Cooling)	Actual Work (kJ kJ^{-1} Cooling)	Cost of Cooling (£ kJ^{-1} Cooling)
−40	0.35	0.59	13.0
−20	0.24	0.40	9.0
0	0.15	0.25	5.5

[a] *Assumptions: Refrigerant evaporation temperature is 5 °C less than process cooling temperature; refrigerant is condensed at 35 °C; ideal work is 0.6 of actual work; cost of electricity is £22.2 GJ^{-1} (£0.08 kW^{-1}h^{-1}).*

temperature of the refrigerant are dictated by the temperature of the heat sink, which accepts heat from the refrigerant condenser and the temperature of the process stream requiring cooling, respectively. Subambient heat exchangers are designed to allow a minimum temperature approach of 1−5 °C. A low temperature approach increases the heat transfer area required, but crucially increases the evaporation temperature and reduces the temperature difference $T_{cond}-T_{evap}$, which reduces the power demand of the refrigeration cycle. If a low-temperature heat sink is available to accept heat from the refrigerant condenser, this can reduce the compression work significantly. Table 6.3 illustrates unit operating costs for various operating conditions in a refrigeration cycle and rejecting heat to cooling water. Figure 6.10 summarizes illustrative utility costs at various temperature levels.

The compressor power demand and the condenser cooling duty may be estimated more accurately using a relatively rigorous model of the refrigeration cycle, taking into account the thermodynamic properties of the refrigerant fluid, the compressor type and efficiency and the pressure drop within the heat exchangers and pipework of the refrigeration cycle, etc. [3].

The cost of absorption refrigeration systems requires the absorption−desorption process to be designed and modeled to predict its energy requirements and to allow

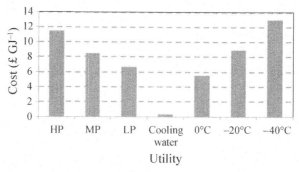

FIGURE 6.10 Unit Costs of Heating and Cooling Utilities Depend on Their Temperatures

Illustrative costs from Tables 6.2 and 6.3.

sizing of the associated equipment. The costs of this technology are not discussed further in this chapter; more details are provided by Ludwig [8].

6.1.2.3 Summary—process economics and energy efficiency

It is clear from Figure 6.10 that the form, quantity and quality (i.e. temperature) of energy required by distillation processes significantly affect distillation operating costs. Therefore, for energy-efficient distillation, the designer or operator should seek technical solutions taking into account the costs of utilities.

6.1.3 Energy efficiency: sustainable industrial development

Sustainable industrial development means that products that contribute to economic activity, satisfy human needs and support or enhance quality of life need to be produced with reduced consumption of materials and with reduced environmental damage [12]. Achieving this aim needs increased efficiency in the use of materials (including feedstocks, water and materials of construction) and of energy. The products being produced as a part of economic activity also need to add to quality of life. Their production should minimize the impact on the environment, in terms of consumption of energy, materials and water and of emissions of carbon dioxide and other damaging substances to the air, including substances that potentially deplete the ozone layer. A distinction must be made between energy derived from fossil fuels and other finite resources and that obtained from renewable sources.

Energy-efficient distillation processes play an important role in sustainable industrial development: decreasing demand for energy, whether in the form of heat, cooling, or work, can effectively reduce consumption of materials, energy and cooling water and decrease carbon dioxide and other atmospheric emissions. When energy efficiency reduces production costs, it also increases the economic sustainability of industrial activity.

A holistic consideration of the environmental impact of distillation processes should take into account the process life cycle [12], considering the extraction and consumption of fuels, extraction of materials of construction of the processing equipment and the final use of the products of the process. In a "cradle-to-grave" analysis, the use of resources and the emissions and wastes for the life cycle of a product consider the materials—their extraction, processing, transport, use (and reuse or recycling) and disposal—for the product and the equipment used to produce and process the product [12]. Note that the results of life cycle analyses depend significantly on where the boundaries are drawn for the analysis.

6.2 Energy-efficient distillation

Distillation columns are applied in the process industries in order to carry out separations for various reasons, including removal of contaminants that have a negative impact on downstream processes or are harmful or toxic and recovery of material that would otherwise be wasted. Distillation is a versatile process that allows a wide range of flow rates, of mixtures, of mixture compositions, etc. to be separated into products of any degree of purity or with any degree of recovery (in the absence of azeotropes). The degree of separation is generally dictated by downstream requirements, e.g. the effect of impurities on a catalyst, or environmental regulation related to transport fuels, but the sharpness of the separation is sometimes a degree of freedom in design or operation, e.g. the recovery of a reactant for recycle.

When considering the energy efficiency of distillation processes, it is important that like-for-like comparisons are made between options. In the following discussion, it is assumed that the separation, specified in terms of product purity or recovery (with respect to a single component or a group of components), is fixed.

6.2.1 Energy-efficient distillation: conceptual design of simple columns

A "simple" distillation column is defined as one in which a single feed is separated into two products, where the column has a single reboiler and a single condenser. Conceptual design of the distillation column means selecting the operating conditions and design parameters that will ensure that the column can carry out the specified separation.

A simple distillation column requires heat in the reboiler, to create a vapor stream that flows upwards through the column and cooling in the condenser, to provide liquid reflux in the column. The column operates over a pressure range, where the pressure at the bottom of the column is greater than that at the top of column; the pressure drop is a result of friction in the column and the hydraulic head of liquid within column. The pressure drop is typically around 0.3−1.0 kPa per distillation stage, depending on the type of internals [13−15]. In the following discussion, uniform pressure in the column will be assumed in order to simplify analysis.

Heat must be provided to the reboiler at the bottom of the column, where the mixture being separated has a relatively high boiling point, being enriched in less volatile components. The condenser, at the top of the column, rejects heat at a lower temperature, as the mixture being condensed is concentrated in more volatile components. Unfortunately, because of this temperature difference, heat rejected by the condenser cannot be reused by the reboiler.

6.2.1.1 Degrees of freedom in design

This section considers the design of a simple distillation column to separate a given feed into products that achieve given separation requirements (e.g. product purity or component recovery to a product). The feed flow rate and composition are taken to be fully defined. There are several degrees of freedom in the design of the column that can be manipulated to promote energy efficiency. The feed temperature and pressure can be manipulated. The column operating pressure must be set (and the associated pressure drop must be estimated). The number of theoretical stages, feed stage location and the type of condenser (to produce an overhead product

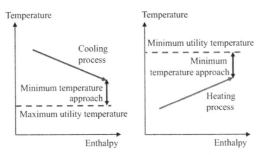

FIGURE 6.11 Utility Temperatures are Set by Heating and Cooling Temperatures

that is in the liquid phase, vapor phase, or both) must be chosen. All of these design decisions impact on the heating and cooling requirements of the column and the associated temperatures.

Heating and cooling in distillation processes often takes place over a range of temperatures. The following discussion makes the conservative assumption that a hot utility or heat source must be hot enough, or a cold utility or heat sink must be cold enough, to provide all of the heating or cooling, as illustrated in Figure 6.11. Opportunities to use more than one heat source or sink, or heating or cooling utilities that operate over a range of temperatures (e.g. hot oil, mixed refrigerants), are not considered in the following discussions.

Table 6.4 presents data relevant to an industrial distillation process separating *iso*-butane from *n*-butane [16] that will be used to illustrate how column design affects heating and cooling requirements.

Table 6.4 Summary of Illustrative Distillation Process[a]

Feed		Column Design	
Flow rate	26,122 kg h^{-1}	Number of stages	76
Pressure	650 kPa	Feed stage (from top)	36
Temperature	55.4 °C	Top pressure	650 kPa
Feed Composition (by Mass)		Bottom pressure	650 kPa
Propane	1.54%	**Specifications**	
iso-Butane	29.5%	*iso*-Butane in distillate	93.5 wt%
n-Butane	67.7%	*n*-Butane in bottom product	98.1 wt%
iso-Butene	0.13%	**Products**	
1-Butene	0.20%	Distillate	8114 kg h^{-1}
neo-pentane	0.11%	Bottom product	18,008 kg h^{-1}
iso-Pentane	0.77%		
n-Pentane	0.08%		

[a] Adapted from on an industrial example [16]; Peng-Robinson equation of state is used to model the column in Aspen HYSYS v. 7.3.

6.2.1.2 Column operating pressure

In the discussion that follows, the influence of the operating pressure on the distillation performance will be considered; the pressure drop is assumed to be negligible. The column operating pressure is the most important distillation design parameter; it affects the temperatures at which heating and cooling are required, the type of heating and cooling utility required, as well as heating and cooling duties. Furthermore, it impacts on the number of theoretical stages needed for the separation and the diameter of the column.

Firstly, the distillation operating pressure affects the temperatures of heating and cooling. The bubble point and dew point temperatures of a mixture of a given composition depend strongly on pressure. For mixtures in which the vapor phase behaves as an ideal gas and the liquid phase behaves as an ideal liquid solution [14], Raoult's law can be used, together with a suitable vapor-pressure model, to predict the pressure–temperature relationships. For less ideally behaved mixtures, other fluid property models, including cubic equations of state and activity coefficient models, are needed.

Higher operating pressures increase heating and cooling temperatures, as illustrated in Figure 6.12(a) for the case presented in Table 6.4. In general, these higher temperatures may mean that a hotter heat source is needed in the reboiler (e.g. MP steam

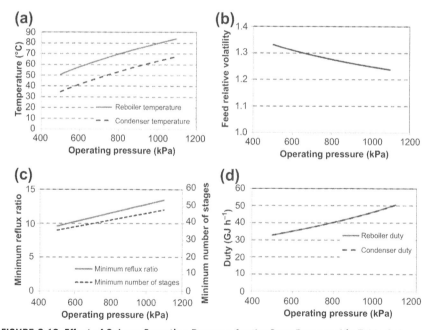

FIGURE 6.12 Effect of Column Operating Pressure for the Case Presented in Table 6.4

(a) Reboiler and condenser temperatures; (b) Relative volatility; (c) Minimum reflux ratio and minimum number of stages; (d) Reboiler and condenser duties.

rather than LP steam) and heat is rejected at a higher temperature in the condenser (e.g. more moderate refrigeration, or cooling water rather than refrigerated cooling).

Secondly, as illustrated in Figure 6.12(b), as temperatures increase with increased pressure, the components in the mixture typically become more similar in volatility [17] and their separation becomes more difficult. With this decrease in the driving force for the separation (the difference in volatilities of the components being separated), the distillation column will need more reflux and more theoretical stages to compensate; Figure 6.12(c) indicates that the minimum reflux ratio and minimum number of stages both increase with pressure.

With a higher operating pressure, the capital cost of the column will increase. A taller column will be needed, to accommodate a greater number of theoretical stages and a thicker shell may be needed to withstand the pressure. A wider column may or may not be needed, as the diameter of the column is dictated mainly by volumetric flow of the vapor; although the molar flow of vapor increases with pressure, so does its density.

The increase in reflux resulting from the pressure increase is likely to increase the reboiler and condenser duties (although this may be offset by the decrease in the latent heat of vaporization as the pressure increases); Figure 6.12(d) provides an illustration. The impact on operating costs depends on both the duty and the utility—if, at the higher temperature, steam at a higher temperature were needed, the cost per unit of heat would increase. However, if the higher temperature raised the temperature at which refrigeration were required, or allowed refrigeration to be avoided completely, the cost per unit of cooling could decrease quite significantly.

In general, therefore, the combination of these effects suggests that it is best to operate at as low a pressure as is practical and which avoids the use of refrigeration. While operating under vacuum makes the separation easier, the low vapor density requires a greater cross-sectional area in the column to accommodate the high volumetric flow of material. More importantly, operating under vacuum increases the complexity of the flowsheet, requiring additional, costly equipment to draw the vacuum and recover material drawn into the vacuum system. Vacuum also introduces safety hazards, related to risk of air ingress and resulting fire or explosion of flammable process fluids.

The logical conclusion is that it is best to operate at atmospheric pressure unless there are good reasons not to [3]. In particular, if:

1. increasing the operating pressure allows refrigeration to be avoided or facilitates the use of refrigeration at more moderate conditions;
2. operating under a vacuum avoids degradation of thermally sensitive materials because of the lower temperatures in the column;
3. changing the pressure (up or down) creates an opportunity for heat recovery within the wider process (see Section 6.4);
4. the cost of compressing the feed products (to meet downstream specifications) outweighs the benefits of increased operating pressure.

In the design of distillation columns, it is important that the operating pressure is considered first, as the other degrees of freedom for design are all strongly influenced by the operating pressure and as the operating pressure influences most strongly which utilities will be required.

6.2.1.3 Pressure drop

Distillation benefits from low operating pressures in terms of heating and cooling duties; the corollary is that high pressure drop in a column is detrimental to energy efficiency, as part of the column will operate at pressures well above the minimum pressure in the column, where volatilities are lower and more reflux is required to compensate. Pressure drops increase column heating and cooling duties, especially in columns operating at atmospheric or subatmospheric pressures [18]. Higher pressure drop is particularly detrimental in vacuum distillation, as the energy required to draw the vacuum is wasted in a context where the separation is particularly sensitive to volatilities. In these cases, additional separation stages may increase, rather than decrease energy requirements, because of the increased pressure drop [18]. Especially in columns operating under vacuum, it is beneficial to use distillation internals and auxiliary equipment (e.g. trays or packing, gas and liquid distributors, mist eliminators, reboiler, condenser) with low pressure drop.

In the example presented in Table 6.4, a pressure drop of 0.5 kPa per stage increases the reboiler duty by up to 2%. An additional pressure drop of 50 kPa each in the condenser and reboiler increases this penalty by a further 3%.

6.2.1.4 Number of theoretical stages

The number of theoretical stages required to carry out a specified separation in a simple distillation column (at a given operating pressure) is not unique. Some minimum number of stages is required to achieve the separation [19]. For different numbers of stages, different reflux ratios will be required, corresponding to different reboiler and condenser duties, as illustrated in Figure 6.13(a). A key design decision is to select the number of stages in the column. There is a trade-off between operating costs and capital investment—increasing the number of stages would increase

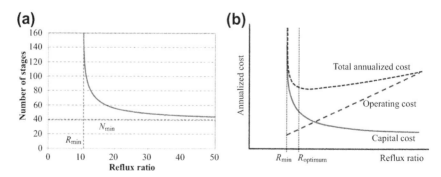

FIGURE 6.13 Number of Theoretical Stages

(a) Number of stages vs. reflux ratio for the example presented in Table 6.4; (b) Cost trade-offs with reflux ratio.

the height of the column and therefore its cost, but would decrease reflux requirements and hence reduce duties and operating costs, as well as costs of heat transfer equipment.

A heuristic rule, based on a wide range of mixtures, separation specifications, operating pressures, etc., is that the reflux ratio should be in excess of the minimum reflux ratio by approximately 10%, as illustrated in Figure 6.13(b). In difficult separations, where high reflux ratios are required, or in cases that refrigeration is required, a lower excess reflux ratio may be applied. However, operating too close to the minimum reflux ratio can make the process highly sensitive to operational fluctuations or errors in the design calculations, e.g. estimated tray efficiencies or predicted phase equilibrium behavior [5]. If heating and cooling are inexpensive, e.g. where recovered heat is used in the reboiler, higher reflux ratios may be suitable.

6.2.1.5 Feed condition

The temperature (or "thermal condition") of the feed is another important degree of freedom. It is generally the case that a feed (at the column pressure) should enter the column as a saturated liquid [20] or at a temperature between its bubble point and dew point temperatures. A superheated feed or a subcooled feed will introduce thermodynamic inefficiencies, as it will need to be cooled or heated, respectively, to saturation conditions to participate in the separation processes within the column.

The thermal condition of the feed entering a distillation column impacts directly on the flow rate of vapor and of liquid within the column, as illustrated in Figure 6.14(a). In turn, the feed condition affects the amount of vaporization required in the reboiler and the amount of condensing required and thus the heating and cooling duties. The feed condition therefore presents an opportunity to manipulate the heating and cooling duties of the column, by using a feed preheater or precooler, as shown in Figure 6.14(b). The total heating and cooling duties tend to be least when the feed enters as a saturated liquid [20], so from the point of view of duty only, a saturated liquid feed is most energy efficient. It is often appropriate to assume that the total heating and cooling duties of the column remain unchanged when the feed condition changes. However, as shown in Figure 6.14(c), the total heating duty for a saturated vapor feed is around 16% more than that for a saturated liquid feed for the example of Table 6.4.

The hot utility suitable for heating the feed may be cooler and cheaper than that used in the reboiler, as the bubble point and dew point temperatures of the feed are usually lower than that in the reboiler. Using feed heating may thus decrease the total cost of providing heat (in both the reboiler and preheater). An analogous argument applies to precooling the feed. Whether preheating (or precooling) the feed reduces operating costs depends mainly on whether a cheaper, more energy-efficient heating medium can be employed. For the illustrative example, Figure 6.14(d) shows that the feed and reboiler temperatures are relatively similar in this near-binary separation—it is likely that heating the feed would actually increase total heating costs.

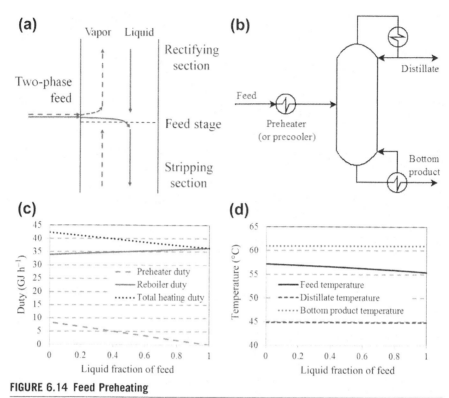

FIGURE 6.14 Feed Preheating

(a) A two-phase feed entering the column affects vapor and liquid flows in the column (feed and column pressure are assumed equal); (b) Preheater and reboiler both require heat; (c) Effect on total heating duty with feed heating (example in Table 6.4); (d) Effect of feed preheating on feed temperature.

6.2.1.6 Feed stage location

When a feed stream enters a distillation column, it mixes with the liquid and vapor streams at the feed stage and participates in the mass transfer processes on the feed stage. The stage onto which the feed is introduced is a degree of freedom in the design. If the feed composition or temperature are very different to those on the feed stage, the mixing of the feed with the material within the column disrupts the composition profile in the column. These "mixing effects" are thermodynamically inefficient and cause the heating and cooling duties of the column to increase [17]. On the other hand, if the composition and temperature of the feed and the feed stage are similar, then the feed scarcely influences the mass transfer taking place on the feed stage, which is more energy efficient.

The above argument is applicable to both single-phase and two-phase feeds; the compositions of both the liquid phase and the vapor phase of the feed should ideally match those on the feed stage [3,5]. For the illustrative example of Table 6.4, selecting a feed stage four stages above or below the best feed stage (35) increases the heating and cooling duty by around 1.5%.

6.2.1.7 Condenser type

The phase of the overhead product of a distillation column may be a design degree of freedom, if, for example, the phase of the distillation product is not specified. Figure 6.15 illustrates that the overhead product may be a liquid and/or a vapor. A "total condenser" produces a liquid product; all the overhead vapor is condensed to its bubble point, so the condenser duty is maximized and the product temperature is minimized. Where the cooling medium is expensive, as in refrigerated condensers, this may lead to high operating costs. Instead, a "partial condenser" (in which additional separation takes place [3]) can provide a vapor-phase product and liquid reflux; as less material is condensed, the condenser duty is lower and cooling is required to a more moderate temperature.

Especially where components are present that are difficult to condense at the column operating pressure, a "mixed condenser" may offer a compromise [3], i.e. providing a liquid product and liquid reflux but also producing a vapor product. If the overhead product is itself the feed to a distillation column, the energy demand of the downstream column will be affected by the type of condenser of the upstream column.

6.2.1.8 Summary—column design for energy efficiency

Key degrees of freedom for column design affect both the quantity and quality of heating and cooling required. The column operating pressure is significant, as it affects heating and cooling duties, as well as the temperatures involved. The number of stages in the column also determines duties; the trade-off between energy demand and capital investment is noteworthy. The feed condition, feed stage location and phase of the distillate product should also be selected considering energy efficiency. This section assumes that all heating and cooling is provided by utilities, such as steam, cooling water and refrigeration. Section 6.4 discusses how exploiting heat recovery opportunities can also enhance energy efficiency. Furthermore, Section 6.5 presents a range of complex column configurations that can improve energy efficiency.

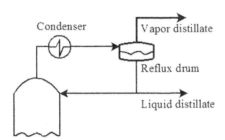

FIGURE 6.15 A Condenser May Produce Vapor-Phase and/or Liquid-Phase Products

It is sometimes assumed that there are strong trade-offs between energy efficiency and capital investment; however, energy-efficient designs may also have reduced capital costs, for example, if a more compact column, with a smaller diameter, is needed because of reduced vapor and liquid flows in the column.

6.3 Energy-efficient distillation: operation and control
6.3.1 Energy-efficient column operation

In practice, distillation columns designed for energy efficiency may not be operated efficiently. Plants need to be monitored, maintained and managed for energy efficiency and causes of inefficiency and solutions to overcome these, need to be identified [21]. A column operated inefficiently, compared to design conditions, may use utilities in excessive amounts or at excessively high (or low) temperatures, may not achieve expected throughputs, or may not meet product specifications, impacting on process economics.

Economic trade-offs exist between reflux, production rate and product purity in an operational distillation column. If a higher-purity product can be sold for a higher price that more than compensates for the additional energy required, it may make sense to use higher reflux in the column. However, higher reflux will increase energy demand and may also lead to flooding, restricting the production capacity of the column. Conservative operating policies, where reflux and product purities are higher than required, thus reduce energy efficiency and potentially production capacity, i.e. increase costs and limit product revenue [21].

A number of operational problems can reduce energy efficiency. Fouling or polymerization in the reboiler can reduce heat transfer rates, increasing steam demand or requiring steam at higher temperatures. Steam flows will also be excessive if some of the steam does not condense in process heaters; appropriate maintenance and operation of the steam system and steam traps can avoid this problem. A high pressure drop may indicate plugging or fouling of column internals or flooding in the column, with associated increases in heating and cooling duties [21]. If column internals are broken, damaged, or poorly installed, there may be effectively fewer theoretical stages in the column, so additional reflux will be needed to compensate. Poorly insulated equipment may have significant heat losses, increased utility demand. If the feed or product specifications change, compared to the design case, column operating conditions, e.g. feed temperature, reflux ratio, reflux temperature, may need to be adjusted accordingly to operate efficiently.

6.3.2 Process control

The control system for a distillation process manipulates operating parameters to meet process specifications. A column that is controlled inappropriately may operate inefficiently with respect to energy consumption and production rates. Further detail on the control of distillation processes is presented in a later chapter [22].

If control of the column is not tight but the set points are appropriate, then the product quality may, on average, meet specifications and the energy demand, on average, should be acceptable. In other words, deviations around the set point are of less importance than the set point itself [23]. Process simulation and optimization can identify the most appropriate set points for control variables to minimize the use of the most costly utilities, given the physical constraints of existing equipment. Control to avoid subcooling the reflux may reduce demand for both hot and cold utilities.

In winter months, it may be possible to reduce the column operating pressure to take advantage of colder ambient temperatures. The higher volatilities at the lower pressure should allow the reflux to decrease and thus reduce the condenser and reboiler duties as well as the feed preheat temperature—the control set points should all be adjusted accordingly.

6.3.3 Summary—operation and control

The energy efficiency of a well-designed distillation process can be promoted by effective management, maintenance, control and operation of the plant. How the column is operated affects the separation performed, production capacity and energy demand and in turn process economics and energy efficiency. Judicious choice of control set points, along with effective control strategies, can facilitate energy-efficient column operation.

6.4 Heat integration of distillation

A heat-integrated process is one in which heat is recovered from process streams requiring cooling and reused by process streams requiring heating. Heat is recovered using heat exchangers, which facilitate heat transfer from one process stream to another. Figure 6.16 illustrates a heat exchanger in which heat is transferred from a "hot" stream, requiring cooling, to a "cold" stream, requiring heating. The hot stream is also known as a heat source and the cold stream as a heat sink.

The recovery and reuse of heat reduces demand for heating and cooling utilities and thus increases the energy efficiency of a process; the improved energy efficiency

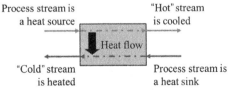

FIGURE 6.16 A Heat Exchanger Facilitates Heat Recovery from a "Hot" Stream to a "Cold" Stream

has a positive impact on operating costs, environmental impact and sustainability, as discussed in Section 6.1.

6.4.1 Heat exchange and heat recovery

For heat to flow from one stream to another in a heat exchanger, the heat source must be at a temperature that is hotter than the heat sink. The greater the temperature difference between the streams, the greater the driving force for heat transfer, so the lower the heat transfer area required. Equation (6.2) presents a simple model for heat exchangers that relates the heat transfer duty $\dot{Q}_{ht}$ to the logarithmic mean temperature difference (ΔT_{lm}), the overall heat transfer coefficient U and the heat transfer area A_{ht} [5]:

$$\dot{Q}_{ht} = U \cdot A_{ht} \cdot \Delta T_{lm} \tag{6.2}$$

The cost of the heat exchanger, which is highly correlated with heat transfer area, is lowest for large temperature differences. However, fewer opportunities for heat recovery will exist if a high minimum temperature approach in the exchanger is specified, which will in turn increase operating costs. In above-ambient processes, this trade-off typically leads to a minimum temperature approach of 10 °C or greater; in subambient processes, the high cost of refrigeration can justify the use of much lower temperature differences, of around 5 °C or even 1 °C [3].

A process may employ several heat exchangers to recover heat from one or more hot streams. Together, these exchangers form a heat exchanger network. The design of the network can be challenging, especially when several process streams requiring heating or cooling are present. Systematic approaches have been developed for design of heat exchanger networks that maximize heat recovery. One such approach, known as pinch analysis, which is based on thermodynamic analysis, is useful for identifying the maximum heat recovery that can be achieved, for a given minimum temperature approach [3–5]. Furthermore, the "pinch" in pinch analysis indicates a constraint for heat recovery: below the pinch temperature, it does not make sense for the process to act as a sink for heat available for recovery and above the pinch temperature it is not sensible for the process to reject heat.

In the context of energy-efficient distillation, it is pertinent to ask under which conditions it is beneficial for heat rejected by a column to be recovered for reuse and under which conditions it makes sense for recovered heat to be reused by a column. The pinch concept helps resolve this question.

For a given process (that is, including various streams requiring heating and cooling but excluding the distillation column), the pinch temperature can be determined for a set value of the minimum temperature approach [3,4]. Compared to the case where utilities supply all the heating and cooling of a distillation column, the total utility demand (of the column and overall process together) will not be reduced if the column rejects heat at a temperature below the pinch and acts as a heat sink at a temperature above the pinch. This condition may be quickly and conveniently applied to identify situations in which there is no benefit of recovering heat between the process and the column. In the converse situation, where the temperatures at which heat is rejected by the column and reused by the column are both above or both below the pinch, the total heating or cooling demand (of the column and process together) may be reduced. A decrease in total utility demand may not lead to a decrease in operating cost, as utility costs depend strongly on temperature. Note that heat may be rejected by the condenser or feed cooler, as discussed in Section 6.4.2 and the reboiler and feed heater are heat sinks. Section 6.5 considers heat integration of complex column configurations.

6.4.1.1 Summary—heat exchange and heat integration

Heat integration can greatly enhance the energy efficiency of distillation processes. For heat recovery, an available heat source needs to be hotter than a process heat sink; furthermore, pinch analysis recommends that heat recovery between a distillation column and an associated process is only implemented if the temperatures of all streams exchanging heat are either above or below the pinch temperature.

6.4.2 Distillation design for heat integration

Heat recovery can increase the energy efficiency of a distillation column of a given design (given operating conditions, number of theoretical stages, feed condition, etc.). Moreover, the design of the column could potentially be manipulated to enhance or create opportunities for heat integration. The key design degrees of freedom discussed in Section 6.2.1 impact on heating and cooling requirements (both quality and quantity) and therefore on opportunities for heat integration.

6.4.2.1 Operating pressure

As discussed in Section 6.2.1, increasing the column operating pressure raises the temperature in the condenser and reboiler but is also likely to increase their duties. If a feed heater or cooler is present and operates at an increased pressure, the temperatures of the feed heater or cooler will also increase. Changing the operating pressure may therefore present an opportunity to manipulate the temperatures of the column heat sinks and sources so they lie entirely above, or entirely below, the pinch temperature. In this case, it may be possible to satisfy column heating and cooling requirements using process heat sinks and sources and therefore reduce overall utility requirements.

In columns operating at subambient temperatures, in which refrigerated condensing is required, the operating pressure impacts significantly on the distillation condenser temperature and therefore on the refrigerant evaporation temperature. Furthermore, the column reboiler may provide a useful heat sink for the refrigeration cycle rejecting heat from its condenser. As Eqn (7.1) shows, the energy efficiency of a refrigeration cycle can be increased by reducing the temperature of the condenser (i.e. allowing heat to be rejected to a colder heat sink). A column reboiler may act as such a heat sink; the column operating pressure can sometimes be manipulated to increase the energy efficiency of a refrigeration cycle by reducing the compression power demand of the cycle. Tables 6.5, 6.6 and 6.7 summarize an illustrative example.

In this example, in which light hydrocarbons are distilled to separate ethene and methane, the column and refrigeration system performance are modeled using simple

Table 6.5 Summary of Light Hydrocarbons Distillation Process

Feed	
Flow rate	6000 kmol h^{-1}
Pressure	1000 kPa
Temperature	−70 °C
Vapor fraction	0.248
Feed composition (mole fraction)	
Methane	0.24
Ethene	0.46
Ethane	0.20
Propene	0.08
Propane	0.02
Specifications	
Methane recovery to distillate	99%
Ethene recovery to bottom product	96%

Table 6.6 Summary of Column Design for Distillation of Light Hydrocarbons[a]

	Column Operating Pressure	
Column Design Parameter	**1000 kPa**	**2000 kPa**
Ratio R/R_{min}	1.10	1.10
Minimum reflux ratio	0.607	0.750
Actual reflux ratio	0.668	0.825
Number of stages	10.9	13.4
Feed stage (from top)	3.1	3.8
Distillate flow rate	1543 kmol h^{-1}	1542 kmol h^{-1}
Bottom product flow rate	4457 kmol h^{-1}	4458 kmol h^{-1}
Feed pressure	1000 kPa	2000 kPa
Feed temperature	−70 °C	−52.4 °C
Condenser duty	12.4 GJ h^{-1}	12.2 GJ h^{-1}
Condenser temperature	−104.8 °C	−91.6 °C
Reboiler duty	16.8 GJ h^{-1}	24.1 GJ h^{-1}
Reboiler temperature	−43.4 °C	−18.3 °C

[a] *Peng-Robinson equation of state is used to model the column in Aspen HYSYS v. 7.3. The shortcut distillation model used to design and simulate the distillation column. The pressure drop within the column is assumed to be negligible.*

Table 6.7 Summary of Refrigeration Requirements for Light Hydrocarbons Separation[a,b]

Refrigeration Cycle[a]	Without Heat Integration		With Heat Integration	
Column Operating Pressure	1000 kPa	2000 kPa	1000 kPa	2000 kPa
Condenser duty	12.4 GJ h^{-1}	12.2 GJ h^{-1}	12.4 GJ h^{-1}	12.2 GJ h^{-1}
Condenser temperature	−104.8 °C	−91.6 °C	−104.8 °C	−91.6 °C
Reboiler duty	16.8 GJ h^{-1}	24.1 GJ h^{-1}	16.8 GJ h^{-1}	24.1 GJ h^{-1}
Reboiler temperature	−43.4 °C	−18.3 °C	−43.4 °C	−18.3 °C
Refrigerant evaporation temperature	163.4 K (−109.8 °C)	176.6 K (−96.6 °C)	163.4 K (−109.8 °C)	176.6 K (−96.6 °C)
Refrigerant cooling medium	Ambient	Ambient	Reboiler	Reboiler
Refrigerant condensing temperature	308 K (35 °C)	308 K (35 °C)	235 K (−38.4 °C)	260 K (−13.3 °C)
Ideal compression power demand	11.0 GJ h^{-1} (3.1 MW)	9.1 GJ h^{-1} (2.5 MW)	5.5 GJ h^{-1} (1.5 MW)	5.8 GJ h^{-1} (1.6 MW)
Actual compression power demand	18.4 GJ h^{-1} (5.1 MW)	15.2 GJ h^{-1} (4.2 MW)	9.1 GJ h^{-1} (2.5 MW)	9.6 GJ h^{-1} (2.7 MW)
Refrigerant condensing duty	30.8 GJ h^{-1}	27.4 GJ h^{-1}	21.5 GJ h^{-1}	21.8 GJ h^{-1}
Feed compression power demand[b]	–	2.0 GJ h^{-1} (0.6 MW)	–	2.0 GJ h^{-1} (0.6 MW)
Suitable refrigerants [3]	Methane + ethene + propene	Ethene + propene	Methane + ethene	Ethene

[a] Refrigerant power demand is estimated using the Carnot model, assuming that the ideal power demand is 60% of the actual power demand and the minimum temperature approach is 5 °C in the evaporator and condenser. In the "With heat integration" cases, heat from the refrigerant condenser is rejected to the column reboiler.
[b] A centrifugal compressor with an isentropic efficiency of 85% is used to compress the vapor fraction of the feed.

models—the short-cut distillation design models and the Carnot model. Refrigeration is required to condense the methane-rich overhead product. Table 6.6 shows how the condenser temperature is affected by operating pressure and Table 6.7 illustrates the effect on the compression power demand. In the case that there is no heat integration, the higher-pressure column has a lower power demand, considering both feed compression and the refrigeration system. Table 6.7 also illustrates the potential benefits of heat integration—rejecting heat from the refrigeration condenser to the column reboiler reduces the power demand of the refrigeration system. (Note, however, that for the column operating at 1000 kPa, the reboiler duty is smaller than the refrigeration condenser duty, so the proposed heat recovery scheme is only partially feasible.)

An additional benefit of using a higher pressure in the distillation column may be obtained if the difference in temperature between the evaporator and condenser temperature of the cycle is small enough to avoid using cascaded refrigeration cycles that are required to overcome the limitations in the operating range of refrigerants [3]. In the example presented in Table 6.7, it would be feasible to use a single refrigerant, ethane, when heat integration is implemented for the column operating at 2000 kPa, while a cascade of two or three refrigeration cycles would be required if the refrigerant were condensed using an ambient cooling medium.

6.4.2.2 Number of theoretical stages

The reflux ratio required to carry out a given separation at a given operating pressure depends on the number of theoretical stages present. When heating and cooling are provided by process utilities, the trade-off between operating and capital costs suggests a reflux ratio of 10% greater than the minimum reflux ratio. In a heat-integrated column, where heating and cooling are provided by recovered heat, the operating costs per unit of heating or cooling are likely to be much lower than for utilities or even negligible. Therefore, a very different trade-off—between the column capital cost, the cost of heat recovery equipment and column operating costs—is likely to exist and the most appropriate number of theoretical stages (and thus reflux ratio) could be very different to that in the case that the heating and cooling needs of the column are met solely by utilities [3].

6.4.2.3 Feed condition

The effect of heating or cooling the feed stream is discussed in Section 6.2.1. Compared to a design in which the column feed is a saturated liquid, preheating the feed can allow the use of heat at a more moderate temperature than that in the reboiler. This more moderate temperature may allow the feed heater to use process heat sources that are not sufficiently hot to serve the reboiler; although the total heating duty may increase slightly as a result of feed heating, energy efficiency may increase because of heat recovery. An analogous effect may be obtained by using a feed cooler, where heat is rejected at a more moderate (warmer) temperature that allows its reuse by the associated process.

6.4.2.4 Summary—distillation design for heat integration

Significant benefits, with respect to energy demand, operating costs and environmental impact, can be achieved by heat integrating distillation columns with an associated process. Direct heat recovery, between heat sinks and sources of a distillation column and an associated process, as well as indirect heat recovery, between a distillation column and a refrigeration cycle, can significantly reduce utility demand. The design degrees of freedom of a distillation column can potentially be exploited to create or enhance opportunities for heat recovery.

6.5 Energy-efficient distillation: advanced and complex column configurations

So far, the discussion has only considered simple distillation columns. Complex column configurations can increase the energy efficiency of distillation. These configurations include those related to a single separation, i.e. the use of intermediate heating and cooling, "double-effect distillation" (also known as two-pressure distillation), open-cycle heat pumping (also known as vapor recompression) and internally heat-integrated column arrangements, as well as those related to two or more separations, including thermally coupled columns and prefractionation arrangements.

6.5.1 Intermediate heating and cooling (side-reboilers and side-condensers)

In a conventional distillation column (and ignoring feed heating), heat is provided to the column in the reboiler, with the highest temperature in the column and heat is rejected from the condenser, with the lowest temperature. Therefore, all the heat is supplied at a temperature above that in the reboiler. Instead, heat can be supplied at an intermediate temperature, at an intermediate point in the column, using a so-called side-reboiler or inter-reboiler, as illustrated in Figure 6.17(a).

Compared to the simple column, introducing some of the heat at a more moderate temperature could improve energy efficiency. On the other hand, the internal reflux is lower between a reboiler and a side-reboiler. As a result, less separation may take place on those stages—the total heating duty may increase, or additional stages may be needed, to compensate for the lower reflux. Thus, there is a trade-off between the quality and the quantity of heat that is needed. Furthermore, there is a limit to the amount of heat that can be provided by the side-reboiler. If the side-reboiler duty is too high, then the internal reflux ratio may be reduced to below the minimum level for the required separation to be feasible (i.e. a new pinch may form in the column) [24]. The use of side-condensers is analogous to that of side-reboilers. Where the temperature of the side-reboiler allows the use of recovered heat, the demand for a hot utility may be reduced, in spite of the increased total

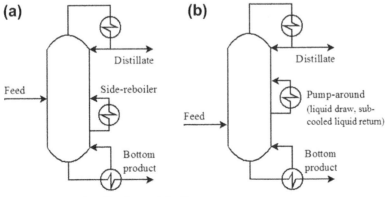

FIGURE 6.17 Intermediate Heating and Cooling

(a) Column with a side-reboiler; (b) Column with a pump-around.

heat duty and so energy efficiency may increase, compared to the case without inter-reboiling. Side-reboilers incur capital expenditure and increase the complexity of the distillation column, however.

Inter-cooling, or side-condensing, similarly, can reduce the amount of heat rejected by the condenser (at the lowest temperature in the column) and provide a potentially useful heat source at a more moderate temperature. In practice, it is difficult to withdraw a vapor stream from a column, condense it and return it to the column. It is often more practical to withdraw a liquid stream, subcool it and return it to the column; on entering the column, the subcooled liquid causes some of the vapor to condense and to flow down the column, increasing the reflux below the return stage.

Such an arrangement is commonly applied in petroleum refining, for example, in the atmospheric crude oil distillation unit, as a pump-around or pump-back loop [25], as shown in Figure 17(b). The pump-around (or pump-back) streams are sub-cooled before being returned to the column; the heat rejected in the pump-around plays an important role in preheating the column feed, reducing the requirement for fired heating.

6.5.2 Double-effect distillation

In a double-effect (or "two-pressure") distillation configuration, a single separation is carried out in two separate distillation columns operating at two different operating pressures, as illustrated in Figure 6.18(a). The feed to the process is split into two streams and fed to columns 1 and 2; the pressure of the feed to column 2 is raised. This high pressure is selected to allow the heat rejected from the condenser

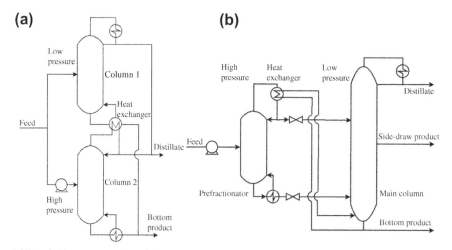

FIGURE 6.18 Double-Effect Distillation Configurations

(a) Simple column; (b) Prefractionator configuration [20].

Adapted and reprinted from Journal of Natural Gas Science and Engineering, Copyright 2012, with permission from Elsevier.

of column 2 to be transferred to the reboiler of the lower-pressure column 1 in a heat exchanger. By exchanging heat between the condenser of column 2 and the reboiler of column 1, the requirement for hot and cold utilities can be significant reduced, to almost half of that required in a single column carrying out the same separation.

Such a substantial reduction in both heating and cooling duties could lead to a significant reduction in operating costs. The drawbacks of this configuration relate firstly to the additional capital expenditure required—the two columns each processing about half of the total feed are likely to cost significantly more than a single column separating the whole feed [5]. Furthermore, the use of high pressure in column 2 is likely to increase the heating and cooling duty per unit of feed processed and the increase in temperature is likely to incur increased costs per unit of heating if a more expensive hot utility is needed. Furthermore, the more extreme reboiler temperature in column 2 is likely to reduce opportunities for heat integration with the associated process.

In spite of these drawbacks, double-effect distillation is applied industrially, e.g. for ethanol–water and acetic acid–water separations, as well as in multiple-stage evaporation processes [26,27]. Figure 6.18(b) illustrates the double-effect concept applied to the prefractionator arrangement discussed in Section 6.5.5.

6.5.3 Heat pumping (vapor recompression)

Heat rejected from the condenser of a simple distillation column cannot be reused directly by the reboiler of the column, because the condenser is colder than the reboiler. Heat pumping may be applied to upgrade the heat rejected by the condenser to a temperature that allows its reuse in the reboiler of the column (or in another heat sink). Vapor compression is the most widely used technology for heat pumping. Other heat pumping technologies, including absorption and compression-resorption heat pumps, are being developed with a view to improving energy efficiency [26].

Figure 6.19 illustrates a column using a compressor to upgrade heat, known as "mechanical vapor recompression". After compression, the temperature of the overhead vapor is sufficiently high to allow heat to be rejected from the column condenser to the reboiler. Therefore, the reboiler does not require a hot utility. (A trim cooler may also be needed to reject heat that cannot be used in the reboiler.) Once the compressed vapor has been cooled in this way and after the distillate product has been withdrawn, the pressure is reduced and the cooled, condensed material returns to the column as reflux.

The drawback of this arrangement is that the overhead vapor requires compression, involving expensive compression equipment and a costly form of energy: power. These costs are least for a small temperature difference between the condenser and reboiler, for example, for close-boiling mixtures [26]. However, if a refrigeration cycle using compression would otherwise be needed to condense the overhead vapor, then this "open loop" heat pumping arrangement may offer advantages because of the inbuilt heat integration and the direct heat transfer in the condenser-reboiler exchanger (compared to the separate evaporator and condenser of the refrigeration cycle). In

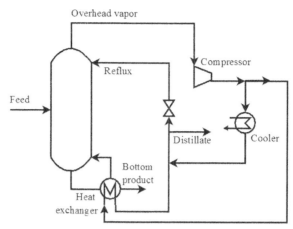

FIGURE 6.19 Distillation Column with Mechanical Vapor Recompression

practice, this arrangement is applied in a wide range of separations, including those involving moderately light hydrocarbons (C_3 to C_8), aromatics, alcohols and high-boiling solvents from aqueous solutions, etc. [28–30].

6.5.4 Internally heat-integrated distillation

Three complex column configurations that can enhance energy efficiency of distillation have been discussed; all of these features are combined in an internally heat-integrated distillation column, known as a HIDiC. Firstly, intermediate heating and cooling in a distillation column, as discussed in Section 6.5.1, can increase energy efficiency and enhance opportunities for heat recovery because of the more moderate temperatures involved. Secondly, as described in Section 6.5.2, double-effect distillation exploits the effect of operating pressure on heating and cooling temperatures to facilitate recovery of heat rejected by a high-pressure column to a low-pressure column. Thirdly, heat pumping, discussed in Section 6.5.3, uses compression work to elevate the temperature of the heat rejected by a column condenser to allow heat recovery.

An internally heat-integrated distillation column, as illustrated in Figure 6.20(a), has a high-pressure rectifying section and a separate low-pressure stripping section. Temperatures in the rectifying section are higher than those in the stripping section, allowing heat transfer between the column sections.

Effectively, the rectifying section uses intermediate cooling along its length and the stripping section uses intermediate heating along its length. The overhead vapor of the stripping section is compressed to raise its pressure to that of the rectifying section. The reboiler and condenser duties can thus be significantly reduced, or even eliminated, especially if a feed preheater is used [31]. As illustrated in Figure 6.20(b), the varying reflux ratio within each column section lends itself to even distribution of mass

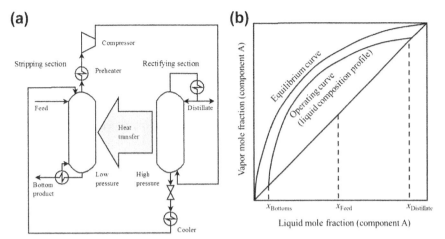

FIGURE 6.20 Internally Heat-integrated Distillation

(a) Internally heat-integrated distillation concept [31]; (b) Driving forces are evenly distributed within the columns [32].

(a) Adapted and reprinted from the Journal of Chemical Engineering of Japan, Copyright 2012, with permission from The Society of Chemical Engineers, Japan; (b) Adapted and reprinted from Chemical Engineering Research and Design, Copyright 2003, with permission from Elsevier.

transfer driving forces through the column, which leads to more thermodynamically reversible behavior and therefore lower energy requirements [32].

Challenges for the HIDiC technology related to conceptual design, mass transfer modeling, separation and heat transfer equipment design and process control, have been addressed by research programs and in pilot plant studies. A variety of hardware has been applied in these studies, including two concentric shells, plate-fin contactors, conventional trays with heat transfer plates and "thin plate" contactors [31].

Energy savings of 30−60% have been demonstrated in these studies at bench- and pilot-plant scale, e.g. a column 27 m high with a diameter of 1.4 m [31]. In spite of intensive research and development efforts, concentrated in Japan since the mid-1980s, internally heat-integrated distillation columns have not to date been applied commercially. Ongoing research addresses heat transfer equipment, batch distillation, azeotropic mixtures, column retrofit, hybrid membrane-HIDiC configurations, reactive HIDiC, etc. [31].

6.5.5 Energy efficiency in distillation sequences

6.5.5.1 Energy-efficient sequences of simple columns

When a multicomponent mixture is to be separated into three or more products, a sequence of two or more simple distillation columns is needed. Once the sequence of separations is selected, each simple column could be designed for energy efficiency, considering column operating conditions, heat integration, etc. Complex column configurations carrying out more than two separations that promote energy efficiency are described further in the following sections.

6.5.5.2 Columns with side-draw products

A side-draw stream (or side-stream) may be withdrawn from a single column, where the purity of the side-draw depends on the composition profile within the column. It is sometimes possible to withdraw a stream that is enriched in an intermediate-boiling component, producing three products from a single column. The capital investment in a single column could be substantially less than that required for two simple columns performing the same separation. However, a relatively pure side-draw product typically requires high reflux ratios in the column, unless the intermediate-boiling components are abundant in the feed. A single column with a relatively pure side-draw product is unlikely to require less heating and cooling than a sequence of two columns producing three products [33].

6.5.5.3 Thermal coupling (side-strippers and side-rectifiers)

In a sequence of simple distillation columns, feed stage mixing gives rise to thermo-dynamic inefficiencies, which translate to low energy efficiency. The exception is the special, but industrially unimportant, case of columns carrying out so-called "preferred" splits, where the aim of the separation is to separate the lightest and heaviest components in the mixture; in this case, mixing effects of the feed stage and where liquid reflux and vapor boil-up reenter the column may be negligible [20].

These mixing effects can be partly avoided by using thermal coupling, where material streams flowing between two distillation columns couple the transfer of material with that of heat. As illustrated in Figure 6.21(a), the vapor stream leaving

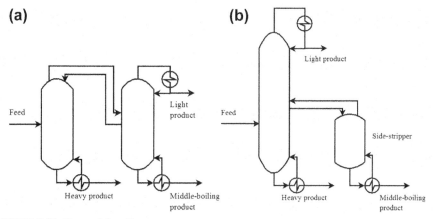

FIGURE 6.21 Thermal Coupling

(a) In thermally coupled indirect sequence of columns, the side-draw from the downstream column provides reflux to the upstream column; (b) The equivalent side-stripper configuration.

Adapted and reprinted with permission from N.A. Carlberg, A.W. Westerberg, Industrial and Engineering Chemistry Research 28 (9) 1379–1386, Copyright 1989, American Chemical Society.

overhead may be fed directly to a downstream column that also provides liquid reflux to the upstream column. By avoiding this thermodynamic inefficiency, the total heating and cooling duties of the thermally coupled column configuration may be less than that of the corresponding sequence of simple columns [34].

As shown in Figure 6.21(b), thermally coupled configurations can also be viewed as side-stream columns where the side-stream undergoes further separation and the remaining material is returned to the main column. In this sense, thermally coupled columns can produce, energy efficiently, a high-purity side-stream product.

Thermally coupled distillation configurations can reduce the quantity of heating and cooling required by a sequence. However, as there is only one reboiler providing heat for both columns, all the heat is needed at a high temperature. Similarly, if there is only one condenser, all the heat rejected is at the lowest temperature in the sequence. Thus, there is a trade-off between the quantity and the quality of the heating and cooling required in the sequence. Especially where two different temperature levels facilitate the use of a cheaper utility or the recovery of heat, thermal coupling may have a detrimental effect on energy efficiency.

Conceptual design of thermally coupled distillation columns is challenging, as these columns present additional degrees of freedom for design and are more complex to model [34]. Important industrial applications of thermal coupling are the use of side-strippers in refinery distillation columns, such as the atmospheric crude oil distillation unit [25] and side-rectifiers in air separation (to purify intermediate-boiling argon) [35,36]. Thermally coupled configurations present practical challenges related to withdrawing part of a vapor stream from a column and transferring a vapor stream from a lower-pressure downstream column to an upstream column at a high pressure [35]. In practice, two thermally coupled columns are constrained to operate at similar pressures; this constraint limits opportunities to manipulate the operating pressure for energy efficiency.

6.5.5.4 Prefractionation arrangements

Instead of introducing a feed directly to a distillation column, where mixing effects will inevitably increase energy requirements, it is possible to partially separate the feed, in a "prefractionation" column, before it undergoes further separation in the column. Furthermore, prefractionation may allow composition profiles to develop in the column that lend themselves to the withdrawal of a relatively pure side-draw product.

Figure 6.22(a) presents such a prefractionation arrangement. By appropriately designing both the prefractionator, which carries out the preliminary separation between the components to be separated and the main column, the feeds to the main column can be introduced with little remixing taking place; the composition profiles in both columns can develop in thermodynamically efficient ways [20]. The total heating and cooling required by the sequence of columns may be significantly lower (around 20–40%, depending on the feed mixture, composition and product specifications) than that of a sequence of simple columns carrying out the same separations [20,27].

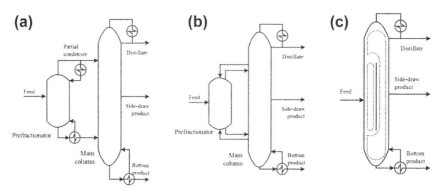

FIGURE 6.22 Prefractionation Arrangements

(a) Prefractionator; (b) Thermally coupled prefractionator (Petlyuk column); (c) Dividing wall column [3].

Reprinted from Chemical Process Design and Integration, Copyright 2005, with permission from John Wiley and Sons.

The prefractionator may also be thermally coupled with the main column, as shown in Figure 6.22(b); this arrangement is also known as a Petlyuk column. The main benefit of introducing thermal coupling is that only a single reboiler and a single condenser are needed, compared to the uncoupled arrangement of Figure 6.22(a). There is little or no benefit in terms of heating and cooling duties, but there is a penalty with respect to quality of heating and cooling, as heating and cooling are required at the most extreme temperatures in the column. Furthermore, thermal coupling reduces opportunities to operate the two columns at different pressures. Operation and control are also more complicated than for the uncoupled arrangement.

The thermally coupled arrangement can be implemented in a single shell, as shown in Figure 6.22(c). The prefractionation is carried out on one side of a vertical wall, or partition, while the final separation takes place on the other side of the wall as well as above and below it. Liquid reflux from the overhead condenser is split across the top of the wall; reboiled vapor from the reboiler distributes across the wall according to the pressure drop on either side of the wall.

This dividing wall column configuration has essentially the same heating and cooling requirements as the Petlyuk column and is as constrained with respect to operating pressure. The use of a single column shell can bring significant advantages with respect to capital investment (around 30% [3]) because of the single shell and reduced costs of piping, foundations, plot space, etc. Specialized internals are needed to accommodate the dividing wall; both trays and packings have been successfully applied in these columns [37–40]. Liquid distribution arrangements include passive (e.g. weirs) and active (e.g. control valves) systems. A key challenge is to design the column and internals to achieve the correct vapor flows; these flows are dictated by the pressure drop on each side of the wall [3,37].

Dividing wall distillation is an established technology, although design and operation remain challenging [38]. These columns have been applied industrially,

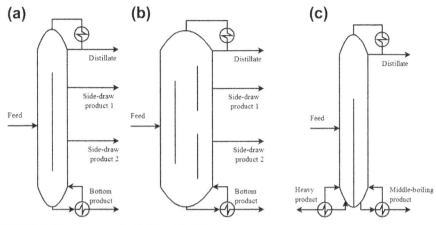

FIGURE 6.23 Extensions to Dividing Wall Columns

(a) Kaibel column; (b) Column with multiple dividing walls [20]; (c) Side-stripper dividing wall column.

Reprinted from Journal of Natural Gas Science and Engineering, Copyright 2012, with permission from Elsevier.

initially in specialty chemicals production (in the late 1980s), but also in petroleum refining and many other areas [38–40]. The range of applications includes operating pressures of 0.2–1000 kPa and a wide range of product purities; column diameters are typically 0.6–4 m, but larger columns have been constructed [38]. Control strategies, equally, are well established [37,38]. Dividing wall columns have been employed in process revamps and for azeotropic, extractive and reactive distillation applications [37,38].

Various extensions of the Petlyuk and dividing wall column concepts have been developed, including the so-called Kaibel column with more than three products, columns with multiple dividing walls and side-stripper (and side-rectifier) arrangements, as shown in Figure 6.23. While these configurations are relatively complex, they can potentially bring additional advantages in terms of energy demand and capital costs [20].

6.5.5.5 Summary—energy-efficient distillation sequences

Many distillation sequences can carry out a required separation; column operating conditions can be selected for energy-efficient operation. The problem of selecting and designing the most energy-efficient sequence can be tackled by developing suitable models and applying optimization techniques [34].

Nonconventional column configurations, such as thermally coupled columns and prefractionation arrangements, can bring significant benefits in terms of energy demand, but at the expense of higher quality energy and with constraints relating to operating pressure. These columns, while complex and specialized, may also require less capital investment than the corresponding sequences of simple columns.

6.6 Energy-efficient distillation: evaluation of energy requirements

Process modeling and analysis tools that assist with the evaluation of distillation energy efficiency can be used to quantify distillation energy requirements. Models for individual columns are outlined; sequences can be evaluated considering each column in turn. However, interactions between the columns need to be captured by models for distillation sequences and complex column configurations. It is assumed in this discussion that the same separation is carried out in all the design alternatives; it is sometimes also relevant to consider interactions between product quality (e.g. purity) or yield and energy requirements.

6.6.1 Shortcut distillation column models

Shortcut distillation models, such as the Fenske, Underwood and Gilliland models, are well established and provide estimates of the minimum reflux ratio, minimum number of stages and actual reflux (or number of theoretical stages) of a column carrying out a specified separation, given feed data and the operating pressure [14,17,41]. These models are useful for developing conceptual column and sequence designs and for comparing them; when used together with simple models for capital and operating cost, the shortcut models allow economic evaluation of alternative design concepts.

The shortcut models assume constant relative volatility and constant molar overflow. The Underwood equations, to calculate the minimum reflux ratio, are of most relevance for evaluating energy requirements. However, care is needed when applying the method in multicomponent separations, especially when some components distribute significantly between the top and bottom products [3]. Also, the results may be sensitive to the relative volatility values selected to represent a column, e.g. whether based on the feed mixture at the feed conditions or on the feed stage, or based on the compositions of the products, where these compositions are themselves estimated using shortcut models [3,17].

The assumption of constant relative volatility may be unrealistic not only for obviously nonideal mixtures, e.g. those forming azeotropes but also for apparently well-behaved mixtures, e.g. light hydrocarbons such as ethene–ethane mixtures. The assumption of constant molar overflow is least valid for mixtures where components have dissimilar latent heats of vaporization; carrying out a separate enthalpy balance over the condenser can help to relate minimum vapor flow at the column pinch to that at the top of the column [17]. Other extensions to the shortcut models allow some complex column configurations to be evaluated, such as those with side-draw products and multiple feeds [17], but the application of the models is less straightforward in these cases.

The shortcut models, together with data related to the thermal and thermodynamic properties of the components and mixtures involved, allow energy requirements to be determined. In particular, the effects on heating and cooling duties and the associated temperatures of design degrees of freedom, namely column operating pressure (and pressure drop), reflux ratio (or number of stages), feed condition and condenser type, can be estimated. However, as the models provide no detail of material or energy flows or temperature profiles within column sections, they cannot be used directly to evaluate intermediate heating and cooling options.

In spite of their shortcomings, shortcut models have been found useful when incorporated into a range of methodologies for conceptual design of columns and sequences (e.g. [34,42]), facilitated by the relatively low computational demands of the shortcut models.

6.6.2 More rigorous distillation models

Instead of applying shortcut distillation models, rigorous models that carry out stage-by-stage calculations of the material balance, phase equilibrium and enthalpy balance may be used to evaluate distillation column and sequence designs. These models involve a relatively large number of equations (depending on the number of components and number of equilibrium stages) and the countercurrent nature of the flows in the column complicates their solution [14,17,41].

Fortunately, robust and efficient algorithms for solving these equations simultaneously are well established and have been implemented in commercial process simulation software. The accuracy of these models is limited by the assumption that phase equilibrium is achieved on every theoretical stage; the use of stage efficiencies is a convenient, if rather empirical, way to represent mass transfer limitations. However, stage efficiencies are not straightforward to predict accurately and often are inadequate for representing mass transfer effects [41,43].

6.6.3 Distillation process performance indicators

Distillation models may be applied to predict and evaluate the performance of design alternatives. The separation specifications should be set appropriately, e.g. in terms of product purity or recovery, rather than in terms of product flow rate and reflux ratio, to permit valid comparisons. When evaluating design alternatives for distillation columns or sequences, it is reasonable to focus on energy demand, as operating costs typically dominate distillation process economics. In above-ambient processes, the cost of heating typically far outweighs the cost of cooling and in processes operating below ambient temperatures, the cost of refrigerated cooling is the overriding cost.

Various indicators have been applied for evaluating the performance of alternative distillation column or sequence designs; these are discussed following:

1. *Minimum or actual vapor load* [3,42]—The molar flow of vapor (known as vapor load) is readily calculated using shortcut distillation models and is correlated with actual heating and cooling duties. It does not account for energy quality, i.e. the temperatures at which heating and cooling are required, nor capital investment.
2. *Total heating or cooling duty* [20,44]—Reboiler, condenser and feed heater and cooler duties can be determined using shortcut or more rigorous models and

heat of vaporization data, but this measure of performance does not recognize the impact of temperature on energy efficiency or capital—energy trade-offs.

3. *Ideal or actual power demand*—In distillation processes involving heat pumping, such as those requiring subambient cooling, the dominant cost is that of providing power to the refrigeration system. Eqn. (6.1) correlates ideal power demand with the process cooling duty, the temperature at which cooling is required and the temperature "lift" ($T_{cond} - T_{evap}$). The impact of heat integration on power demand, where heat is rejected from a refrigeration cycle to a process stream, can also be captured. Capital investment for the distillation and heat pumping processes is neglected by this indicator.

4. *Operating cost*—The operating cost can be determined from the heating and cooling duties and refrigeration power demand, using refrigeration process models and given the cost of utilities, but it does not account for capital investment.

5. *Total annualized cost*—Capital costs can be approximated using more or less sophisticated approaches for capital cost estimation [5,11]. Capital cost models are inevitably inaccurate and generally less established for novel technologies, such as dividing wall columns. Trade-offs between capital costs and operating costs can be taken into account by considering total annualized costs [5]. The impact of changes in product yield or quality can be accounted for if product revenue is also considered.

6. *Simple payback*—Payback is the ratio of annual income, profit, or savings to capital investment and thus provides a measure of benefit to cost. It can therefore compare processes with different throughputs or yields but cannot reflect the scale of the investment.

6.6.4 Thermodynamic analysis of distillation columns

The concept of thermodynamically reversible distillation has proven useful for analyzing and comparing distillation design alternatives [45,46]. In a reversible distillation process, a simple column with infinitely many stages carries out a specified separation, as illustrated in Figure 6.24. The driving force is reduced to zero on each stage by continuous addition of heat below the feed stage and continuous removal of heat above the feed stage.

The total heat provided to a reversible distillation column is the minimum energy demand for a specified separation at a given operating pressure. Minimum energy demand is an important benchmark for evaluating energy efficiency of distillation columns. The temperatures at which heating and cooling are required are also of interest.

6.6.4.1 Reversible and near-reversible distillation profiles

Various methods have been developed to generate and interpret reversible and near-reversible column profiles to quantify ideal heat flows to the column. These heat flows relate to opportunities to apply side-reboilers and side-condensers at more moderate temperatures than in the reboiler and condenser.

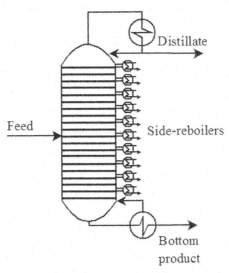

FIGURE 6.24 Reversible Distillation Column [46]

Reprinted from Chemical Engineering Science, Copyright 2011, with permission from Elsevier.

Reversible distillation is possible in theory for binary separations; in this case, the liquid composition profile is identical to the equilibrium curve. A stage-by-stage enthalpy balance can identify the amount of heat that needs to be added to each stage, as well as the associated temperature. Simulating the reversible column, to generate composition, temperature and enthalpy profiles, is relatively straightforward [45−47].

Real, irreversible distillation columns require more energy than thermodynamically ideal columns, because driving forces for separation and for heat transfer are non-negligible and because mixing streams of different composition or temperature generates entropy. To account for real columns, with a finite number of stages, "near-reversible" profiles can be constructed and interpreted to identify opportunities to increase energy efficiency by reducing the reflux ratio, modifying the feed condition and employing side-reboilers and side-condensers [47].

However, for multicomponent separations, only preferred splits can be carried out reversibly—in these separations, the lightest and heaviest components can be recovered to any degree to the overhead and bottom products, respectively; the remaining components distribute between the products. Most multicomponent splits of practical importance cannot be analyzed with respect to reversible operation. As the splits are not preferred splits, it is unavoidable that entropy is generated by feed-stage mixing. Several approaches have been developed to evaluate energy demand for multicomponent columns operating "as reversibly as possible" [46].

A well-established approach, known as "column targeting", treats multicomponent separations as pseudo-binary separations to generate a temperature−enthalpy profile or "column grand composite curve" [24,47]. The procedure is straightforward, but the validity of the analysis is severely restricted by several key assumptions: that reversible behavior is possible throughout the column, that minimum flows depend only the phase equilibrium behavior of the two key components (or groups of key components) and that molar enthalpies of the liquid and vapor streams are insensitive to composition [46]. Furthermore, the method does not compensate for the reduced reboil or reflux on some stages of a finite distillation column with a side-reboiler or side-condenser and the resulting penalty—to meet the separation specifications, either the total heating and cooling duty or the number of equilibrium stages must increase. The duties and temperatures of side-reboilers and side-condensers predicted using this temperature–enthalpy analysis are typically unrealistic with respect to both quantity and quality of heat flows [46].

An alternative approach uses the Underwood equations or rigorous models to identify the minimum vapor load in a distillation column separating a known feed [42]. The "Vmin diagram" identifies the minimum energy demand, corresponding to minimum reflux, for *all* separations of the feed and therefore for any particular separation. The energy demand of a reversible column is easily computed; the Vmin diagram is particularly powerful for determining the minimum energy demand of Petlyuk arrangements [20], where each subsection of the Petlyuk column carries out a preferred split.

6.6.4.2 Minimum driving force profiles

Other methodologies consider driving forces for separation within the column, recognizing that nonzero driving forces are needed in practice. Strictly speaking, chemical potential is the driving force for distillation, but defining an overall mass transfer driving force in terms of chemical potential is not straightforward [46]. Therefore, exergy, Ex, as defined in Eqn (6.3), has been proposed as a measure of irreversibility, as it accounts for both mass and heat transfer driving forces on a separation stage.

$$Ex = H - T_0 S \qquad (6.3)$$

where H is enthalpy and S is entropy for a stream and T_0 is the ambient (reference) temperature.

The exergy loss of a stage is easily computed from converged simulation results, by applying an exergy balance over the stage [46].

This methodology considers design alternatives with side-reboilers and side-condensers of various duties and with additional stages, such that the total heat load is fixed and the exergy loss over any stage may not exceed a specified value. These designs can be systematically generated and evaluated with respect to exergy, operating cost, or capital—energy trade-offs [46].

6.6.4.3 Summary—evaluation of distillation energy requirements

To evaluate distillation design alternatives with respect to energy efficiency, quantitative models are needed to represent the distillation process and its energy requirements, considering energy quality and quantity. The validity of the evaluation depends on the quality of the models employed and the measure of performance applied. Finding the most energy-efficient designs needs the set of design alternatives to be explored thoroughly; tools based on thermodynamic analysis of the columns can assist with this time-consuming task. Heat-integration opportunities should be considered simultaneously.

6.7 Conclusions

Distillation is an energy-intensive process, where heating and cooling needs (duty and temperature) dominate process economics, environmental impact and sustainability. A distillation column carrying out a specified separation has several degrees of freedom for design that affect energy efficiency. The column operating pressure is the most significant design variable, especially when it allows energy-intensive refrigeration to be avoided. Heat recovery between a distillation column and an associated process or refrigeration cycle can enhance energy efficiency; a challenge for distillation process design is to create and exploit opportunities for heat recovery. Well-designed distillation processes also need to be operated and controlled with energy efficiency in mind; generally, over-refluxing results in excessive energy demand and also constrains throughput.

Simple distillation columns are ubiquitous. Complex column configurations, such as dividing wall columns, are becoming more established and can offer substantial benefits with respect to energy efficiency as well as capital investment. Especially because these configurations cannot fully exploit the effects of operating pressure on energy efficiency, quantitative benefits are often diminished by qualitative penalties.

Although distillation is a mature technology, economic and environmental drivers continue to motivate further development of energy-efficient distillation processes that are not yet commercially proven, such as internally heat-integrated distillation columns. Research interest in hybrid distillation processes that can reduce energy consumption is vigorous; recent research publications address distillation—pervaporation, adsorption—distillation, liquid—liquid extraction—distillation and crystallization—distillation processes, as well as novel heat-pumped configurations and short-path distillation.

In order to design and operate energy-efficient column configurations, research in computer-aided process engineering continues to develop comprehensive, reasonably accurate and computationally robust tools for process synthesis, modeling, analysis and optimization.

References

[1] IEA, Key World Energy Statistics 2013, 2013 (accessed 09.11.2013), www.iea.org/publications/freepublications/publication/name,31287,en.html.
[2] IEA, CO_2 Emissions from Fuel Combustion: Highlights, 2013 (accessed 9.11.2013), www.iea.org/publications/freepublications/publication/name,32870,en.html.

[3] R. Smith, Chemical Process Design and Integration, John Wiley & Sons Ltd., UK, 2005, ISBN 0-471-48681-7.

[4] I.C. Kemp, Pinch Analysis and Process Integration: A User Guide on Process Integration for the Efficient Use of Energy, second ed., Elsevier Ltd., 2006, ISBN 978-0-7506-8260-2.

[5] R. Sinnott, G.P. Towler, Chemical Process Design, fifth ed., Elsevier Ltd, 2009, ISBN 978-0-7506-8551-1.

[6] P. Mullinger, B. Jenkins, Introduction, in: Industrial and Process Furnaces: Principles, Design and Operation, Elsevier, Oxford, 2008, ISBN 978-0-7506-8692-1, pp. 1–30, http://dx.doi.org/10.1016/B978-0-7506-8692-1.00001-6.

[7] C.R. Branan, Refrigeration, in: Rules of Thumb for Chemical Engineers, fourth ed., Elsevier Inc., 2005, ISBN 978-0-7506-7856-8, pp. 176–200, http://dx.doi.org/10.1016/B978-075067856-8/50011-6.

[8] E. Ludwig, Refrigeration systems, in: Applied Process Design for Chemical & Petrochemical Plants, vol. 3, Elsevier, 2001, ISBN 978-0-88415-651-2, pp. 289–367, http://dx.doi.org/10.1016/S1874-8635(01)80004-2.

[9] NREL, Use Low-grade Waste Steam to Power Absorption Chillers, 2006. Industrial Technologies Program (ITP) Steam Tip Sheet # 14 (accessed 30.11.2012), http://www.nrel.gov/docs/fy06osti/39316.pdf.

[10] E. Ludwig, Heat transfer, in: Applied Process Design for Chemical & Petrochemical Plants, vol. 3, Elsevier, 2001, ISBN 978-0-88415-651-2, pp. 1–288, http://dx.doi.org/10.1016/S1874-8635(01)80003-0.

[11] A.M. Gerrard, Guide to Capital Cost Estimating, fourth ed., IChemE, 2000, ISBN 0-85295-399-2.

[12] A. Azapagic, S. Perdan, Sustainable Development in Practice: Case Studies for Engineers and Scientists, second ed., John Wiley & Sons Inc., 2010, ISBN 978-0-470-71872-8.

[13] E. Ludwig, Distillation, in: Applied Process Design for Chemical & Petrochemical Plants, vol. 2, Elsevier, 1997, pp. 1–107, http://dx.doi.org/10.1016/S1874-8635(97)80002-7.

[14] J.D. Seader, E.J. Henley, Separation Process Principles, John Wiley and Sons, Inc., 1998, ISBN 0-471-58626-9.

[15] H.Z. Kister, Distillation Design, McGraw-Hill, 1992, ISBN 978-0-07-034909-4.

[16] K.T. Klemola, J.K. Ilme, Distillation efficiencies of an industrial-scale i-butane/n-butane fractionator, Ind. Eng. Chem. Res. 35 (12) (1996) 4579–4586, http://dx.doi.org/10.1021/ie960390r.

[17] C.J. King, Separation Processes, second ed., McGraw-Hill, 1980, ISBN 0070346127.

[18] W.L. Luyben, Effect of tray pressure drop on the trade-off between trays and energy, Ind. Eng. Chem. Res. 51 (2012) 9186–9190, http://dx.doi.org/10.1021/ie300634a.

[19] R. Agrawal, V. Shah, Conceptual design of zeotropic distillation processes, in: Gorak, et al. (Eds.), Distillation: Fundamentals and Principles, vol. 1, Elsevier, 2014.

[20] I. Halvorsen, S. Skogestad, Energy efficient distillation, J. Nat. Gas Sci. Eng. 3 (4) (2011) 571–580, http://dx.doi.org/10.1016/j.jngse.2011.06.002.

[21] ETSU, Efficiency Improvements to Existing Distillation Equipment, Good Practice Guide 269, ETSU, Oxfordshire, 1999.

[22] W.L. Luyben, Distillation control, in: Gorak, et al. (Eds.), Distillation: Fundamentals and Principles, vol. 3, Elsevier, 2014.

[23] D. White, Optimize energy use in distillation, Chem. Eng. Prog. (March 2012) 35–41.

[24] S. Bandyopadhyay, R.K. Malik, U.V. Shenoy, Temperature-enthalpy curve for energy targeting of distillation columns, Comput. Chem. Eng. 22 (12) (1998) 1733−1744, http://dx.doi.org/10.1016/S0098-1354(98)00250-6.

[25] S. Fraser, Distillation in refining, in: Gorak, et al. (Eds.), Distillation: Fundamentals and Principles, vol. 3, Elsevier, 2014.

[26] A.A. Kiss, S.J. Flores Landaeta, C.A. Infante Ferreira, Towards energy efficiency distillation technologies − making the right choice, Energy 47 (1) (2012) 531−542, http://dx.doi.org/10.1016/j.energy.2012.09.038.

[27] L.C.B.A. Bessa, F.R.M. Batista, A.J.A. Meirelles, Double-effect integration of multicomponent alcoholic distillation columns, Energy 45 (1) (2012) 603−612, http://dx.doi.org/10.1016/j.energy.2012.07.038.

[28] D. Hänggi, I. Meszaros, Vapor recompression: distillation without steam, Sulzer Tech. Rev. (1999), 3/1/1999 (accessed 29.10.2012), http://www.sulzer.com/en/-/media/Documents/Cross_Division/STR/1999/1999_01_32_haenggi_e.pdf.

[29] EERE, Steam Technical Brief: Industrial Heat Pumps for Steam and Fuel Savings, 2003 (accessed 4.11.2012), http://www1.eere.energy.gov/manufacturing/tech_deployment/pdfs/heatpump.pdf.

[30] A. Cooper, M. Lyon, Make the most of MVR, Chem. Eng. ISSN: 0009-2460 111 (7) (2004).

[31] K. Matsuda, K. Iwakabe, M. Nakaiwa, Recent advances in internally heat-integrated distillation columns (HIDiC) for sustainable development, J. Chem. Eng. Jpn. 45 (6) (2012) 363−372, http://dx.doi.org/10.1252/jcej.11we127.

[32] M. Nakaiwa, K. Huang, A. Endo, T. Ohmori, T. Akiya, T. Takamatsu, Internally heat-integrated distillation columns: a review, Chem. Eng. Res. Des. 81 (1) (2003) 162−177, http://dx.doi.org/10.1205/026387603321158320.

[33] N. Doukas, W.L. Luyben, Economics of alternative distillation configurations for the separation of ternary mixtures, Ind. Eng. Chem. Res. 17 (3) (1978) 272−281.

[34] J.A. Caballero, I.E. Grossmann, Optimization of distillation processes, in: Gorak, et al. (Eds.), Distillation: Fundamentals and Principles, vol. 1, Elsevier, 2014.

[35] R. Agrawal, Thermally coupled distillation with reduced number of intercolumn vapor transfers, AIChE J. 46 (11) (2000) 2198−2210, http://dx.doi.org/10.1002/aic.690461112.

[36] Linde, Cryogenic Air Separation: History and Technological Progress, 2012 (last accessed 26.10.2012), http://www.linde-le.de/process_plants/air_separation_plants/documents/L_2_1_e_09_150dpi.pdf.

[37] O. Yildirim, A.A. Kiss, E.Y. Kenig, Dividing wall columns in chemical process industry: a review on current activities, Sep. Purif. Technol. 80 (3) (2011) 403−417, http://dx.doi.org/10.1016/j.seppur.2011.05.009.

[38] N. Asprion, G. Kaibel, Dividing wall columns: fundamentals and recent advances, Chem. Eng. Process. 49 (2) (2010) 139−146, http://dx.doi.org/10.1016/j.cep.2010.01.013.

[39] A.A. Kiss, Advanced Distillation Technologies: Design, Control and Applications, John Wiley & Sons, 2013, ISBN 978-1-119-99361-2.

[40] B. Kaibel, Dividing wall columns, in: Gorak, et al. (Eds.), Distillation: Fundamentals and Principles, vol. 2, Elsevier, 2014.

[41] E. Kenig, S. Blagov, Modeling of distillation processes, in: Gorak, et al. (Eds.), Distillation: Fundamentals and Principles, vol. 1, Elsevier, 2014.

[42] I.J. Halvorsen, S. Skogestad, Minimum energy consumption in multicomponent distillation. 1. Vmin diagram for a two-product column, Ind. Eng. Chem. Res. 42 (3) (2003) 596−604, http://dx.doi.org/10.1021/ie010863g.

[43] R. Taylor, Mass transfer in distillation, in: Gorak, et al. (Eds.), Distillation: Fundamentals and Principles, vol. 1, Elsevier, 2014.

[44] A. Lucia, B.R. McCallum, Energy targeting and minimum energy distillation column sequences, Comput. Chem. Eng. 34 (6) (2010) 931—942, http://dx.doi.org/10.1016/j.compchemeng.2009.10.006.

[45] Z. Fonyo, Thermodynamic analysis of rectification: 1. Reversible model of rectification, Int. Chem. Eng. 14 (1) (1974) 18—27.

[46] F. Soares Pinto, R. Zemp, M. Jobson, R. Smith, Thermodynamic optimisation of distillation columns, Chem. Eng. Sci. 66 (13) (2011) 2920—2934, http://dx.doi.org/10.1016/j.ces.2011.03.022.

[47] V.R. Dhole, B. Linnhoff, Distillation column targets, Comput. Chem. Eng. 17 (5—6) (1993) 549—560, http://dx.doi.org/10.1016/0098-1354(93)80043-M.

Conceptual Design of Zeotropic Distillation Processes

7

Vishesh H. Shah[1], Rakesh Agrawal[2]

Engineering and Process Sciences Laboratory, The Dow Chemical Company, Midland, MI, USA[1],
School of Chemical Engineering, Purdue University, West Lafayette, IN, USA[2]

CHAPTER OUTLINE

7.1 Introduction

A distillation column typically separates a feed mixture into two product streams. If an ideal binary mixture is introduced to a distillation column, the two components in the mixture can be separated to arbitrarily high purities in the two product streams. However, real-world separation problems often involve multicomponent feed streams. Moreover, components in such mixtures can sometimes have significant nonideal interactions with each other. This chapter will focus only on distillation of near-ideal mixtures. The next chapter addresses the distillation of nonideal mixtures. By near-ideal mixtures, we mean mixtures that do not form azeotropes and the component relative volatilities allow their separation through distillation. Consider the separation of an ideal three-component mixture (ABC) into three product streams, each enriched in one of the components, using a single distillation column (see Figure 7.1).

In Figure 7.1, and throughout this chapter, components have been arranged alphabetically with decreasing relative volatilities. Reboilers have been represented

Distillation: Fundamentals and Principles. http://dx.doi.org/10.1016/B978-0-12-386547-2.00007-7
Copyright © 2014 Elsevier Inc. All rights reserved.

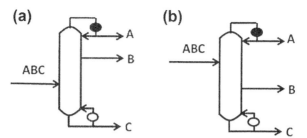

FIGURE 7.1 Separation of an Ideal Three-Component Mixture Using a Single Distillation Column

by nonfilled circles, and condensers have been represented by filled circles. In Figure 7.1(a) and (b), it can be seen that the side-stream (B) will always be contaminated by the lightest component A or the heaviest component C, respectively. The intermediate boiling component B thus cannot be obtained with a high purity for practical values of reflux ratios using a single distillation column. On the other hand, inclusion of a second distillation column enables easy separation of the three components to any arbitrarily high purity. Three arrangements that use two distillation columns to separate a feed mixture into three product streams are shown in Figure 7.2. The three arrangements carry out the same overall separation. Such multicolumn distillation arrangements are also often referred to as distillation sequences, distillation configurations, distillation trains, or distillation schemes.

A split represents separation of a mixture into two product streams. A distillation configuration is thus a collection of splits. For instance, the configuration shown in Figure 7.2(a) carries out the split ABC−A/BC, followed by BC−B/C. Such splits are referred to as sharp splits because the product streams have no (or acceptably low levels of) overlapping components. On the other hand, one of the splits in the configuration shown in Figure 7.2(c) is ABC−AB/BC. Such a split is referred to as a nonsharp split because the product streams have significant amounts of overlapping intermediate volatility components. Accordingly, distillation configurations are classified in the literature as either sharp split configurations (if all their splits are sharp),

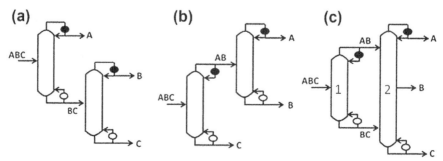

FIGURE 7.2 Multicolumn Distillation Arrangements for Separating a Three-Component Mixture

or nonsharp split configurations (if at least one split is nonsharp). The configurations shown in Figure 7.2(a) and (b) are thus sharp split configurations, while the configuration shown in Figure 7.2(c) is a nonsharp split configuration.

In our nomenclature, the components A, B, etc. can actually represent multiple species, with lumping providing a convenient way to address the design of multicolumn arrangements. For instance, atmospheric distillation of petroleum crude involves a feed stream with thousands of components [1], but it can be treated as a five-component mixture from the point of view of conceptual design of multicolumn configurations [2]. Also, in streams such as AB, trace amounts of other components such as C will always be present. Therefore, another convenient feature of our nomenclature is that when we describe a stream such as AB, we imply that the other components such as C are present at acceptably low levels.

Consider separation of a saturated liquid mixture containing 56.6 wt% benzene (A), 28.4 wt% toluene (B), and 15 wt% p-xylene (C) into three product streams with at least 99 wt% purity of each component, using the configurations shown in Figure 7.2. These configurations are simulated in ASPEN Plus v7.1 using the Peng–Robinson equation of state with the condenser pressures specified as $1.013 \cdot 10^5$ Pa and with sufficient theoretical stages in each distillation column. The total reboiler duty of each configuration is minimized by varying appropriate variables such as the reflux ratios in each distillation column, and the overhead and bottom stream flow rates. The in-built local optimization solver in ASPEN Plus is used. The configuration in Figure 7.2(c) is found to have the least total reboiler duty. The configurations in Figure 7.2(a) and (b) are found to have total reboiler duties that are 23% and 29% higher, respectively. This is a fictitious example, because an actual aromatics feed stream contains other impurities and usually requires advanced distillation techniques like extractive distillation to purify the aromatic compounds [3]. However, this example clearly shows that even though different multicolumn distillation configurations carry out the same overall separation, they can differ significantly in critical performance parameters like heat demand, motivating the need to select the optimal scheme for a given distillation need.

Several examples of multicolumn distillation sequences that separate mixtures into more than two product streams can be found. For separation of a mixture into four product streams using multicolumn distillation, one of the largest scale examples involves the purification of ethylene from a stream obtained by cracking naphtha. The conventional configuration to carry out this separation is shown in Figure 7.3, and it has been the preferred choice in industry for over 50 years [4]. A major fraction of the cost of this process is associated with the total heat demand of the configuration and the cryogenic utility demand for condensing the overhead product of the first distillation column.

A well-known example of separating a mixture into five product streams is the atmospheric distillation unit in refineries, which uses four distillation columns to separate crude oil into useful fractions [2]. Total world petroleum production in 2012 was 86.152 million bbl of crude oil/day [5]. Petroleum crude distillation roughly consumes energy at the equivalent of 2% of the crude produced [2], working

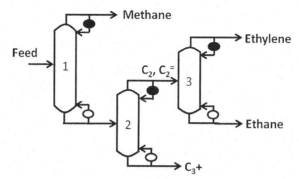

FIGURE 7.3 Distillation Configuration to Purify Ethylene

out to nearly 1.7 million bbl/day. Many other multicomponent distillation examples can be found throughout the chemical and petrochemical industries.

This chapter will allow readers to answer questions such as:

1. How many possible configurations exist for separating a mixture into n product streams?
2. How does one enumerate all these configurations?
3. How does one find the optimal configuration from among these options?

7.2 Synthesizing all possible distillation configurations

The term "search space" has been used to refer to the set of all possible configurations for a given distillation need [6]. The concept of a search space is based on the premise that if the optimal configuration is not "defined", it will not be found. Therefore, a good search space must include all configurations that can potentially be optimal for any given feed. At the same time, it must not include configurations that are never optimal for any separation. For instance, Figures 7.1 and 7.2 represent a combined search space for three-component separations, but a good search space for three-component separations is probably limited to Figure 7.2. Early attempts to synthesize all possible distillation configurations were lacking in one of the two characteristics of a good search space. In other words, these attempts either missed potentially optimal configurations or included several nonoptimal configurations. On the other hand, recent techniques to generate the search space have both these characteristics, but they differ from each other significantly in terms of computational demand.

7.2.1 A brief history

The first attempt to synthesize optimal distillation schemes probably dates back to 1947 and provided heuristics to guide the selection of distillation sequences [7].

Subsequently, additional heuristics were developed by other researchers [8−11], and a summary of some key heuristics is provided by Seader and Westerberg [12]. In 1972, Thompson and King [13] presented a computer programing approach for synthesizing sharp split configurations and provided a formula to estimate the number of sharp split distillation configurations for any n component separation:

$$\frac{[2(n-1)]!}{n!(n-1)!}$$

Other computer-aided techniques for synthesizing optimal and near optimal sharp split distillation sequences were also developed [14−16]. These techniques did not include nonsharp split distillation sequences in the search space.

In 1976, a superstructure flowsheet approach was proposed (see Figure 7.4), which was purported to contain several useful distillation configurations [18]. The idea behind the superstructure was that optimization of a single process flowsheet would provide the optimal distillation scheme by allowing streams and column sections to have nonzero as well as zero flows. It can be seen from Figure 7.4 that the superstructure included several nonsharp split configurations.

In 1996, satellite distillation configurations were invented [17] (see Figure 7.5). It was demonstrated that the four-component superstructure flowsheet did not include the four-component satellite configuration shown in Figure 7.5. Importantly,

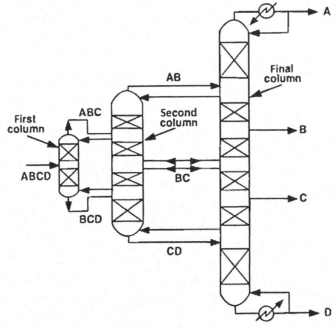

FIGURE 7.4 Superstructure Flowsheet [17]

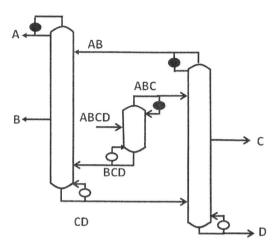

FIGURE 7.5 An Example of a Satellite Configuration

the four-component satellite configuration was shown to be attractive for some applications, thereby demonstrating a shortcoming of the superstructure flowsheet.

In 1998, the notion of "states" and "tasks" for distillation sequence synthesis was introduced by Sargent [19]. States basically represent streams in a configuration, and tasks represent splits. In 2003, a systematic set of rules was proposed for the first time to synthesize all possible distillation configurations that use $n - 1$ distillation columns for an n-component separation [20]. These rules could synthesize sharp split as well as nonsharp split configurations (including satellite-type configurations) for any number of components in the feed. Subsequent developments in the synthesis of all possible distillation configurations were mostly mathematical in nature, whereby different mathematical representations were developed to apply a set of rules to a numerical representation of states and/or tasks. For instance, a mathematical programing approach based on the state-task network concept was developed by Caballero and Grossmann [21]. Binary integer variables that can take values of either 0 or 1 were assigned to states and tasks to mathematically describe distillation configurations. Another mathematical formulation based on the notion of a "supernetwork" was developed in 2010, where binary integer variables were associated with rectifying and stripping sections (which act like half-tasks or half-splits) [22].

While this list of prior work is by no means complete, it summarizes some of the notable contributions over the years. Recent mathematical formulations have been successful in synthesizing the three distillation configurations shown in Figure 7.2 for a three-component separation, and also 18 distillation configurations, 203 distillation configurations, and 4373 distillation configurations for four-component, five-component, and six-component separations, respectively. From the large number of configurations, it can be seen that synthesis of a search space needs to be tackled using computers to avoid tedious manual calculations. However, these mathematical formulations were computationally demanding and were unable to generate the search space for more than six component separations within a practical amount of computational time.

A novel and easy-to-use mathematical framework was developed to improve the computational efficiency of synthesizing the search space [23,24]. This framework was based on the observation that while the main feed stream and the final product streams are present in all distillation configurations (stream ABC and streams A, B, and C in Figure 7.2), each distillation configuration is unique because of the set of transfer streams it utilizes. This framework involves six steps that have been summarized in the following section.

7.2.2 Computationally efficient search space synthesis

The six steps of the method developed by Shah and Agrawal [24] are explained and illustrated with a four-component separation example.

Step 1: Obtain the value of n, the number of product streams that the feed mixture is to be separated into.

For our illustration, $n = 4$.

Step 2: Generate an $n \times n$ upper triangular matrix, i.e. all elements below the diagonal are assigned numerical values of 0.

For our four-component separation example, we obtain the matrix shown in Figure 7.6.

Step 3: Define a unique correspondence between upper triangular elements of the matrix to streams that are encountered in a configuration.

If the components are numbered according to their volatility (A: component 1, B: component 2, etc.), then the following convention can be used:

1. The stream(s) corresponding to the matrix element(s) of row i begins with component i, and when a stream contains multiple components, the subsequent components are $i + 1$, $i + 2$, etc.
2. The stream(s) corresponding to the matrix element(s) of column j contains "$n + 1 - j$" components.

For the four-component separation example, we thus obtain the correspondence shown in Figure 7.7. We can thus see that for any n-component separation problem, the (1,1) element of a matrix will always correspond to the feed stream, while the elements in the nth column of the matrix will always correspond to the final product streams. The remaining upper triangular elements correspond to "submixtures" that

$$\begin{bmatrix} - & - & - & - \\ 0 & - & - & - \\ 0 & 0 & - & - \\ 0 & 0 & 0 & - \end{bmatrix}$$

FIGURE 7.6 Upper Triangular Matrix for a Four-Component Separation

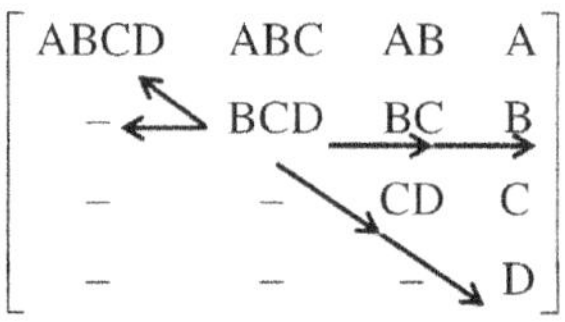

$$
\begin{bmatrix}
ABCD & ABC & AB & A \\
 & BCD & BC & B \\
 & & CD & C \\
 & & & D
\end{bmatrix}
$$

FIGURE 7.7 Associating Upper Triangular Matrix Elements to Streams in a Configuration

are transferred between two distillation columns of a multicolumn distillation arrangement. These are also referred to as "transfer streams".

It can also be seen that all possible top products of the distillation of any feed stream can only lie on a horizontal path to the right of the feed stream in the matrix. Similarly, all possible bottom products of the distillation of any feed stream can only lie on a diagonal path to the right of the feed stream in the matrix. Conversely, all possible feed streams that can produce a particular stream in the matrix by distillation will only lie on horizontal or diagonal paths to the left of the particular stream. These observations are illustrated in Figure 7.7 for stream BCD, whereby it can be formed only by distilling stream ABCD; and distillation of stream BCD can result only in streams BC or B as top products and streams CD or D as bottom products.

Step 4: Using binary integer variables for the upper triangular matrix elements, generate candidate matrices that represent all possible combinations of the presence and absence of transfer streams.

Presence of streams in a configuration can be indicated by numerical values of 1, and absence of streams in a configuration can be indicated by numerical values of 0. In an $n \times n$ matrix, there are $n(n + 1)/2$ upper triangular elements. Among these, $n + 1$ elements correspond to the main feed stream and the final product streams. Therefore $d = n(n + 1)/2 - (n + 1)$, elements correspond to submixtures and can take values of either 0 or 1, where d refers to the degrees of freedom. For instance, the matrix shown in Figure 7.8(i) has values of 1 at the elements associated with the main feed stream (ABCD) and the final product streams (A, B, C, D). The elements associated with the submixture streams (ABC, BCD, AB, BC, CD) have values of 0. Therefore, the matrix shown in Figure 7.8(i) corresponds to a configuration that does not transfer any submixture streams between distillation columns. The matrix shown in Figure 7.8(ii) describes the transfer of stream CD between distillation columns, while the matrix shown in Figure 7.8(iii) describes transfer of stream BC between distillation columns. By continuing with such combinations, it can be seen that the $2^d = 2^5 = 32$ candidate matrices shown in Figure 7.8 for a four-component separation represent all possible combinations of the presence and absence of transfer streams.

Step 5: Eliminate matrices that correspond to infeasible configurations.
The following rules are used:

1. For every stream that exists in the matrix (except the main feed stream), ensure that at least one corresponding stream that can act as its feed also exists within the matrix. This rule basically states that for a submixture or a product stream to

$$
\text{(i)}\begin{bmatrix}1&0&0&1\\0&0&0&1\\0&0&0&1\\0&0&0&1\end{bmatrix}\quad
\text{(ii)}\begin{bmatrix}1&0&0&1\\0&0&0&1\\0&0&1&1\\0&0&0&1\end{bmatrix}\quad
\text{(iii)}\begin{bmatrix}1&0&0&1\\0&0&1&1\\0&0&0&1\\0&0&0&1\end{bmatrix}\quad
\text{(iv)}\begin{bmatrix}1&0&0&1\\0&0&1&1\\0&0&1&1\\0&0&0&1\end{bmatrix}
$$

$$
\text{(v)}\begin{bmatrix}1&0&1&1\\0&0&0&1\\0&0&0&1\\0&0&0&1\end{bmatrix}\quad
\text{(vi)}\begin{bmatrix}1&0&1&1\\0&0&0&1\\0&0&1&1\\0&0&0&1\end{bmatrix}\quad
\text{(vii)}\begin{bmatrix}1&0&1&1\\0&0&1&1\\0&0&0&1\\0&0&0&1\end{bmatrix}\quad
\text{(viii)}\begin{bmatrix}1&0&1&1\\0&0&1&1\\0&0&1&1\\0&0&0&1\end{bmatrix}
$$

$$
\text{(ix)}\begin{bmatrix}1&0&0&1\\0&1&0&1\\0&0&0&1\\0&0&0&1\end{bmatrix}\quad
\text{(x)}\begin{bmatrix}1&0&0&1\\0&1&0&1\\0&0&1&1\\0&0&0&1\end{bmatrix}\quad
\text{(xi)}\begin{bmatrix}1&0&0&1\\0&1&1&1\\0&0&0&1\\0&0&0&1\end{bmatrix}\quad
\text{(xii)}\begin{bmatrix}1&0&0&1\\0&1&1&1\\0&0&1&1\\0&0&0&1\end{bmatrix}
$$

$$
\text{(xiii)}\begin{bmatrix}1&0&1&1\\0&1&0&1\\0&0&0&1\\0&0&0&1\end{bmatrix}\quad
\text{(xiv)}\begin{bmatrix}1&0&1&1\\0&1&0&1\\0&0&1&1\\0&0&0&1\end{bmatrix}\quad
\text{(xv)}\begin{bmatrix}1&0&1&1\\0&1&1&1\\0&0&0&1\\0&0&0&1\end{bmatrix}\quad
\text{(xvi)}\begin{bmatrix}1&0&1&1\\0&1&1&1\\0&0&1&1\\0&0&0&1\end{bmatrix}
$$

$$
\text{(xvii)}\begin{bmatrix}1&1&0&1\\0&0&0&1\\0&0&0&1\\0&0&0&1\end{bmatrix}\quad
\text{(xviii)}\begin{bmatrix}1&1&0&1\\0&0&0&1\\0&0&1&1\\0&0&0&1\end{bmatrix}\quad
\text{(xix)}\begin{bmatrix}1&1&0&1\\0&0&1&1\\0&0&0&1\\0&0&0&1\end{bmatrix}\quad
\text{(xx)}\begin{bmatrix}1&1&0&1\\0&0&1&1\\0&0&1&1\\0&0&0&1\end{bmatrix}
$$

$$
\text{(xxi)}\begin{bmatrix}1&1&1&1\\0&0&0&1\\0&0&0&1\\0&0&0&1\end{bmatrix}\quad
\text{(xxii)}\begin{bmatrix}1&1&1&1\\0&0&0&1\\0&0&1&1\\0&0&0&1\end{bmatrix}\quad
\text{(xxiii)}\begin{bmatrix}1&1&1&1\\0&0&1&1\\0&0&0&1\\0&0&0&1\end{bmatrix}\quad
\text{(xxiv)}\begin{bmatrix}1&1&1&1\\0&0&1&1\\0&0&1&1\\0&0&0&1\end{bmatrix}
$$

$$
\text{(xxv)}\begin{bmatrix}1&1&0&1\\0&1&0&1\\0&0&0&1\\0&0&0&1\end{bmatrix}\quad
\text{(xxvi)}\begin{bmatrix}1&1&0&1\\0&1&0&1\\0&0&1&1\\0&0&0&1\end{bmatrix}\quad
\text{(xxvii)}\begin{bmatrix}1&1&0&1\\0&1&1&1\\0&0&0&1\\0&0&0&1\end{bmatrix}\quad
\text{(xxviii)}\begin{bmatrix}1&1&0&1\\0&1&1&1\\0&0&1&1\\0&0&0&1\end{bmatrix}
$$

$$
\text{(xxix)}\begin{bmatrix}1&1&1&1\\0&1&0&1\\0&0&0&1\\0&0&0&1\end{bmatrix}\quad
\text{(xxx)}\begin{bmatrix}1&1&1&1\\0&1&0&1\\0&0&1&1\\0&0&0&1\end{bmatrix}\quad
\text{(xxxi)}\begin{bmatrix}1&1&1&1\\0&1&1&1\\0&0&0&1\\0&0&0&1\end{bmatrix}\quad
\text{(xxxii)}\begin{bmatrix}1&1&1&1\\0&1&1&1\\0&0&1&1\\0&0&0&1\end{bmatrix}
$$

FIGURE 7.8 Candidate Matrices for a Four-Component Separation

exist in a configuration, there must exist a distillation column with a feed that can produce the stream of interest.

2. Disallow disappearance of components in a split (such as splitting of BCDE to B and DE, since component C cannot "disappear").

These rules can be converted to mathematical constraints and implemented in a computer program [24]. For the four-component problem, the matrices shown in Figure 7.8(i)−(v), (vii)−(ix), (xiii), (xvii)−(xviii) can be eliminated based on rule 1, while the matrices shown in Figure 7.8(i)−(v), (vii), (ix), (xiii), (xvii)−(xviii), (xxv)−(xxvi), (xxix) can be eliminated on the basis of rule 2. Each matrix that remains after this elimination corresponds to a unique feasible distillation configuration.

The use of these rules to eliminate infeasible matrices is illustrated in Figure 7.9 for the matrix shown in Figure 7.8(ii). The (2,4) element of the matrix corresponds to stream B and can be obtained only from a distillation of either stream AB or stream BCD or stream BC, none of which are present in the matrix, i.e. the corresponding

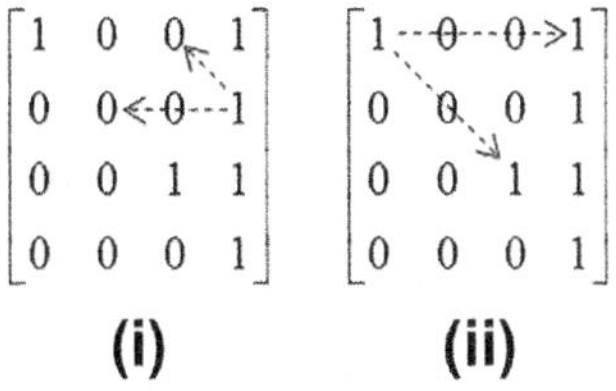

$$
\text{(i)}\begin{bmatrix}1&0&0&1\\0&0&0&1\\0&0&1&1\\0&0&0&1\end{bmatrix}\qquad
\text{(ii)}\begin{bmatrix}1&0&0&1\\0&0&0&1\\0&0&1&1\\0&0&0&1\end{bmatrix}
$$

FIGURE 7.9 Application of Rules to Eliminate Infeasible Matrices

(i) all possible feed mixtures of stream B are absent and (ii) component B disappears during the distillation of ABCD to A and CD.

matrix elements have values of 0 as shown in Figure 7.9(i). Therefore, this matrix violates rule 1 and is classified as an infeasible matrix. At this point, it is not even necessary to apply rule 2 to the matrix. However, as Figure 7.9(ii) illustrates, this matrix also happens to violate rule 2 by describing a split of ABCD to A and CD with component B disappearing.

Step 6: Convert each remaining matrix to a distillation configuration by placing splits that make common product streams in the same distillation column.

This is illustrated in Figure 7.10 for the configuration shown in Figure 7.8(xxx). In Figure 7.8(xxx) (which corresponds to a feasible matrix), the elements that have values of 1 are replaced with the corresponding streams as shown in Figure 7.10 prior to systematically enumerating all the splits described by the matrix (ABCD−ABC/ BCD, ABC−AB/C, BCD−B/CD, AB−A/B, CD−C/D). Each split can be represented by a pseudo-distillation column as shown in Figure 7.10. Pseudo-columns producing a common stream are then placed in the same actual distillation column. When a pseudo-column producing a product stream from its bottom is combined with another pseudo-column producing the same product stream from its top, then the associated reboiler and condenser are eliminated and the product stream is now produced from an intermediate location of the resulting actual distillation column. Products B (pseudo-columns marked 3) and C (pseudo-columns marked 2) are such examples.

The six-step procedure results in the synthesis of 18 distillation configurations shown in Figure 7.11 for a four-component separation. The transfer streams in Figure 7.11 are associated with simple reboilers and condensers. Other reboiler and condenser arrangements can also be used when appropriate [25,26].

The n-component distillation configurations shown in Figures 7.2 and 7.11 that use $n - 1$ distillation columns, with each distillation column using one reboiler and condenser, and produce each of the n products only once, are called basic configurations [20]. Furthermore, any n-component configuration that uses $n - 1$ distillation columns is referred to as a regular-column configuration. As we will see later in the section on thermal coupling in this chapter, a regular-column configuration need not be a basic configuration (but the reverse is always true).

The six steps can easily be automated using a computer program, for any value of n. The degrees of freedom (d) for the six-step formulation are of the order of n^2 for an n-component separation. For a state-task network formulation and a supernetwork formulation, the degrees of freedom are of the order of n^3. Therefore, these methods have to explore $2^{\circ(n^3)}$ candidates to synthesize all possible schemes, while the six-step formulation has to explore $2^{\circ(n^2)}$ candidates. An alternative matrix-based approach developed by Ivakpour and Kasiri is based on classifying upper triangular matrix elements as being associated with either a reboiler or a condenser or a side-stream [27]. These three choices for upper triangular matrix elements result in the exploration of $3^{\circ(n^2)}$ candidates to synthesize all possible schemes. The six-step formulation described in this chapter is thus preferred due to its greater computational efficiency. In fact, the six-step method can elucidate the exact number of distillation configurations for even seven- and eight-component separations as shown in Table 7.1 (see only configurations without thermal

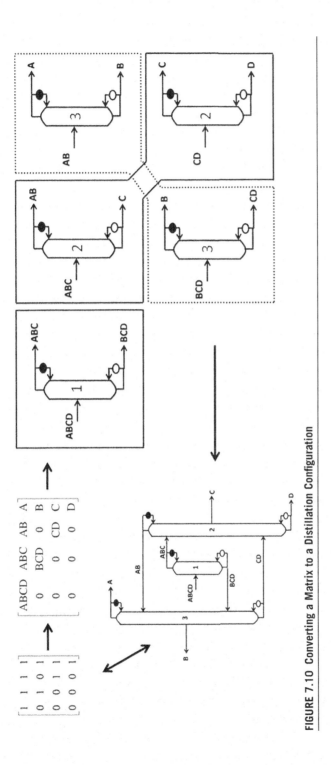

FIGURE 7.10 Converting a Matrix to a Distillation Configuration

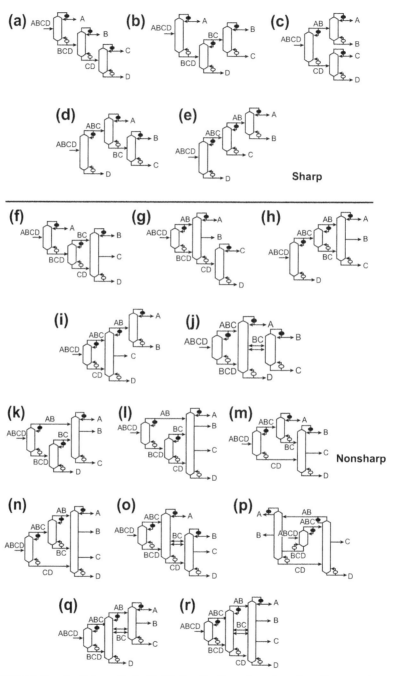

FIGURE 7.11 Regular-Column Basic Distillation Configurations for a Four-Component Separation

Table 7.1 Number of Regular-Column Distillation Configurations

Number of Components	Number of Configurations Without Thermal Coupling			Number of Configurations with Thermal Coupling			Total Number of Regular-Column Configurations
	Sharp	Nonsharp	Total	Partially Thermally Coupled	Completely Thermally Coupled	Total	
3	2	1	3	2	3	5	8
4	5	13	18	116	18	134	152
5	14	189	203	5722	203	5925	6128
6	42	4331	4373	498,166	4373	502,539	506,912
7	132	185,289	185,421	84,845,350	185,421	85,030,771	85,216,192
8	429	15,766,778	15,767,207	28,991,159,474	15,767,207	29,006,926,681	29,022,693,888

coupling). From the second, third, and fourth columns in Table 7.1, we can see that the search space contains mostly nonsharp split configurations. Giridhar and Agrawal carried out a thorough numerical analysis of 120 different four-component separations and showed that nonsharp split configurations must be included in a search space [6]. Therefore, in spite of the large size of the search space, each sharp and nonsharp split configuration must be included to ensure successful identification of the optimal scheme for a given distillation need.

7.3 Thermal coupling

An often-seen feature in several multicolumn distillation configurations is the presence of thermal coupling links. An introduction to thermal coupling links is provided in the previous chapter. These links replace condensers or reboilers associated with transfer streams by two-way liquid vapor communications between distillation columns. Figure 7.12 shows the introduction of thermal coupling links in the configurations of Figure 7.2(a) and (c). It is not necessary to replace all the transfer stream heat exchangers in a configuration by thermal coupling links. For instance, Figure 7.13(b)–(p) shows all possible thermally coupled configurations that can be derived from the configuration shown in Figure 7.13(a).

In Figure 7.2(c), a portion of the vapor at the top of the first distillation column is condensed to provide reflux to the first distillation column. Replacement of condenser AB by a thermal coupling link in Figure 7.12(b) shows that thermal coupling allows an augmented flow of vapor from the first distillation column to the second distillation column because no vapor is condensed at the eliminated intermediate temperature level condenser (AB). The augmented vapor flow is condensed at the top of the second distillation column at the lowest temperature in condenser A. The condensate provides reflux and supports the separation in the top section of the second distillation column before a portion of it is returned as

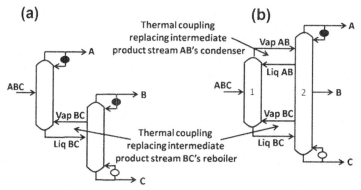

FIGURE 7.12 Thermal Coupling Links [24]

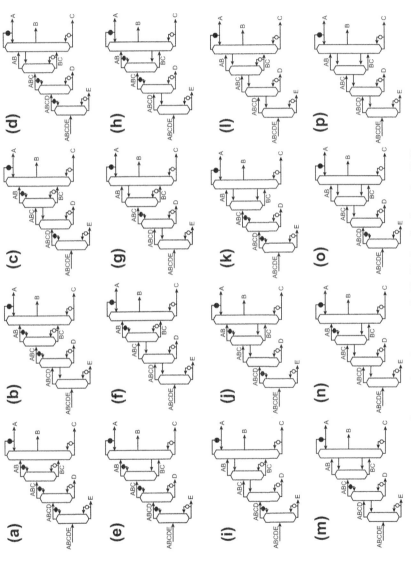

FIGURE 7.13 Thermally Coupled Configurations Derived from a Basic Distillation Configuration [28]

reflux to the first distillation column. Similarly, thermal coupling at reboiler BC uses augmented liquid flow to generate vapor flow and drive some of the separation in the bottom section of the second distillation column. Therefore, it can be seen that thermal coupling permits the use of already generated vapor and liquid to drive a greater amount of separation by eventually condensing the vapor or boiling the liquid at a lower or higher temperature, respectively. Thermal coupling links can thus reduce the total heat demand or total cooling demand for distillations. The lower utility consumption of thermally coupled configurations generally (but not necessarily always) occurs at the expense of utility temperature level [28]. Some well-known examples of commercial distillation processes that utilize thermal coupling links are the atmospheric distillation train for petroleum crude distillation and argon production from cryogenic air distillation.

If all the transfer stream heat exchangers in a distillation configuration are replaced by thermal coupling links, the resulting scheme is referred to as a "completely" thermally coupled configuration. However, if one or more of the transfer stream heat exchangers are retained, the scheme is referred to as a "partially" thermally coupled configuration. The configurations shown in Figure 7.13(b)−(o) are thus partially thermally coupled configurations, while the configuration shown in Figure 7.13(p) is a completely thermally coupled configuration. It can be shown that a completely thermally coupled configuration will have the lowest heating and cooling demand from among all the thermally coupled configurations that can be derived from a basic configuration (without thermal coupling links). Therefore, completely thermally coupled configurations should be included in a search space. However, a completely thermally coupled configuration will demand more heat and cooling at the higher temperature reboiler(s) and lower temperature condenser(s), respectively, as compared to partially thermally coupled configurations. Therefore, a tradeoff exists between total utility demand and utility temperature levels, with partially thermally coupled configurations well poised to strike an appropriate balance.

7.3.1 Impact of thermal coupling on utility demand and temperature levels

Most studies on thermally coupled configurations have focused on completely thermally coupled configurations, but a recent study compared the performance of partially thermally coupled configurations to completely thermally coupled configurations [28]. This study found that:

1. Some thermal coupling links can unnecessarily incur a temperature level penalty without reducing the utility demand. In such situations, partially thermally coupled configurations are preferred over completely thermally coupled configurations. Therefore, such thermal coupling links should not be introduced unless other constraints in a plant warrant their use. Importantly, partially thermally coupled configurations should be included in a search space.

2. Sometimes, thermal coupling links can reduce the utility demand without incurring any temperature level penalty. These types of thermal coupling links

should certainly be included in a distillation configuration. A qualitative example of this phenomenon is illustrated in Figure 7.14.

Each split in a distillation configuration has a vapor flow requirement (based on minimum reflux calculations). If the actual vapor flow is lower than the vapor flow requirement of a split, the split will not achieve its desired separation. In Figure 7.14, the lengths of the vertical arrows inside the distillation columns represent the vapor flow requirements of the corresponding splits. Therefore, for the separation we are considering, split CD−C/D has the largest vapor flow requirement, followed by splits ABC−A/BC and BC−B/C, respectively. In fact, let us assume that the vapor flow requirement for split CD−C/D is greater than the sum of the vapor flow requirements of splits ABC−A/BC and BC−B/C. Since splits CD−C/D and BC−B/C belong to the same distillation column, the vapor generated at reboiler D corresponds to the larger of their vapor flow requirements (i.e. split CD−C/D) to ensure that both splits have sufficient vapor. The column sections that separate BC−B/C will thus have surplus vapor flow. Introduction of a thermal coupling link to replace reboiler BC can utilize a portion of this surplus vapor flow that equals the requirement of the ABC−A/BC split, to drive the separation in the second distillation column without requiring the generation of extra vapor at reboiler D. Therefore, the total utility demand is lowered by elimination of reboiler BC, but importantly there is no increase in the heat demanded at the high temperature reboiler D. This is illustrated in Figure 7.14(b). Such special separations thus allow lower utility demand without temperature level penalties.

This discussion shows that partially as well as completely thermally coupled configurations can be optimal configurations for a given distillation need depending on the characteristics of the mixture to be separated, and both types of configurations should be included in a search space. These thermally coupled configurations can easily be synthesized using the mathematical frameworks mentioned in the previous section. Table 7.1 lists the number of thermally coupled distillation configurations for up to eight component separations and was generated using the six-step matrix-based framework. It can be seen that the number of thermally coupled

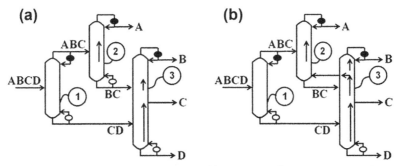

FIGURE 7.14 Thermal Coupling that Reduces Utility Demand Without a Temperature Level Penalty [28]

configurations is significantly larger than the number of basic configurations without thermal coupling. This is because a basic distillation configuration corresponds to exactly one completely thermally coupled configuration but can potentially lead to several partially thermally coupled configurations (see Figure 7.13). Since the objective of a search space is to include all potentially optimal configurations, all thermally coupled configurations should be included in spite of their large numbers.

7.3.2 Elimination of any utility temperature penalty due to thermal coupling

An interesting observation is that if we introduce liquid–vapor communications between distillation columns while retaining heat exchangers, we can avoid the temperature level penalty altogether for any separation [28,29]. These types of thermal coupling links are not pure thermal coupling links as defined earlier. Any retained transfer stream heat exchangers associated with these thermal coupling links will be smaller than the original heat exchangers and will have lower utility demand. This concept is illustrated in Figure 7.15. The lengths of the vertical arrows inside the distillation columns once again represent the vapor flow requirements for the corresponding splits. We are thus considering a separation where split ABC–A/BC has the largest vapor flow requirement followed by splits CD–C/D and BC–B/C, respectively. Therefore, introduction of a pure thermal coupling link at reboiler BC (like Figure 7.14(b)) will require generation of more vapor at reboiler D than that demanded by split CD–C/D. On the other hand, transferring only the available amount of surplus vapor to the bottom of the second distillation column together with retention of a smaller reboiler BC still reduces the total utility demand of the configuration without any temperature level penalty.

Several options are available for these types of thermal coupling links [28,29]. For instance in Figure 7.15(a), a portion of the bottom liquid BC from the second column is totally vaporized and fed to the column, while allowing the vapor transfer from the third distillation column to be introduced sufficiently above the bottom of the second distillation column to take advantage of the rectified composition of the transfer vapor stream. Alternatively, in Figure 7.15(b), the partial reboiler BC is configured such that the boiled-up vapor is in equilibrium with the liquid BC,

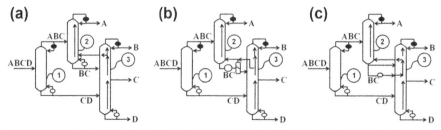

FIGURE 7.15 Partial Retention of Heat Exchangers during Thermal Coupling to Avoid Temperature Penalty [28]

thereby allowing it to be combined with the vapor being transferred from the third distillation column. Another option is shown in Figure 7.15(c), where the vapor generated by reboiler BC is introduced below liquid BC in the third distillation column. This indirectly augments the transfer vapor stream flow to fully meet the vapor demand of split ABC−A/BC.

7.3.3 Operability of thermally coupled columns

A known concern of thermally coupled configurations with multiple thermal coupling links is the potential transfer of vapor streams in opposite directions between distillation columns. For instance, in the thermally coupled configuration shown in Figure 7.16(a), the vapor associated with stream BC is transferred from the second distillation column to the first distillation column, while the vapor associated with stream AB is transferred from the first distillation column to the second distillation column. To allow spontaneous vapor flow and to avoid using compressors, we would require the bottom portion of the first distillation column to be at a lower pressure than the corresponding pressure in the second distillation column, while we would require the top portion of the first distillation column to be at a higher pressure than the corresponding pressure in the second distillation column. Operating such a distillation process would thus require very careful design and adjustment of the pressure drop in each distillation column, along with appropriate process control. However, thermal coupling links allow rearrangement of column sections to obtain distillation configurations that are equivalent to the original configuration [31]. For instance, the configurations shown in Figure 7.16 are thermodynamically equivalent. The configurations shown in Figure 7.16(b) and (c) are thus preferred, since they can alleviate operational difficulties associated with the vapor transfers. Using the technique of rearranging distillation sections, the operational difficulty associated with an intercolumn vapor transfer can thus be alleviated in thermally coupled configurations and configurations can be made operational [32]. Other techniques have also been developed for mitigating operational difficulties while allowing a sequence to realize the benefit of thermal coupling links [33]. From the synthesis perspective, these rearranged configurations can be treated as a single entity. The appropriate rearrangements can be considered after an optimal configuration is identified.

7.4 Identifying optimal configurations

As the previous sections illustrate, mathematical frameworks can successfully synthesize all possible regular-column distillation configurations with and without thermal coupling. These distillation configurations use $n - 1$ distillation columns for an n-component separation. Each of the synthesized configurations can potentially be optimal for a given distillation need. The number of candidate configurations is extremely large, and computational tools are required to identify optimal distillation

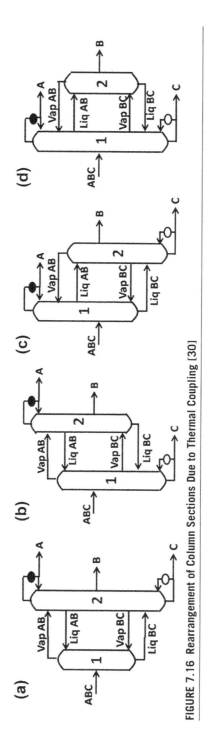

FIGURE 7.16 Rearrangement of Column Sections Due to Thermal Coupling [30]

schemes from the search space. Even with the tremendous advances in computational and optimization techniques, identifying optimal schemes is not a trivial task.

From a chemical plant's perspective, an optimal distillation configuration generally has the least total cost (operating plus capital cost). Estimating the operating cost of a multicolumn distillation arrangement is fairly straightforward for most separations. On the other hand, the procedure for estimating the capital cost of a multicolumn distillation arrangement can vary significantly depending on the separation under consideration, because of factors like the material of construction for instance. There are several other important features in a chemical plant that cannot be quantified as easily, such as the safety aspects of a configuration, its operational robustness in the face of process disturbances, etc. The objective function or criterion for identifying the optimal scheme thus depends greatly on the separation under consideration and other plant and market-driven constraints.

Regardless of the choice of the objective function, several variables can be manipulated to optimize a given distillation configuration as described in Chapter 12. The vast majority of these variables are mathematically continuous variables such as the flow rates and compositions of transfer streams, reflux ratios for the distillation columns, heights and diameters of the distillation columns, operating pressures in the distillation columns, heating and cooling utility temperatures, etc. Discrete variables like number of stages and feed stage location also exist for distillation columns that use trays. The variables generally have nonlinear dependencies on each other, and also affect most objective functions in a nonlinear fashion. Therefore, optimization of a single distillation configuration generally requires the use of a nonlinear programing solver. The nonlinear programing formulation for a distillation configuration strongly depends on the objective function under consideration and the desired level of accuracy in the calculations. Therefore, details of nonlinear programing formulations will not be discussed here but are addressed in Chapter 12.

Two approaches can be used for the problem of identifying the optimal sequence for a given distillation need from among several distillation configurations: (1) a simultaneous optimization approach, and (2) a sequential optimization approach. These approaches are summarized in Figure 7.17.

We have used the sequential approach for up to five-component separation problems and have found it to work well (reliable convergence to globally optimal solutions within reasonable computational time). We believe that it should be feasible to satisfactorily extend the sequential approach to six-component distillations, especially through the use of parallel computing. The simultaneous approach has also been used by researchers [21,34,35] but has currently not developed to a level where global optimality of the solution can be guaranteed. The use of the optimization solver BARON [36] can help overcome such limitations because of its ability to certify global optimality.

It is challenging to perform an optimization exercise with real thermodynamic phase equilibrium parameters when the search space includes thousands to millions of configurations. To overcome this problem, we recommend the use of

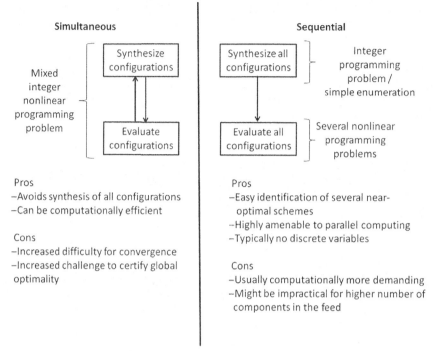

FIGURE 7.17 Strategies for Identifying Optimal Distillation Configurations

Underwood's method [37] using constant relative volatilities for a first-pass screening. Such an exercise provides the top candidates, which can then be simulated in detail using a commercial process simulator with real thermodynamic phase equilibrium parameters. We have found this technique to work reliably. For four- and five-component mixtures, we used the sequential approach to rank list all the configurations according to their heat duty using constant relative volatilities and Underwood's equations to calculate vapor flow rates. The top few candidates were then re-optimized using ASPEN Plus, and the rank listing was found to hold fairly well.

In summary, the choice of either the sequential or the simultaneous approach can be made by practitioners based on the guidelines presented in Figure 7.17. The sequential approach rank lists all the configurations, and quite often, multiple configurations have similar heat duties and costs. This provides a process engineer a choice to examine these similar configurations according to other criteria such as operability, manufacturability, retrofitability, etc. To our knowledge, no commercial software or simulator currently provides either of these two approaches for synthesizing optimal distillation configurations. However, several tools based on these approaches are being developed by academic research groups with a focus on improving convergence and speed [21,38].

7.5 **An example: petroleum crude distillation**

Crude oil is typically distilled into five fractions using the configuration shown in Figure 7.18(a). This configuration consists of a main column with side-strippers and has been in use for more than 75 years [39]. The thermal coupling links allow rearrangement of the column sections, and it can be seen that the configuration of Figure 7.18(a) is equivalent to the more familiar version shown in Figure 7.18(b). According to our nomenclature, the conventional configuration for crude distillation is thus a completely thermally coupled sharp split configuration that separates a mixture into five product streams. It is worth noting that an actual crude distillation configuration uses pump-arounds. Also, steam is introduced at the bottom of the columns in lieu of reboilers, and heat integrations with the plant are employed. However, as a first pass, the main features of the configuration used are captured by the one shown in Figure 7.18 [39].

From Table 7.1, we can see that a total of 6128 configurations with and without thermal coupling exist for separating a mixture into five product streams. The conventional configuration is just one of these 6128 possible options.

The minimum total vapor duty requirement of a distillation configuration has frequently been used in the literature as a popular substitute for the total cost of a configuration because it is proportional to the heat demand of a configuration and to the diameters of the distillation columns in the configuration [40]. The minimum total vapor duty requirement of a configuration is basically the sum of the vapor flows generated at all the reboilers in the configuration, with the calculations being done under minimum reflux conditions.

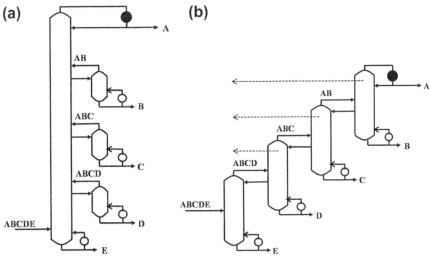

FIGURE 7.18 Petroleum Crude Distillation [24]

Underwood's equations can be used to estimate the minimum vapor duty requirement of any split in a distillation configuration [37]. One approach for optimizing a distillation configuration using Underwood's equations utilizes the concept of a "transition split" [41] and involves sequential optimization of each split in the distillation configuration. This sequential optimization of a single distillation configuration can be carried out analytically [25]. Another approach involves simultaneous optimization of all the splits in a configuration. This approach cannot be carried out analytically and requires the use of a nonlinear programing optimization solver. Through a comprehensive numerical study of four- and five-component separations, Nallasivam et al. showed that the analytical sequential method is unable to estimate the global minimum total vapor duty requirement of all distillation configurations [38]. In fact, in some cases, the analytical sequential method was observed to give vapor duty requirements that were 83% higher than the globally optimal vapor duty requirements estimated using the simultaneous optimization of splits using nonlinear programing. Therefore, the nonlinear programing-based simultaneous optimization strategy is needed for accurately optimizing a given distillation configuration.

The minimum total vapor duty requirements of each of the 6128 candidate configurations for crude distillation were obtained using the sequential optimization strategy shown in Figure 7.17 coupled with the simultaneous optimization strategy for each distillation configuration as described above [24]. The petroleum crude distillation search space was thus systematically explored by solving 6128 nonlinear programing problems to global optimality to obtain a rank-ordered list of configurations with respect to their total vapor duty requirements [24]. Interestingly, this study showed that the conventional configuration was the best among the historically known sharp split configurations, in spite of the lack of computational tools in the era in which it was designed. However, the study also showed that thousands of non-sharp split configurations had significantly lower minimum total vapor duty requirements than the conventional configuration. Some of these configurations were structurally quite similar to the conventional configuration, thereby indicating possible retrofit opportunities for the existing refineries. The top candidates after optimization of the complete space of 6128 configurations were estimated to have vapor duty reductions of nearly 48%! See Figure 7.19 for an example of such a configuration with estimated savings of 48% and 17% for representative light and heavy crude oil mixtures, respectively [24]. If such configurations are successfully implemented in all refineries around the world, there is a potential to save nearly 292 million bbl oil/year, with annual savings running in billions of dollars. In fact, a giant oil field is defined as one that has produced or will produce at least 100 million bbl of oil [42]. These numbers thus potentially translate to the discovery of a new giant oil field every 4 months.

This example illustrates the value of using computer-aided mathematical tools for conceptual design of multicolumn distillation arrangements. Significant benefits can be realized not just for new distillation needs but also for "mature" distillation processes.

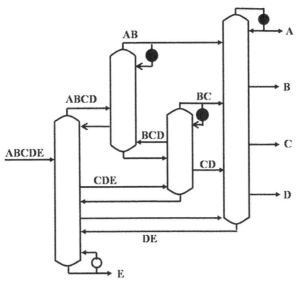

FIGURE 7.19 An Attractive Candidate for Petroleum Crude Distillation [24]

7.6 Additional multicolumn configurations

Most of the literature on the synthesis of distillation configurations has focused on configurations that use $n - 1$ distillation columns for separating a mixture into n product streams. These distillation configurations are thus referred to as "regular-column" configurations. Some configurations with more than $n - 1$ columns and less than $n - 1$ columns have also been studied in the literature. We shall refer to them as "plus-column" and "subcolumn" configurations, respectively. An example of a plus-column configuration is shown in Figure 7.20, whereas examples of sub-column configurations are shown in Figure 7.1.

Sometimes feed streams are split and processed in parallel distillation columns that perform the same separation task (for structural or asset utilization purposes). Alternatively, distillation columns are occasionally split into different shells for tall tower designs. Such situations are not considered in our nomenclature of plus-column, regular-column, and subcolumn configurations. Plus-column configurations such as the one shown in Figure 7.20 use more distillation columns than regular-column configurations to carry out the same overall separation and are expected to have higher capital costs. Moreover, a recent study compared the minimum total vapor duty requirements of all possible four-component plus-column and regular-column configurations without thermal coupling for 120 different feed conditions and found that an optimal regular-column configuration always had lower minimum total vapor duty requirement than any plus-column configuration [6]. A similar result is expected to be valid for separations involving any number of components,

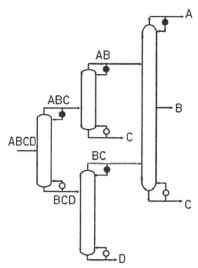

FIGURE 7.20 A Plus-Column Configuration that Uses Four Distillation Columns to Separate a Mixture into Four Product Streams [6]

making it acceptable to eliminate plus-column configurations from a search space. Table 7.2 shows the number of plus-column configurations without thermal coupling for up to six component separations. For a feed containing more than four components, a comparison with Table 7.1 shows that the number of plus-column configurations without thermal coupling is much larger than the number of regular-column configurations without thermal coupling. This difference will be even more pronounced if we include thermally coupled derivatives of these configurations. Therefore, elimination of plus-column configurations from the search space is quite beneficial for identification of the optimal distillation schemes in a practical amount of computational time.

Table 7.2 Additional Multicolumn Configurations

Number of Components	Number of Plus-Column Configurations Without Thermal Coupling	Number of Subcolumn Configurations	
		Without Thermal Coupling	With Thermal Coupling
3	0	1	0
4	6	12	18
5	1149	198	1279
6	4,070,490	5142	124,346
7	–	224,257	20,168,590
8	–	17,056,898	5,739,609,045

On the other hand, subcolumn configurations are expected to have lower capital costs than regular-column configurations. Subcolumn configurations have actually been studied at least since 1942 [43,44]. However, methods to synthesize subcolumn configurations have been published only recently [45,46]. Errico et al. published a method to generate sharp split subcolumn configurations [45]. Shenvi et al. developed a mathematical framework to synthesize all possible subcolumn configurations, including nonsharp split configurations, by extending the six-step formulation for regular-column configurations described in this chapter [46]. Their development of a systematic framework led to the synthesis of several novel subcolumn configurations even for a four-component separation (shown in Figure 7.21). For a case study involving purification of four n-alkanes, it was shown that one of the novel four-component subcolumn configurations had only slightly higher total reboiler duty than an optimal regular-column configuration, while

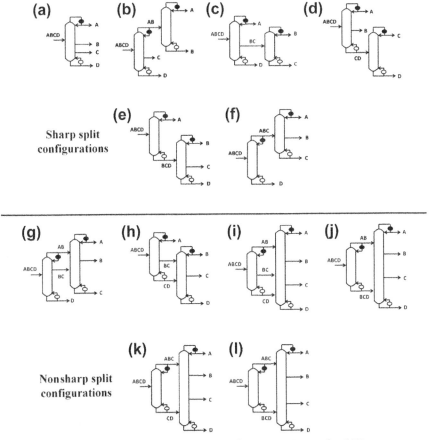

FIGURE 7.21 Subcolumn Configurations for a Four-Component Separation [46]

utilizing one lesser distillation column. Subcolumn configurations can thus be potentially optimal for a given distillation need, if we consider both capital and operating costs. Therefore, they should be included in a search space.

The number of subcolumn configurations with and without thermal coupling are listed in Table 7.2 for up to eight component separations. From Table 7.2, it can be seen that only a single subcolumn configuration exists for a three-component separation. However, Figure 7.1 shows two different arrangements of a subcolumn configuration for a three-component separation. For the purpose of synthesizing a search space, these two configurations can be treated as a single entity with the location of the side-stream subject to optimization. This avoids the introduction of multiplicity and limits the computational burden for synthesizing a search space. The 12 subcolumn configurations without thermal coupling for a four-component separation (as indicated in Table 7.2) are shown in Figure 7.21.

Subcolumn configurations are often believed to be unable to produce all product streams with arbitrarily high purity. For instance, the configurations in Figure 7.1 will need very large reflux ratios to increase the purity of side-stream B. However, several of the newly synthesized subcolumn configurations were found to not have such constraints on purity. In other words, these configurations can make product streams of arbitrarily high purity just like the regular-column configurations. For instance, if the configuration of Figure 7.21(j) is optimized to produce high purity product streams, stream BCD will have a very low concentration of component B. The resulting configuration is shown in Figure 7.22(b). This configuration, and other subcolumn configurations that have no limitations on product purities, are found to contain a heat-and-mass integrated section. The heat-and-mass integrated section is so named because these subcolumn configurations can be synthesized from regular-column configurations by introducing simultaneous heat-and-mass integration between two distillation columns (see Figure 7.22(a)).

It can be seen that a heat-and-mass integrated section is bounded by two streams. If these two streams have overlapping components, the overlapping components will go through a maximum in the concentration profile in the heat-and-mass integrated

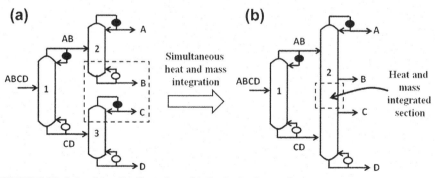

FIGURE 7.22 Subcolumn Configuration with No Purity Constraint [47]

section. A recent finding shows that a natural amount of the overlapping components can be withdrawn from the heat-and-mass integrated section to exploit the peak in the concentration profile, thereby leading to a significantly lower utility demand for the overall separation [47]. This discovery thus led to an entirely new class of subcolumn configurations. These configurations should also be included in the search space. Two examples of these novel configurations are shown in Figure 7.23. In the configuration shown in Figure 7.23(a), the middle section of the second distillation column is bounded by streams BC and CD. If sufficient theoretical stages are present, the composition profile in that section can exhibit a high purity of component C at some intermediate location. This allows the withdrawal of some amount of product-grade stream C from the second distillation column itself and results in a lower utility demand by exploiting a separation that is already achieved inside a distillation column in the sequence. Similarly, the configuration shown in Figure 7.23(b) allows for the withdrawal of a second CD stream from the middle section of the second distillation column.

Plus-column configurations are not used in the industry for normal applications. On the other hand, subcolumn configurations are often used in the industry and are referred to as side-stream columns. Columns having single side-streams have been classified as pasteurization columns or heavy-ends columns and have been used for applications such as ethylene fractionation and ethanol-water separation [48].

Finally, another important class of distillation configurations is characterized by the presence of dividing wall columns. The concept of dividing wall columns was invented in 1949 [49], and currently more than a hundred applications of this concept exist in the chemical industry [50]. Dividing wall columns are created by combining two or more thermally coupled distillation columns in a single shell, thereby providing the energy benefit of thermal coupling with a reduced capital cost. Pendergast et al. [51], Kaibel [52], and Kiss [53] have provided overviews

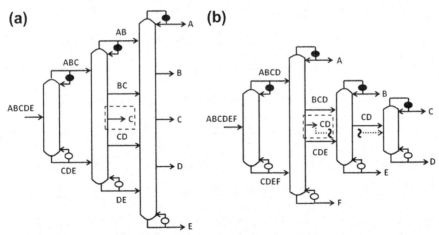

FIGURE 7.23 Strategic Side-Stream Withdrawals that Exploit Peaks in Composition Profiles [47]

of this technology. Because of their additional capital benefits, configurations involving dividing wall distillation columns should also be included in a search space. These configurations can be easily synthesized from partially or completely thermally coupled configurations. Some unconventional dividing wall columns with multiple reboilers and condensers have also been suggested for better control of product purities [54].

7.7 Summary and thoughts toward the future

Real world separations often involve multicomponent feed streams. Multicolumn distillation arrangements can be used to separate these mixtures into product streams of acceptable purity. This chapter was focused on designing optimal multicolumn arrangements for separating near-ideal (nonazeotropic) mixtures.

Mathematical tools for synthesizing all possible configurations (with and without thermal coupling) and identifying optimal schemes were described. These tools were focused on distillation configurations that used $n - 1$ distillation columns for an n-component separation. Other types of multicolumn distillation configurations were also described, with recommendations on whether they should be included or excluded from a search space. These ideas will help readers understand how many configurations exist for a given distillation need and will also help them select optimal distillation configurations. Some of the tools described in this chapter are excellent for a first-pass screening, whereby practicing engineers can quickly narrow down the number of distillation options from thousands and even millions to literally a handful.

Going forward, aspects such as heat integration are expected to play a significant role in the selection of optimal distillation configurations. This includes not just heat integration of a scheme with the rest of the plant but also heat integration within a configuration itself. In fact, heat integration strategies for individual distillation columns in a configuration can also be implemented [55].

Most of the literature for synthesizing and designing optimal distillation schemes has focused on ideal separations. As summarized in this chapter, excellent progress in this area has been made in recent years, with reasonably reliable tools available for use. The extension of these tools to include nonideal distillations is a very promising area for future research. We believe that significant breakthroughs can be realized in this area in the coming years.

References

[1] N. Aske, Characterisation of crude oil components, asphaltene aggregation and emulsion stability by means of near infrared spectroscopy and multivariate analysis (Ph.D. thesis), Norwegian University of Science and Technology, 2002.

[2] M. Bagajewicz, S. Ji, Rigorous procedure for the design of conventional atmospheric crude fractionation units. Part I: targeting, Ind. Eng. Chem. Res. 40 (2) (2001) 617–626.

[3]   W.A. Sweeney, P.F. Bryan, BTX Processing. Kirk-Othmer Encyclopedia of Chemical Technology, 2007.

[4]   A. Shenvi, Synthesis of energy efficient distillation configurations (Ph.D. thesis), Purdue University, 2012.

[5]   BP, Statistical Review of World Energy, June 2012. Available at: www.bp.com/statisticalreview.

[6]   A. Giridhar, R. Agrawal, Synthesis of distillation configurations: I. Characteristics of a good search space, Comp. Chem. Eng. 34 (1) (2010) 73−83.

[7]   F.J. Lockhart, Multi-column distillation of natural gasoline, Pet. Refin. 26 (8) (1947) 104−108.

[8]   W.D. Harbert, Which tower goes where? Pet. Refin. 36 (3) (March 1957) 169−174.

[9]   C.J. King, Separation Sequences, McGraw-Hill, New York, 1971.

[10] H. Nishimura, Y. Hiraizumi, Optimal system pattern for multicomponent distillation systems, Int. Chem. Eng. 11 (1) (1971) 188−193.

[11] D.F. Rudd, G.J. Powers, J.J. Siirola, Process Synthesis, Prentice-Hall, Englewood Cliffs, New Jersey, 1973.

[12] J.D. Seader, A.W. Westerberg, A combined heuristic and evolutionary strategy for synthesis of simple separation sequences, AIChE J. 23 (6) (1977) 951−954.

[13] R.W. Thompson, C.J. King, Systematic synthesis of separation schemes, AIChE J. 18 (5) (1972) 941−948.

[14] J.E. Hendry, R.R. Hughes, Generating separation process flowsheets, Chem. Eng. Prog. 68 (6) (1972) 71−76.

[15] F. Rodrigo, J.D. Seader, Synthesis of separation sequences by ordered branch search, AIChE J. 21 (5) (1975) 885−894.

[16] A.W. Westerberg, G. Stephanopoulos, Studies in process synthesis − I. Branch and bound strategy with list techniques for the synthesis of separation schemes, Chem. Eng. Sci. 30 (8) (1975) 963−972.

[17] R. Agrawal, Synthesis of distillation column configurations for a multicomponent separation, Ind. Eng. Chem. Res. 35 (4) (1996) 1059−1071.

[18] R.W.H. Sargent, K. Gaminibandara, Optimum design of plate distillation columns, in: L.W.C. Dixon (Ed.), Optimization in Action, Academic Press, New York, 1976, pp. 267−314.

[19] R.W.H. Sargent, A functional approach to process synthesis and its application to distillation systems, Comp. Chem. Eng. 22 (1−2) (1998) 31−45.

[20] R. Agrawal, Synthesis of multicomponent distillation configurations, AIChE J. 49 (2) (2003) 379−401.

[21] J.A. Caballero, I.E. Grossmann, Structural considerations and modeling in the synthesis of heat-integrated-thermally coupled distillation sequences, Ind. Eng. Chem. Res. 45 (25) (2006) 8454−8474.

[22] A. Giridhar, R. Agrawal, Synthesis of distillation configurations: II. A search formulation for basic configurations, Comp. Chem. Eng. 34 (1) (2010) 84−95.

[23] V.H. Shah, A.V. Giridhar, R. Agrawal, A Computationally Efficient Method to Generate a Complete Search Space for Multicomponent Distillation Sequences, AIChE Annual Meeting, Philadelphia, PA, Fall 2008.

[24] V.H. Shah, R. Agrawal, A matrix method for multicomponent distillation sequences, AIChE J. 56 (7) (2010) 1759−1775.

[25] Z.T. Fidkowski, Distillation configurations and their energy requirements, AIChE J. 52 (6) (2006) 2098−2106.

[26] R. Agrawal, Z.T. Fidkowski, Ternary distillation schemes with partial reboiler or partial condenser, Ind. Eng. Chem. Res. 37 (8) (1998) 3455−3462.

[27] J. Ivakpour, N. Kasiri, Synthesis of distillation column sequences for nonsharp separations, Ind. Eng. Chem. Res. 48 (18) (2009) 8635−8649.

[28] V.H. Shah, R. Agrawal, Are all thermal coupling links between multicomponent distillation columns useful from an energy perspective? Ind. Eng. Chem. Res. 50 (3) (2011) 1770−1777.

[29] R. Agrawal, Z.T. Fidkowski, Thermodynamically efficient systems for ternary distillation, Ind. Eng. Chem. Res. 38 (5) (1999) 2065−2074.

[30] V.H. Shah, R. Agrawal, Multicomponent distillation configurations with large energy savings, in: Proceedings of Distillation and Absorption, Eindhoven, The Netherlands, 2010.

[31] R. Agrawal, Z.T. Fidkowski, More operable arrangements of fully thermally coupled distillation columns, AIChE J. 44 (11) (1998) 2565−2568.

[32] R. Agrawal, More operable fully thermally coupled distillation column configurations for multicomponent distillation, Trans. IChemE 77 (Part A) (1999) 543−553.

[33] R. Agrawal, Thermally coupled distillation with reduced number of intercolumn vapor transfers, AIChE J. 46 (11) (2000) 2198−2210.

[34] J.A. Caballero, I.E. Grossmann, Generalized disjunctive programming model for the optimal synthesis of thermally linked distillation columns, Ind. Eng. Chem. Res. 40 (10) (2001) 2260−2274.

[35] J.A. Caballero, I.E. Grossmann, Design of distillation sequences: from conventional to fully thermally coupled distillation systems, Comp. Chem. Eng. 28 (11) (2004) 2307−2329.

[36] M. Tawarmalani, N.V. Sahinidis, A polyhedral branch-and-cut approach to global optimization, Math. Prog. 103 (2) (2005) 225−249.

[37] A.J.V. Underwood, Fractional distillation of multicomponent mixtures, Chem. Eng. Prog. 44 (8) (1948) 603−614.

[38] U. Nallasivam, V.H. Shah, A. Shenvi, M. Tawarmalani, R. Agrawal, Global optimization of multicomponent distillation configurations: 1. Need for a reliable global optimization algorithm, AIChE J. 59 (3) (2013) 971−981.

[39] K. Liebmann, V. Dhole, Integrated crude distillation design, Comp. Chem. Eng. 19 (S1) (1995) 119−124.

[40] V. Rod, J. Marek, Separation sequences in multicomponent rectification, Collect. Czechoslov. Chem. Commun. 24 (1959) 3240−3248.

[41] M.F. Doherty, M.F. Malone, Conceptual Design of Distillation Systems, McGraw-Hill, NY, 2001.

[42] M.T. Halbouty, Giant oil and gas fields in United States, Am. Assoc. Pet. Geol. Bull. 52 (7) (1968) 1115−1151.

[43] A.J. Brugma, Process and Device for Fractional Distillation of Liquid Mixtures, More Particularly Petroleum, 1942, US Patent 2,295,256.

[44] J.K. Kim, P.C. Wankat, Quaternary distillation systems with less than N-1 columns, Ind. Eng. Chem. Res. 43 (14) (2004) 3838−3846.

[45] M. Errico, B.-G. Rong, G. Tola, I. Turunen, A method for systematic synthesis of multicomponent distillation systems with less than N-1 columns, Chem. Eng. Process. Process Intensif. 48 (4) (2009) 907−920.

[46] A.A. Shenvi, V.H. Shah, J.A. Zeller, R. Agrawal, A synthesis method for multicomponent distillation sequences with fewer columns, AIChE J. 58 (8) (2012) 2479−2494.

[47] A.A. Shenvi, V.H. Shah, R. Agrawal, New multicomponent distillation configurations with simultaneous heat and mass integration, AIChE J. 59 (1) (2013) 272−282.

[48] S. Ochiai, F.G. Shinskey, Distillation: calculations of relative gains, in: Control and Optimization of Unit Operations, fourth ed.Instrument Engineers' Handbook, vol. 2, CRC Press, September 2005.

[49] R.O. Wright, Fractionation Apparatus, 1949, U.S. Patent 2,471,134.

[50] G. Parkinson, Dividing-wall columns find greater appeal, Chem. Eng. Prog. 103 (5) (2007) 8−11.

[51] J.G. Pendergast, D. Vickery, P. Au-Yeung, J. Anderson, Consider dividing Wall columns, Chem. Process. (December 2008).

[52] B. Kaibel, Dividing wall columns, in: A. Górak, Z. Olujic (Eds.), Distillation Book, vol. IIElsevier, Amsterdam, 2014.

[53] A.A. Kiss, Advanced Distillation Technologies: Design, Control and Applications, John Wiley & Sons, Ltd, Chichester, UK, 2013.

[54] R. Agrawal, Multicomponent distillation columns with partitions and multiple reboilers and condensers, Ind. Eng. Chem. Res. 40 (20) (2001) 4258−4266.

[55] A.A. Shenvi, D.M. Herron, R. Agrawal, Energy efficiency limitations of the conventional heat integrated distillation column (HIDiC) configuration for binary distillation, Ind. Eng. Chem. Res. 50 (1) (2011) 119−130.

Conceptual Design of Azeotropic Distillation Processes

8

Mirko Skiborowski, Andreas Harwardt, Wolfgang Marquardt

AVT-Process Systems Engineering, RWTH Aachen University, Aachen, Germany

CHAPTER OUTLINE

Distillation: Fundamentals and Principles. http://dx.doi.org/10.1016/B978-0-12-386547-2.00008-9

8.1 Introduction

8.1.1 Azeotropic distillation processes

According to Kiva et al. [1] "azeotropic distillation is defined as distillation that involves components that form azeotropes." In fact, many industrially relevant liquid mixtures are strongly nonideal and exhibit azeotropic behavior. This has a tremendous influence on the design of distillation processes, since the composition space is divided into regions with different volatility order of the components. Therefore, in contrast to zeotropic distillation (see Chapter 7), the determination of a feasible separation cannot be easily derived by the boiling order of the pure components, and inherent limitations have to be considered due to the nonideality of the mixture to be separated.

Although the separation of an azeotropic mixture into its pure components is often impossible using a sequence of simple distillation columns (with a single feed and two product streams), it often becomes possible if more complex flowsheets with recycle streams are chosen. Frequently an additional component, the so-called entrainer (or solvent), is added to facilitate the separation. Therefore, we consider entrainer-free processes like curved-boundary and pressure-swing distillation as well as entrainer-based extractive distillation as instances of homogeneous azeotropic distillation. Heteroazeotropic distillation refers to heterogeneous distillation processes comprising a combination of (at least) a distillation column and a decanter; it can be viewed as a hybrid separation process because it combines distillation (in the column) and liquid phase split (in the decanter). We deliberately do not want to distinguish cases where the azeotropic behavior is inherently present in the mixture or where it is introduced by an added entrainer.

Figure 8.1 shows some examples of different types of azeotropic distillation processes. Curved-boundary distillation exploits the curvature of the distillation

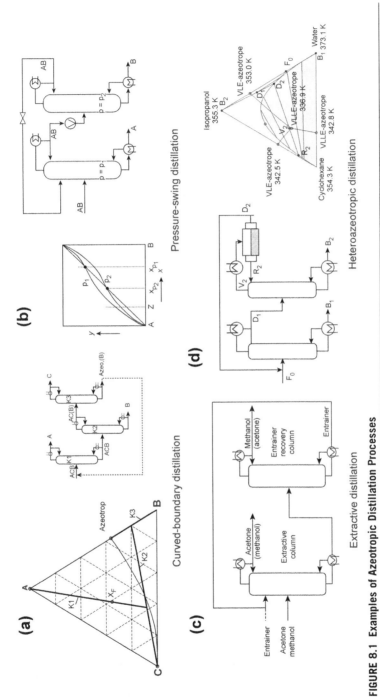

FIGURE 8.1 Examples of Azeotropic Distillation Processes

(a) Curved-boundary distillation, (b) Pressure-swing-distillation, (c) Extractive distillation and (d) Heteroazeotropic distillation.

boundary and employs a recycle to enable a complete separation. In pressure-swing distillation the sensitivity of the azeotrope with respect to pressure variation is used to accomplish the separation in a two-column process. The addition of an entrainer can increase process performance. Extractive distillation refers to an entrainer-based process, where the entrainer interacts with the azeotropic mixture to locally alter the concentration-dependent relative volatilities [2]. The process generally consists of a multiple-feed separation column and another column for entrainer recovery. If the entrainer induces a miscibility gap, a so-called heteroazeotropic distillation process is established.

Distillation can also be combined with other unit operations like extraction, crystallization, or membrane separations, which are not limited by azeotropic behavior, either as a prefractionation step or as part of an integrated hybrid separation process (see Chapter 9). Reactive distillation has also been suggested to separate azeotropic mixtures. Nevertheless, these processes are not in the scope of this chapter.

8.1.2 Conceptual process design

Conceptual design has to rely on creativity, solid technological knowledge, and broad experience of the design engineer. It therefore has been referred to as rather an art than a science [3]. The inherent limitations regarding split feasibility and the variety of design options render the conceptual design of azeotropic distillation processes a complex task. The space of possible designs is tremendous. The design decisions include the selection of the individual separation steps including the possible choice of an eligible entrainer, their sequencing, and the design of the complete flowsheet mostly showing inevitable recycle structures. The consideration of other unit operations and integrated hybrid processes expands the design space even further. Hence, it is just about impossible to determine an adequate design without the support of systematic, model-based methods and tools to determine a feasible process variant of high performance.

To this end, Marquardt et al. [4] have suggested a process synthesis framework to guide the process of designing a separation process for a given azeotropic mixture operating at minimal cost. Their three-step approach, depicted in Figure 8.2, starts with a variant-generation step followed by two evaluation steps relying on models of different levels of detail.

The variant-generation step is of major importance since the best process can only be determined if it is included in the generated variant tree, whereas the size of the variant tree is usually constrained by the resources available in the design process. Variant generation should rely on a broad knowledge base, including thermodynamics of the mixture, heuristic design guidelines, as well as the designers' experience and intuition. Screening methods are preferably used to assess split feasibility and separation effort. However, a final assessment is often only possible by means of rigorous modeling and optimization in the last step.

While it is common practice to utilize process simulation in order to evaluate feasibility and performance of different process variants, this approach has the

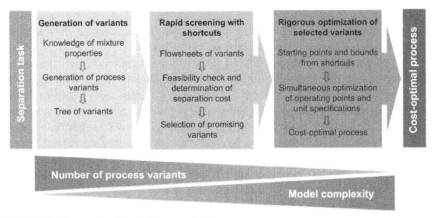

FIGURE 8.2 Process Synthesis Framework [4]

distinct drawback that all degrees of freedom have to be specified in advance. Hence, numerous simulations have to be performed to determine a feasible and well-performing design for any of the process variants. The quality of selected design can only be quantified in relation to the simulation results available. Consequently, there is no way to assess the gap to the best possible design. The formulation of a superstructure and the subsequent application of mathematical programming present an alternative approach to determine an optimal design. However, due to a large number of design decisions and the nonlinearity and the high dimensionality of the required mathematical models, a direct optimization of large superstructures incorporating all possible process variants together with all detailed design options is not promising. Therefore, Marquardt et al. [4] propose the decomposition of the problem into a first one to evaluate candidate design variants by means of shortcut methods, followed by a second one to rigorously optimize specific and detailed superstructures for each of the promising process variants identified before by means of mathematical programming.

The exhaustive screening of a large number of variants in the first problem requires computationally efficient shortcut methods. In contrast to simulation-based approaches, these methods do not require the detailed specification of the design. They provide characteristic performance indicators like minimum energy demand (MED) or minimum number of column trays, which can be used to assess the economic potential and rank the different process variants. However, reliable results can only be obtained if these shortcuts rigorously account for the nonideality of azeotropic mixtures as well as the induced limitations due to distillation boundaries to quantify split feasibility and performance of a process variant. The screening with such shortcut methods also reduces the need to limit the number of possible variants and hence the engineering effort prior to model-based evaluation by heuristic rules.

Although a first ranking of the process variants can be accomplished, the shortcut methods only provide an estimate on performance and split feasibility without determining all detail of the design. Therefore, the most promising variants are investigated further in order to determine the detailed conceptual design of the most economic process variant by rigorous optimization.

This chapter follows the individual steps of the process synthesis framework and provides an overview of the different methods capable of supporting the design engineer in each of the steps. To this end, different tools for entrainer selection, for the analysis of split feasibility, and for variant generation are presented first. Next, an elaborate review of thermodynamically motivated shortcut methods is given. Then, a short introduction to rigorous superstructure optimization for a cost-optimal design of azeotropic distillation processes is provided. The chapter closes with illustrative examples of different distillation processes facilitating the separation of azeotropic mixtures. As long as not stated differently, adiabatic and isobaric distillation columns with equilibrium trays are assumed throughout the chapter.

8.2 Generation of distillation process variants

The first step in the conceptual design process refers to the generation of candidate variants. Relevant unit operations have to be selected and combined into a flowsheet to build a coarse process structure, which needs to be refined in subsequent steps. Especially distillation boundaries (DB), emanating from azeotropes, severely complicate distillation process synthesis. Recycle streams as well as entrainer compounds need to be considered to facilitate the separation. Thermodynamic insight is crucial for this step. Consequently, reliable thermodynamic property models are essential.

The generation of design variants is commonly based on heuristic rules. These rules represent the experience collected during the solution of similar design problems and often involve thermodynamic properties as governing decision criteria [5–7]. Computer-aided approaches, like rule-based expert systems [8] or case-based reasoning [9,10], have been developed to automate variant generation. While previous experience can be coded and systematically reused, all these approaches lack the potential to generate truly innovative designs and to resolve contradicting suggestions. Thus, the design of cost-optimal separation processes should rely on both the knowledge and experience of a creative design team and model-based decision support to identify the best from a broad spectrum of possible variants.

Several methodologies have been proposed to support the design engineer in the generation of process variants for azeotropic distillation including the selection of suitable entrainers and the assessment of split feasibility.

8.2.1 Entrainer selection

The selection of an appropriate entrainer comprises a major design decision in the case of extractive distillation and heteroazeotropic distillation [11]. Candidate compounds can be selected from established databases based on their selectivity [12]. Further restrictions on their applicability, like the boiling or melting point or the formation of additional azeotropes or even miscibility gaps, can be used for the evaluation of entrainer candidates [13,14]. Alternatively, computer-aided molecular design (CAMD) methods [15] can be employed to systematically identify compounds with desirable properties. The application of a generate-and-test procedure [16] or the formulation and solution of a mixed-integer nonlinear programming (MINLP) problem [17] constitute two options with individual strengths and weaknesses. Group contribution methods are typically employed for the prediction of thermodynamic properties from molecular structure [18]. While CAMD constitutes a vital possibility to innovative entrainer selection, the limited accuracy of the property prediction models limits the power of CAMD and necessitates experimental validation. Although the absolute accuracy of the predictions might be low, the qualitative ranking of solvent candidates has been confirmed experimentally at least in some reported case studies [15,19].

Depending on the properties of the entrainer additional distillation boundaries may be introduced, which can have a major effect on split feasibility and process performance [20]. Entrainer selection should not only be based on thermodynamic properties and heuristics, which may even contradict each other [21], rather, the selection of promising entrainer candidates should be integrated with the evaluation of candidate process structures [22]. Recently reported simultaneous approaches to the design of entrainer-based processes include property clustering [23] and hypothetical molecules [24]. In the latter approach, for example, a thermodynamic model, which links molecular structure to the properties of a hypothetical entrainer molecule, allows for a simultaneous optimization of process structure and entrainer. Hence, instead of selecting entrainer candidates prior to process design, the optimal entrainer is determined to match the properties of the hypothetical molecule resulting from integrated process and product design.

8.2.2 Split feasibility

For a feasible separation, the product compositions

- have to obey the overall mass balance and therefore need to be on a straight line with the feed composition in the composition space,
- need to be connected by a concentration profile throughout the column.

In order to determine the concentration profile in a simple packed or tray distillation column, the height of packing or the number of trays in the rectifying and stripping sections (N_R,N_S) as well as the operating point (reflux ratio RR and distillate-to-feed ratio D/F) have to be defined.

In order to reduce the number of degrees of freedom, strongly simplifying abstractions of simple distillation columns are utilized, including the so-called total reflux distillation columns and reversible distillation columns (see Figure 8.3). Both abstractions do not properly match a real distillation column, since the

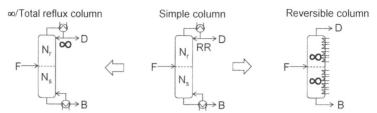

FIGURE 8.3 Different Abstractions of a Distillation Column

separation is assumed to be performed at an infinitely high energy demand in the first case or at infinitely high capital cost in the second case [25]. However, both abstractions allow for a qualitative analysis of column behavior in face of the distillation boundaries and distillation regions of a given mixture, and hence support the assessment of split feasibility.

8.2.2.1 Columns operated at total reflux

In case of total reflux columns, the vapor stream is completely condensed at the top of the column and fully redirected to the column. Note that this case is often also called infinite reflux column, referring to an infinite reflux ratio (corresponding to total reflux). In either case, the internal streams and their concentration profile are not affected by the feed streams, such that the entire column can be considered as a single section. Either residue curves (RC) or distillation lines (DL) describe the composition profile in a packed or tray column, respectively, which is operated at total reflux (or at an infinite reflux ratio).

Residue curves correspond to the time evolution of the concentration of the liquid residue in an open evaporation still as illustrated in Figure 8.4. The mass balances of the still result—after some simple transformation—in the well-known Rayleigh equation in dimensionless time τ [26],

$$\frac{dx_i}{d\tau} = x_i - y_i, \quad x_i(\tau = 0) = x_i^0, \quad i = 1...n_c, \tag{8.1}$$

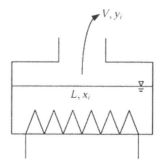

FIGURE 8.4 Open Evaporation

where x_i denotes the liquid and y_i the vapor composition of all components $i = 1,...n_c$ in the mixture at vapor–liquid equilibrium (VLE). These concentrations are related by the VLE conditions

$$y_i = K_i(x, y, p, T)x_i, \quad i = 1...n_c. \tag{8.2}$$

In contrast to ideal mixtures, the vapor–liquid distribution ratios K_i cannot be calculated simply by Raoult's law but need to account for the nonidealities in both phases by means of some suitable thermodynamical model such as an equation of state (EOS) or G^E model.

The numerical solution of Eqns (8.1) and (8.2) results in a RC originating at the initial concentrations (x_i^0, $i = 1...n_c$). The dimensionless parameter τ cannot only be interpreted as dimensionless time in an open evaporation but also as dimensionless height in a packed column operated at total reflux (or infinite reflux ratio) [21,27]. Hence, the solution of Eqns (8.1) and (8.2) also describes the change of the liquid compositions over the column height. This interpretation can be easily derived from a differential equation model of a column section assuming constant molar overflow (CMO) and total reflux [28].

By simply specifying different initial concentrations, the so-called residue curve map (RCM) can be constructed. The RCM of the ternary mixture of acetone, chloroform, and benzene, which exhibits a binary azeotrope between acetone and chloroform, is exemplarily depicted in Figure 8.5. The three pure components and the azeotrope constitute the so-called singular points of the mixture, i.e. the stationary

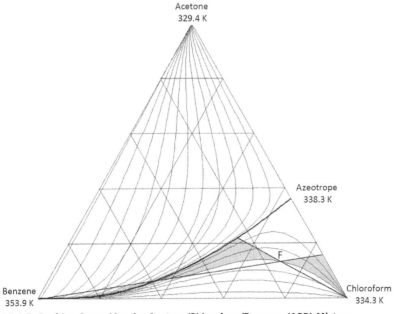

FIGURE 8.5 Residue Curve Map for Acetone/Chloroform/Benzene (ACB) Mixture

points of Eqn (8.1), where the differentials are zero. The singular points can be classified as unstable or stable node or as saddle [26]. Any RC follows a distinct path from an unstable to a stable node and can further be attracted by one or more saddles along its path. In addition, the uniqueness of the solutions of ordinary differential equations implies that the RCs may never intersect or touch each other. The topological consistency of a given multicomponent mixture can be checked by a generalized rule of azeotropy reported by Zharov [29]. Kiva et al. [1] present an extensive review on the structural properties of VLE diagrams and RCM as well as their topological classification according to Serafimov [30].

The RCM presents a simple means of visually analyzing the boiling properties of a ternary mixture and allows for a first assessment of the feasibility of a ternary distillative split [31]. Furthermore, the RCM can also be used to identify feasible separations and recycle structures for ternary separations [32,33]. Especially the limiting RCs that originate from saddle azeotropes are of special interest for conceptual design. They separate the RCM into different distillation regions that lead to different possible products. The limiting RCs connect different pairs of a saddle and an unstable or stable node and are termed simple distillation boundaries (SDB) [34]. Consequently, the feasible product region of a simple packed distillation column at total reflux for a given feed (F) can be determined from the RCs through the feed concentration and the mass balance lines, as depicted by the shaded regions in Figure 8.5.

The use of RCM for a graphical inspection of the behavior of ternary mixtures has evolved to an industrial standard [35]. It is also applicable to quaternary mixtures, but mixtures with more than four components can only be inspected visually through projection techniques [36] or after component lumping [37], which is, however, only telling in exceptional cases [38].

It is often assumed that the RCs are also an accurate representation of the liquid composition profiles at total reflux in a column with equilibrium trays [39]. However, an RC is at best an approximation to the liquid composition profile [40]. Even at total reflux there is a distinct difference between RCs and the liquid composition profiles in tray columns, which are termed distillation lines (see below) and were, according to Kiva et al. [1], first introduced by Zharov [41,42]. Therefore, the type of equipment has to be considered if a quantitative evaluation of split feasibility is to be performed [43].

Distillation lines (DL) can be derived from a mass balance around a section of a tray distillation column. Under the CMO assumption, the mass balance of the rectifying section (see Figure 8.7) results in [28]

$$x_{n,i} = \frac{RR+1}{RR}y_{n+1,i} - \frac{1}{RR}x_{D,i}, \quad i = 1...n_c, \quad n = 1...N, \tag{8.3}$$

where $x_{n,i}$ denotes the liquid and $y_{n,i}$ the vapor composition of all components $i = 1...n_c$ on trays $n = 1...N$. In the limiting case of a reflux ratio $RR = \frac{L}{D} \to \infty$ this equation results in

$$x_{n,i} = y_{n+1,i}, \quad i = 1...n_c, \quad n = 1...N \tag{8.4}$$

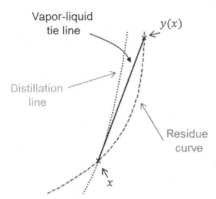

FIGURE 8.6 Sketch of Residue Curves and Distillation Line Characteristics

which is the equation for the operating line of a multistage column at total reflux (or infinite reflux ratio). While this difference equation represents a finite set of vapor−liquid tie-lines, the DL is defined as the continuous trajectory of liquid compositions, for which the vapor−liquid tie-lines represent chords for every composition on the line [40]. According to Petlyuk [44], this point of view dates back to the Russian literature of the 1970s, where the DLs were termed continuous c-lines.

Although RCs and DLs are very similar, they also feature distinct differences. While the RC proceeds in a distillation region in the direction of increasing temperature and therefore links the light boiler at the top to the heavy boiler at the bottom of the column, the DL proceeds from the bottom to the top of a column. But more importantly, the vapor−liquid tie-lines represent chords of the DL, but they are tangent to the RC. Hence, RC and DL are only the same if they are linear. For most mixtures, they are nonlinear and deviate from each other. The sketch in Figure 8.6 illustrates this geometric behavior (see also [40]). Since the difference between RC and DL is often small, either one can be used for a (rough) qualitative analysis of mixture behavior and possible separations.

8.2.2.2 Columns operated at a finite reflux ratio

While a column operated at total reflux can be represented as a single section, the effect of the feed cannot be neglected, and rectifying and stripping sections have to be distinguished at finite reflux ratios. Therefore, the regions of feasible product compositions of columns operated at finite reflux ratios do not match those of columns operated at total reflux. At finite reflux ratios, it is possible to find feasible designs that are infeasible at total reflux [11,25], while other designs are only feasible at total reflux or at very high reflux ratios.

In order to determine the course of tray-to-tray profiles in a column section, a balance envelope is formed as exemplarily depicted in Figure 8.7 for the stripping section. Assuming that the bottom product (B, x_B) and the reboiler duty $(\dot{Q}_B)$ are

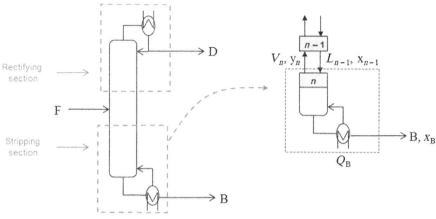

FIGURE 8.7

Column sections (left) and balance envelope for tray-to-tray calculations in stripping section (right).

fully specified, the liquid composition on each tray can be calculated by solving the nonlinear system of equations,

$$0 = L_{n-1} - V_n - B, \tag{8.5}$$

$$0 = L_{n-1}x_{n-1,i} - V_n y_{n,i} - Bx_{B,i}, \quad i = 1...n_c, \tag{8.6}$$

$$0 = L_{n-1}h_L(x_{n-1}, T_{n-1}, p) - V_n h_V(y_n, T_n, p) - Bh_B + \dot{Q}_B, \tag{8.7}$$

$$0 = y_{n-1,i} - K_i(x_{n-1}, y_{n-1}, T_{n-1}, p)x_{n-1,i}, \quad i = 1...n_c, \tag{8.8}$$

$$0 = \sum_{i=1}^{n_c} x_{n,i} - 1. \tag{8.9}$$

with appropriate models for the calculation of liquid and vapor enthalpies (h_L, h_V) and an appropriate model for the VLE in Eqn (8.8) starting at the bottom tray. A similar equation system can be constructed for the rectifying section. The condenser duty in the rectifying section ($\dot{Q}_C$) and the reboiler duty in the stripping section ($\dot{Q}_B$) are not independent but are linked by the overall energy balance of the column. Therefore, the equations are fully determined for any given feed specification (F, z_F, h_F) and a fully specified product and an energy duty at either the top or the bottom of the column. The other product follows from the overall mass balance, and the energy duty at either the reboiler or the condenser is determined from the overall energy balance. For any feasible separation at finite reflux, the concentration profiles of stripping and rectifying section need to intersect at the feed tray.

8.2.2.3 Reversible distillation
The concept of a reversible distillation column is a hypothetical construct that describes a distillation column, which—in correspondence with the second law of thermodynamics—operates at zero entropy production. This requires (1) isobaric

and adiabatic operation, (2) heat transfer at vanishing temperature difference along the column, and (3) no non-equilibrium contact between liquid and vapor streams throughout the column, neither at the condenser, the reboiler, or the feed tray locations [44,45]. As a consequence, reversible columns are of infinite height or have an infinite number of trays.

In contrast to columns operating at total reflux but similar to columns operating at finite reflux ratios, the concentration profile in a reversible column is affected by the feed, i.e. there is always a kink at the feed location. Since phase equilibrium is assumed throughout the column, the concentration profiles in the rectifying and the stripping sections of a reversible column are similar to the so-called pinch lines through the product compositions [45]. Pinch lines and pinch points, i.e. points on the pinch lines, are also important for the calculation of the MED as will be shown in Section 8.3.

Single pinch points can be calculated as the solution of the pinch equation system that comprises the tray-to-tray equation system Eqns (8.5)–(8.9) extended by the constraints of vanishing driving force, i.e. $x_{n+1} = x_n = x_{\text{pinch}}$.[1] The pinch points can be calculated for each section of the distillation column; they only depend on product compositions as well as on the reflux or reboil ratio. In order to calculate all pinch points as a function of energy duty, the solution branches of the pinch equation system can be calculated efficiently starting from all the singular points of the mixture by a continuation approach [46,47]. The solution branches are called pinch lines, pinch branches, or locus of pinch points [34]. Secondary pinch branches can be detected by means of bifurcation analysis embedded into the continuation. While the calculation of all pinch branches connected to the singular points is mostly sufficient, additional isolated pinch branches may occur that can only be detected by computationally more involved methods [48,49].

For a feasible reversible distillation, the products have to be connected by a continuous path constituted of the product pinch lines and intersecting in the feed concentration [50].

8.2.2.4 Feasible product specifications

If distillation boundaries exist that separate the composition space in different distillation regions, not all product specifications can be reached—a specification can only be feasible if both product compositions are situated within the same distillation region.

Common distillation boundaries are the SDBs, constituted by the separatrices of the RC map, which represent the distillation boundaries of packed columns operated at infinite reflux ratio and the total reflux boundaries (TRBs) [34] corresponding to the separatrices of the DL map, which represent the distillation boundaries of tray columns. Similar to SDBs and TRBs, the concentration simplex can also be divided

---

[1] Only the fixed points of pinch equation system that are in the path of the column profile are called pinch points.

into different distillation regions using the pinch lines, for which the product boundary of a reversible distillation column, the so-called pinch distillation boundary (PDB), can be constructed. In case of ternary mixtures, the PDB can be constructed by an involved geometrical procedure as the envelope of the tangents to RCs in their inflection points [25]. However, a set of equations can be derived defining the PDB as the locus of pitchfork bifurcation points of the pinch equation system [51] allowing for an algorithmic assessment of split feasibility by means of PDBs [52].

While SDBs and TRBs, both assuming total reflux separations, result in very similar distillation boundaries, the PDB, assuming a reversible separation, may deviate significantly from those boundaries at total reflux. The degree of deviation depends on the curvature of the boundaries. Though split feasibility of finite reflux distillation is different from both, total reflux and reversible distillation [11,34], PDBs and SDBs or TRBs can be used to bound the feasible products of a real distillation column. In particular, achievable products in a real distillation column are confined by those reached in a separation at minimum work (reversible distillation, PDB) and those reached in a separation at maximum work (total reflux distillation, SDB or TRB) [25,53].

Product compositions situated between those two limiting distillation boundaries can in principle be reached from both adjacent distillation regions at certain reflux ratios, crossing either the total or the reversible distillation boundaries [34]. This characteristic is illustrated for the acetone, chloroform, benzene (ACB) mixture in Figure 8.8, where the bottom product B can be reached from the chloroform-rich region (i.e. the one where the chloroform vertex constitutes the unstable node of the RCM) at $RR = 15$ crossing the PDB or from the acetone-rich region (i.e. the one where the acetone vertex constitutes the unstable node of the RCM) at $RR = 5$ crossing the SDB/TRB. While the qualitative behavior of distillation boundaries differs only slightly, there is a significant quantitative difference between the total reflux boundaries (SDB/TRB) and the reversible column boundary (PDB); in contrast to the former, the latter does not end in the pure benzene vertex but runs into the binary edge connecting chloroform and benzene at a benzene ratio of about 80% (see insert in Figure 8.8).

Since precise product specifications are usually not necessary for the generation of design variants (cf. first step in Figure 8.2), the consideration of one of the limiting distillation boundaries is typically sufficient. However, the evaluation of the feasibility of the separation has to be assured in the later steps of the design framework (cf. Figure 8.2). Obviously, shortcut methods, employed in the second step, are candidates for this task. However, they rely on simplified models that do not require computation of complete concentration profiles throughout the column. While these methods are computationally efficient, they cannot reliably assess the feasibility of product specifications. Although the differences between SDB/TRB and PDB are often negligible, there are cases where the difference is sufficient to result in false conclusions in case of a quantitative evaluation of feasibility.

This problem is accounted for by utilizing both types of distillation boundaries simultaneously to assess split feasibility. For ternary mixtures so-called operation

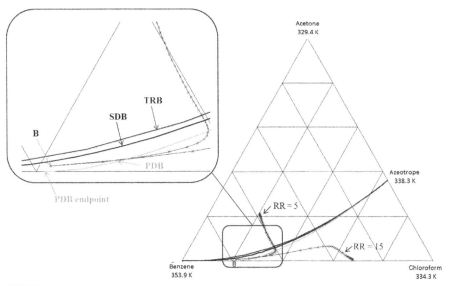

FIGURE 8.8

Different distillation boundaries and tray-to-tray profiles for stripping section at different reflux ratios for the acetone, chloroform, benzene (ACB) mixture. SDB, simple distillation boundaries; TRB, total reflux boundaries; PDB, pinch distillation boundary.

leaves of a column section can be constructed from distillation and pinch lines to contain all possible composition profiles from minimum to total reflux [25,53,54]. If the operation leaves of both column sections overlap at some finite reflux ratio, the separation is expected to be feasible. Unfortunately, this approach is limited to ternary mixtures. In order to at least estimate the feasibility of a separation of a multicomponent mixture, Thong et al. [55] propose to use structurally linear approximations of the equivalent manifolds. Alternatively, the SDB/TRB and PDB can be calculated to assess split feasibility; a separation can be expected to be infeasible, if the product concentrations are located in regions separated by both types of distillation boundaries; in contrast, the assessment is inconclusive if (one of) the product concentrations lies in the region between SDB/TRB and PDB (cf. the ACB example in Figure 8.8).

If a desired separation in a single or a sequence of single-feed columns is not possible, some azeotropic distillation process, either extractive or heteroazeotropic distillation, offer alternatives for the separation of an azeotropic mixture.

8.2.2.5 Extractive distillation
The location of the feed streams in multi-feed distillation columns does not make any difference regarding the assessment of split feasibility in case of zeotropic multi-component mixtures, but the location of the feed trays and the feed ratio may have a significant impact on achievable products in case of azeotropic distillation [21]. A

separation that is infeasible in a single-feed column may be feasible in a two-feed column. Therefore, the concentration profiles in the extractive section, i.e. the section between the two feed trays, have to connect the concentration profiles of the rectifying with those of the stripping section [56]. Since the concentration profiles at total reflux are independent of the feed distribution, split feasibility of extractive distillation can never be assessed at total reflux [11]. However, specific RCM topologies indicate the feasibility of extractive distillation at finite reflux ratios and allow for a first evaluation of suitable entrainer candidates at least for ternary mixtures [14,21,57]. The possible distillate products can, for example, be determined from an analysis of the isovolatility curve(s) [58].

Feasibility of extractive distillation strongly depends on the entrainer flow rate (or rather the ratio of both feed streams) and the reflux ratio. Extractive distillation is only feasible above a minimum entrainer flow rate for a given reflux ratio limited by a minimum and a maximum reflux ratio [59]. For ternary mixtures, the possible concentration profiles in the extractive section can be analyzed by constructing the balance envelope for the extractive section [56,60]; Wahnschafft and Westerberg [56] calculate the operation leaves for the rectifying and stripping sections and the pinch lines of the extractive section to analyze the feasibility of extractive distillation. An example is depicted in Figure 8.9.

A separation is feasible if the operation leaves reach into the region bounded by the pinch lines of the extractive section. The size of the region depends on the entrainer feed rate, while the actual composition profiles of the rectifying and stripping sections depend on the reflux and reboil ratios. Therefore, feasibility of extractive distillation is limited by some minimum entrainer flow rate and a minimum and

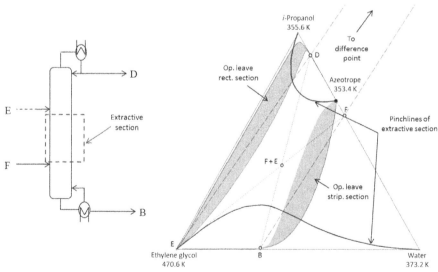

FIGURE 8.9 Analysis of Split Feasibility for Ternary Extractive Distillation

(Similar to Ref. [56]).

a maximum reflux ratio [56]. These limits can also be computed applying concepts from bifurcation theory [59,61] to extend the assessment of feasibility to mixtures with more than three components.

8.2.2.6 Heteroazeotropic distillation

Heteroazeotropic distillation exploits differences in volatility and liquid—liquid phase split by linking a distillation column and a decanter. Therefore, it may also be interpreted as a hybrid separation process [62,63]. Liquid phase splitting in the decanter facilitates breaking of the azeotrope to reach high purity products. Heteroazeotropic distillation is therefore used for the separation of heterogeneous azeotropic mixtures; it also offers a favorable option for the separation of homogeneous azeotropic mixtures if an entrainer is added to induce a liquid—liquid phase split.

For ternary mixtures, the feasibility of heteroazeotropic distillation can directly be assessed by an analysis of the according RCM. Similar to homogeneous systems, the RC cannot cross the SDB. However, if a point on an RC inside the heterogeneous region results in two separate liquid phases with equilibrium concentrations x^I and x^{II} in two different distillation regions, liquid—liquid phase separation can be exploited to cross the distillation boundary. An example for such a process is depicted in Figure 8.10.

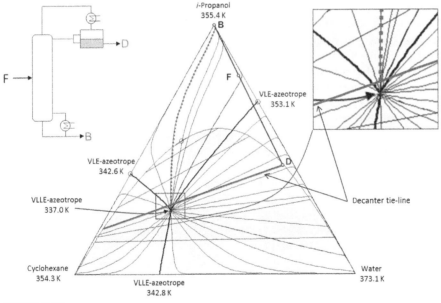

FIGURE 8.10

Residue curve map of the heterogeneous mixture of isopropanol, water, and cyclohexane at $1.013 \cdot 10^5$ Pa. VLE, vapor—liquid equilibrium; VLLE, vapor—liquid—liquid equilibrium.

In contrast to homogeneous systems, it is not trivial to construct the RCM. While VLE computed from Eqn (8.2) or (8.8) applies in the homogeneous region, vapor—liquid—liquid equilibrium (VLLE) has to be considered in the heterogeneous region by

$$x_i = \phi x_i^{\mathrm{I}} + (1 - \phi)x_i^{\mathrm{II}}, \quad i = 1\ldots n_{\mathrm{c}}, \tag{8.10}$$

$$y_i = K_i\left(x^{\mathrm{I}}, y, p, T\right)x_i^{\mathrm{I}}, \quad i = 1\ldots n_{\mathrm{c}}, \tag{8.11}$$

$$y_i = K_i\left(x^{\mathrm{II}}, y, p, T\right)x_i^{\mathrm{II}}, \quad i = 1\ldots n_{\mathrm{c}}, \tag{8.12}$$

$$0 = \sum\nolimits_{i=1}^{n_{\mathrm{c}}} x_i^{\mathrm{I}} - 1. \tag{8.13}$$

$$0 = \sum\nolimits_{i=1}^{n_{\mathrm{c}}} x_i^{\mathrm{II}} - 1. \tag{8.14}$$

where ϕ represents the liquid phase ratio. However, it is not trivial to computationally distinguish between the metastable (heterogeneous liquid) and absolutely stable (homogeneous liquid) regions because of the trivial solution of Eqns (8.10)—(8.14). Reliable and sound techniques for assessing phase stability are required [64]. Global and local phase stability tests have been reviewed by Cairns et al. [65] and Zhang et al. [66]. While only global methods assure a correct solution, especially the combination of local methods, like the negative flash [67], and continuation methods [68,69] or rate-based methods [70], allow for computationally efficient solutions with high reliability.

Due to the combination of a column and a decanter, the design of a heteroazeotropic distillation column offers additional degrees of freedom and reveals different requirements for feasible separations. In the standard configuration of a heteroazeotropic distillation column, the decanter separates the condensed liquid at the top of the column. For a feasible separation, a continuous concentration profile needs to connect the bottom product and the decanter tie-line, where the top product is situated. Therefore, the top vapor composition must be in the same distillation region as the bottom product [71]. An exemplary specification is depicted in Figure 8.10, where the dotted RC represents a feasible column profile. Other configurations are also possible to exploit the topology of the phase diagram [72—74]. The design of such processes relies on a creative use of RCM analysis to identify a suitable configuration for the separation of the azeotropic mixture into desired products. The differences of split feasibility at total and finite reflux also need to be considered for simple heteroazeotropic distillation columns. For a detailed analysis of the different distillation boundaries for heterogeneous mixtures, we refer the reader to the work of Krolikowski et al. [75].

8.2.3 Algorithmic approaches for the generation of distillation process variants

Although significant effort has been spent on the automatic generation of separation process variants, only the generation of distillation sequences for zeotropic mixtures can be fully automated [76–78]. The occurrence of distillation boundaries in azeotropic mixtures, independent of their type, severely complicates the determination of possible products and hence the structure of a distillation process. This is especially true for mixtures with more than four components, where those geometric methods that heavily rely on visualization are not applicable anymore. For the identification of possible splits, the type of split is also of special importance, since some of the methods only apply to certain types. In case of a *sharp split*, both products contain at least one component less than the feed (e.g. A,B,C,D), meaning that each product contains at least one of the feed components only in a very small quantity (e.g. A: B,C,D). In case of a *semi-sharp split*, one product contains all components of the feed (e.g. A,B:A,B,C,D). If all components are present in both products, the split is termed *nonsharp* or *sloppy split* (e.g. A,B,C,D:A,B,C,D) [79]. While splits with both products on distillation boundaries are sometimes also called sharp splits [80], we will term them *highest purity splits*, independent of the number of components present in the products.

Several attempts have been made to at least facilitate the systematic generation of distillation process variants for the separation of azeotropic mixtures [79–88]. There are two major problems an algorithmic generation of process variants is faced with: (1) the determination of feasible product compositions, and (2) the consideration of recycles. In contrast to zeotropic distillation processes, the product specifications are limited by distillation boundaries, and recycles are often required to break azeotropes in azeotropic distillation (cf. Figure 8.1).

Rooks et al. [80] propose an algorithm to determine the reachability matrix [89], which approximately represents the structure of the RCM. This square matrix involves only binary elements to encode if two singular points can be connected by an RC in the direction of increasing temperature. Based on the reachability matrix, distillation regions can be identified as subsets of the singular points. Rooks et al. [80] also propose a simple test to determine the feasibility of highest purity splits based on a common saddle, which attracts composition profiles calculated from both specified product compositions at high reflux ratio. Although this test is sufficient, it excludes a wide range of potentially feasible splits [55]. Thong and Jobson [55] therefore extend the methodology of Rooks et al. by a more general feasibility test. These authors suggest to construct envelopes of the concentration profiles in the rectifying and stripping sections using the concept of operation leaves. These curved bodies are, however, approximated by polyhedrons to facilitate computation, in particular for separations involving more than three components. Product regions and feasible splits can be identified based on this test. Thong and Jobson further extend their approach toward an algorithmic procedure to generate column sequences [85,86]: first, all possible splits are identified, and then a superstructure covering (preferably all) recycling options is generated and reduced in complexity by applying a set of heuristic rules. The resulting superstructure is used to determine the compositions of product and recycle streams in an iterative procedure. This strategy could however be automated, casting it into an optimization problem with mass balances and appropriate feasibility conditions as equality constraints. While this approach allows for an almost complete automation of generation of distillation process variants, it does not account for extractive or heteroazeotropic distillation. Furthermore, a possible shift of the distillation boundary by an appropriate choice of the operating pressure in a column to facilitate pressure-swing distillation is also not included.

Instead of aiming at an automation of the variant generation step, Wahnschafft et al. [83] decompose it into different tasks. They propose a sequential procedure for the synthesis of complex separation schemes, including extractive and heteroazeotropic distillation. The procedure can be interpreted as a code of practice for a manual design and guides the design engineer toward a systematic synthesis of variants.

Recently, Petlyuk et al. [79,88] presented a computational simple method for the identification and analysis of possible splits and product regions for sharp splits. The method is based on an analysis of the pinch branch terminals and can also be used for the synthesis of distillation sequences.

While the design of distillation sequences for single-feed homogeneous distillation columns can be largely automated, the design of azeotropic distillation processes, including the choice of operating pressures and the selection of entrainers for extractive or heteroazeotropic distillation processes, can so far only be guided and supported by an educated application of a combination of the methods presented in this section.

8.3 Shortcut evaluation of distillation processes

Once flowsheet candidates and feasible product specifications for each of the separation steps have been identified, shortcut methods can be applied to evaluate process performance according to some target. While total annualized cost is the ultimate evaluation criterion from an industrial perspective, the information required for a reasonably detailed cost estimate might not be available in this early design stage. Alternatively, the MED of the distillation sequence is an attractive target. It not only provides a good estimate of operational cost but also correlates well with investment cost because of its strong dependence on heat exchanger areas and column diameters. Another alternative is a minimum number of trays (MTN).

Different types of shortcut methods have been proposed, which differ mainly in the underlying assumptions, computational demand, and range of applicability. CMO and constant relative volatilities (CRVs) are common assumptions that are generally appropriate for zeotropic mixtures with similar components. They should, however, be used with caution when dealing with azeotropic mixtures.

The remainder of this section presents an overview of different shortcut methods for simple distillation columns and their extension to extractive and heteroazeotropic distillation. Advantages and limitations of the different shortcut methods are highlighted to foster basic understanding and to lay the foundations for appropriately selecting and applying these methods to the design of azeotropic distillation processes.

8.3.1 **Simple columns**

Shortcut methods to assist the design of simple columns have been available for several decades. The general concept of any shortcut method is to directly address the design problem as an inverse problem, i.e. by specifying the desired effect, the product compositions, and determining its cause, the design parameters, for which the products are achievable for a given feed. The column is decomposed in the rectifying and stripping section; the tray numbers in both sections and the energy demand are determined to allow for an intersection of the composition profiles of both sections, which are calculated starting from the product compositions. The limiting MED and MTN can be calculated if either an infinite number of trays or an infinite energy input is assumed. The best-known shortcut methods include the McCabe–Thiele and Underwood methods.

The *McCabe–Thiele method* [90] is based on a graphical evaluation of the VLE line and the operating lines of both column sections in the x–y-diagram for binary separations assuming CMO. The MED as well as the MTN can be determined as depicted in Figure 8.11. The McCabe–Thiele method is the simplest available shortcut method, but it can only be applied for binary mixtures and its accuracy is limited if the CMO assumption is not valid.

Underwood's method [91] allows a numerical evaluation of the MED for separations involving any number of components. The calculations are based on CRVs of the components, which can be derived from pure component vapor pressures, and hence limits the application to ideal (or at least zeotropic) mixtures. Extensions to azeotropic mixtures have been suggested [92,93] where azeotropes are treated as pseudo-components.

However, similar to the McCabe–Thiele method, the CMO and in addition the CRV assumptions do not hold in this case and therefore severely limit the accuracy of Underwood's method in case of azeotropic distillation. Nevertheless, due to its

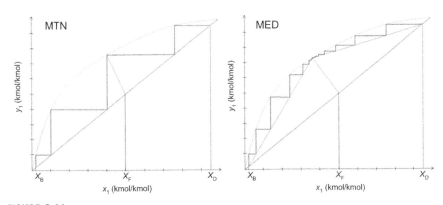

FIGURE 8.11

McCabe–Thiele diagram for the determination of minimum number of trays (MTN) (left) and minimum energy demand (MED) (right).

simplicity the Underwood's method has found wide application in zeotropic distillation. Implementations are available in commercial process simulators.

In the last decades, more sophisticated shortcut methods have been proposed for MED calculation of azeotropic distillation processes. Similar to the McCabe—Thiele method, these methods try to trace the column profiles in both column sections in order to determine the energy demand, for which both section profiles first intersect. The different methods either rely on rigorous and computationally involved tray-to-tray calculations or utilize pinch points to approximate the course of the tray-to-tray profile at MED.

8.3.1.1 Shortcut methods relying on concentration profile calculations

The *boundary value method* (BVM) [20] has been the first shortcut method designed for the treatment of multicomponent azeotropic mixtures. It calculates tray-to-tray profiles in both column sections, starting from the product compositions. The specified separation is identified to be feasible if the concentration profiles of both column sections intersect. The MED is the lowest energy at which this intersection occurs. Unfortunately, the concentration profiles and hence the MED strongly depend on the distribution of trace impurities in the product specifications [46]. Since only three out of $2n_c$ product specifications (B, D, $x_{B,i}$, $x_{D,i}$, $i = 1,...n_c - 1$) can be fixed in the BVM for a given feed and column pressure. The determination of MED for a mixture of four or more components requires the simultaneous variation of energy input and of $n_c - 3$ product specifications [94]. This property together with the need of a visual inspection of the intersection of the concentration profiles limits the practical applicability of the BVM largely to ternary mixtures.

If an affine mapping from column height to a dimensionless bubble-point temperature is introduced [95], the transformed boundary value problem can be cast into an optimization problem, which is solved by the so-called *temperature collocation*. The bubble-point distance can be used to verify the intersection of the column profiles and hence to determine the MED for any number of components in the mixture.

In contrast to this approach, the *shortest stripping line method* (SSLM) [96,97] and related approaches [46,98] try to overcome these limitations by a unidirectional calculation of the concentration profile. The SSLM relies on the observation that for a feasible separation the column with the (geometric) shortest stripping profile in the concentration space operates at MED. Instead of calculating the concentration profiles in each column section starting from the product composition and trying to make them intersect, the stripping profile is first calculated starting from the bottom product for a large number of trays (e.g. 300 [97]), then the rectifying profile is calculated starting from the last tray of the stripping profile toward the distillate. The separation is expected to be feasible if the calculated rectifying profile reaches a vicinity of the specified distillate product. The MED is calculated as the solution of a nonlinear programming (NLP) problem minimizing the length of the stripping line. If the SSLM is incorporated into a two-level design method, the distribution of nonkey components in the bottom product can also be optimized [99]. Both the BVM and the SSLM are very accurate and result in an exact column profile if the profiles of the BVM intersect at a common tray or the SSLM profile runs exactly in the defined distillate composition.

While the SSLM is applicable to any kind of separation, it comes very close to a direct minimization of the energy demand by means of a rigorous model as presented in Section 8.4. In particular, the MED can be computed at comparable effort by solving an NLP with the energy input as the objective function, a column model with large tray numbers in both column sections and product purity constraints.

8.3.1.2 Pinch-based shortcut methods

In distillation columns operated at MED, the concentration profile usually exhibits regions in which there is only little change. The concentrations in these regions are very close to the stationary points of the tray-to-tray equation system (8.5)–(8.9) and hence to the pinch points. At pinch points the down-coming liquid and the up-flowing vapor streams approach equilibrium [100] because the operating lines come very close to the equilibrium surface, such that the driving forces for mass transfer tend to vanish [45].

Pinch points, as any solution of a set of difference equations, can be classified according to their stability properties [101]. The $n_c - 1$ positive real eigenvalues and real eigenvectors of the Jacobian of the tray-to-tray equation system describe the local behavior of the concentration profiles in the vicinity of the pinch points [94]. If all the eigenvalues are bigger than one, the concentration profiles leave the corresponding pinch in the direction of one of the eigenvectors. In contrast, if all the eigenvalues are smaller than one, the concentration profiles approach the pinch [94,102]. Therefore, pinch points with only stable eigenvalues ($\lambda_i < 1$) represent a stable node, and pinch points with only unstable eigenvalues ($\lambda_i > 1$) represent an unstable node, respectively. Pinch points with stable and unstable eigenvalues represent saddle points.

Pinch-based shortcut methods characterize the MED by some geometrical criteria regarding the location of certain pinch points. In case of a binary separation, the MED is described by a single pinch point, defined by the intersection of the operating lines and the equilibrium line as depicted in Figure 8.11. In ternary systems this scenario is also possible, but corresponds to a specific set of products, the so-called "preferred split" [103]. Figure 8.12 shows the pinch points and the associated eigenvectors in case of a sharp split for a ternary separation operated at MED. In addition, several tray-to-tray profiles, which are calculated from a concentration in very close proximity of the product, are depicted. It is obvious that the calculated concentration profiles qualitatively follow the path of the pinch points. All the concentrations profiles end in the stable pinch of the rectifying and stripping sections; they are more or less attracted and deflected by the saddle pinch points. For a sharp split, the profiles of the rectifying and stripping sections traverse very close to the saddle points (e.g. r2 and s2 in Figure 8.12).

Based on the theory of discrete dynamical systems [94], most pinch-based shortcut methods also relate to the geometric interpretation of Underwood's method for ideal mixtures, showing that the MED relates to the collinearity of certain pinch

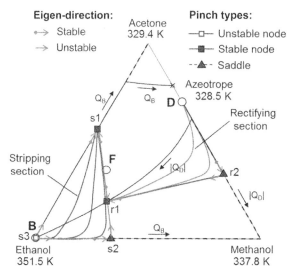

FIGURE 8.12 Tray-to-Tray Profiles and Pinch Points at Minimum Energy Demand

points [20,94,104]. Koehler et al. [105] give an excellent survey of the relations between Underwood's method restricted to ideal mixtures and pinch-based methods applicable to general (homogeneous) mixtures. In case of the ternary mixture in Figure 8.12, the saddle pinch s2 as well as the stable pinches r1 and s1 become approximately collinear at MED. In this configuration, the stable pinch r1 marks the switch from the rectifying to the stripping section, which corresponds to the concentration on the feed tray and is therefore termed feed pinch.

If the stable pinch point s1 is replaced by the concentration F of a saturated liquid feed for the assessment of the collinearity condition, the accuracy of the MED estimate can be improved [20]. The *zero volume criterion* (ZVC) [94] allows for a direct evaluation of such a collinearity condition for three-component mixtures. It can be extended to a coplanarity condition for multicomponent mixtures by requiring that the volume defined by the $n_c - 1$ compositions of the feed, the feed pinch, and $n_c - 2$ pinch points in the adjacent column section becomes zero at MED [94].

Instead of directly applying a collinearity condition to ternary mixtures, which—even if CRV and CMO assumptions hold—is only valid for sharp splits [104], the *minimum angle criterion* (MAC) [45] aims at the minimization of the angle between the composition vectors formed by the concentrations of the feed F and of a pinch point in both sections, e.g. the feed pinch r1 and the saddle pinch s2 in the example shown in Figure 8.12. If the angle is zero, the criterion reduces to the collinearity criterion used by the ZVC [45].

Similar to the MAC, the *feed angle method* (FAM) [106] aims at the minimization of an angle between two composition vectors, formed, however, by the feed pinch r1 with the saddle pinch s2 and with the composition on the tray following the feed pinch in the adjacent section. The FAM can also be applied to multicomponent mixtures if the hyperplane defined by the feed pinch and the saddle pinches in the adjacent section is formed and if the angle between this hyperplane and the vector defined by the compositions of the feed pinch and a tray in the adjacent section is minimized. A modified FAM is available for cases where there is no feed pinch [106]. It has been shown that the FAM results in very good MED estimates, also in particular for strongly nonideal mixtures [106].

One major limitation of all three methods is the need to determine the relevant pinch points, since this task can be difficult and is at least time-consuming, especially in case of multicomponent mixtures with four and more components and additional azeotropes. The *rectification body method* (RBM) [107] overcomes these problems by calculating and utilizing all existing pinch points. In a first step, all the pinch points are calculated and classified. Afterward, possible paths connecting the pinch points are generated using information on their stability properties and on the number of stable eigenvectors. Additional criteria related to thermodynamic consistency (e.g. check of monotonic entropy production) or geometrical constraints (e.g. check for tangential pinch) are used to identify thermodynamically infeasible paths [107]. Based on the generated paths of pinch points, linear rectification bodies (RBs) are constructed from the product compositions and the relevant pinch points. The RBs can be interpreted as linear approximations of the envelopes of all tray-to-tray profiles calculated from concentrations in close proximity of the product compositions for a given energy duty [55,107]. MED is identified as the lowest energy demand at which the RB for stripping and rectifying sections intersect. In case of ternary mixtures, this is equivalent to the collinearity criteria. An additional advantage of the RBM is that the intersection of the polyhedrons can easily be checked computationally in a multidimensional space. This also facilitates the determination of MED in those cases where no feed pinch is present [107]. The method has been fully automated for an—in principle—arbitrary number of components.

The accuracy of the MED estimate is limited in case of strongly nonideal mixtures. It can be improved, however, if the RBM is complemented by a subsequent application of the MAC or the FAM.

One major advantage of pinch-based shortcut methods is their insensitivity toward the specification of trace impurities. While the specification of impurities strongly determines the results of shortcut methods relying on tray-to-tray calculations like the BVM and the SSLM, the amount and distribution of trace components has only little effect on the location of the pinch points and hence the results obtained from pinch-based shortcut methods [46,94]. Figure 8.13 displays the geometrical criteria of different pinch-based shortcut methods.

8.3.1.3 Hybrid shortcut methods
Another possibility to enhance the quality of MED predictions, especially in case of strongly nonideal mixtures, is the combination of tray-to-tray calculations and pinch-based methods.

The *eigenvalue criterion* (EC) [102], for example, can be interpreted as an extension of the BVM if applied to sharp splits. Instead of calculating the concentration

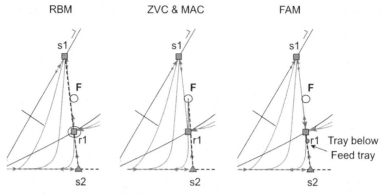

FIGURE 8.13

Geometrical criteria for minimum energy demand illustrated for the separation presented in Figure 8.12; rectification body method (RBM), zero volume criterion (ZVC), minimum angle criterion (MAC) and feed angle method (FAM).

profiles starting from the product compositions, the calculation is started at a concentration in the proximity of the relevant saddle pinch in the direction of the unstable eigenvector. While this approach avoids the precise specification of trace impurities in the products and works well for ternary mixtures, the choice of the relevant saddle pinch and the assessment of the intersection of the concentration profiles complicate its use for mixtures with four or more components.

The *continuous distillation region method* (CDRM) [108] improves the accuracy of the RBM, using tray-to-tray calculations to replace the linear connection of the pinch points by the concentration profiles. The resulting envelopes are constructed solely from pinch point information and are therefore equivalent to the continuous distillation regions (CDRs) introduced by Fidkowski et al. [27]. A separation is feasible, if the CDR of the column sections containing the product compositions intersect. Though much better estimates of the MED can be determined for strongly nonideal separations, the CDRM is limited to ternary mixtures.

The *feed pinch method* (FPM) [106] is similar to the SSLM, but instead of calculating the complete column profile, the FPM employs tray-to-tray calculations for the non-pinched section starting at the feed pinch concentration. While the FPM can improve the MED estimate for strongly nonideal separations, it suffers from the same limitations as the SSLM and, in addition, can only be applied to separations exhibiting a feed pinch.

8.3.1.4 Tangent pinch

Separations of azeotropic mixtures may be controlled by the presence of so-called tangent pinches [109]. These special pinch points limit the applicability of all pinch-based methods and need particular attention to properly estimate the MED of the separation. For binary separations, a tangent pinch is reflected by multiple

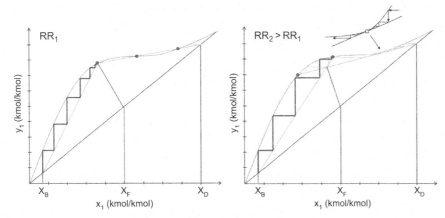

FIGURE 8.14 Illustration of a Tangent Pinch for a Binary Separation

intersections of the equilibrium curve and the operating line of one of the column sections, as depicted in Figure 8.14. Though both operating lines intersect the equilibrium curve for the reflux ratio RR_1, the separation is not feasible, since the composition profile of the rectifying section ends prematurely at the first intersection of the rectifying section operating line and the equilibrium curve. The reflux ratio therefore needs to be increased to a value of RR_2 at which the operating line of the rectifying section intersects only once with the equilibrium curve, being the tangent to the equilibrium curve at the highlighted position. At this reflux ratio, a tangent pinch occurs, because the composition profiles of both column sections intersect for the first time; hence, a feasible separation at MED is established.

The concept of tangent pinch can also be extended to multicomponent mixtures [109]. In this case, the tangent pinch is reflected by the existence of multiple pinch points on a single pinch line for a given reflux ratio. The equivalence of both conditions is given by the fact that the equilibrium curve corresponds to the reversible operating line [45] or in other words the pinch line for binary systems. The tangent pinches are turning points in the map from reflux ratio to pinch compositions, which correspond to saddle-node bifurcations of the pinch equation system [110]. They are also characterized by a local maximum of reflux ratio or energy demand along a pinch line [105].

8.3.1.5 Restrictions of pinch-based shortcuts

The use of pinch-based shortcut methods for the determination of the MED requires the concentration profiles to pass close to the pinch points, which only happens for a large or—in the limit—an infinite number of trays in both column sections. Pinch-based shortcut methods are therefore limited to separations that at least exhibit one pinched region in each column section. Whether such pinched regions occur primarily depends on the product specifications and therefore on the type of split. While pinched regions typically occur in case of sharp splits, they can occur for semisharp splits, if a feed pinch, a tangent pinch, or a product composition very close to a distillation boundary is present. Especially the last case has to be carefully analyzed considering the possible difference between SDB/TRB and PDB (cf. Figure 8.8). Koehler et al. [105] conclude that in such cases rigorous calculations are unavoidable to get reliable predictions of MED. In case of sloppy or nonsharp splits, pinch-based shortcut methods usually result in inaccurate estimates of the MED.

8.3.1.6 Summary
While the restrictions regarding product specification discussed above are certainly limiting the applicability of pinch-based shortcuts, the opportunity of applying them to separations with sharp splits without the need to specify trace impurities makes them advantageous over any shortcut method relying on tray-by-tray calculations. In these practically relevant cases, they constitute an efficient tool for MED estimation, even in case of multicomponent separations involving up to 10 components [111]. However, if the presence of pinched regions cannot be guaranteed, shortcut methods relying on tray-to-tray calculations should rather be preferred to determine the MED. Therefore, the choice of a suitable shortcut method should always consider the type of split and the topology of the mixture in order to provide reliable MED estimates.

8.3.2 Extractive distillation

Extractive distillation is performed in a sequence of at least two columns, a two-feed column with an extractive middle section followed by a single-feed entrainer recovery column (cf. Figure 8.1). Shortcut methods are of special interest for extractive distillation, since they allow for a fast evaluation of process performance for different entrainer candidates [61]. While split feasibility and MED of the entrainer recovery column can be assessed by the methods discussed so far, two-feed columns require an adjustment of the shortcut methods to account for the distinct feed positions. In addition, the entrainer flowrate introduces another degree of freedom for the design, which strongly influences separability and MED. The minimum entrainer flowrate as well as the minimum and the maximum reflux ratio, which limit the operating window of the column, can be calculated from a bifurcation analysis of the pinch lines of the middle section [59]. An increase in entrainer flowrate and, accordingly, the maximum reflux rate might, however, be necessary to render the separation in the complete column feasible; this is also preferable to improve the operational stability of the process [59,61].

For a fixed entrainer flowrate, the MED can be calculated by means of several shortcut methods. Those methods relying on tray-by-tray calculations, like the BVM, have to cope with the problem that the column profile exhibits two transitions, one at either feed stage, making the search for a consistent composition profile at MED even more difficult [112]. Under the assumption of a feed pinch occurring at one of the intersections of the profiles of the column sections of a multi-feed column [112], a collinearity criterion can be applied to determine the MED. Levy and

Doherty [112] state that collinearity of the feed pinch of the controlling column section, the feed composition, and a saddle pinch of the extractive column section provides an excellent approximation of the MED. However, the controlling section and the relevant pinches have to be determined somehow to apply this criterion. In addition, mixtures with four or more components do not necessarily exhibit a feed pinch, and more than one saddle pinch may occur [61]. Brüggemann et al. [61] extended the RBM for extractive distillation columns to overcome these restrictions. A bifurcation analysis of the pinch equation system reveals that the MED is determined by an intersection of the RB of all three column sections for an entrainer flowrate above the minimum (e.g. for $E \approx 1.1E_{min}$). In contrast to the RB of the rectifying and stripping sections, the RB of the middle section is generated by a linear extension of the eigenvectors of the saddle pinches of the extractive section. This extension of the RBM is designed to properly handle mixtures with four or more components and to be independent of the occurrence of a feed pinch.

The application of pinch-based methods for two-feed columns relies on requirements regarding product specifications that are similar to those of single-feed columns. If the controlling pinch points are identified, MAC or FAM can again be applied to increase the accuracy of the MED estimate. It also has to be mentioned that two-feed column designs are not only essential for extractive distillation but can also offer energy savings if a separation of two feed streams with the same components at different compositions is to be performed [112].

8.3.3 Heteroazeotropic distillation

Shortcut methods for the design of heteroazeotropic distillation columns not only have to account for the liquid phase splitting in the decanter but also need to consider the possibility of heterogeneous trays in the column. Therefore, a phase stability test (see Section 8.2.2.6) needs to be carried out for every tray during the computations.

The BVM can be applied to determine the MED of a heteroazeotropic distillation column if a decanter model is simply added [113]. In contrast to simple distillation columns, the rectifying section profile has to be calculated starting from the liquid composition of the reflux leaving the decanter to the top stage of the column and not from the top product. Therefore, the phase ratio of the decanter reflux presents an additional degree of freedom for the design of a heteroazeotropic distillation column. In case the decanter tie-line intersects the vapor line (cf. Figure 8.15), the reflux from the decanter with a specific phase ratio $\phi_1 = \phi_1^0$ results in a heterogeneous tray at the top of the column. Additional heterogeneous trays are possible if their respective liquid–liquid tie-line intersects with the vapor line. Hence, in contrast to a simple column, the composition profile depends not only on the product specification and the energy duty but also on the number of heterogeneous trays (k) in the column and the phase ratio on the last heterogeneous tray (ϕ_k) [108], rendering the BVM design even more tedious for such columns.

The sensitivity of the rectifying profile with respect to the phase ratio of the decanter reflux is exemplarily illustrated in Figure 8.15. Any phase ratio $\phi_1 \neq \phi_1^0$

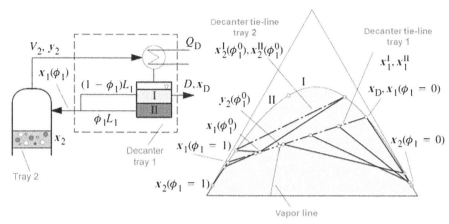

FIGURE 8.15

Balance envelope for the decanter of a heteroazeotropic distillation column (left) and illustration of the sensitivity of tray-to-tray calculations (right).

Reprinted (adapted) with permission from Ref. [108]. Copyright 2002 American Chemical Society.

results in a liquid composition x_2 outside the miscibility gap. The calculation of the complete column profile starting from the bottom product avoids the specification of any additional degrees of freedom. A design is considered feasible if the vapor composition of the topmost tray is located on the decanter tie-line. Therefore, methods working from the bottom product upwards, like the SSLM or the FPM [106], are more favorable in case of heteroazeotropic distillation than the BVM.

Pinch-based shortcut methods can also be applied to heteroazeotropic distillation if the pinch points are properly calculated either for the homogeneous or the heterogeneous pinch equation system, depending on the state of the liquid composition on a pinched tray. The pinch points only depend on the product specifications and the energy duty. For the RBM, the RB of the rectifying section needs to be constructed with the liquid composition of the reflux from the decanter, instead of that of the top product (cf. discussion on BVM above). The ZVC, the MAC, or the FAM can likewise be applied after the relevant pinch points are determined.

However, most heterogeneous separations exhibit strongly nonlinear composition profiles, for which collinearity criteria are not of sufficient accuracy to determine the MED [108,112]. While the RBM and the ZVC lead to poor estimates, the CDRM and the FAM have been shown to provide accurate MED estimates [106,108]. Therefore, either these pinch-based shortcuts or unidirectional profile calculations starting at the bottoms product or a feed pinch (like SSLM or FPM) should be used to determine the MED for a heteroazeotropic distillation column.

Complex column configurations, other than with a decanter at the top of the column, are possible but may require appropriate modifications of the shortcut methods [73,74]. One additional degree of freedom for the design of heteroazeotropic distillation columns that most shortcuts do not consider is the operating temperature of the decanter. It has a direct influence on the miscibility gap and consequently on the feasibility of product specifications and should therefore be considered.

8.4 Optimization-based conceptual design of distillation processes

The results obtained from shortcut methods provide a preliminary ranking of design alternatives but do not provide a complete conceptual design of a distillation process. Rather, detailed MESH models, based on mass balances, equilibrium calculation, summation constraints, and energy balances, have to be used to determine a "rigorous" conceptual design of distillation processes. Total annualized cost, comprised of annual operating costs and annualized investment costs, constitute a suitable performance measure to evaluate various design alternatives. While the operating costs are accounting for energy costs resulting from the energy duty of the columns in the sequence, investment costs are estimated by simple yet nonlinear cost correlations, including material correction factors that account for economy of scale (e.g. [114–116]).

The "optimal" separation process design, including the decision on structural and operational degrees of freedom, is typically determined by a sequence of simulation studies accompanied by cost estimation in tedious trial-and-error procedures in industrial practice (and often even in academia). The column design, i.e. the determination of energy duty higher than the MED, the number of theoretical trays, and the optimal feed tray location, can also be determined by means of an extension of the ZVC [46].

The most versatile and efficient approach to conceptual design of distillation processes, however, relies on rigorous numerical optimization of detailed tray-to-tray models [4,117]. Here, the manual search is replaced by a numerical procedure that relies on mathematical programming. In particular, a superstructure of the separation flowsheet is developed to account for the various structural alternatives of the separation process, and an optimization problem is formulated to allow for a simultaneous determination of all the design degrees of freedom by directly minimizing an appropriate objective function, which typically measures total annualized cost. Discrete decisions accounting for the position of the feed tray and the number of trays in each column section can be encoded by binary decision variables as introduced and worked out by Grossmann and coworkers (see [117] for the original contribution and the Chapter 12 "Optimization of Distillation Processes" for a detailed overview on problem formulations and solution strategies). The resulting superstructure models are generally of large scale and present strongly non-convex MINLP problems that are particularly hard to solve [118].

This is especially true for the numerical optimization of azeotropic distillation processes, which has only been addressed more recently [119–121]. Special emphasis has to be placed on appropriate property models for thermodynamic quantities, in particular for equilibrium concentrations, which comprise complex nonlinear equations. Hence, the computational complexity of the MINLP problems has been the limiting factor to really use this strategy successfully in the design of azeotropic distillation processes. A careful initialization and bounding of all variables has been shown to be indispensable [119]. The insight collected during the application of a shortcut method as well as the numerical values of key quantities derived from the results of the shortcut calculations can be used to facilitate a first feasible design [119,122]. These ideas have been instrumental for formulating the process synthesis framework [4] reviewed in Section 8.1.2, where only the most promising variants identified during shortcut-based screening are further evaluated and refined by means of rigorous design optimization. The use of the shortcut results for a reasonable initialization of the concentration and temperature profiles has shown to tremendously improve the robustness of the solvers. In addition to this favorable initialization, a stepwise solution with incremental model refinement can further improve robustness [122].

Since the computational effort increases exponentially with the number of binary decision variables, the set of possible feed tray and reboiler/condenser locations should be constrained (based on the results of the shortcut step or experience) to facilitate a robust and efficient optimization. Bauer and Stichlmeier [119] propose to first solve a relaxed MINLP, where binary variables are allowed to take every value between zero and one and to subsequently heuristically project the continuous solution to binary decisions. More recently, the reformulation of the original MINLP problem to a continuous problem has shown great potential compared to a direct solution of the MINLP problem in case of azeotropic distillation processes [122−124]. While the solution of the continuous optimization problem tends toward a solution with discrete feed and recycle tray locations in case of single-feed zeotropic columns [117,125], additional nonlinear constraints may be necessary to enforce discrete solutions in case of multi-feed columns or other complex separation processes [123,124,126]. Especially the use of nonlinear complementary constraints [127] has resulted in excellent performance in the optimization of azeotropic distillation processes with multiple columns, recycle streams, and multi-feed columns: a solution of the MINLP problem can be achieved by the solution of a series of NLP problems with subsequently tightened relaxations [126].

Optimization models of curved-boundary and pressure-swing distillation processes can be generated by coupling single-column models. Similarly, extractive distillation processes can directly be modeled as the combination of a two-feed and a single-feed column, connected with a recycle of the entrainer. For complex columns with multiple feed or side streams, additional constraints can be added to improve the convergence [126].

Distillation processes involving heteroazeotropic multicomponent mixtures require special attention with respect to the assessment of phase stability in order to correctly distinguish between trays with two or three phases. Automatic switching between VLE and VLLE equilibrium models has to be enabled on every single tray (cf. Section 8.2.2.6). Since the separation performance may strongly depend on the phase ratio in the decanter and on the number of heterogeneous trays in the column [108], these degrees of freedom have to be determined during the optimization. The correct consideration of phase equilibrium is computationally involved. Reliable methods are essential to ensure reliable optimization results. In addition, the model discontinuities stemming from switching between heterogeneous and homogeneous trays may complicate or even impede the optimization. The solution of the equilibrium problem including a phase stability test as a subproblem has been shown to improve the robustness of the solution algorithms in dynamic simulation [128]. This computational strategy has been shown to improve robustness also in rigorous design optimization [129].

8.5 Design studies for different types of azeotropic distillation processes

The last section of this chapter presents a set of illustrative case studies for the design of a number of the azeotropic distillation processes following the process synthesis framework and employing a combination of shortcut methods and rigorous optimization methods. The cases are selected to cover the most common options for breaking azeotropes in distillation processes, i.e. curved-boundary, pressure-swing, extractive, and heteroazeoptropic distillation processes are covered subsequently.

8.5.1 Curved-boundary distillation process

In order to demonstrate the design of a curved-boundary distillation process, the separation of a quaternary mixture of acetone, chloroform, benzene, and toluene (ACBT mixture) with a single binary azeotrope between acetone and chloroform is investigated. The feed is considered to be an equimolar mixture at a flowrate of 10 mol/s and a pressure of $1.013 \cdot 10^5$ Pa.

8.5.1.1 Generation of process variants

Different process variants can be generated that exploit the curvature of the distillation boundary (see Figure 8.16). In fact, a comparison of the different distillation boundaries under total reflux (SDB) and reversible (PDB) operation shows that the distillation boundary bends toward the benzene/toluene/chloroform plane at finite reflux ratios (cf. Figure 8.8 for an illustration for the ternary acetone/benzene/chloroform plane). This allows for a complete separation of acetone if the feed composition is shifted toward the acetone/benzene/toluene plane. This can be accomplished by appropriate recycle of benzene and toluene [130].

   If the necessary recycle is considered as product and the feed is modified to account for the recycle, a state-task network [131] can be built to generate all possible distillation sequences [111]. Figure 8.16 depicts a selection of possible distillation sequences.

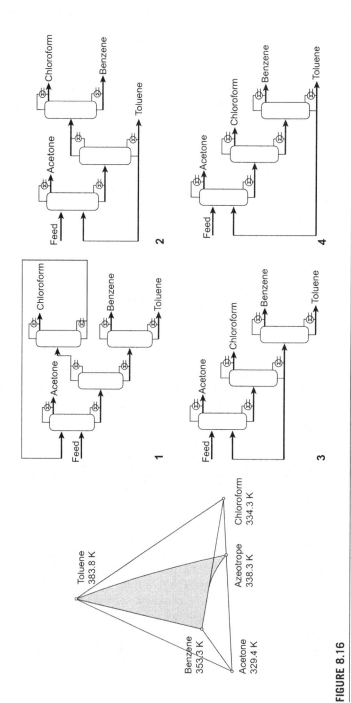

FIGURE 8.16

Concentration simplex with distillation boundary (left) and exemplary process variants for the separation (right) of the acetone, chloroform, benzene, and toluene mixture.

8.5.1.2 Shortcut-based screening

In order to evaluate the different column sequences, any of the presented shortcut methods for simple distillation columns can be applied to determine the MED of each of the single separation tasks first, which can then be summed up to obtain the MED of the sequence.

The MED of any sequence can further be decreased by internal heat integration (HI) of the reboiler and condenser duties if the temperature differences are appropriate. The temperature levels can be altered by changing the operating pressure within the different distillation columns. In order to design an optimal heat-integrated distillation process, the MED of each process sequence can be directly minimized by either keeping the operating pressure of each column as a degree of freedom in the optimization problem embedding shortcut models for every column, or, a two-step approach can be chosen where the MED of each column is evaluated first at different discrete pressure levels and a combinatorial problem is solved afterward [132]. Although the first step of the two-step approach requires an extensive application of the computationally efficient shortcut methods, it is numerically more stable and cannot get trapped in local minima. Note that feasibility of the single separation tasks at different pressure levels needs to be checked, since the position of the azeotrope and the course of the distillation boundary change with the pressure.

The MED for each of the process variants depicted in Figure 8.16, evaluated with the RBM as the shortcut method, is listed for sequences with and without heat integration in Table 8.1. The differences in energy demand between the best and the worst case are significant. In any case, sequence 2 has the lowest MED and will therefore be investigated closer.

8.5.1.3 Rigorous design optimization

A rigorous optimization problem is formulated for sequence 2 in order to determine a more detailed design that minimizes the TAC, considering both operational and investment costs rather than only operational costs (reflected by MED). The results of the shortcut evaluation can be used to initialize the MINLP model of the three-column process, which is afterward optimized in a stepwise solution strategy after a continuous reformulation of the MINLP as described in Section 8.4. The results of the optimization of sequence 2 with and without heat integration are listed in Table 8.2.

While the rigorous design of the heat-integrated process shows about 50% savings in reboiler energy, similar to the result obtained from the shortcut evaluation, the savings in TAC are only about 11%. The lower savings in terms of cost are

Table 8.1 MED for the Sequences Depicted in Figure 8.16 [126,130]

	Sequence 1	Sequence 2	Sequence 3	Sequence 4
Without HI	1399 kW	1103 kW	1201 kW	1209 kW
With HI	615 kW	475 kW	574 kW	501 kW

Table 8.2 Results of the Rigorous Design Optimization of Sequence 2 [130]

	Process Without Heat Integration			Heat-Integrated Process		
	Acetone Column	Toluene Column	Chloroform/ Benzene Column	Acetone Column	Toluene Column	Chloroform/ Benzene Column
Capital cost (€/a)	110,089	76,819	83,340	116,925	76,712	82,355
Operating cost (€/a)	55,781	57,696	33,887	82,258	12,046	6345
TAC (€/a)	417,612			376,641		
Reboiler duty (kW)	453	511	266	539	75 + 297	0 + 320
Cond. Duty (kW)	462	339	417	0 + 297	102 + 320	479
Number of trays	51	25	40	70	28	34
Feed tray	14	13	17	17	14	15
Recycle feed tray	8			8		
Diameter (cm)	68.6	72.5	64.3	49.7	70.7	76.5
Recycle (flow and concentration)	8.27 mol/s [0, 0, 0.01, 0.99]			5.86 mol/s [0, 0, 0.01, 0.99]		

The underlined values highlight the amount of energy, which can be saved for heating and cooling in the heat-integrated design. For example the 297 kW cond. duty for the acetone column are used as heating steam for the toluene column and therefore listed as part of the reboiler duty of that column. In addition to that "internal" heat 75 kW of external heat need to be provided.

due to the higher investment cost and the higher operational cost due to a necessary utility change for the acetone column, which is operated at elevated pressure in case of the heat-integrated design. Although the savings are still significant, the results show that a decision on heat integration should only be made on the basis of a rigorous conceptual design.

8.5.2 Pressure-swing distillation process

The separation of a binary mixture of 95.8 mol% water and 4.2 mol% ethanol is investigated as a case study for the design of a pressure-swing distillation process. The mixture exhibits a minimum boiling azeotrope. A saturated feed stream of 10 mol/s at a pressure of $1.013 \cdot 10^5$ Pa is assumed. This section summarizes the results of the case study presented by Brüggemann and Marquardt [133]

8.5.2.1 Generation of process variants

The pressure sensitivity of the azeotrope can be exploited to facilitate a separation in a simple two-column process. In addition to the simple pressure-swing process, an enhanced process using acetone [134] as an entrainer is also considered. Operating pressures of $1.013 \cdot 10^5$ Pa and $10.13 \cdot 10^5$ Pa are assumed to provide a sufficiently wide pressure-swing area, as depicted in Figure 8.17. Note that at an elevated pressure of $10.13 \cdot 10^5$ Pa three additional azeotropes occur.

8.5.2.2 Shortcut-based evaluation

If the feed and the bottom products of both columns of the simple pressure-swing process are specified as (nearly) pure water and ethanol, there is still one remaining

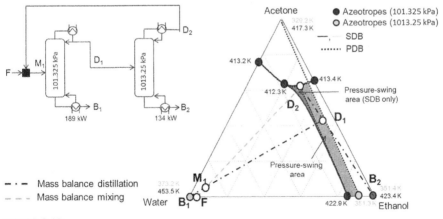

FIGURE 8.17

Process structure (left) and composition simplex (right) with mass balance lines of the separation and distillation boundaries for shortcut evaluation of the mixture. SDB, simple distillation boundaries; PDB, pinch distillation boundary.

Adapted from Ref. [133].

degree of freedom, e.g. the D/F ratio in the first column. For the entrainer-enhanced process, the amount of entrainer in the process constitutes another degree of freedom.

In order to determine the process with the lowest energy demand, the MED has to be evaluated for all possible distillate compositions, which are limited to the shaded pressure-swing area in Figure 8.17. This evaluation can be automated by means of an optimization problem, which incorporates both, a shortcut model to evaluate the MED of each of the columns, and a feasibility test that constrains the possible product specifications. For this purpose, Brüggemann and Marquardt [133] combined the RBM with a feasibility test based on the PDB [52] to determine the optimal product specifications, which are also depicted in Figure 8.17. The MED of the complete process is determined as 323 kW. Since the simple pressure-swing process requires more than 400 kW, the use of acetone as an entrainer significantly reduces the MED of the process. Note that the first column exhibits a tangential pinch that needs to be accounted for in the shortcut evaluation. However, by means of manually changing the specification and visual analysis of the results of a BVM calculation, Knapp and Doherty [134] reported a slightly lower MED of 306 kW. In their process specification, the distillate of the second column is positioned in the part of the pressure-swing area, which is labeled as "SDB only" in Figure 8.17, since it is separated from the bottom product by a PDB. Although feasible, this separation is rejected according to the PDB-based feasibility test and therefore excluded in the optimization approach of Brüggemann and Marquardt. In addition, as these authors report, an evaluation of such a specification by means of pinch-based shortcut methods is of limited value due to the pure approximation of the column profile by straight lines connecting the pinch points and by complex pinch topologies [133]. As mentioned in Section 8.3, pinch-based shortcut methods need to be applied with caution if the MED for columns with semisharp splits is to be evaluated.

8.5.2.3 Rigorous design optimization

The results of the shortcut evaluation can then be used to initialize a rigorous optimization problem. Kossack et al. [135] therefore use the results of the RBM-based optimization of Brüggemann and Marquardt to generate linear concentration and temperature profiles based on the pinch points. After an optimization of each of the columns at a fixed recycle flowrate taken from the shortcut evaluation, the complete two-column process is optimized for TAC with a free recycle flowrate. The resulting process is depicted in Figure 8.18.

It is interesting to note that the rigorous optimization shifts the distillate composition of the second column further toward the ternary azeotrope at $10.13 \cdot 10^5$ Pa, reducing the recycle flowrate as well as the D/F ratio in the first column. Even though the optimization minimizes total annualized cost rather than energy demand, the computed energy demand is only 277 kW and hence well below the MED estimate. The identification of this process design by means of the BVM is possible, in principle. However, it necessitates tedious evaluations, varying the product specifications and the reflux ratio. Though the quality of the results obtained from the

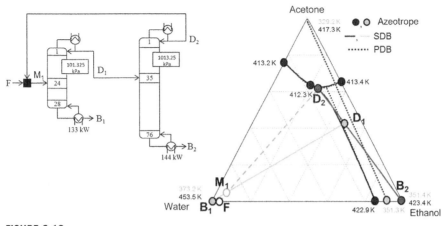

FIGURE 8.18

Optimized process: flowsheet (left) and concentration simplex with balance lines and distillation boundaries (right). SDB, simple distillation boundaries; PDB, pinch distillation boundary.

pinch-based shortcut method may be limited, they can be used successfully to initialize and efficiently solve the rigorous MINLP problem and therefore solve the design problem in the spirit of the process synthesis framework.

Additional design decisions, such as introducing the recycle from the second column as a second feed to the first column, enhancing the simple pressure-swing distillation process by an extractive agent, or changing the pressure levels, can be integrated in the optimization model and can further reduce the necessary energy demand and process cost. However, possibly necessary utility changes need to be considered when changing the operating pressure.

8.5.3 Extractive distillation process

The separation of a binary azeotropic mixture of acetone and methanol, which is typically separated in an extractive distillation process using water as an entrainer [33], is investigated to demonstrate the application of the synthesis framework to such a process, including methods for entrainer selection. A saturated feed of 10 mol/s at $1.013 \cdot 10^5$ Pa with an almost azeotropic composition of 77.74 mol% acetone and 22.26 mol% methanol is assumed. This section summarizes the results of the case study presented by Kossack et al. [22].

8.5.3.1 Generation of process variants

A classical two-column process structure as depicted in Figure 8.19 in combination with a set of different entrainer compounds is considered as a superstructure for process optimization. Depending on the association behavior of the entrainer, either acetone or methanol can be obtained as the distillate of the extractive column.

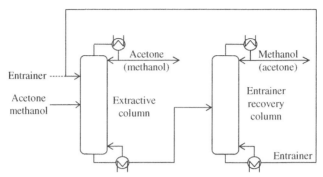

FIGURE 8.19 Extractive Distillation Process for Acetone/Methanol Separation

The selection of candidate entrainer compounds is based on a literature study and a CAMD procedure employing a generate-and-test strategy as implemented in the ICAS software package [16]. The physical properties are predicted from molecular structure using group contribution methods [18]. Based on the literature survey, water, ethanol, isopropanol, chlorobenzene, and ethylene glycol are selected as entrainer candidates [58,136]. For the generation of candidate entrainer molecules either acyclic alkanes and alcohols are considered if methanol is the solute, or chlorine-containing compounds, if acetone is the solute. The entrainer candidates are limited to molecules with less than nine functional groups. According to heuristic guidelines, the boiling temperature is restricted to a range of 350−500 K. Molecules are discarded if they introduce additional azeotropes or miscibility gaps or if they cannot be found in common thermodynamic property databases. In addition, a minimum selectivity of $S_{ij}^{\infty} > 2.5$ is requested. The CAMD procedure largely reproduces the entrainer candidates from the literature and suggests several new molecules. However, due to the inaccuracies of the UNIFAC predictions and the requested design specifications, in particular the minimum selectivity, some molecules are discarded in the generate-and-test procedure. Finally, molecules from both literature survey and CAMD have been chosen as summarized in Table 8.3 to be evaluated in the following steps.

8.5.3.2 Shortcut-based evaluation

The RBM and its extension for extractive distillation [61,107] are utilized to determine the minimum entrainer flowrate and energy demand for the extractive column and the MED for the entrainer recovery column. The results of the calculations are shown in Table 8.3. No feasible entrainer flowrate could be determined for ethylene glycol and benzyl ethyl ether.

Although chlorobenzene has the highest selectivity at infinite dilution and its choice results in the lowest energy demand for the extractive column, it is only the second best candidate concerning the MED of the total process. The results show that a ranking based on selectivities is a useful but not a reliable means to assess the performance of the complete extractive distillation process. It does not

Table 8.3 Estimates of Selectivities and MED Based on RBM for Different Entrainer Candidates [22]

Entrainer Candidate	S_{ij}^{∞}	$\frac{E_{min,bit}}{F}$	$\frac{E}{F}$	$\dot{Q}_{B,min}/F$ Extractive Column	$\dot{Q}_{B,min}/F$ Total Process
Distillate Acetone					
DMSO	2.89	0.651	0.771	5.42	7.20
Water	2.42	0.814	0.933	9.866	11.803
Ethanol	1.65	0.721	0.760	11.077	16.981
Isopropanol	1.71	2.187	2.923	15.97	26.101
Ethylene glycol	4.19	–	–	–	–
Distillate Methanol					
Chlorobenzene	5.84	1.615	2.015	3.75	7.93
p-Xylene (UNIFAC)	4.40	0.81	1.67	3.82	8.71
Mesitylene	3.75	0.82	1.55	4.01	9.562
p-Xylene (UNIQUAC)	2.32	0.77	2.38	4.32	9.94
1,2,3-Trimethylbenze	3.59	0.83	1.63	3.93	10.11
Benzyl ethyl ether	2.23	1.13	–	–	–

consider the effort for the recovery of the entrainer, which can be critical if it comes to the selection of an optimal solvent [137].

8.5.3.3 Rigorous design optimization

A rigorous design optimization with the most promising entrainer candidates is performed to complete the conceptual design. Again, the results of the shortcut calculation can be used for reasonable initialization of the temperature and concentration profiles in the columns. The results of the optimization are listed in Table 8.4. A comparison with the results obtained with the shortcut method shows that only the position of DMSO in the ranking has changed. Although the DMSO process still has the lowest energy requirement, the chlorobenzene process is the most economic one. This is due to the high boiling temperature of DMSO of about 464 K, which necessitates the use of a more expensive heating utility. The number of trays and the location of the feed tray differ significantly for the various designs. The comparison of the results for p-xylene also shows the significance of the quality of the property model on the resulting process design.

8.5.4 Heteroazeotropic distillation process

The separation of the binary mixture water and isopropanol is investigated to illustrate the design of heteroazeotropic distillation processes. The miscibility gap exploited to break the azeotrope is induced by the addition of an entrainer, in this

Table 8.4 Results of the Rigorous Design Optimization for the Most Promising Entrainer Candidates [22]

Entrainer Candidate	$\frac{E}{F}$	$\frac{Q_{B,total}}{F}$	$\frac{TAC}{F}$	Number of Trays Extr. Col.	Number of Trays Rec. Col.	Feed Tray E1	Feed Tray F1	Feed Tray F2
Distillate Acetone								
DMSO	0.992	7.57	34.20	35	13	2	27	5
Water	1.886	11.11	44.64	54	26	33	50	15
Distillate Methanol								
Chlorobenzene	2.382	8.37	32.56	42	16	10	26	7
p-Xylene (UNIFAC)	2.249	9.33	35.62	43	13	10	24	6
Mesitylene	1.900	10.38	38.92	37	9	4	22	4
p-Xylene (UNIQUAC)	2.597	10.63	39.67	47	11	8	30	6

case cyclohexane. A saturated feed of 10 mol/s at $1.013 \cdot 10^5$ Pa with a composition of 32.5 mol% isopropanol and 67.5 mol% water is assumed.

8.5.4.1 Generation of process variants

Once a suitable entrainer has been chosen, two different flowsheet variants exist as shown in Figure 8.20. In variant A, the feed is mixed with the recycle and fed to the conventional column, where the vaporous top product is fed to the heteroazeotropic column. Both pure components are recovered at the bottom of the respective columns. In variant B, the feed is mixed with the recycle stream and fed to the heteroazeotropic column.

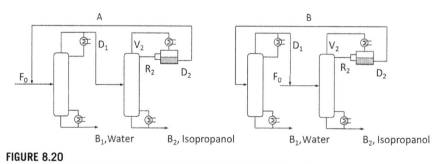

FIGURE 8.20

Variants A (left) and B (right) of the heteroazeotropic distillation process for the separation of isopropanol and water.

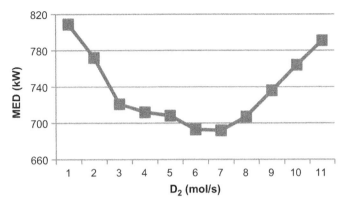

FIGURE 8.21

Minimum energy demand (MED) of flowsheet variant A (cf. Figure 8.20) for different recycle flowrates.

8.5.4.2 Shortcut-based evaluation

The separation performance is measured by an estimation of the MED by means of the FPM. The recycle flowrate constitutes an additional design degree of freedom with significant impact on feasibility and energy demand. The dependence of the MED of variant A on the recycle flowrate is displayed in Figure 8.21 and the results of the MED estimation for the suggested flowsheet variants at their respective optimal recycle flowrate D2 are given in Table 8.5.

Based on the shortcut evaluation, variant A has a significantly lower energy demand and is therefore further investigated using rigorous optimization.

8.5.4.3 Rigorous design optimization

A more detailed design is determined by rigorous design optimization minimizing TAC, which in this case also includes entrainer replacement cost. The cost-optimal process has a solvent replacement rate of 2.0 mmol/s at a cost of 4.8 k€/a. The total energy demand of the separation system is 700 kW, only slightly above the MED estimated by the shortcut method. The recycle flowrate of the cost-optimal design

Table 8.5 Results of the Shortcut Evaluation for Two Different Flowsheet Variants

	Variant A	Variant B
Flowrate D2	5.0 mol/s	18.2 mol/s
MED, column 1	487 kW	775 kW
MED, column 2	205 kW	1368 kW
MED, process	692 kW	2143 kW
MED, minimum energy demand		

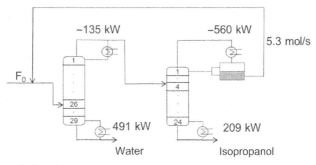

FIGURE 8.22 Optimized Flowsheet

is 5.3 mol/s, compared to 5.0 mol/s for the design at minimum MED determined by the shortcut method. The TAC is 179 k€/a, where energy cost accounts for 69% and solvent replacement cost accounts for 2.8% [129]. Figure 8.22 provides an overview on the optimized flowsheet.

The computed cost-optimal design is very close to the one operated at MED identified by the shortcut method. Hence, the selection of promising process variants based on shortcut evaluation provides very meaningful results in this case.

8.6 Summary and conclusions

This chapter summarizes the state of the art in model-based design of azeotropic distillation processes. The available design methods are organized in a three-step synthesis framework tailored to the design of distillation processes. After the generation of separation process variants, largely based on an analysis of the properties of the mixture to be separated, shortcut methods can be used to evaluate the process alternatives with respect to split feasibility and separation effort. The shortcut methods not only provide the performance indicators of interest but also help to gain insight in the inherent thermodynamic limitations of a separation. While all the various shortcut methods serve this purpose, they have individual strengths and weaknesses regarding their application to different types of mixtures and separations. An educated use of the shortcut methods is indispensable to guarantee a proper valuation of the results. There is still room for improvement in the area of shortcut methods for distillation processes. In particular, a quantification of the error in the MED estimates and in the split feasibility tests would be very desirable. Furthermore, a better understanding of the effect of finite reflux ratio and the choice of specifications on split feasibility, in particular, in combination with nonsharp splits, deserves further attention.

Though the shortcut methods provide a value in themselves, they are particularly useful as a preprocessing step toward rigorous optimization of a conceptual distillation process design. The significant progress in this field in recent years is largely due to the combination of shortcut methods for proper initialization of the resulting large-scale and non-convex MINLP and of continuous reformulation methods to improve the robustness and efficiency of the numerical solution. Design problems of industrial interest can be solved rigorously with the optimization technology available. Despite the fact that flat optima are to be expected, global optimization methods would be very desirable, though a significant increase in computational complexity is to be expected.

Future research should concentrate on improving the capabilities of the existing methods to successfully tackle larger separation problems involving a much larger number of components and more complex separation flowsheets. Finally, these powerful design methods will only be rolled out into industrial practice if a suitable business model is implemented and if reliable and versatile software implementations become available.

References

[1] V.N. Kiva, E.K. Hilmen, S. Skogestad, Azeotropic phase equilibrium diagrams: a survey, Chem. Eng. Sci. 58 (2003) 1903–1953.

[2] V. Gerbaud, I. Rodriguez-Donis, Extractive distillation [Chapter 6], in: A. Gorak, Z. Olujic (Eds.), Distillation: equipment and processes, vol. 2, Elsevier, 2014.

[3] A.W. Westerberg, A retrospective on design and process synthesis, Comp. Chem. Eng. 28 (2004) 447–458.

[4] W. Marquardt, S. Kossack, K. Kraemer, A framework for the systematic design of hybrid separation processes, Chin. J. Chem. Eng. 16 (2008) 333–342.

[5] J.M. Douglas, A hierachical decision procedure for process synthesis, AIChE J. 31 (1985) 353–362.

[6] S.D. Barnicki, J.R. Fair, Separation system synthesis: a knowledge-based approach. 1. Liquid mixture separations, Ind. Eng. Chem. Res. 29 (1990) 421–432.

[7] S.D. Barnicki, J.R. Fair, Separation system synthesis: a knowledge-based approach. 2. Gas/vapor mixtures, Ind. Eng. Chem. Res. 31 (1992) 1679–1694.

[8] R.L. Kirkwood, M.H. Locke, J.M. Douglas, A prototype expert system for synthesizing chemical process flowsheets, Comp. Chem. Eng. 12 (1988) 329–343.

[9] E. Pajula, T. Seuranen, T. Koiranen, M. Hurme, Synthesis of separation processes by using case-based reasoning, Comp. Chem. Eng. 25 (2001) 775–782.

[10] T. Seuranen, M. Hurme, E. Pajula, Synthesis of separation processes by case-based reasoning, Comp. Chem. Eng. 29 (2005) 1473–1482.

[11] L. Laroche, N. Bekiaris, H.W. Andersen, M. Morari, Homogeneous azeotropic distillation – separability and flowsheet synthesis, Ind. Eng. Chem. Res. 31 (1992) 2190–2209.

[12] E. Blass, Entwicklung Verfahrenstechnischer Prozesse, Springer, Berlin, 1997.

[13] J.G. Stichlmair, J.R. Herguijuela, Separation regions and processes of zeotropic and azeotropic ternary distillation, AIChE J. 38 (1992) 1523–1535.

[14] E.T. Foucher, M.F. Doherty, M.F. Malone, Automatic screening of entrainers for homogeneous azeotropic distillation, Ind. Eng. Chem. Res. 30 (1991) 760–772.

[15] R. Gani, E.A. Brignole, Molecular design of solvents for liquid extraction based on UNIFAC, Fluid Phase Equilib. 13 (1983) 331–340.

[16] P.M. Harper, R. Gani, A multi-step and multi-level approach for computer aided molecular design, Comp. Chem. Eng. 24 (2000) 677−683.

[17] A.T. Karunanithi, L.E.K. Achenie, R. Gani, A new decomposition-based computer-aided molecular/mixture design methodology for the design of optimal solvents and solvent mixtures, Ind. Eng. Chem. Res. 44 (2005) 4785−4797.

[18] R.C. Reid, J.M. Prausnitz, B. Poling, The Properties of Gases and Liquids, fourth ed., McGraw-Hill, New York, 1987.

[19] M. Peters, M. Zavrel, J. Kahlen, T. Schmidt, M. Ansorge-Schumacher, W. Leitner, J. Büchs, L. Greiner, A. Spiess, Systematic approach to solvent selection for biphasic systems with a combination of COSMO-RS and a dynamic modeling tool, Eng. Life Sci. 8 (2008) 546−552.

[20] S.G. Levy, D.B. van Dongen, M.F. Doherty, Design and synthesis of homogeneous azeotropic distillations. 2. Minimum reflux calculations for nonideal and azeotropic columns, Ind. Eng. Chem. Fund 24 (1985) 463−474.

[21] L. Laroche, N. Bekiaris, H.W. Andersen, M. Morari, The curious behavior of homogeneous azeotropic distillation − implications for entrainer selection, AIChE J. 38 (1992) 1309−1328.

[22] S. Kossack, K. Kraemer, R. Gani, W. Marquardt, A systematic synthesis framework for extractive distillation processes, Chem. Eng. Res. Des. 86 (2008) 781−792.

[23] M.R. Eden, S.B. Jorgensen, R. Gani, M.M. El-Halwagi, A novel framework for simultaneous separation process and product design, Chem. Eng. Process. 43 (2004) 595−608.

[24] A. Bardow, K. Steur, J. Gross, Continuous-molecular targeting for integrated solvent and process design, Ind. Eng. Chem. Res. 49 (2010) 2834−2840.

[25] O.M. Wahnschafft, J.W. Koehler, E. Blass, A.W. Westerberg, The product composition regions of single-feed azeotropic distillation-columns, Ind. Eng. Chem. Res. 31 (1992) 2345−2362.

[26] M.F. Doherty, J.D. Perkins, Dynamics of distillation processes. 1. Simple distillation of multicomponent non-reacting, homogeneous liquid-mixtures, Chem. Eng. Sci. 33 (1978) 281−301.

[27] Z.T. Fidkowski, M.F. Doherty, M.F. Malone, Feasibility of separations for distillation of nonideal ternary mixtures, AIChE J. 39 (1993) 1303−1321.

[28] D.B. Vandongen, M.F. Doherty, Design and synthesis of homogeneous azeotropic distillations. 1. Problem formulation for a single column, Ind. Eng. Chem. Fund 24 (1985) 454−463.

[29] V.T. Zharov, Non-local relations in vapor-liquid equilibrium diagrams for multicomponent systems, Russ. J. Phys. Chem. (1969) 2784−2791.

[30] L.A. Serafimov, The azeotropic rule and the classification of multicomponent mixtures VII. Diagrams for ternary mixtures, J. Phys. Chem. (USSR) (1970) 567−642.

[31] G.J.A.F. Fien, Y.A. Liu, Heuristic synthesis and shortcut design of separation processes using residue curve maps − a review, Ind. Eng. Chem. Res. 33 (1994) 2505−2522.

[32] J. Stichlmair, J.R. Fair, Distillation: Principles and Practice, Wiley-VCH, Chichester, 1998.

[33] M.F. Doherty, M.F. Malone, Conceptual Design of Distillation Systems, McGraw-Hill, Boston, 2001.

[34] L.J. Krolikowski, Determination of distillation regions for non-ideal ternary mixtures, AIChE J. 52 (2006) 532−544.

[35] W.E. de Villiers, R.N. French, G.J. Koplos, Navigate phase equilibria via residue curve maps, Chem. Eng. Prog. 98 (2002) 66−71.

[36] C. Wibowo, K.M. Ng, Visualization of high-dimensional phase diagrams of molecular and ionic mixtures, AIChE J. 48 (2002) 991−1000.

[37] C.A. Jaksland, R. Gani, K.M. Lien, Separation process design and synthesis based on thermodynamic insights, Chem. Eng. Sci. 50 (1995) 511−530.

[38] S. Brüggemann, W. Marquardt, Rapid screening of design alternatives for nonideal multiproduct distillation processes, Comp. Chem. Eng. 29 (2005) 165−179.

[39] N. Bekiaris, G.A. Meski, C.M. Radu, M. Morari, Multiple steady states in homogeneous azeotropic distillation, Ind. Eng. Chem. Res. 32 (1993) 2023−2038.

[40] S. Widagdo, W.D. Seider, Azeotropic distillation, AIChE J. 42 (1996) 96−130.

[41] V.T. Zharov, Phase transformations and rectifcation of multicomponent solutions, Russ. J. Appl. Chem. 41 (1968) 2530−2536.

[42] V.T. Zharov, Phase representations and rectifcation of many-component solutions—II, Russ. J. Appl. Chem. 42 (1969) 94−98.

[43] M. Jobson, D. Hildebrandt, D. Glasser, Attainable products for the vapour-liquid separation of homogeneous ternary mixtures, Chem. Eng. J. Biochem. Eng. J. 59 (1995) 51−70.

[44] F. Petlyuk, Distillation Theory and Its Application to Optimal Design of Separation Units, Cambridge University Press, Cambridge, 2004.

[45] J. Köhler, P. Aguirre, E. Blass, Minimum reflux calculations for nonideal mixtures using the reversible distillation model, Chem. Eng. Sci. 46 (1991) 3007−3021.

[46] V. Julka, M.F. Doherty, Geometric nonlinear analysis of multicomponent nonideal distillation: a simple computer-aided design procedure, Chem. Eng. Sci. 48 (1993) 1367−1391.

[47] J. Bausa, Näherungsverfahren für den konzeptionellen Entwurf und die thermodynamische Analyse von destillativen Trennprozessen, in: Fortschrittberichte VDI, VDI Verlag, Reihe 3, Nr.692, Düsseldorf, 2001.

[48] N. Felbab, D. Hildebrandt, D. Glasser, A new method of locating all pinch points in nonideal distillation systems, and its application to pinch point loci and distillation boundaries, Comp. Chem. Eng. 35 (2011) 1072−1087.

[49] D.A. Beneke, S.B. Kim, A.A. Linninger, Pinch point calculations and its implications on robust distillation design, Chin. J. Chem. Eng. 19 (2011) 911−925.

[50] P. Pöllmann, E. Blass, Best products of homogeneous azeotropic distillations, Gas Sep. Purif. 8 (1994) 194−229.

[51] A. Davydian, M.F. Malone, M.F. Doherty, Boundary modes in a single feed distillation column for the separation of azeotropic mixtures, Theo. Found. Chem. Eng. 31 (1997) 327−338.

[52] S. Brüggemann, W. Marquardt, Conceptual design of distillation processes for mixtures with distillation boundaries: I. Computational assessment of split feasibility, AIChE J. 57 (2011) 1526−1539.

[53] F.J.L. Castillo, D.Y.-C. Thong, G.P. Towler, Homogeneous azeotropic distillation. 1. Design procedure for single-feed columns at nontotal reflux, Ind. Eng. Chem. Res. 37 (1998) 987−997.

[54] F.J.L. Castillo, D.Y.C. Thong, G.P. Towler, Homogeneous azeotropic distillation. 2. Design procedure for sequences of columns, Ind. Eng. Chem. Res. 37 (1998) 998−1008.

[55] D.Y.C. Thong, M. Jobson, Multicomponent homogeneous azeotropic distillation 1. Assessing product feasibility, Chem. Eng. Sci. 56 (2001) 4369−4391.

[56] O.M. Wahnschafft, A.W. Westerberg, The product composition regions of azeotropic distillation-columns 2. Separability in 2-feed columns and entrainer selection, Ind. Eng. Chem. Res. 32 (1993) 1108−1120.

[57] M.F. Doherty, G.A. Caldarola, Design and synthesis of homogeneous azeotropic distillations. 3. The sequencing of columns for azeotropic and extractive distillations, Ind. Eng. Chem. Fund 24 (1985) 474−485.

[58] L. Laroche, N. Bekiaris, H.W. Andersen, M. Morari, Homogeneous azeotropic distillation − comparing entrainers, Can. J. Chem. Eng. 69 (1991) 1302−1319.

[59] J.P. Knapp, M.F. Doherty, Minimum entrainer flows for extractive distillation − a bifurcation theoretic approach, AIChE J. 40 (1994) 243−268.

[60] M. Tapp, S.T. Holland, D. Hildebrandt, D. Glasser, Column profile maps. 1. Derivation and interpretation, Ind. Eng. Chem. Res. 43 (2004) 364−374.

[61] S. Brüggemann, W. Marquardt, Shortcut methods for nonideal multicomponent distillation: 3. Extractive distillation columns, AIChE J. 50 (2004) 1129−1149.

[62] M. Franke, A. Gorak, J. Strube, Design and optimization of hybrid separation processes, Chem. Ing. Tech. 76 (2004) 199−210.

[63] M. Skiborowski, A. Harwardt, W. Marquardt, Conceptual design of distillation-based hybrid separation processes, Annu. Rev. Chem. Biomol. Eng. 4 (2013) 45−68.

[64] H.N. Pham, M.F. Doherty, Design and synthesis of heterogeneous azeotropic distillations—I. Heterogeneous phase diagrams, Chem. Eng. Sci. 45 (1990) 1823−1836.

[65] B.P. Cairns, I.A. Furzer, Multicomponent 3-phase azeotropic distillation. 2. Phase-stability and phase-splitting algorithms, Ind. Eng. Chem. Res. 29 (1990) 1364−1382.

[66] H. Zhang, A. Bonilla-Petriciolet, G.P. Rangaiah, A review on global optimization methods for phase equilibrium modeling and calculations, Open Thermodyn. J. 5 (2011) 71−92.

[67] C.H. Whitson, M.L. Michelsen, The negative flash, Fluid Phase Equilib. 53 (1989) 51−71.

[68] J. Bausa, W. Marquardt, Quick and reliable phase stability test in VLLE flash calculations by homotopy continuation, Comp. Chem. Eng. 24 (2000) 2447−2456.

[69] F. Jalali-Farahani, J.D. Seader, Use of homotopy-continuation method in stability analysis of multiphase, reacting systems, Comp. Chem. Eng. 24 (2000) 1997−2008.

[70] F. Steyer, D. Flockerzi, K. Sundmacher, Equilibrium and rate-based approaches to liquid-liquid phase splitting calculations, Comp. Chem. Eng. 30 (2005) 277−284.

[71] S.K. Wasylkiewicz, L.C. Kobylka, F.J.L. Castillo, Synthesis and design of heterogeneous separation systems with recycle streams, Chem. Eng. J. 92 (2003) 201−208.

[72] H. Pham, M.F. Doherty, Design and synthesis of heterogeneous azeotropic distillations III. Column sequences, Chem. Eng. Sci. 45 (1990) 1845−1854.

[73] R.Y. Urdaneta, Targeting and conceptual design of heteroazeotropic distillation processes, in: Fortschrittberichte VDI, VDI Verlag, Reihe 3, Nr.841, Düsseldorf, 2005.

[74] P. Prayoonyong, M. Jobson, Flowsheet synthesis and complex distillation column design for separating ternary heterogeneous azeotropic mixtures, Chem. Eng. Res. Des. 89 (2011) 1362−1376.

[75] A.R. Królikowski, L.J. Królikowski, S.K. Wasylkiewicz, Distillation profiles in ternary heterogeneous mixtures with distillation boundaries, Chem. Eng. Res. Des. 89 (2011) 879−893.

[76] R.W. Thompson, C.J. King, Systematic synthesis of separation schemes, AIChE J. 18 (1972) 941−948.

[77] R.N.S. Rathore, K.A. van Wormer, G.J. Powers, Synthesis strategies for multicomponent separation systems with energy integration, AIChE J. 20 (1974) 491−502.

[78] V.H. Shah, R. Agrawal, A matrix method for multicomponent distillation sequences, AIChE J. 56 (2010) 1759−1775.

[79] F. Petlyuk, R. Danilov, S. Skouras, S. Skogestad, Identification and analysis of possible splits for azeotropic mixtures—1. Method for column sections, Chem. Eng. Sci. 66 (2011) 2512−2522.

[80] R.E. Rooks, V. Julka, M.F. Doherty, M.F. Malone, Structure of distillation regions for multicomponent azeotropic mixtures, AIChE J. 44 (1998) 1382−1391.

[81] F. Friedler, K. Tarjan, Y. Huang, L. Fan, Graph-theoretic approach to process synthesis: polynomial algorithm for maximal structure generation, Comp. Chem. Eng. 17 (1993) 929−942.

[82] G. Feng, L.T. Fan, P.A. Seib, B. Bertok, L. Kalotai, F. Friedler, Graph-theoretic method for the algorithmic synthesis of azeotropic-distillation systems, Ind. Eng. Chem. Res. 42 (2003) 3602−3611.

[83] O.M. Wahnschafft, J.P. Lerudulier, A.W. Westerberg, A problem decomposition approach for the synthesis of complex separation processes with recycles, Ind. Eng. Chem. Res. 32 (1993) 1121−1141.

[84] R.W.H. Sargent, A functional approach to process synthesis and its application to distillation systems, Comp. Chem. Eng. 22 (1998) 31−45.

[85] D.Y.C. Thong, M. Jobson, Multicomponent homogeneous azeotropic distillation 3. Column sequence synthesis, Chem. Eng. Sci. 56 (2001) 4417−4432.

[86] D.Y.-C. Thong, G. Liu, M. Jobson, R. Smith, Synthesis of distillation sequences for separating multicomponent azeotropic mixtures, Chem. Eng. Process. 43 (2004) 239−250.

[87] G.L. Liu, M. Jobson, R. Smith, O.M. Wahnschafft, Recycle selection for homogeneous azeotropic distillation sequences, Ind. Eng. Chem. Res. 44 (2005) 4641−4655.

[88] F. Petlyuk, R. Danilov, S. Skouras, S. Skogestad, Identification and analysis of possible splits for azeotropic mixtures. 2. Method for simple columns, Chem. Eng. Sci. 69 (2012) 159−169.

[89] J.R. Knight, M.F. Doherty, Systematic approaches to the synthesis of separation schemes for azeotropic mixtures, in: J.J. Siirola, I.E. Grossmann, G. Stephanopoulos (Eds.), Foundations of computer-aided process design, 1990, p. 417.

[90] W.L. Mccabe, E.W. Thiele, Graphical design of fractionating columns, Ind. Eng. Chem. 17 (1925) 605−611.

[91] A. Underwood, Fractional distillation of multicomponent mixtures, Chem. Eng. Prog. 44 (1948) 603−614.

[92] A. Vogelpohl, Die näherungsweise Berechnung der Rektifikation von Gemischen mit binären Azeotropen, Chem. Ing. Tech. 46 (1974) 195.

[93] G.L. Liu, M. Jobson, R. Smith, O.M. Wahnschafft, Shortcut design method for columns separating azeotropic mixtures, Ind. Eng. Chem. Res. 43 (2004) 3908−3923.

[94] V. Julka, M.F. Doherty, Geometric behavior and minimum flows for nonideal multicomponent distillation, Chem. Eng. Sci. 45 (1990) 1801−1822.

[95] L.B. Zhang, A.A. Linninger, Temperature collocation algorithm for fast and robust distillation design, Ind. Eng. Chem. Res. 43 (2004) 3163−3182.

[96] A. Lucia, A. Amale, R. Taylor, Energy efficient hybrid separation processes, Ind. Eng. Chem. Res. 45 (2006) 8319−8328.

[97] A. Lucia, A. Amale, R. Taylor, Distillation pinch points and more, Comp. Chem. Eng. 32 (2008) 1342−1364.

[98] D.Y.C. Thong, M. Jobson, Multicomponent homogeneous azeotropic distillation 2. Column design, Chem. Eng. Sci. 56 (2001) 4393−4416.

[99] A. Amale, A. Lucia, A two-level distillation design method, AIChE J. 54 (2008) 2888−2903.

[100] C.D. Holland, Multicomponent Distillation, Prentice-Hall, Englewood Cliffs, NJ, 1963.

[101] D. Kaplan, L. Glass, Understanding Nonlinear Dynamics, Textbooks in Mathematical Sciences, Springer-Verlag, New York, 1995.

[102] P. Pöllmann, S.B. Glanz, E. Blass, Calculating minimum reflux of non-ideal multicomponent distillation using eigenvalue theory, Comp. Chem. Eng. 18 (1994) S49−S53.

[103] J. Stichlmair, H. Offers, R.W. Potthoff, Minimum reflux and minimum reboil in ternary distillation, Ind. Eng. Chem. Res. 32 (1993) 2438−2445.

[104] N. Franklin, J. Forsyth, The interpretation of minimum reflux conditions in multicomponent distillation, Trans. Inst. Chem. Eng. 31 (1953) 363−388.

[105] J. Köhler, P. Pöllmann, E. Blass, A review on minimum energy calculations for ideal and nonideal distillations, Ind. Eng. Chem. Res. 34 (1995) 1003−1020.

[106] K. Kraemer, A. Harwardt, M. Skiborowski, S. Mitra, W. Marquardt, Shortcut-based design of multicomponent heteroazeotropic distillation, Chem. Eng. Res. Des. 89 (2011) 1168−1189.

[107] J. Bausa, R. von Watzdorf, W. Marquardt, Shortcut methods for nonideal multicomponent distillation: 1. Simple columns, AIChE J. 44 (1998) 2181−2198.

[108] R.Y. Urdaneta, J. Bausa, S. Bruggemann, W. Marquardt, Analysis and conceptual design of ternary heterogeneous azeotropic distillation processes, Ind. Eng. Chem. Res. 41 (2002) 3849−3866.

[109] S.G. Levy, M.F. Doherty, A simple exact method for calculating tangent pinch points in multicomponent nonideal mixtures by bifurcation theory, Chem. Eng. Sci. 41 (1986) 3155−3160.

[110] Z.T. Fidkowski, M.F. Malone, M.F. Doherty, Nonideal multicomponent distillation: use of bifurcation theory for design, AIChE J. 37 (1991) 1761−1779.

[111] A. Harwardt, S. Kossack, W. Marquardt, Optimal column sequencing for multicomponent mixtures, in: B. Braunschweig, X. Joulia (Eds.), 18th European Symposium on Computer Aided Process Engineering − ESCAPE 18, 2008, pp. 91−96.

[112] S.G. Levy, M.F. Doherty, Design and synthesis of homogeneous azeotropic distillations. 4. Minimum reflux calculations for multiple-feed columns, Ind. Eng. Chem. Fund 25 (1986) 269−279.

[113] H. Pham, P. Ryan, M.F. Doherty, Design and minimum reflux for heterogeneous azeotropic distillation columns, AIChE J. 35 (1989) 1585−1591.

[114] K.M. Guthrie, Capital cost estimating, Chem. Eng. Tech. (1969) 114−142.

[115] J.M. Douglas, Conceptual Design of Chemical Processes, McGraw-Hill, New York, 1988.

[116] W.D. Seider, J.D. Seader, D. Levin, Process Design Principles, John Wiley & Sons Inc, New York, 1999.

[117] J. Viswanathan, I.E. Grossmann, Optimal feed locations and number of trays for distillation-columns with multiple feeds, Ind. Eng. Chem. Res. 32 (1993) 2942−2949.

[118] J. Kallrath, Mixed integer optimization in the chemical process industry - experience, potential and future perspectives, Chem. Eng. Res. Des. 78 (2000) 809−822.

[119] M.H. Bauer, J. Stichlmair, Synthesis and optimization of distillation sequences for the separation of azeotropic mixtures, Comp. Chem. Eng. 19 (1995) S15−S20.

[120] H. Yeomans, I.E. Grossmann, Optimal design of complex distillation columns using rigorous tray-by-tray disjunctive programming models, Ind. Eng. Chem. Res. 39 (2000) 4326−4335.

[121] M. Barttfeld, P.A. Aguirre, I.E. Grossmann, Alternative representations and formulations for the economic optimization of multicomponent distillation columns, Comp. Chem. Eng. 27 (2003) 363−383.

[122] S. Kossack, K. Kraemer, W. Marquardt, Efficient optimization based design of distillation columns for homogenous azeotropic mixtures, Ind. Eng. Chem. Res. 45 (2006) 8492−8502.

[123] Y.-D. Lang, L.T. Biegler, Distributed stream method for tray optimization, AIChE J. 48 (2002) 582−595.

[124] F.J. Neves, D.C. Silva, N.M. Oliveira, A robust strategy for optimizing complex distillation columns, Comp. Chem. Eng. 29 (2005) 1457−1471.

[125] G. Dünnebier, C.C. Pantelides, Optimal design of thermally coupled distillation columns, Ind. Eng. Chem. Res. 38 (1999) 162−176.

[126] K. Kraemer, S. Kossack, W. Marquardt, Efficient optimization-based design of distillation processes for homogeneous azeotropic mixtures, Ind. Eng. Chem. Res. 48 (2009) 6749−6764.

[127] O. Stein, J. Oldenburg, W. Marquardt, Continuous reformulations of discrete-continuous optimization problems, Comp. Chem. Eng. 28 (2004) 1951−1966.

[128] S. Brüggemann, J. Oldenburg, P. Zhang, W. Marquardt, Robust dynamic simulation of three-phase reactive batch distillation columns, Ind. Eng. Chem. Res. 43 (2004) 3672−3684.

[129] M. Skiborowski, A. Harwardt, W. Marquardt, Efficient optimization based design for the separation of heterogeneous azeotropic mixtures, Computers and Chemical Engineering. doi:10.1016/j.compchemeng.2014.03.012.

[130] K. Kraemer, A. Harwardt, W. Marquardt, Design of heat-integrated distillation processes using shortcut methods and rigorous optimization, in: 10th International Symposium, 2009.

[131] R.W.H. Sargent, K. Gaminibandara, Optimum design of plate distillation columns: optimization in action, in: L.C.W. Dixon (Ed.), Academic Press, London, 1976, pp. 267−314.

[132] A. Harwardt, K. Kraemer, W. Marquardt, Optimization based design of heat integrated distillation processes, in: Proceedings of the 8th world congress of chemical engineering (WCCE), August 23−27, 2009. Montreal/Canada.

[133] S. Brüggermann, W. Marquardt, Conceptual design of distillation processes for mixtures with distillation boundaries. II. Optimization of recycle policies, AIChE J. 57 (2011) 1540−1556.

[134] J.P. Knapp, M.F. Doherty, A new pressure-swing-distillation process for separating homogeneous azeotropic mixtures, Ind. Eng. Chem. Res. 31 (1992) 346−357.

[135] S. Kossack, K. Kraemer, W. Marquardt, Combining shortcut methods and rigorous MINLP optimization for the design of distillation processes for homogeneous azeotropic mixtures, in: E. Sorensen (Ed.), Distillation & Absorption 2006, IChemE, London, United Kingdom, 2006, pp. 122−131.

[136] Z.G. Lei, C.Y. Li, B.H. Chen, Extractive distillation: a review, Sep. Pur. Rev. 32 (2003) 121−213.

[137] S.O. Momoh, Assessing the accuracy of selectivity as a basis for solvent screening in extractive distillation processes, Sep. Sci. Technol. 26 (1991) 729−742.

Hybrid Distillation Schemes: Design, Analysis, and Application

9

Deenesh K. Babi, Rafiqul Gani

Department of Chemical and Biochemical Engineering,
Technical University of Denmark, Lyngby, Denmark

CHAPTER OUTLINE

9.1 Introduction

Hybrid distillation in this chapter refers to the use of more than one operation, where at least one is conventional distillation. Conventional distillation (CD) operations refer to distillation columns operating with one feed stream and two product streams. Hybrid distillation is employed when a conventional distillation is unable to separate a feed mixture into two high-purity products. For example, binary azeotropic mixtures cannot be separated into two pure compounds in a single conventional distillation column. Also, it may be difficult to separate mixtures with close boiling compounds into high-purity products through a single conventional distillation column. One of the alternatives in these cases is to employ a hybrid distillation scheme (HDS).

Distillation: Fundamentals and Principles. http://dx.doi.org/10.1016/B978-0-12-386547-2.00009-0

Many types of HDSs can potentially be designed and/or configured to perform a specific separation task. In this chapter, the following HDSs are considered: membrane-based operations combined with conventional distillation column (see for example [1]); two conventional distillation columns operating at different pressures (see for example [2]); solvent-based schemes involving two distillation columns (see for example [3]); and a combination of reactive and non-reaction distillation stages in one intensified column (see for example [4]). Each of the above types of HDSs is closely related to the mixture to be separated, that is, the separation task. The following types of mixtures are considered in this chapter: binary azeotropic mixtures; binary close-boiling mixtures; multicomponent mixtures with at least one binary pair that forms an azeotrope or is close-boiling; and reactive mixtures with difficult downstream separations.

A typical separation task could be defined as: *Given*—a mixture to be separated into two or more almost pure compounds. *Required*—the identification of an appropriate hybrid distillation scheme that can achieve the design targets (specified products).

The required HDS can be determined in many ways. For example, the trial-error experimental approach would be reliable but could be time consuming and expensive, even if the experimental facilities are available. Another example is the application of a model-based approach where the initial design is determined, analyzed, and verified through a collection of model-based methods and tools, and where adjustment of the initial design and final design verification is made through focused experimental work. In this way, calculation intensive work covering the search over a wide area is performed quickly and efficiently with reliable model-based (computer-aided) methods and tools to establish a small set of (feasible) design alternatives, and the expensive and time consuming, but more accurate, work is done at the end to obtain the final design.

The objective of this chapter is to present and discuss some of the important issues related to the use of model-based methods and tools for selection, design and application of HDSs. First, a rule-based procedure for selection of a specific HDS for a given separation task is presented. This procedure takes into account the relationship between HDSs and separation tasks represented by the feed mixtures. Next, the model-based (computer-aided) methods and tools that are needed for synthesis/design/analysis/verification of HDSs are presented. Four types of methods and tools are considered, classified broadly in terms of analysis; selection; synthesis/design; and verification. Next, application examples highlighting the use of different HDSs for various feed mixtures are presented. The chapter ends with conclusions and suggestions for future work.

9.2 Selection of HDS: rule-based procedure

The types of HDSs covered in this chapter are listed in Table 9.1, where the main characteristics (use of external agent) of each HDS are also given.

Table 9.1 List of Hybrid Distillation Schemes Covered in this Chapter

Type of HDS	First Operation	Second Operation	External Agent
Pressure-swing (PS-HDS)	Distillation at pressure p_1	Distillation at pressure p_2	Pressure
Solvent-based (S-HDS)	Extract solute	Solvent recovery	Solvent
Membrane-based (M-HDS)	Permeate one compound	Product recovery	Membrane
Reactive distillation (RD)	Reaction	–	Heat of reaction
Reactive agent-based (R-HDS)	React away one compound	Regenerate reactive agent	Reactive agent

A rule-based selection procedure is described next. Note that this set of rules is not supposed to be comprehensive and does not cover all separation tasks. These rules should be considered as an initial set that can be further extended.

- **Rule 1**: If the mixture is binary and forms an azeotrope, which is pressure sensitive (PS),
  Then, select a pressure-swing HDS (PS-HDS)

Figure 9.1(a) shows the vapor-liquid equilibrium (VLE) phase diagram of a typical pressure-dependent binary azeotrope, and Figure 9.1(b) shows a typical PS-HDS.

- **Rule 2**: If the mixture is binary and azeotropic, but not pressure sensitive,
  Then, select solvent-based HDS (S-HDS)
  or membrane-based HDS (M-HDS)
  or reactive agent-based HDS (R-HDS)

Figure 9.2(a) shows the VLE phase diagram of a typical pressure-independent binary azeotrope, and Figure 9.2(b)−(d) show typical S-HDS, M-HDS, and R-HDS, respectively.

- **Rule 3**: If the mixture is multicomponent, non-azeotropic for all pairs except for one pair that form an azeotrope or are close-boiling,
  Then, if the azeotrope is not sensitive to pressure, select S-HDS (only for the close-boiling/azeotrope pair)
  or M-HDS (only for the close-boiling/azeotrope pair)
  or reactive distillation (RD) (only for the close-boiling/azeotrope pair)
  If the azeotrope is sensitive to pressure, select PS-HDS.

Figure 9.3 illustrates the case of first applying a series of conventional distillation columns to separate the non-azeotropic compounds and then applying solvent-based extraction for the pressure insensitive azeotrope pair. This rule is based on the principle of performing the most difficult separation task last.

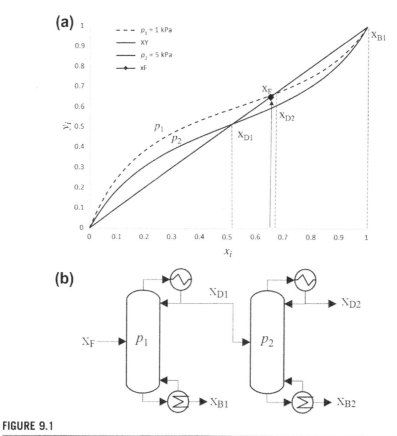

FIGURE 9.1

(a) VLE phase diagram as a function of pressure. (b) Configuration of a PS-HDS.

- **Rule 4**: If the mixture is multicomponent and at least one binary pair forms an azeotrope,
  Then, first isolate the pairs that form azeotropes,
  Then follow the appropriate rules (Rule 1 or Rule 2)
  or remove the azeotropes if any of the azeotrope compounds are not part of the specified product

Figure 9.4 highlights a case where an azeotropic pair is removed first because it is not part of the desired products (compound C is the desired product) through a sequence of conventional distillation operations. Note that in this case, HDS as defined here is not necessary.

- **Rule 5**: If the mixture is reactive, an equilibrium reaction is present and products form azeotropes with unconverted reactants,
  Then, select RD

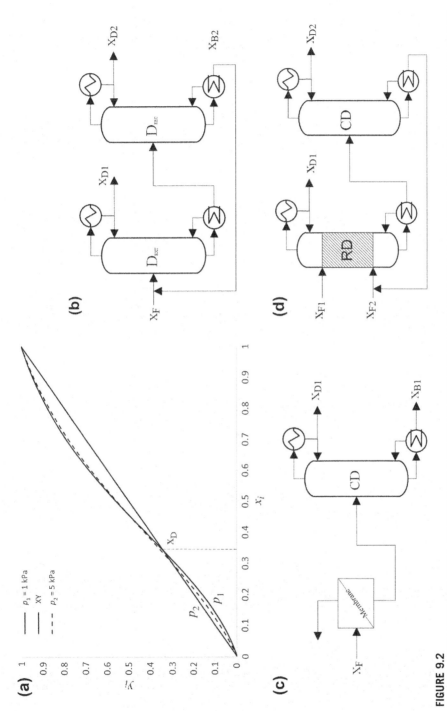

FIGURE 9.2

(a) VLE phase diagram where pressure has a minor effect. (b) Configuration of an S-HDS (column CD regenerates the reactive agent for column RD). (c) Configuration of an M-HDS. (d) Configuration of an R-HDS (column CD regenerates the reactive agent for column RD).

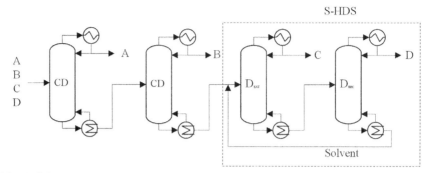

FIGURE 9.3

Configuration for separation of a multicomponent mixture with a pressure-insensitive binary azeotrope (compounds C-D form a pressure-insensitive binary azeotrope).

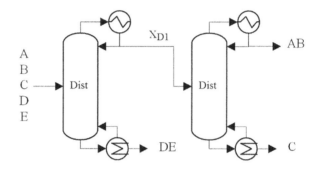

FIGURE 9.4

Configuration for separation of a multicomponent mixture with several azeotropes that does not require HDS (note: pairs DE and AB form azeotropes; compound C is the product).

Figure 9.5(a) shows a superstructure of alternative schemes of RD with or without non-reactive stages. An excellent review on R-HDS is given by Lutze and Gorak [5].

Figure 9.5(b) shows a superstructure of HDS alternatives for separation of binary azeotropic mixtures and includes all possible hybrid separation schemes covered in this chapter. For example, setting the binary variables, Y2, Y4, Y7, Y10, Y14, and Y17, to one and all other binary variables to zero, an S-HDS is obtained, as also highlighted through the dark lines in Figure 9.5. In this way, by setting a sub-set of binary variables to one (meaning that the corresponding streams do exist) and the remaining binary variables to zero (meaning that the corresponding streams do not exist), different HDS can be generated and then evaluated through simulation. Also, the optimal HDS can be determined through solving a mixed integer linear (or non-linear) programming problem. If, however, the number of feasible combinations of binary variables is not large, feasible HDS can be generated by enumeration and evaluated by simulation to find the optimal (see Ref. [6]).

Applying the above rules, an appropriate HDS (one feasible alternative) can be selected for preliminary design and analysis. Based on these rules, relationships between mixtures to be separated and HDS types can be established, as given in Table 9.2.

9.3 **Model-based computer-aided methods and tools**

Design and analysis of HDSs can be performed through the following methods and tools:

- Methods and tools for mixture analysis
- Methods and tools for selection (solvent, membrane, reactive agent, operating pressure)

(a)

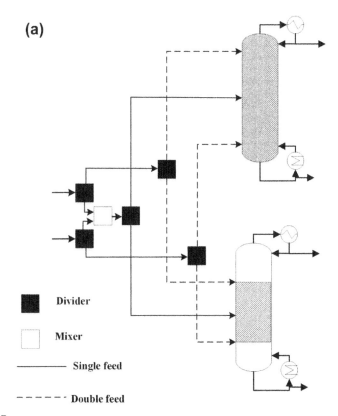

Divider

Mixer

—— **Single feed**

----- **Double feed**

FIGURE 9.5

(a) Superstructure of RD configurations. (b) Superstructure of all possible HDS (Note: Ex is extraction column; D1 and D2 are distillation columns with or without reactive stages; DEC is decanter or membrane separator).—Adopted from Ref. [5].

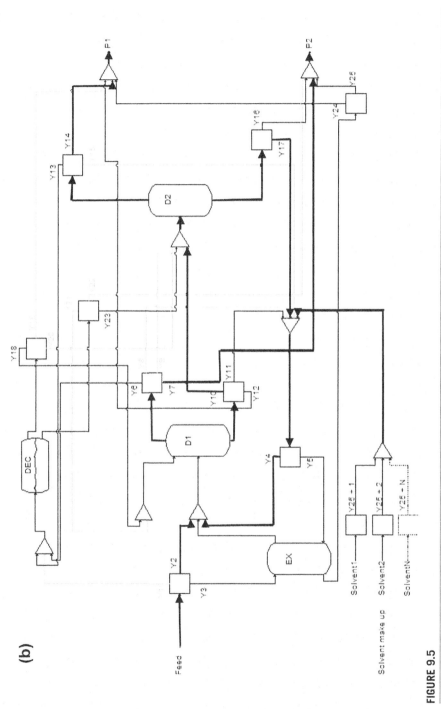

FIGURE 9.5

(continued)

Table 9.2 Mixtures to be Separated versus Type of HDS

Type	Azeotrope	Pressure Sensitive	Close-Boiling	Reactive	HDS Scheme	Reference to the Chapter
Binary	Yes	No	No	No	S-HDS, M-HDS	Figure 9.2(b) and(c)
Binary	Yes	Yes	No	No	PS-HDS, S-HDS, M-HDS	Figures 9.1(b) and 9.2(b, c)
Binary	No	–	Yes	No	S-HDS, M-HDS, R-HDS	Figure 9.2(b)–(d)
Multicompound	Yes*	No	No	No	CD + S-HDS	Figure 9.2(b)
Multicompound	Yes*	Yes*	No	No	CD + S-HDS (or PS-HDS or M-HDS)	Figure 9.2(b) (Figures 9.1(b) and 9.2(c))
Multicompound	Yes*	No	Yes	No	CD + S-HDS (or M-HDS)	Figure 9.2(b) (Figure 9.2(c))
Multicompound	Yes*	Yes*	Yes*	No	CD + S-HDS (or PS-HDS or M-HDS)	Figure 9.2(b) (Figures 9.1(b) and 9.2(c))
Multicompound	No	No	No	Yes	RD	Figure 9.5(a)
Multicompound	Yes*	No	No	Yes	RD + M-HDS	Figure 9.5(a) + Figure 9.2(c)
Multicompound	Yes*	Yes*	No	Yes	RD + PS-HDS	Figure 9.5(a) + Figure 9.1(b)
Multicompound	Yes*	Yes*	Yes*	No	CD + S-HDS (or PS-HDS or M-HDS or R-HDS)	Figure 9.2(b) (Figures 9.1(b) and 9.2(c, d))

Note: * indicates at least one pair of compounds forms an azeotrope or are close-boiling.

- Methods and tools for synthesis and design (configuration of HDSs, design of operations)
- Methods and tools for HDS analysis

Table 9.3 gives a selected list of methods and tools that have been used for the calculations and results given in this chapter and that can be used for various calculations related to synthesis, design, and analysis of HDSs.

Table 9.3 Summary of Methods and Tools for Synthesis, Design, and Analysis of HDS

Objective	Method	Tool-Name/Tool-Type	Features
Phase diagram generation	Property model based	ICAS-utility/Analysis[a]	Group contribution based property models used for VLE, LLE, SLE, distillation boundary, residue curve, etc., calculations
Solvent selection	CAMD; database search	ProCAMD[a]/Selection[a]	Searches for solvents for various types of solvent based separation processes
Membrane selection	Database search	CAPEC database[a]/Selection[a]	A database with known membrane assisted separations has been created with collected data
Distillation (with or without reaction)	Driving force based; equilibrium based	PDS-ICAS[a]/Design, Analysis[a]	Based on generated phase diagrams and driving force diagrams design of distillation columns to be used in HDS can be synthesized and simulated
Modeling	Equation oriented problem solution	MoT-ICAS[a]/Analysis[a]	Process and property models can be quickly generated and solved without spending time on programming
Process simulation	Model based calculations	Aspen Plus/PROII/Analysis[a]	Models for distillation and reactive distillation are available for use in analysis of HDS

[a] *ICAS Analysis-Tools [17].*

9.3.1 **Methods and tools for mixture analysis**

The feed mixture needs to be analyzed in terms of the presence of azeotropes and/or close-boiling compounds; the effect of pressure on the azeotropes; the effect of addition of solvents on the azeotropes; and the available driving forces for specific separations. Without the solvent, mixtures analyzed have two or more compounds whilst with the solvent, the mixtures analyzed will have three or more compounds. Driving force, D_{ij}, for the separation of compound i from compound j is given by Gani and Bek-Pedersen [7],

$$D_{ij} = y_i - x_i = \frac{x_i \cdot \alpha_{ij}}{1 + x_i \cdot (\alpha_{ij} - 1)} - x_i \tag{9.1}$$

where x_i is the liquid composition of compound i, y_i is the vapor composition of compound i, and α_{ij} is the relative volatility of compound i related to compound j. Choice of the model for relative volatility is important as it will be reflected in the resulting phase diagram. For non-ideal systems, the relative volatility cannot be assumed to be a constant.

Accurate vapor–liquid equilibrium needs to be predicted for all mixtures so that they can be analyzed. Figure 9.6(a) highlights the driving force diagram of a binary azeotropic mixture that is pressure insensitive, and Figure 9.6(b) highlights the driving force diagram of a pressure-sensitive binary azeotrope. Figure 9.6(c) highlights the driving force diagram for a ternary mixture on a solvent-free basis. In order to generate these driving force diagrams, a bubble point calculation program where, given the liquid phase composition and pressure, the vapor phase composition and temperature is calculated.

Repeating the bubble point calculations for different values of the liquid composition of compound i (light boiling compound) will generate the needed data points for the entire phase (driving force) diagram as shown in Figure 9.6(a)–(c) (for all these calculations, the UNIFAC-VLE model was used for the liquid-phase activity coefficient, the vapor phase was assumed ideal, and data fitted to the Antoine correlation were used for the pure component vapor pressure). Note, however, how the selected property models for the pure component vapor pressures and the equilibrium constants (influenced by the liquid phase activities and the vapor phase fugacities) play a major role in the accuracy of these calculations.

When solvents are added to binary mixtures, the sequencing and design of the extractive and solvent recovery columns need to consider the resulting distillation boundaries and residue curves on ternary diagrams [8,9]. Figure 9.7(a) and (b) highlight the distillation boundaries and residue curves for homogeneous and heterogeneous azeotropic systems. In the case of a homogeneous azeotrope (see Figure 9.7(a)), the HDS consists of an extractive distillation column, followed by a solvent recovery column (as shown in Figure 9.2(b)). In the case of heterogeneous azeotropic systems (see Figure 9.7(b)), an azeotropic distillation column with one vapor phase and two liquid phases is employed for extraction, followed by a solvent recovery column. Numerous publications can be found on solvent-based separation

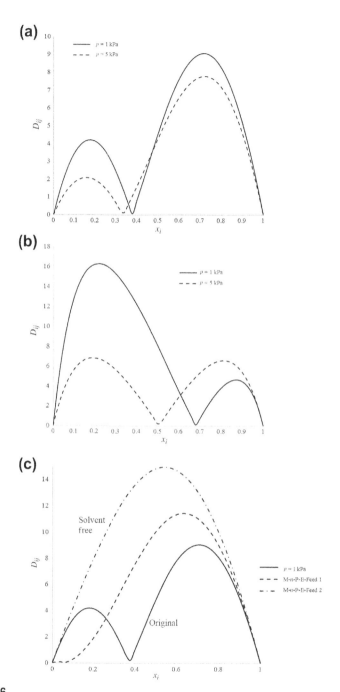

FIGURE 9.6

(a) Driving force diagrams at two pressures for acetone–chloroform system. (b) Driving force diagram at two pressures for methanol–methyl acetate system. (c) Driving force diagram on solvent-free basis for the acetone–chloroform system with methyl-n-pentyl-ether (M-n-P-E) as the solvent.

(a)

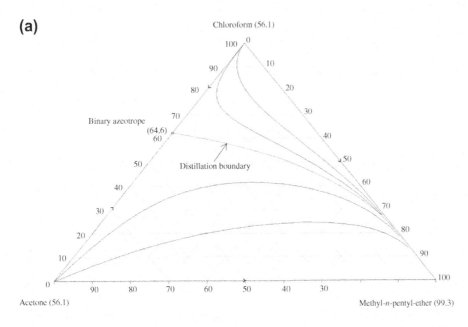

(b)

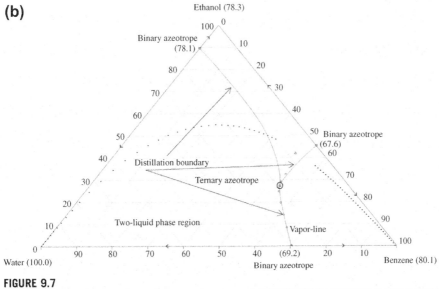

FIGURE 9.7

(a) Ternary homogeneous azeotropic system (acetone, chloroform, methyl-*n*-pentyl-ether) at 1 kPa. Temperature (°C). (b) Ternary heterogeneous azeotropic system (ethanol, water, benzene) at 1 kPa. Temperature (°C).

Table 9.4 Formula Matrix for the MeOAc System

Element/Component	MeOH	HOAc	MeOAc	H$_2$O
A = C$_2$H$_2$O	0	1	1	0
B = CH$_4$O	1	0	1	0
C = H$_2$O	0	1	0	1

of ethanol from water, for example, S-HDS with ionic liquids [10,11]; organic-solvent based S-HDS [12,13]; and M-HDS [14]. Feasibility of the HDS depends on the selected solvent and the distillation boundary. To generate these figures, a batch distillation simulation at total reflux, coupled with a method to locate azeotropes and bubble point calculations, needs to be employed [8].

For reactive systems, since they will usually involve three or more compounds, the element-based method of Perez-Cisneros et al. [15] can be used to generate similar types of diagrams as in Figures 9.1(a), 9.6(a) and 9.7(a). Consider the methyl acetate system consisting of the following four compounds: methanol (MeOH); acetic acid (HOAc); methyl acetate (MeOAc); and water (H$_2$O). Using the element-based method, this four-compound system can be represented by three elements (see Table 9.4). Performing reactive bubble point and reactive flash calculations [4], reactive VLE phase diagrams for the ternary element system (see Figure 9.8(a)) and reactive residue curves and distillation boundaries (see Figure 9.8(b)) can be generated.

For membrane based separation, permeability through the membrane of the permeate needs to be calculated and plotted. Figure 9.9 shows the calculated permeation of water through a membrane (PERVAP 2201-Sulzer) as a function of mass percentage of water and compared with experimental data [16].

The calculations needed to generate the data points can be done directly or indirectly through options found in commercial simulators. The ICAS (Integrated Computer Aided System [17]) provides special toolboxes to generate these figures. Therefore, it is a very useful tool in the analysis of mixtures to be separated by HDS.

9.3.2 Methods and tools for selection

As highlighted through Tables 9.1 and 9.2, design and application of HDS need one or more of the following: operating pressure for PS-HDS, solvent for S-HDS, membrane for M-HDS, or a reactive agent for R-HDS. Methods and tools for their selection are briefly discussed here (see also Table 9.3).

Selection of operating pressure: The objective here is to stay as close as possible to the atmospheric pressure in order to keep the operating costs down and to select two operating pressures that give the largest driving forces for the two regions of the azeotropes. The simple rule to select the operating pressure p_1 for the first column is to select the corresponding section of the azeotrope that has the largest driving force,

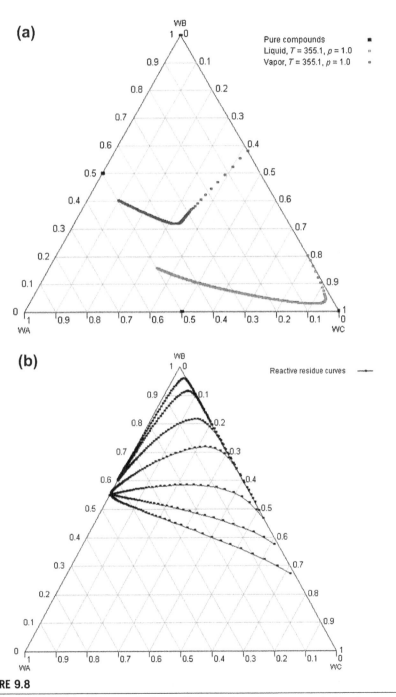

FIGURE 9.8

(a) Reactive VLE phase diagram for the methyl acetate system represented by three elements. Temperature (K), Pressure (kPa). (b) Reactive residue curves for the methyl acetate system represented by three elements.

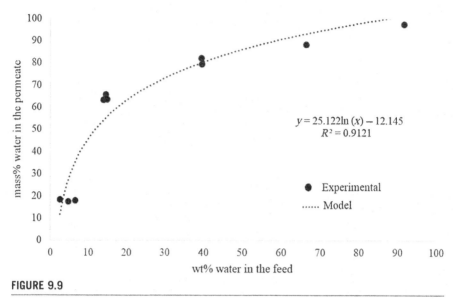

FIGURE 9.9

Permeability of water through a membrane (model-generated data compared with experimental data from Ref. [16]).

provided the feed mixture composition is also within the composition range of this azeotrope. The operating pressure for p_2 corresponds to the separation with the driving force for the other region of the azeotrope but at pressure p_2.

Selection of solvent: The type of solvent selected and/or designed defines the nature of the solvent-based extraction column. If the selected solvent is totally miscible (in the liquid phase) with the binary mixture and selectively dissolves one compound more than the other, then a two-phase vapor–liquid system is obtained and an extractive distillation operation can take place in the solvent-based extraction column (see Figures 9.6(c) and 9.7(a)). On the other hand, if the selected solvent creates two liquid phases with the binary azeotropic mixture, a three-phase vapor–liquid–liquid system is obtained and an azeotropic distillation operation takes place in the solvent-based extraction column (see Figure 9.7(b)).

The selection of the solvent is usually based on a set of essential, desirable, and environmental-health-safety (EHS) properties [18]. An example is given here for a two-phase extractive distillation column.

The solvent must (essential properties):

- Be a liquid at the feed condition (defined by its normal boiling point and normal melting point)
- Selectively dissolve one of the compounds in the mixture (for example, dissolves compound A)
- Not form azeotropes with compounds A or B
- Not react with any of the compounds

- Have a sufficient difference in boiling point and vapor pressure with respect to compound A (so it can be easily recovered and recycled)

The solvent could have (desirable properties):

- Higher selectivity for A than B (so solvent loss would be low)
- Higher solubility of A (compared to other solvents)
- Low heat of vaporization (so cost of reboiler heat supply will be low)

The solvent should have the following EHS properties:

- Non-toxic
- Not contributing to global warming potential
- Not contributing to ozone depletion
- Obtained from renewable sources

ICAS has a special tool for selection of solvents for separation of azeotropes by extractive distillation [6]. See also, Refs [19,20,21] for a latest version of solvent selection method and computer aided tool; Lek-utaiwan et al. [20] for solvent selection and verification studies and [21] on the method employed for computer-aided solvent design and selection.

Selection of membranes: For M-HDS, there needs to be a membrane assisted operation for which a membrane needs to be selected. To our knowledge, a systematic model-based membrane selection procedure has not yet been proposed. In most cases, membranes are selected through expert advice and/or database search. The CAPEC database contains a small section for data on membranes, where reported data on membrane assisted operations have been stored [22,23].

Selection of reactive agents: For R-HDS, the objective is to react away a compound that needs to be recovered or removed. To our knowledge, a systematic model-based selection procedure has not yet been reported in the open literature. Castier et al. [24] have reported this kind of R-HDS, and CO_2 capture with solvents is another example. In both cases, a second column is necessary for regeneration of reactive agent or solvent.

9.3.3 Methods and tools for synthesis and design of HDS

The different types of HDS require their scheme-specific design decisions. Since in all schemes, a conventional distillation column is also used, a simple but efficient graphical design procedure based on available driving force [7] for these columns is discussed next.

According to this method, first, data of two coexisting phases (liquid and vapor in equilibrium or not) are collected and plotted (see Figure 9.6(a)−(c)) in terms of driving force (see Eqn (9.1)) on the y-axis and the liquid composition of the light-boiling compound on the x-axis. The design of the distillation column is illustrated through Figure 9.10, where it is shown that in order to utilize the largest available driving force, the two operating lines need to intersect at the maximum point of the driving force. The number of stages, reflux rate, and feed stage location can

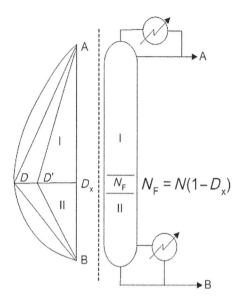

FIGURE 9.10

Driving force based design of distillation columns (on the left is the driving force diagram and on the right is the corresponding design of the distillation column).

then be calculated by a McCabe-Thiele method (as shown by Ref. [7]). In a similar manner, the distillation columns for solvent-based extractive distillation [6] and reactive distillation [4] can also be designed. The PDS toolbox in ICAS [17] provides options for these calculations. Note that the results obtained at this point are not final, but rather very good initial designs that need to be verified and adjusted through detailed rigorous simulation.

It can be noted that the driving force diagram is always concave with a unique maximum. In a single conventional column with one feed stream and two products, this means that the triangle formed by the two operating lines intersecting at the feed stage (where the maximum driving force is located) gives a measure of the maximum available driving force for the column operation. As the driving force is inversely proportional to the energy required, the design decisions related to feed stage location, reflux rate, number of stages, etc., can be worked out for a column separating a given feed mixture into two specified products. Bek-Pedersen and Gani [25] have shown that as long as the two specified products are on either side of the maximum driving force, a near-optimal design of the distillation column design and operation is obtained. This procedure can be applied for design and analysis of all the different distillation columns used in the various types of HDS considered in this chapter. Note that for pressure-sensitive azeotropic systems, the procedure is applied once for each column and the corresponding azeotropic region. For a solvent-based extraction column, the solvent-free driving force is used together with the distillation boundaries from the ternary diagrams. For reactive systems, converting to the element-based method provides the reactive driving force with respect to elements and again the same procedure is applied (See [4,15]). For multi-component mixtures, separation tasks are ordered in terms of the driving forces of binary pairs (key compounds) formed from the mixture compounds. The easiest separation task is the one having the largest driving force, which is performed first.

9.3.4 Methods and tools for HDS analysis

The most common method and tool in model-based computer-aided analysis of HDSs is the process simulator where specialized models for distillation column operations with or without reaction are available. The results reported in this chapter were obtained through such models available in Aspen Plus, PRO/II, and ICAS [17].

9.4 Application of HDS

9.4.1 Pressure-swing distillation (separation of methanol—methyl acetate azeotrope)

Pressure-swing distillation is feasible when the driving force is large enough for a single conventional distillation column to be employed for each region of the azeotrope and the azeotrope (composition) moves sufficiently with respect to pressure. The important design decisions are the operating pressures for each conventional distillation column. Since each column operates at different pressures, they can be configured to obtain one pure product and the binary azeotrope, which is recycled (see Figure 9.1(b)). As the two columns operate at two different pressures, opportunities for heat integration between the two columns may also exist. Figure 9.1(a) shows the pressure dependence of the azeotrope of the binary mixture of methanol—methyl acetate. Column 1 is operated at $5 \cdot 10^5$ Pa pressure, while column 2 is operated at $1 \cdot 10^5$ Pa pressure. Driving force diagrams (see Figure 9.6(b)) are used for the design and operation of each conventional column. Table 9.5 gives the design and simulations results of PS-HDS for the separation of the pressure-sensitive methanol—methyl acetate binary azeotropic mixture.

Table 9.5 Design and Simulation Results of PS-HDS for Separation of Methanol–Methyl Acetate ($x1$: mole fraction of Methyl-Acetate)

| | Azeotrope | | Pressure | Top Product | | Bottom Product | | Feed Stage | Total Stages | $\dot{Q}_R$ | $\dot{Q}_C$ |
	x_1	T (K)	(Pa)	x_1	T (K)	x_1	T (K)	(–)	(–)	(GJ/h)	(GJ/h)
Column1	0.4993	378.13	$5 \cdot 10^5$	0.5658	376.64	0.9998	385.01	7	18	77.59	68.76
Column2	0.6656	326.67	$1 \cdot 10^5$	0.5783	326.55	0	337.35	7	18	68.8	75.53

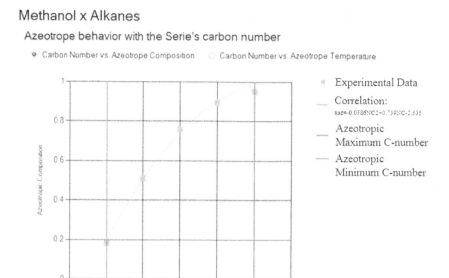

Methanol x Alkanes

Azeotrope behavior with the Serie's carbon number

FIGURE 9.11

Azeotropic composition of methanol as a function of carbon number of alkanes (C5–C9). Beyond this range, there are no azeotropes (figure generated through AzeoPro-tool in ICAS).

The column configuration will change depending on whether the azeotrope is low-boiling or high-boiling. More details on pressure-swing distillation can be found in Refs [2,6,26]. A list of homogeneous azeotropes that are pressure sensitive can be found through the AzeoPro-tool in ICAS [17].

For azeotropic mixtures, another interesting feature to note is that for a series of binary mixtures (where one compound is fixed, and the molecular type of the other compound is fixed but not its carbon number), the behavior of the mixture is similar within a range of carbon numbers of the second compound. Figure 9.11 shows the trend for mixtures of methanol with *n*-alkanes. This means that the hybrid distillation scheme obtained for one binary system may be easily adopted for all the other mixtures in the same series (in this case from C5 to C9).

9.4.2 Solvent-based extraction (separation of acetone–chloroform)

Solvent-based extraction becomes necessary when the dependence of the azeotrope on pressure is insignificant, or the driving force is not large enough to employ a conventional distillation column. Also, for close-boiling mixtures, which by definition have very low driving forces, solvent-based extraction is an alternative worth considering. This hybrid distillation scheme also involves two distillation columns, one for

Table 9.6 Design and Simulation Results of S-HDS for Solvent-Based Separation of Acetone–Chloroform (x1: mole fraction of Acetone)

	Azeotrope		Pressure	Top Product		Bottom Product		Feed Stage	Total Stages	$\dot{Q}_R$	$\dot{Q}_C$
	x_1	T (K)	(Pa)	x_1	T (K)	x_1	T (K)	(–)	(–)	(GJ/h)	(GJ/h)
Column1	0.3409	337.6	$1 \cdot 10^5$	0.9695	330.52	0.0061	365.39	20	35	12.2	6.8
Column2				0.0306	334.99	0	372.11	12	30	30.06	26.52

extraction of compound A (for example, the solute) with the solvent and one for the recovery and recycling of the solvent. In this case, compound B is obtained as a product from column 1 while compound A is obtained as a product from column 2. Subject to the availability of an appropriate solvent, this hybrid scheme is the most common application of hybrid distillation. Once the solvent has been found, the two distillation columns can be designed through the driving force method by employing solvent-free driving force diagrams.

Figure 9.2(a) showed the phase diagram for the acetone–chloroform mixture, Figure 9.6(a) showed the driving force diagram as a function of pressure, and Figure 9.6(c) showed the solvent-free driving force diagram when the solvent is methyl-*n*-pentyl-ether. From Figure 9.6(c), it can be noted that the azeotrope is clearly broken and in terms of the solvent-free diagram, the design outlined in Figure 9.10 can be applied. The S-HDS was shown in Figure 9.2(b) and the process design and simulation results are given in Table 9.6. See Lek-utaiwan et al. (2011) and Ref. [3] for other examples of S-HDS. A list of homogeneous azeotropic systems for which a solvent may be necessary in an S-HDS can be obtained through the AzeoPro-tool in ICAS [17].

9.4.3 Membrane-based HDS

Consider the mixture HOAc, MeOH, MeOAc, and H_2O coming out of a reactor. We know that this mixture forms three binary azeotropes, two of them with H_2O. Therefore, removing water from the system by a membrane-based operation leaves us the simple task of removing HOAc through a conventional distillation operation followed by a PS-HDS for the separation of MeOAc and MeOH. As shown in Figure 9.9, removal of water by a membrane-assisted operation is feasible in this case. The final M-HDS is shown in Figure 9.12. Note that Figure 9.12 shows just one M-HDS plus PS-HDS. In principle, many other alternatives can also be generated.

9.4.4 Combination of reactive-nonreactive stages in one column

This type of hybrid distillation scheme is employed when the downstream separation of reactor effluent becomes difficult because of the presence of azeotropes and/or

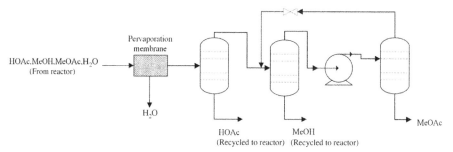

FIGURE 9.12 Membrane-Assisted Operation Plus Distillation

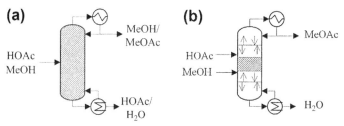

FIGURE 9.13

(a) RD column with one mixed feed and no non-reactive stages. (b) RD column with separate feed streams and non-reactive stages.

close-boiling compounds. The design of HDSs involves design issues that are related to the specific hybrid distillation scheme employed. As pointed out, it means using the conventional distillation column design under special conditions. The methyl acetate system (Refs [27,28]) is taken here as an example. This system has three binary azeotropes between MeOH/MeOAc, MeOAc/H$_2$O, and HOAc/H$_2$O. By employing RD, a top product containing the MeOH/MeOAc azeotrope and a bottom product of HOAc/H$_2$O are obtained as shown in Figure 9.13(a). However, instead of one mixed feed of the two reactants, if two feed streams are used, then it is possible to obtain nearly pure products (MeOAc at the top and H$_2$O at the bottom), as shown by Huss et al. [29].

For the RD-column shown in Figure 9.13(b), the column details and the rigorous simulation details are given in Tables 9.7 and 9.8. Babi et al. [30] reproduced these results using the RD column models in Ref. [31].

The equilibrium constant correlation used for the rigorous simulation is shown in Eqn (9.2) [27]. The thermodynamic model used is UNIQUAC model with binary interaction parameters obtained from Ref. [32].

$$\ln(K_i) = 0.83983 + \frac{782.98}{T[K]} \tag{9.2}$$

The stream summary for the feasible RD column is given in Table 9.8 (results obtained using reactive distillation equilibrium model in RadFrac in Aspen Plus).

Table 9.7 Column Specifications for the Feasible RD Column Design

Variable	Value
Reflux ratio, RR	2.2
Distillate rate (kmol/h)	280
Stages	45
Feed stage HOAc/1st reactive stage	7
Feed stage MeOAc/2nd reactive stage	40
Rectifying section stages	5
Stripping section stages	4

Table 9.8 Stream Summary for the Feasible RD Column

	HOAc Feed	MeOH Feed	MeOAc Product	H₂O Product
Component Mole Fraction				
MeOH	0	1	9.52E-04	2.00E-03
HOAc	1	0	4.36E-06	2.95E-03
MeOAc	0	0	0.9970462	1.45E-10
H_2O	0	0	2.00E-03	0.9950487
Mole flow (kmol/h)	230.28	230.28	230.28	230.28
Component Mass Fraction				
MeOH	0	1	4.13E-04	3.53E-03
HOAc	1	0	3.54E-06	9.75E-03
MeOAc	0	0	0.9990971	5.93E-10
H_2O	0	0	4.87E-04	0.9867199
Mass flow (kg/h)	13828.9	7378.669	17024	4183.576
Temperature (K)	328.15	328.15	329.7771	372.4841
Pressure (Pa)	$1 \cdot 10^5$	$1 \cdot 10^5$	$1 \cdot 10^5$	$1 \cdot 10^5$

Other variations of the MeOAc system include an R-HDS as highlighted through Figure 9.14 [30], where a membrane reactor is used followed by non-reactive distillation columns. The objective of the membrane reactor is in situ removal of H_2O from the reaction system using a pervaporation membrane. Since water forms azeotropes with MeOAc and HOAc, removing H_2O from the reactor effluent simplifies the downstream separation problem. In this case, first the HOAc is removed through a CD column followed by a PS-HDS (see Figure 9.1(b)). Compared to RD shown in Figure 9.13(b), it can be noted that this R-HDS gives a purer product and there is less loss of raw materials (compare Figure 9.14 with Table 9.8).

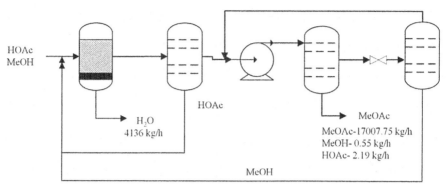

FIGURE 9.14 The R-HDS Option for MeOAc System [28]

9.5 Conclusions and future perspectives

A class of hybrid distillation schemes has been presented in terms of their design, analysis, and application to separation tasks that conventional distillation operations are not able to perform. The use of model-based computer-aided methods and tools has been highlighted for different applications of hybrid distillation schemes. A number of methods and tools already exist for performing efficient and reliable calculations needed for design, analysis, and application of the considered hybrid distillation schemes. However, as more and more difficult separation tasks become known, these methods and tools will also need to be extended and improved. Models will continue to play a very important role—not only in providing reliable calculations but also in providing truly innovative solutions that can be obtained through the use of predictive models with wide application ranges, which also expands the scope significance of the employed methods and tools. However, more work is needed in increasing database and knowledge of separation processes employing external agents such as membranes, adsorbents, reactive agents, etc., that cannot currently be explored to their full potential. Finally, techniques such as process intensification could also be incorporated so that more innovative hybrid distillation schemes could be generated and investigated.

References

[1] C. Buchaly, P. Kries, A. Gorak, Chem. Eng. Process. 46 (2007) 790−799.

[2] J.P. Knapp, M.F. Doherty, Ind. Eng. Chem. Res. 31 (1992) 346.

[3] C. Kossack, K. Kreamer, R. Gani, W. Marquardt, Chem. Eng. Res. Des. 86 (2008) 781−792.

[4] O.S. Daza, E.S. Perez-Cisneros, E. Bek-Pedersen, R. Gani, AIChE J. 49 (2003) 2822−2841.

[5] P. Lutze, A. Gorak, Chem. Eng. Res. Des. 91 (2013) 1978−1997.

[6] M. Hostrup, P.M. Harper, R. Gani, Comput. Chem. Eng. 23 (1999) 1395−1414.

[7] R. Gani, E. Bek-Pedersen, AIChE J. 46 (2000) 1271−1274.

[8] B.S. Bossen, S.B. Jørgensen, R. Gani, Ind. Eng. Chem. Res. 32 (1993) 620−633.

[9] H.N. Pham, M.F. Doherty, Chem. Eng. Sci. 45 (1990) 1823−1836.

[10] B.C. Roughton, B. Christian, J. White, K.V. Camarda, R. Gani, Comput. Chem. Eng. 42 (2013) 248−262.

[11] M. Seiler, C. Jork, A. Kavarnu, W. Arlt, R. Hirsch, AIChE J. 50 (2004) 2439−2454.

[12] A.R. Królikowski, L.J. Królikowski, Wasylkiewicz, Chem. Eng. Res. Des. 89 (2011) 879−893.

[13] P.J. Ryan, M.F. Doherty, AIChE J. 35 (1989) 1592−1601.

[14] U. Sander, P. Soukop, J. Membr. Sci. 36 (1988) 463−475. http://www.sciencedirect.com/science/article/pii/037673888880036X.

[15] E.S. Perez-Cisneros, R. Gani, M.L. Michelsen, Chem. Eng. Sci. 52 (1997) 527−543.

[16] D. Van Baelen, B. Van der Bruggen, K. Van den Dungen, J. Degreve, C. Vandecasteele, Chem. Eng. Sci. (60) (2005) 1583−1590.

[17] R. Gani, G. Hytoft, C. Jaksland, A.K. Jensen, Comput. Chem. Eng. 21 (1997) 1135−1146.

[18] P. Harper, R. Gani, P. Kolar, T. Ishikawa, Fluid Phase Equilibr. 158−160 (1999) 337−347. http://www.sciencedirect.com/science/article/pii/S0378381299000898.

[19] I. Mitrofanov, S. Sansonetti, J. Abildskov, G. Sin, R. Gani, Comput.-Aided Chem. Eng. 30 (2012) 762−766.

[20] P. Lek-utaiwan, B. Suphanit, P.L. Douglas, N. Mongkolsiri, Comput. Chem. Eng. 35 (2011) 1088−1100. http://www.sciencedirect.com/science/article/pii/S0098135410003662.

[21] P. Harper, R. Gani, Comput. Chem. Eng. 24 (2000) 677−683.

[22] P. Lutze, An innovative synthesis methodology for process intensification (Ph.D. thesis), Technical University of Denmark, Lyngby, Denmark, 2011.

[23] P.T. Mitkowski, Computer aided design and analysis of reaction-separation and separation-separation systems (Ph.D. thesis), Technical University of Denmark, Lyngby, Denmark, 2008.

[24] M. Castier, P. Rasmussen, A. Fredenslund, Chem. Eng. Sci. 44 (1989) 237−248. http://www.sciencedirect.com/science/article/pii/0009250989850614.

[25] E. Bek-Pedersen, R. Gani, Chem. Eng. Process. Process Intensifi. 43 (2004) 251−262.

[26] W.L. Luyben, Comput. Chem. Eng. 50 (2013) 1−7.

[27] W. Song, G. Venimadhavan, J.M. Manning, M.F. Malone, M.F. Doherty, Ind. Eng. Chem. Res. 37 (1998) 1917−1928.

[28] V.H. Agreda, L.R. Partin, W.H. Heise, Chem. Eng. Prog. 86 (1990) 40.

[29] R.S. Huss, F. Chen, M.F. Malone, M.F. Doherty, Comput. Chem. Eng. 27 (2003) 1855−1866.

[30] D.K. Babi, J.M. Woodley, R. Gani, Chem. Eng. Process. Process Intensifi. (submitted for publication).

[31] Aspen, Aspen PLUS, Aspen Technology, USA, 2011.

[32] T. Pöpken, L. Götze, J. Gmehling, Ind. Eng. Chem. Res. (39) (2000) 2601−2611.

Modeling of Distillation Processes

10

Eugeny Y. Kenig[1],[*], Sergei Blagov[2]

Department of Mechanical Engineering, University of Paderborn, Pohlweg 55, Paderborn, Germany[1],
BASF SE, GT/SI, Ludwigshafen, Germany[2]
*Corresponding author: e-mail: eugeny.kenig@upb.de.

CHAPTER OUTLINE

Distillation: Fundamentals and Principles. http://dx.doi.org/10.1016/B978-0-12-386547-2.00010-7

Essentially, all models are wrong, but some are useful
George E.P. Box

10.1 Introduction

Modeling is actually the way to understand the world around us. Based on our senses and supported by available equipment, we are able to perceive the complex picture of natural objects and processes. However, we can understand it only *via and during the modeling*, that is, via creating a reasonably simplified image of these objects and processes. The more complexity is concentrated in the object or process under consideration, the more skillful and careful the modeling must be. Of course, this general reasoning applies to chemical engineering, and, in particular, to distillation.

Distillation is a very complex technological process; on the other hand, it is one of the most widespread and important separation operations applied in numerous technological chains. Adequate distillation modeling is, therefore, both challenging and essential. Despite the fact that distillation represents one of the oldest separation processes and its modeling deserves to be judged as one of the most advanced, its rigor and reliability are still insufficient, and no unified approach can be established. Significant difficulties are related to the capturing of the complex counter-current, two-phase flow typical for distillation units. The column internals applied in distillation columns have to provide a large specific contact area in order to facilitate mass transfer, and this results in intricate flow patterns that cannot be resolved properly. As a solution, substantially simplified fluid dynamic representation is often used in distillation models together with gross parameters that are supposed to establish a link to real flow conditions in the equipment. Moreover, in multicomponent separations, one has to account for diffusional interactions resulting in complex coupled mass-transport equations. Distillation thermodynamics is also not simple, especially in highly nonideal azeotropic systems and systems with miscibility gaps. Further complexities are related to simultaneous heat and mass transfer as well as to intricate column configurations, with multiple feeds, side draws, and recycles. Because of these reasons, distillation equipment design and optimization still often require experimental correlations and previous experience [1].

The traditional way to design a distillation unit is to use the two-step modeling approach, starting with *conceptual design* and continuing with *rigorous design*. The former is based on a simplified set of equations and data, and the latter uses the conceptual design information as an input and is based on more sophisticated models. For each of these methods, different modeling ways can be found.

In this Chapter, we give a comprehensive overview of the existing modeling approaches for distillation related to both conceptual and detailed design, and covering both simplified and rigorous methods. In addition to common distillation processes, some more complex operations based on distillation principle, e.g., extractive, azeotropic, thermally coupled, and reactive distillation, as well as microdistillation, are considered. A particular attention is devoted to Underwood's method, since it provides analytical ways to highlight basic regularities of distillation processes.

10.2 Classification of distillation models

Because of the important role of distillation operations in process industries and its extreme energy consumption, is has for years been a focus of intensive research activities. Great achievements in the description of physical backgrounds, fluid dynamics, heat and mass transfer, and thermodynamics have opened new opportunities for distillation modeling. This has been supported by the revolutionary progress in the computer technology, allowing both advanced experimental and numerical process analysis. Nevertheless, the overall complexity of distillation mentioned above still remains very high, thus requiring significant simplification in the process description. The level and degree of specific simplifications depend on several factors, among others, on the complexity of the system to be separated, on the flow pattern, on the column design, as well as on the particular goals of the modeling and the availability of physicochemical data. All this results in a high model diversity. In Figure 10.1 we give a classification of distillation models, most of which are considered in detail in this Chapter.

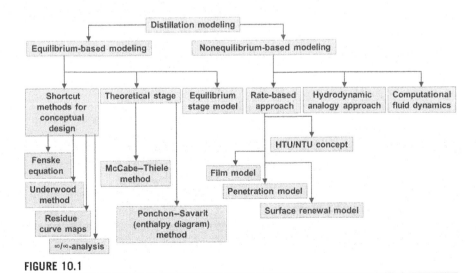

FIGURE 10.1

Classification of modeling methods for distillation.

Two large groups can be recognized: the equilibrium-based and the nonequilibrium-based methods. In contrast to the latter group, the first one does not include mass and heat transfer kinetics directly into the modeling framework, taking advantage of the fact that distillation is often dominated by the thermodynamic equilibrium of the contacting phases. Moreover, many important questions regarding the process feasibility and design can indeed be answered with the aid of just equilibrium information. This can partly be attributed to the fact that liquid present in a distillation column is always close to a boiling point, which can be directly determined from the thermodynamic equilibrium relationships.

The treatment of a distillation column as a cascade of similar segments (the so-called *stages*) is a common feature of distillation modeling, as it helps to reduce the description of a typically very large object (column) to the sequence of repeating smaller objects (stages). The stages are identified with real trays or segments of a packed column (cf. Figure 10.2). This concept has resulted in several different theoretical models, applying mass and heat balances in combination with the thermodynamic equilibrium information, with a wide range of physicochemical assumptions and accuracy. For instance, in *the theoretical stage concept*, it is assumed that the liquid and vapor phases exchange components and energy fast enough to reach thermodynamic equilibrium within the stage, so that the leaving streams *are at thermodynamic equilibrium*. This method is largely used for many practical tasks [2,3].

The main disadvantage of the equilibrium methods is their insufficient link to actual column design. For instance, it is often difficult to predict the column performance accurately without involving mass transfer kinetics and fluid dynamics. This

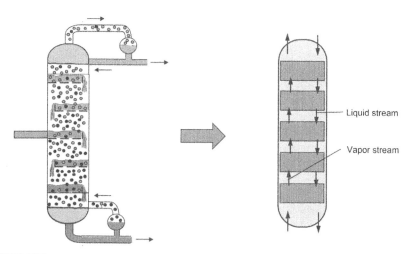

FIGURE 10.2

A simplified representation of a distillation column (left) and a column as a stage cascade with liquid and vapor streams (right).

information is particularly important when column internals should be optimized. Equilibrium-based methods usually employ lumped coefficients, which roughly estimate the deviation between real (nonequilibrium) and ideal process behavior, for instance, stage efficiencies. However, this way is not always sufficient, especially in the case of complex systems with several components.

Nonequilibrium-based models directly include process kinetics description and thus avoid the problem described in the previous paragraph. They can also be used within the stage concept; in this case, the streams leaving a stage are not assumed to reach thermodynamic equilibrium. Such an approach is also known as *the rate-based approach* (RBA) [4]. Moreover, nonequilibrium modeling can also be applied outside the stage concept, either by considering the whole column unit or by using periodic representative elements other than stages. Such methods may have a very rigorous background, as, for instance, partial differential equations of fluid mechanics and advanced numerical methods in the so-called *computational fluid dynamics* (CFD). Alternatively, a simplified flow patterns can be used within the *hydrodynamic analogy* (HA) approach. Yet, at the moment, the application of such methods to large-scale distillation columns cannot be considered realistic, so that the idea of complementary modeling using a combination of rigorous nonequilibrium methods appears more promising [5].

In the subsequent sections, these different modeling methods are considered in more detail.

10.3 Equilibrium-based modeling

Information on vapor—liquid equilibrium is crucial for distillation modeling. It is probably the main reason why numerous equilibrium-based methods have been developed for distillation columns and column configurations. A deep qualitative analysis of possible separation cuts, numerical evaluation of the separation possibilities and difficulties as well as simplified graphical methods are now available, so that the methodical spectrum is quite large. Hence, distillation can be considered by far the best understood equilibrium-based separation operation.

In this Section, we go through the most significant equilibrium-based methods, starting from simple shortcut techniques and finishing with the theoretical stage concept.

10.3.1 Shortcut methods

During the last 30 years, the so-called *shortcut methods* of distillation process design have partially lost their importance for practical design problems. This could have happen due to significant progress in the development of commercial process simulators, which are now capable of governing not only single column units, but also complex process flow sheets. However, shortcut methods are often based on a deep theoretical analysis of distillation process behavior, and, hence, they can still serve useful purposes; for instance, when accurate phase equilibrium and enthalpy data are missing or when ideal or nearly ideal solutions are considered. Besides, these methods often represent excellent tools for various theoretical studies of distillation processes.

In general, shortcut modeling methods of distillation are not aimed at obtaining full process information, e.g., temperature and composition profiles along the column, as more rigorous methods do. Instead, they provide a fast calculation of several particular process characteristics. These characteristics mainly represent special limiting cases with respect to:

- Separation limit
  - Maximum attainable separation
  - Feasibility study for a selected desired separation
- Investment costs limit
  - Minimal number of stages required for a specified separation
- Operating costs limit
  - Minimum energy consumption, minimum reflux ratio

10.3.1.1 Separation limit

For the separation limit analysis, both investment and operating costs are usually excluded from the consideration, so that only thermodynamics limits attainable separation. As a consequence, the mixture might be assumed to be separated in a column of infinite height (infinite number of stages) operating under an infinitely high reflux ratio. This type of consideration is most suitable for conceptual distillation process design [6] and is known in the literature as ∞/∞-*analysis of distillation* [7−10]. In this case, there exists only one process parameter governing the separation through the material balance relationships, for instance, the ratio between distillate- and bottom-product flow rates (distillate-to-bottom ratio).

For ideal mixtures as well as for most mixtures without azeotropes, the thermodynamic relationships are trivial: the components are distributed between the distillate and bottom products in accordance with their volatilities, which, in turn, are related to the boiling temperatures. For example, for a quaternary mixture **ABCD** in which components **A**, **B**, **C**, and **D** are ordered according to rising boiling temperature, the following separations are possible, when the distillate-to-bottom ratio increases from zero to one:

Distillate	Bottom
A	ABCD
AB	BCD
ABC	CD
ABCD	D

Within the context of the ∞/∞-analysis, there is always only one common component distributed between the two products, whereas all the other components are fully located either in distillate or in bottom product. The presented sequence of product pairs does not depend on a particular feed location. The latter only influences the critical values of the distillate-to-bottom ratio, i.e., the values at which the separation type changes. Such product sequences for mixtures without azeotropes result from the analysis of residue curve or distillation line maps.

Residue curves are usually defined as the liquid-phase composition trajectories of an open evaporation process [11]. By balancing the residue in the still, a set of differential equations can be derived for the residue curve:

$$\frac{d\mathbf{x}}{d\tau} = \mathbf{y}^*(\mathbf{x}) - \mathbf{x} \tag{10.1}$$

where $\mathbf{x}$ is the liquid residue composition, $\mathbf{y}^*(\mathbf{x})$ is the vapor composition in equilibrium with the liquid composition $\mathbf{x}$ in the still, and τ is a transformed time parameter (see Ref. [11]).

Plotting several residue curves within the composition triangle referred to as *the composition simplex* hereafter, results in a residue curve map (see Figure 10.3). Such a map allows a visualization of vapor−liquid equilibria for ternary and quaternary systems [6,12,13]. Using the well-known film model (cf. Section 10.4.2.1), Zharov and Serafimov [14] and Vogelpohl [15] showed that the composition profiles in packed distillation columns operating at total reflux coincide with the residue curves if mass transfer coefficients for all components are equal and mass transfer resistance is fully located in the vapor phase.

Distillation lines originate directly from the description of distillation columns operated at total reflux when theoretical stage modeling is used (cf. Section 10.2). They consist of a sequence of phase equilibrium states and vapor liquid equilibrium

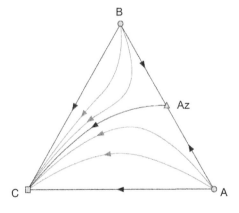

FIGURE 10.3

Example of a residue curve map for a ternary mixture with a binary azeotrope Az. Separatrix C−Az divides the composition triangle into two distillation regions, A−C−Az and B−C−Az. The arrows on the residue curves indicate boiling temperature increase.

tie-lines connecting them [6,14]. The equilibrium state sequence is described by the following recursion formula:

$$\mathbf{x}_{i+1} = \mathbf{y}^*(\mathbf{x}_i) \tag{10.2}$$

The index i can be regarded as the stage index in the column. Plotting several distillation lines in the composition simplex results in distillation line diagrams similar to residue curve maps.

Residue curves give a continuous, and distillation lines a discrete, representation of the same vector field formed by vapor–liquid equilibrium tie-lines, so that their main characteristics are very similar. For instance, they both have the same set of singular points (pure components and azeotropes) and they show the same behavior in the vicinity of these points [16].

In azeotropic mixtures, azeotropes may cause a subdivision of the composition simplex into several distillation regions, each containing a separate bundle of trajectories that originate and end at the same pair of singular points [3,16]. For instance, there are two distillation regions in Figure 10.3: **A–C–Az** and **B–C–Az**. The boundary residue curves or distillation lines are called separatrices (i.e., a separatrix **C–Az** in Figure 10.3).

Under assumptions of infinite column height and total reflux, the maximum separation capacity is attained, thus resulting in the so-called *limiting separations* [7–10]. To investigate these limiting separations, information on the topology of the vapor–liquid equilibrium is required (singular points–pure components and azeotropes, and their links). In other words, only information on the geometry of the separatrices in the composition space has to be provided, which is given in structural residue curve maps or distillation line diagrams [16].

The ∞/∞-analysis states that distillate and bottom products always have to lie within the same distillation region and on the same residue curve (distillation line). If the connecting residue curve lies inside the distillation region, then one of the product points must coincide with one of the *terminus* points (origin or end) of the residue curve, whereas the second product is located inside the region. For example, in Figure 10.3, if the feed is located in the distillation region **A–C–Az**, either the distillate can be withdrawn as pure **A** or the bottom product as pure **C**. If the connecting residue curve lies on the boundary of the distillation region, then both distillation products must lie on this boundary rather than inside the region. For example, in Figure 10.3, the distillate product may lie on the segment **A–Az** and the bottom product on the separatrix **C–Az**, while the corresponding residue curve then passes through the azeotrope **Az**.

As mixtures without azeotropes are characterized by a single distillation region, the identification of all possible separations is trivial and, as mentioned above, does not depend on the feed point location. For azeotropic mixtures with several distillation regions, the determination of the separation product sequence becomes more complicated and generally depends on the feed point locations. Examples of ternary and quaternary vapor–liquid equilibrium diagrams with four distillation regions are presented in Figures 10.4 and 10.5.

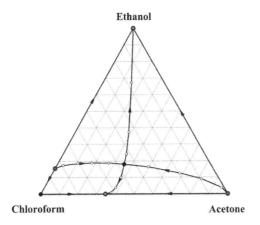

FIGURE 10.4

Structure of the residue curve map with four distillation regions for a ternary mixture of acetone, ethanol and chloroform at 1 bar [17]. Arrows on the residue curves indicate boiling temperature increase.

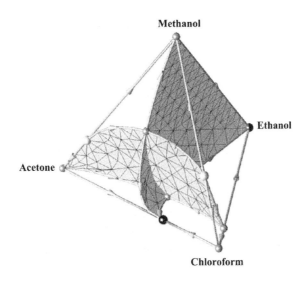

FIGURE 10.5

Structure of the residue curve map with four distillation regions for a quaternary mixture of acetone, methanol, ethanol and chloroform at 1 bar [17]. Arrows on the residue curves indicate boiling temperature increase. Separatrix surfaces are piece-wise approximated by triangular facets.

Despite the outlined simplifications in ∞/∞-analysis, the basic process characteristics are usually retained, allowing the method to support various tasks in the conceptual design of distillation processes like feasibility, multiplicity studies, and bifurcation analysis [10]. The ∞/∞-analysis has been systematically used for studying homogeneous azeotropic and heteroazeotropic distillation. Further, it has been shown that this method can also be applied to distillation sequences [9,10,18].

The lack of commercial software tools for vapor−liquid equilibrium diagram synthesis for more than three components hindered the use of ∞/∞-analysis in many practical problems. Because of the highly nonlinear nature of the phase equilibrium behavior, the determination of the topology of multicomponent residue curve or distillation line maps of arbitrary complexity is generally a challenging task [19−21]. For mixtures with more than three components, the multidimensional separatrices of the vapor−liquid equilibria (distillation boundaries) can only be determined numerically [22,23]. In this case, it is possible to identify discrete sets of points belonging to multidimensional separatrices, and obtain the separatrix surface by a suitable piece-wise linear approximation (cf. Figure 10.5).

For the conceptual design of homogeneous distillation sequences, graphical methods based on residue curves or distillation lines have proven to be very useful, and presently they are used as standard tools [3,6]. Residue curve maps superimposed on the liquid−liquid equilibrium diagram also allow for analysis of heterogeneous systems. The application of these graphical methods to the design of distillation processes is thoroughly explained in the textbooks of Doherty and Malone [6] and Stichlmair and Fair [3]. Most commercial process simulators are now able—at least for ternary mixtures—to build residue curve maps, distillation line diagrams, and liquid−liquid equilibrium diagrams as graphical supporting information illustrating phase equilibrium behavior. However, if distillation sequences with recycles are considered, the graphical approach for the conceptual design of distillation processes becomes cumbersome even for ternary systems, and computer-based numerical analysis is required [17,24].

10.3.1.2 Investment cost limit

In the investment cost limit analysis, the operating costs are not accounted for. Therefore, it is sufficient to consider a distillation column operating under total reflux, i.e., infinitely high reflux ratio; this provides a maximum separation capacity for any given column height. Similar to the previous study, this is a limiting case in which the required height of the column (number of stages) for performing the desired separation has to be minimal. For ideal or nearly ideal mixtures that can be satisfactory described by a constant relative volatility model, Fenske [25] found the following analytical expression:

$$\frac{x_{1,j}}{x_{1,k}} = \left(\frac{\alpha_j}{\alpha_k}\right)^N \frac{x_{N,j}}{x_{N,k}} \tag{10.3}$$

where N is the number of stages in the column, j and k denote appropriate components in the mixture, and $x_{1,j}$ and $x_{N,j}$ are mole fractions of component j on the bottom and top stages, respectively. Relative volatility of component j, α_j, is an adjustable parameter of the constant relative volatility model that relates the compositions of vapor and liquid being at equilibrium in accordance with:

$$y_j^* = \frac{\alpha_j x_j}{\sum\limits_{k=1}^{n} \alpha_k x_k} \tag{10.4}$$

The application of the Fenske equation is thoroughly discussed in numerous textbooks (e.g., Refs [2,26]). For nonideal mixtures, there is no explicit analytical relationship for the determination of the minimal required height. However, this task can always be solved by an iterative numerical procedure.

10.3.1.3 Operational cost limit

In contrast to the investment cost limit, in the operation cost limit analysis, the reflux ratios are finite, whereas both rectifying and stripping sections are considered to be of infinite height. The aim here is to find the minimal reflux ratio sufficient for the desired separation task. Since the reflux ratio is proportional to the required heat duty, this method allows minimal operating costs for the separation to be determined.

Underwood [27] succeeded in developing, in fact, the only known analytical model for hypothetical processes that would occur in a distillation column with infinitely high sections. Although such processes cannot be realized in practice, and, hence, are considered to be a limiting case, they clearly highlight fundamentals of distillation processes. Underwood's equation is derived with two main assumptions:

- The constant relative volatility model (Eqn (10.4)) is applicable.
- The internal molar flow rates of liquid, L, and vapor, V, along any section of a distillation column are constant.

The latter, usually referred to as *constant molar overflow* assumption, excludes heat balancing from the consideration and implies that vaporization heats for all components of the mixture are equal or close, thus leading to equimolar mass transfer between liquid and vapor phases at each cross-section of an apparatus.

Commonly, Underwood's equation is used to evaluate the minimum reflux ratio. The relevant guidelines are described in detail in original works by Underwood [27–29], as well as in numerous textbooks (e.g., Refs [2,3,26]). It should be noted that, due to the fact that both column sections are assumed to be of infinite height, the feed point location has no meaning for Underwood's method. An important limitation of the method is related to the assumption of constant molar flow rates. As discussed by Henley and Seader [2], this assumption can lead to considerably underestimated minimum reflux values. Another important limitation is the very simple vapor–liquid equilibrium treatment, which largely fails for mixtures with significant deviations from ideality, as, e.g., formation of azeotropes.

Several methods have been developed to overcome these limitations for nonideal mixtures. The first computer-based shortcut method utilizing the residue curve concept (Eqn (10.1)), *the boundary value method* [30], was developed to determine the minimum energy demand for separating nonideal ternary mixtures. This method was gradually improved and extended to quaternary systems [31−33]. A more recent method for the determination of the minimum energy requirements for single-feed two-product distillation columns, which are referred hereafter as simple distillation columns, is *the rectification body method* [34] that has been continuously evolved over the years [35,36]. Although both these methods allow for reasonable estimates in ternary and quaternary systems, their application to mixtures with more than four components is hardly possible. Hence, for most practical problems, the limiting minimum reflux ratio can only be determined by trials. In each trial, a different column height is chosen for given process specifications, and an optimization with respect to both the reflux ratio and feed point location is performed. The minimum reflux ratio is then estimated by extrapolation toward infinite column height.

Using this trial-and-error approach, the determination of minimum reflux for individual distillation columns can today be routinely done with common commercial simulation software tools. However, full optimization of complex distillation sequences with recycles still remains difficult. The Underwood method offers a unique opportunity to perform the distillation sequence analysis based on an exact analytical model free of any numerical uncertainty. It can be directly used for ideal and nearly ideal mixtures and, what is even more important, this method provides a deeper insight into the phenomena that occur in real distillation processes.

In the Appendix of this Chapter, an illustrative example of the Underwood method application is given. The parametric study for a simple distillation column is presented using an excellent interpretation of Underwood's equation proposed by Franklin [37−39].

10.3.2 Theoretical stage

In the last decades, the modeling and design of distillation processes has usually been based on the equilibrium stage model. Since 1893, as the first equilibrium stage model was published by Sorel [40], numerous publications discussing various aspects of model development, application, and solution have appeared in the literature (see Ref. [2]). As mentioned in Section 10.2, the equilibrium stage model assumes that each vapor stream leaving a tray or a packing segment is in thermodynamic equilibrium with the correspondent liquid stream leaving the same tray or segment.

In practice, thermodynamic equilibrium can seldom be reached within a single stage, and, therefore, the column consisting of equilibrium stages represents a theoretical rather than a practical column. Consequently, the equilibrium stages are often called *theoretical stages*. By combining the equilibrium relationships between the leaving streams with material and energy balances, concentration and temperature profiles along such a theoretical distillation column can be determined. With a desired product purity given, the number of theoretical stages can be calculated. Such calculations are, however, seldom straightforward, because of the inherent counter-current flow regime of distillation and flow coupling at the column top, bottom, and feed stage(s). In this regard, many different numerical methods have been proposed, most of them being based on stage-by-stage algorithms [26,41]. More recently, a few methods have been developed that use simultaneous and homotopy continuations approaches [26,42,43].

Parallel to numerical methods, a few elegant and simple graphical methods have been developed, allowing a fast estimation of the number of theoretical stages required for a particular distillation task as well as some limiting values, e.g., minimum reflux ratio. Such methods are available for binary systems or systems that can be reduced to binary ones. If constant overflow in both vapor and liquid phase is assumed, the McCabe–Thiele method [44] can be applied. In this case, the heat balance is excluded, implying that the component latent heats of vaporization are close. A more complex and more rigorous graphical method that in addition includes the heat balance is called Ponchon–Savarit graphical method [45,46]. These methods are detailed in many textbooks as well as in monographs dedicated to distillation (see, e.g., Refs [2,26]).

The link between a theoretical and a real column must, however, be found. Such a link could be established if an adequate physical description were possible. However, this is usually not the case, otherwise the theoretical stage concept would probably become redundant. In practice, the adjustment of the equilibrium-based theoretical description to reality is achieved with the help of additional experimental data, which are usually represented as tray or overall column efficiencies or HETP (*height equivalent to a theoretical plate*) values [2]. Most frequently, *the overall tray efficiency* introduced by Murphree in 1925 [47] and, hence, known as *the Murphree efficiency*, is used. It relates the actual change in vapor composition when the vapor passes through the liquid on a tray (plate) to the maximum possible composition change that would be reached if the leaving vapor were in vapor–liquid equilibrium with the tray liquid. For a tray number k:

$$E_{MV} = \frac{y_k - y_{k+1}}{y_k^*(x_k) - y_{k+1}} \tag{10.5}$$

In Eqn (10.5), top-to-bottom tray numbering is applied. In a similar way, the Murphree efficiency is defined based on the liquid-phase composition:

$$E_{ML} = \frac{x_k - x_{k-1}}{x_k^*(y_k) - x_{k-1}} \tag{10.6}$$

Although the application of tray efficiency concept is straightforward for binary systems [48,49], it may be ambiguous for multicomponent systems, because of the cross-effects caused by diffusional interactions of several components [50,51]. These effects cause an unpredictable behavior of the efficiency factors, which vary along the column height, show a strong dependency on the component concentration, and are often different for different components [51–53].

10.4 Nonequilibrium-based modeling

As can be inferred from the notion *nonequilibrium-based*, this modeling way loosens the strong dependence on equilibrium data by including a consideration of process rates directly into the modeling concept. In this regard, it may appear as a method opposite to the equilibrium-based modeling. However, it is not the case, since the equilibrium data continue to be an important factor within the nonequilibrium modeling methods. Rather, nonequilibrium-based modeling should be thought of as an extension of the process consideration toward more rigor and physically consistent treatment of complex phenomena in separation units.

The process kinetics means, above all, mass and heat transfer rates. They depend on many factors, e.g., column internals, fluid dynamic regime, and physicochemical system properties. Very often, these factors cannot be captured explicitly because of their complexity, and some assistance is required from experimental measurements, for instance, in the form of correlations for mass and heat transfer coefficients and specific interfacial contact area, which are then used for similar process conditions and thus for solving unit design tasks [48,54,55].

In the last decades, some kinetics-based methods have been developed that do not require mass and heat transfer coefficient correlations [5,56,57]. Such methods (CFD, HA approach) go beyond the stage concept, considering, where possible, either an entire separation unit or a periodic element of it.

In the literature, the notion of nonequilibrium modeling is usually applied to the *stage-based consideration* of a distillation column. In this Chapter, however, we use the term nonequilibrium modeling in a different, broader manner, covering all modeling methods directly including mass and heat transfer kinetics. To refer to the truly stage-based nonequilibrium modeling, another widely used notion, *the rate-based approach*, is applied.

In this Section, we will briefly consider the diversity of all nonequilibrium-based modeling methods, starting from the very simple HTU–NTU concept.

10.4.1 HTU–NTU concept

The concept of transfer units refers to the need to determine the height of the distillation tower H_C. In the case of tray columns, with their stepwise changes in the column profiles, it can be done using the geometrical consideration of a selected tray type in combination with the known tray efficiency values (cf. Section 10.3.2). However, for the so-called continuous-contact distillation equipment (packed column), efficiencies do not apply. Instead, transfer units are introduced.

In packed columns, flow rates are related to the cross-sectional area of the unit, A_c. In any arbitrary differential column volume (Figure 10.6) with height dl, there is an interfacial contact area

$$dA_{\text{packing}} = a^I A_c \, dl \qquad (10.7)$$

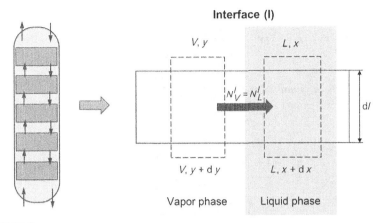

FIGURE 10.6

A distillation column as a stage cascade (left) and a stage balance envelope used for the derivation of the HTU-NTU concept (right).

available. Here, it is assumed that the total packing surface is wetted, otherwise a wetting degree appear in the right part as a multiplier. Further assuming constant molar overflow, the differential species balance in the gas phase (Figure 10.6) yields:

$$V(y + dy) - Vy = N_V^I dA_{\text{packing}} \tag{10.8}$$

or

$$N_V^I a^I A_c dl = V dy \tag{10.9}$$

Considering the mass transfer relation

$$N_V^I = \beta_V (y - y^I) \tag{10.10}$$

we obtain from Eqn (10.9):

$$H_C = \int_O^{H_C} dl = \int_{y_{\text{bottom}}}^{y_{\text{top}}} \frac{V}{\beta_V a^I A_C} \cdot \frac{dy}{(y - y^I)} \tag{10.11}$$

The left-hand term under the integral has the dimension of height and is designated the *height of a transfer unit*, HTU_V:

$$\text{HTU}_V = \frac{V}{\beta_V a^I A_c} \tag{10.12}$$

It can be assumed independent of concentration [49], otherwise, average values can be used [49,58]. Equation (10.11) then reduces to

$$H_C = \text{HTU}_V \int_{y_{\text{bottom}}}^{y_{\text{top}}} \frac{dy}{(y - y^I)} = \text{HTU}_V \cdot \text{NTU}_V \qquad (10.13)$$

where NTU_V is *the number of transfer units*.

Similar expressions can be derived based on the liquid-phase consideration:

$$H_C = \text{HTU}_L \int_{y_{\text{bottom}}}^{y_{\text{top}}} \frac{dx}{(x^I - x)} = \text{HTU}_L \cdot \text{NTU}_L; \quad \text{HTU}_L = \frac{L}{\beta_L a^I A_C} \qquad (10.14)$$

or, based on the overall mass transfer coefficients, K_{OV} and K_{OL}:

$$H_C = \text{HTU}_{OV} \cdot \text{NTU}_{OV}; \quad \text{HTU}_{OV} = \frac{V}{K_{OV} a^I A_c}; \quad \text{NTU}_{OV} = \int_{y_{\text{bottom}}}^{y_{\text{top}}} \frac{dy}{(y - y^*)} \qquad (10.15)$$

$$H_C = \text{HTU}_{OL} \cdot \text{NTU}_{OL}; \quad \text{HTU}_{OL} = \frac{L}{K_{OL} a^I A_c}; \quad \text{NTU}_{OL} = \int_{x_{\text{top}}}^{x_{\text{bottom}}} \frac{dy}{(x^* - x)} \qquad (10.16)$$

If the vapor−liquid equilibrium condition is linearized, vapor and liquid concentrations are related by the equilibrium constant K^{eq}. In this case, the following relationships are valid:

$$\frac{1}{K_{OV}} = \frac{1}{\beta_V} + \frac{K^{\text{eq}}}{\beta_L}; \quad \frac{1}{K_{OL}} = \frac{1}{K^{\text{eq}} \beta_V} + \frac{1}{\beta_L} \qquad (10.17)$$

and

$$\text{HTU}_{OV} = \text{HTU}_V + S \cdot \text{HTU}_L; \quad \text{HTU}_{OL} = \text{HTU}_L + \frac{\text{HTU}_V}{S} \qquad (10.18)$$

where factor $S = K^{\text{eq}} V/L$

It is possible to obtain the relationships for HTU and NTU in a similar form as in Eqns (10.12) and (10.16), even when the molar overflow changes along the column. In this case, however, an appropriate averaging over the column height has to be applied and Eqn (10.10) should be written for the diffusional flux rather than for the overall mass flux [53]:

$$J_V^I = \beta_V (y - y^I) \qquad (10.19)$$

The mass transfer coefficient definition in this case is somewhat different. In most textbooks, Eqn (10.10) and constant molar overflow are applied.

In the literature, correlations for HTU values for various internals and process conditions can be found (see, e.g., Refs [48,54,55]). However, it should be mentioned that the interfacial area and mass-transfer coefficients depend upon the *mass flow rates*, which may vary significantly, even if the *molar flow rates* remain constant. Thus, the constancy of the HTU-values should be used with care [58]. Further details to the application of the HTU–NTU concept can be found elsewhere (see, e.g., Refs [48,54,58]).

The performance of continuous distillation units can also be expressed in a somewhat simpler way in terms of the HETP value introduced above. This value is linked to the equilibrium stage concept providing the number of theoretical stages, N_t, whereupon

$$\text{HETP} = \frac{H_C}{N_t} \tag{10.20}$$

If the equilibrium line can be approximated by a straight line, a simple relations are valid [48,49]:

$$\text{HETP} = \text{HTU}_{OV} \frac{\ln S}{S - 1} \quad \text{for } S \neq 1 \tag{10.21}$$

$$\text{HETP} = \text{HTU}_{OV} \quad \text{for } S = 1 \tag{10.22}$$

The HETP must be determined experimentally for each packing. However, this value varies substantially, both with flow rates and with composition, and, thus, its application is not straightforward [58]. Nevertheless, because of its simplicity, HETP remains a popular means for practical distillation tasks [48].

10.4.2 Rate-based stage modeling

The notion rate-based approach (RBA) is often attributed to the method to model a column stage by involving kinetic (rate) equations for mass and heat transfer [4]. It is also known in literature as the nonequilibrium model (cf. [53]). However, as mentioned above, in this Chapter, we use the term *nonequilibrium* in a more general way, i.e., for all modeling methods directly including mass and heat transfer kinetics, and also beyond the stage concept, for instance, for CFD. Hence, we will use the term *rate-based stage* approach (or simply RBA) for the treatment of *the stage-based distillation modeling* only.

Compared to the equilibrium-based stage description, the RBA represents a more physically consistent modeling way and provides a direct link to the column and internals design [5,53,59]. This is possible because of the inclusion of mass and heat transfer equations directly into the modeling framework and by using process and model parameters governing the process fluid dynamics.

10.4.2.1 Mass transfer models

Mass transfer at the vapor–liquid interface can be described using different theoretical concepts [53,60]. Most often *the film model* introduced by Lewis and Whitman

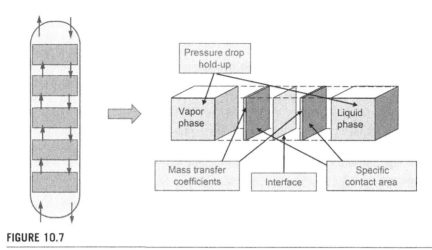

FIGURE 10.7

A distillation column as a stage cascade (left) and a stage represented by the film model (right).

[61] is applied. This model, illustrated by Figure 10.7, assumes that total resistance to mass transfer is concentrated in thin stagnant films adjacent to the phase interface and that transfer occurs within these films by steady-state molecular diffusion alone. Thus, in the film region, one-dimensional diffusion transport normal to the interface takes place. This results in constant values of the component fluxes across the interface and, after integration of the steady-state diffusion equation over the vapor-phase film thickness, one obtains:

$$N_{Vi}^I = c_{Vt} \frac{D_{Vi}}{\delta_V} \left(y_i - y_i^I \right) = \beta_{Vi} \left(y_i - y_i^I \right); \quad i = 1, \dots, n \qquad (10.23)$$

A similar equation is valid for the liquid phase, while the liquid-phase and vapor-phase fluxes are equal at the interface. Thus, the mass transfer coefficient indicates the first-power dependence on the diffusion coefficient:

$$\beta_i = c_t \frac{D_i}{\delta}; \quad i = 1, \dots, n \qquad (10.24)$$

Outside the films, in the bulk fluid phases, the level of mixing is so high that the concentration gradients are absent.

The film model has some similarity with phenomena taking place at many real fluid elements in two-phase systems, e.g., fast change of main process variables (composition, temperature) close to the interface and their much slower change far from the interface. However, the flow pattern in the film model is very simplified, and no reasonable momentum equation can be used here. Thus, the films represent purely the model elements rather than any kind of real film flow. Their thicknesses cannot be measured; they are estimated using the mass transfer coefficient correlations (or *Sherwood correlations*).

Another simplified representation is given by the *penetration model* suggested by Higbie [62]. It is assumed that the contact times of two phases at the interface is fairly short, and thus this contact can be modeled by a series of fluid elements, each of those approaching the interface, remaining there for a short fixed time, and afterward returning back in the bulk phase and being replaced by another similar element. During the exposure time, the mass transfer takes place. This process can be described by a one-dimensional diffusion equation with the corresponding initial and boundary conditions. The solution of this system provides the concentration profiles and, consequently, the mass flux expression

$$N_{Li}^I = c_{Lt}\sqrt{\frac{D_{Li}}{\pi t}}(x_i^I - x_i); \quad i = 1, \dots, n \tag{10.25}$$

Assuming that the exposure time t is the same for all the repeated contacts, Higbie [62] derived an expression for the average flux over the contact time

$$N_{Li}^I = 2c_{Lt}\sqrt{\frac{D_{Li}}{\pi t_e}}(x_i^I - x_i); \quad i = 1, \dots, n \tag{10.26}$$

from which the time-average mass transfer coefficient could be obtained as

$$\beta_{Li} = 2c_{Lt}\sqrt{\frac{D_{Li}}{\pi t_e}}; \quad i = 1, \dots, n \tag{10.27}$$

In contrast to the film model, the penetration model predicts the square root dependence of the mass transfer coefficient on the diffusion coefficient.

The surface renewal model was proposed by Danckwerts [63] and can be considered as a further development of the penetration modeling concept. The main difference here is in another distribution of the residence time of fluid elements at the interface, based on the assumption that the probability of an element being replaced by a new one is independent of the actual time for which it has been exposed. This results in a slightly different expression for the fluxes and mass transfer coefficient

$$N_{Li}^I = c_{Lt}\sqrt{D_{Li}s}(x_i^I - x_i); \quad \beta_{Li} = c_{Lt}\sqrt{D_{Li}s}; \quad i = 1, \dots, n \tag{10.28}$$

In both the penetration and the surface renewal model, mass transfer is described using a combination of a deterministic (nonsteady state diffusion) and stochastic (residence time distribution of fluid element) principles.

An attempt to combine the film and the penetration model was undertaken by Toor and Marchello [64]. They developed the so-called *film-penetration model*, in which elements of a film of thickness δ adjacent to the interface are continuously replaced by the elements having the bulk fluid composition. The model gives a more general expression for the mass transfer coefficient, which reduces to

Eqn (10.24) for large $D_{Li}/(\delta_L^2 s)$ and to Eqn (10.28) for small $D_{Li}/(\delta_L^2 s)$ values. Further modifications of the models described above are reviewed in [49,65].

Basic equations describing the film, penetration, surface renewal, and film-penetration models are usually given for binary systems as scalar equations. However, they can be extended to multicomponent systems by using vector-type composition description and matrix algebra operations (see, e.g., Refs [53,60]).

All four presented models have much in common: they try to replace real phenomena at the interface by a one-dimensional simplified picture, and they exclude real flow and hence momentum transport equations from consideration. As a consequence, they are not able to estimate their main parameters (e.g., the film thickness or average residence time) without additional information, and thus are forced to use experimental correlations for the mass transfer coefficients. In this respect, all four models have the same weak point and hence can be considered as equivalent with respect to mass transfer prediction. However, the film model appears advantageous, since there is a broad spectrum of correlations available in the literature, for all types of internals and systems, whereas for the penetration/surface renewal model, the choice is limited. Moreover, many researchers have used the film model as the basis for the evaluation of mass transfer coefficients. For these reasons, in this Chapter we focus on the film model.

10.4.2.2 Multicomponent diffusion

Within the RBA, diffusion represents one of the central phenomena. This is primarily due to of its slow character: commonly being the slowest process step, diffusion dominates the overall process rate.

For binary systems, diffusion description is given by scalar equations based on Fick's law:

$$J_A = -c_t D_{AB} \nabla x_A \tag{10.29}$$

In contrast, diffusion in systems with more than two diffusing components (multicomponent diffusion) represents a more complex phenomenon characterized by the presence of the so-called *cross effects*, when diffusion of a certain component is caused not only by its own concentration gradient, but also by concentration gradients of all other components [50,51,66]. Such phenomena cannot be adequately covered by simple Fick's law and call for the application of coupled diffusion equations.

Multicomponent diffusion can be rigorously described by the Maxwell–Stefan equations derived from the kinetic theory of gases [66] and connecting diffusion fluxes of the components with the gradients of their chemical potential. With some modification, these equations take a generalized form in which they can be used for the description of real gases and liquids [53]:

$$d_i = \sum_{j=1}^{n} \frac{x_i J_j - x_j J_i}{c_t Ð_{ij}}; \quad i = 1, \dots, n \tag{10.30}$$

where d_i is the generalized driving force:

$$d_i = \frac{x_i}{RT}\frac{d\mu_i}{dz}; \quad i = 1, \ldots, n \tag{10.31}$$

Similar equations can be also written for the vapor phase, and hence, the vapor–liquid mass transfer within the films can be modeled along these lines. The equilibrium state is here assumed at the interface only.

For practical tasks, the application of the Maxwell–Stefan equations is not always desirable or even possible, as it results in a coupled matrix-form and requires the full set of binary diffusion coefficients of the system. In this case, an alternative simplified way is represented by the *effective diffusivity approach*, by which a multicomponent system is treated as if it were a binary one. Here the so-called *effective diffusion coefficients* or *pseudo-binary diffusion coefficients D_i* are used in accordance with the Fick's law description, thus replacing real binary diffusion coefficients. We used these coefficients in the previous Section. The effective diffusion coefficients can be obtained, for instance, via a relevant averaging of the Maxwell–Stefan diffusivities [53]. This simplified method brings good results, provided that the effective diffusion coefficients are estimated properly. However, its application must be done with care.

10.4.2.3 Steady-state modeling

The mass balance equations of the conventional multicomponent rate-based model (see, e.g., Refs [53,67],) are formulated separately for each phase:

$$0 = -\frac{d}{dl}\left(Lx_i^B\right) + N_{Li}^B a^I A_c; \quad i = 1, \ldots, n \tag{10.32}$$

$$0 = \frac{d}{dl}\left(Vy_i^B\right) - N_{Vi}^B a^I A_c; \quad i = 1, \ldots, n \tag{10.33}$$

In Eqns (10.32) and (10.33), it is assumed that transfers from the vapor to the liquid phase are positive.

Equations (10.32) and (10.33) represent the mass balances for continuous systems (packed columns). For discrete systems (tray columns), the differential terms transform to finite differences and the balances are reduced to algebraic equations [53,67].

The bulk-phase balances are completed by the summation equation for the liquid and vapor bulk mole fractions:

$$\sum_{i=1}^{n} x_i^B = 1 \tag{10.34}$$

$$\sum_{i=1}^{n} y_i^B = 1 \tag{10.35}$$

For the determination of axial temperature profiles, differential energy balances are formulated including the product of the liquid molar hold-up and the specific enthalpy as energy capacity. The energy balances written for continuous systems are as follows:

$$0 = -\frac{d}{dl}\left(Lh_L^B\right) + Q_L^B a^I A_c \tag{10.36}$$

$$0 = \frac{d}{dl}\left(Vh_V^B\right) - Q_V^B a^I A_c \tag{10.37}$$

If the dynamic process behavior has to be considered, Eqns (10.32), (10.33), (10.36) and (10.37) become partial differential equations including derivatives of the hold-up in respect to time (see more details in Section 10.4.2.6).

The molar fluxes N_i are related to the diffusional fluxes by

$$N_i = J_i + x_i N_t; \quad i = 1, \ldots, n-1 \tag{10.38}$$

while the latter can be evaluated using the Maxwell–Stefan equations, Eqn (10.30). Here, different specific approaches can be applied (see Ref. [53]), for instance, the method in which the diffusional fluxes are calculated with averaged parameters. For the liquid phase, the corresponding equation is as follows:

$$\mathbf{J}_L = \left[\beta_L^{av}\right]\left[\Gamma_L^{av}\right]\left(\mathbf{x}^B - \mathbf{x}^I\right) \tag{10.39}$$

where the matrix of mass transfer coefficients is defined as:

$$\left[\beta_L^{av}\right] = \left[R_L^{av}\right]^{-1} \tag{10.40}$$

with

$$R_{Lii}^{av} = \frac{x_i^{av}}{\beta_{in}} + \sum_{\substack{k=i \\ k \neq j}}^{n} \frac{x_k^{av}}{\beta_{ik}}; \quad i, j = 1, \ldots, n-1 \, (j \neq i); \tag{10.41}$$

$$R_{Lij}^{av} = -x_i^{av}\left(\frac{1}{\beta_{ij}} - \frac{1}{\beta_{in}}\right); \quad i = 1, \ldots, n-1$$

Equations (10.38)–(10.41) can also be written for the vapor phase, for which $\left[\Gamma_G^{av}\right]$ is usually taken as unit matrix. The binary mass transfer coefficients β_{ij} can be extracted from suitable binary mass transfer correlations, using the appropriate Maxwell–Stefan diffusion coefficients $Đ_{ij}$ (cf. Section 10.4.2.2). According to the linearized theory [50,51], matrices $[R^{av}]$ and $[\Gamma^{av}]$ are evaluated based on some averaged molar fraction, for instance

$$x_i^{av} = \frac{x_i^I + x_i^B}{2}; \quad i = 1, \ldots, n \tag{10.42}$$

All required physical properties, for example diffusivities, are calculated on the same basis. At the interface, phase equilibrium is assumed:

$$y_i^I = K_i^{eq} x_i^I; \quad i = 1, \ldots, n \tag{10.43}$$

where vapor–liquid equilibrium constants K_i^{eq} are determined using appropriate thermodynamic models.

Furthermore, the continuity relation of the interfacial mass fluxes must be fulfilled:

$$N_{Li} = N_{Vi}; \quad i = 1, \ldots, n \tag{10.44}$$

In a similar way, the heat flux continuity is satisfied:

$$Q_V^B = \alpha_V^T \left(T_V^B - T^I \right) + \sum_{i=1}^{n} N_i h_{Vi} = Q_L^B = \alpha_L^T \left(T^I - T_L^B \right) + \sum_{i=1}^{n} N_i h_{Li} \tag{10.45}$$

where α_L^T and α_V^T are heat transfer coefficients in liquid and vapor films, respectively.

10.4.2.4 Model parameters

A complete system of equations describing the film model for distillation of multi-component systems can be found elsewhere (see, e.g., Refs [52,53]) Along with the equations, this model requires some parameters which must be known a priori. Among these parameters are pressure drop, liquid hold-up, specific interfacial area, mass and heat transfer coefficients and, in some cases, residence time distribution. Commonly, these parameters are represented as functions (correlations) of physicochemical properties, operational conditions and column geometry. They usually result from comprehensive and expensive experimental studies. The correlations for different columns and internals have been published in numerous papers and are collected in several reviews and textbooks (see, e.g., Refs [48,54,55,68]).

The film thickness represents a model parameter which can be estimated via the mass transfer coefficient correlations that govern the mass transport dependence on physical properties and process fluid dynamics. A number of mass transfer coefficient correlations for various column internals are available in the literature (see, e.g., Refs [54,68–72] and a recent comprehensive review in [73]).

Similarly, individual heat transfer coefficients correlations can also be found, e.g., in Refs [74–76]. However, in practice, the heat transfer coefficients are usually evaluated using the Chilton–Colburn analogy [77].

A very important and sensitive parameter is the specific vapor–liquid interfacial area. Correlations for this parameter can be found, e.g., in Refs [48,71,72,78].

As long as the relevant process and model parameter correlations are available for the specific distillation column under consideration, the film model basically yields predictive and reliable results (e.g., Refs [52,53,79]).

10.4.2.5 Nonideal flow behavior

The mass balances (see Eqns (10.8), (10.32) and (10.33)) assume plug flow behavior for both the vapor and the liquid phase. However, real flow behavior in distillation units can be much more complex, and, in some cases, it may degrade the column performance, for instance, through back-mixing of liquid and vapor phases or through the formation of stagnant zones.

In the horizontal direction on the vapor side, it may be assumed that either vapor is totally mixed before it enters the stage, or that, after being separated from the liquid on the stage below, the vapor does not mix at all. The real situation obviously lies between these two limiting cases. In small-diameter columns, it is very close to the complete mixing, whereas in large-diameter columns, the vapor is less mixed.

A horizontal liquid flow pattern is very complicated, because of the mixing by vapor, dispersion, and the round cross-section of the column. On single-pass trays, for example, the latter results in the flow path, which first expands and then contracts. A rigorous modeling of this flow pattern is very difficult, and usually the situation is simplified by assuming that the liquid flow is unidirectional and the major deviation from plug flow is the turbulent mixing or eddy diffusion.

For packed columns, a promising approach is represented by differential models like the axial dispersion model [80] and the piston flow model with axial dispersion and mass exchange [81]. When applying the axial dispersion model to cover this nonideality, the liquid-phase mass balances (Eqn (10.32)) transform to the following equation:

$$0 = \frac{D_{ax}}{u_L} \frac{\partial^2}{\partial l^2} \left(L x_i^B \right) - \frac{\partial}{\partial l} \left(L x_i^B \right) + N_{Li}^B a^I A_c; \quad i = 1, \ldots, n \tag{10.46}$$

In this case, correlations for axial dispersion coefficient D_{ax} must also be available [82].

The overall flow pattern becomes even more intricate when the liquid falls into a miscibilty gap. Whereas, in terms of the equilibrium-based modeling, only additional thermodynamic complexity arises, the rate-based modeling requires development of special techniques, as two-phase models, e.g., the film model, are no longer applicable. These issues are considered in detail in, e.g., Ref. [83].

10.4.2.6 Dynamic modeling

Steady-state modeling is not sufficient for batch and semibatch distillation processes, or if one tries to optimize the start-up and shut-down phases of the process. In this case, knowledge of dynamic process behavior is necessary. Furthermore, dynamic information is crucial for the process control, as well as for safety issues and training.

In the dynamic rate-based stage model, molar hold-up terms have to be considered in the mass balance equations, in which the change of both, the specific molar component hold-up and the total molar hold-up, are taken into account. For the liquid phase, these equations are as follows:

$$\frac{\partial}{\partial t} U_{Li} = -\frac{\partial}{\partial l} \left(L x_i^B \right) + N_{Li}^B a^I A_c; \quad i = 1, \ldots, n \tag{10.47}$$

$$U_{Li} = x_i^B U_{Lt}; \quad i = 1, \ldots, n \tag{10.48}$$

The vapor hold-up can often be neglected because of the low vapor-phase density, and the component balance equation reduces to Eqn (10.33). However, this should be avoided if columns are operated at high pressures [84].

The dynamic formulation of the model equations requires a careful analysis of the whole system in order to prevent high index problems during the numerical solution [85]. As a consequence, a consistent set of initial conditions for the dynamic simulations and a suitable description of the hydrodynamics have to be introduced. For instance, pressure drop and liquid hold-up must be correlated with the vapor and liquid flows [82].

10.4.3 Hydrodynamic analogy approach

The very simplified fluid dynamic pattern in the context of the film model may be considered as an advantage that permits avoiding the consideration of real complex flows on, and in, the column internals. On the other hand, it makes this modeling approach highly dependable on the quality of the experimental parameters like the specific interfacial area and mass transfer coefficients. Deviations between the correlations of different authors available in literature are quite significant, and, considering that these correlations are usually determined experimentally for a certain packing, chemical system, and within a definite range of operating conditions, their extrapolation to other conditions is not a trivial task (cf. [86]).

Further difficulties arise when the film model is applied to complex multicomponent and reactive separations. Being initially developed for binary mass transfer [61], this method directly relates the film thickness with the binary diffusivity. However, in a multicomponent mixture, different diffusivities related to different component binary pairs are available, thus resulting in several thicknesses. To obtain the unique film thickness for each phase, this model parameter has to be estimated as an averaged value. Moreover, a formal contradiction is often related to the consideration of convective mass transfer by the film model, since the main elements of this model represent stagnant films. As a consequence, the straightforward application of the film model becomes difficult.

An attempt to avoid using rough fluid dynamic simplifications brought about a method called the hydrodynamic analogy (HA) approach [56,60]. This approach is an alternative way to describe the hydrodynamics and transport phenomena in processes in which the exact location of the phase boundaries is not possible, yet the fluid pattern possess some regularity or structure which can be mirrored by an analogy with more simple flow elements. The basic idea of the approach, illustrated by Figure 10.8, is a reasonable *replacement* of the actual complex hydrodynamics in a column by a combination of *geometrically simpler flow patterns*. Such a geometrical simplification has to be done in agreement with *experimental observations* of fluid flow, which play a crucial part for the successful application of this approach. Once the observed complex flow is reproduced by a sequence of the simplified flow patterns, the partial differential equations of momentum, energy, and mass transfer can be applied to govern the transport phenomena in an entire separation column.

The idea of the HA method was put forward in [56]. It has been successfully applied for different separation operations [87–89], among others for distillation processes in columns equipped with corrugated sheet structured packings. The

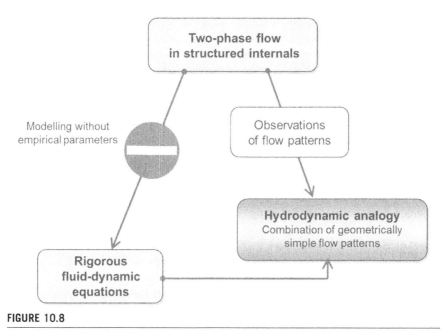

FIGURE 10.8

Illustration of the hydrodynamic analogy approach.

corrugated sheets are installed counter-course in such a way that they form channels, each being formed by the two wall sides and one open side shared between two neighboring channels (see Figure 10.9). In line with the HA concept, the basis for the physical model is provided by the observations of fluid flow together with the geometric characteristics of structured packing. According to experimental studies (e.g., Refs [71,90−93]), these observations can be summarized as follows:

- Gas flow takes place in channels built up by the counter-course arrangement of the corrugated sheets.
- There is a strong interaction between the gas flows in adjacent channels through the open channel side (see Figure 10.9) responsible for small-scale mixing (that is, via turbulence) of the gas phase.
- Liquid generally tends to flow in the form of films over the packing surface, whereas the wave formation is largely suppressed because of the corrugations.
- Liquid flows at the minimal angle built by the packing surface and the vertical axis (the so-called gravity-flow angle).
- Abrupt flow redirection at the corrugation ridges, together with the influence of intersection points with the adjacent corrugated sheets, cause mixing and lateral spreading of the liquid phase.
- Side-effects (abrupt flow redirection at the column wall and at transitions between the packing layers) result in large-scale mixing of the gas phase.

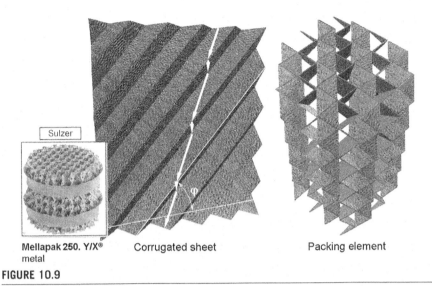

Sulzer

Mellapak 250. Y/X® metal   **Corrugated sheet**   **Packing element**

FIGURE 10.9

A piece of corrugated metal sheet packing (left), a corrugation sheet with observed liquid flow path (middle), and a packing element (right).

The physical model represented in Figure 10.10 comprises all these effects. The pronounced channel flow of the gas phase makes it possible to consider the packing as a bundle of parallel inclined channels with identical cross sections. For simplicity, the circular channel shape is adopted. The number of the channels, as well as their diameter, is determined from the packing geometric specific area and corrugation geometry, respectively. The gas-flow behavior depends on the operating conditions and varies from laminar to turbulent flow. The liquid flows counter-currently to the gas flow in the form of laminar, nonwavy films over the inner surface of the channels. In addition, a uniform distribution of both phases in the radial direction is assumed, i.e., no maldistribution is taken into account in the model. The ratio of wetted to total number of channels is the same as the ratio of effective (interfacial) specific area to geometric specific area of the packing. The periodical ideal mixing approximation is necessary to account for real mixing caused by the abrupt change in the flow direction. The length of the laminar flow interval for the liquid phase corresponds to the distance between the two neighboring corrugation ridges, whereas for the gas phase, it is set to be equal to an average channel length [57].

The system of equations describing this arrangement comprises partial differential equations written for the gas and the liquid phases, coupled through the boundary conditions at the phase interface [5]. Numerical solution of this system (finite difference implicit scheme) yields the local temperature and composition fields. The computed averaged compositions over the packing height were compared with measured data for distillation at total reflux for different structured packings under

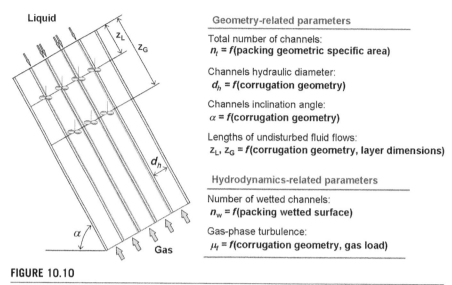

FIGURE 10.10

Physical model of structured packing.

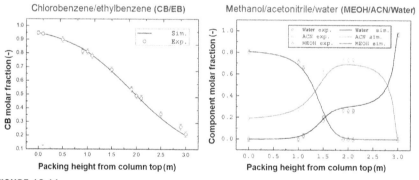

FIGURE 10.11

Comparison of simulated (lines) and experimentally measured (circles) liquid composition profiles in the column supplied with the Montz-Pak A3-500 structured packing: a binary system ethylbenzene/chlorobenzene/(EB/CB) (left), a ternary system methanol/acetonitril/water (MEOH/ACN/WATER) (right). Experimental data from Ref. [94].

a variety of operating conditions, and an excellent agreement was established (see, e.g., Figure 10.11).

The HA method was applied to several separation and reactive separation units [87,89,95] and proved its potential. Recently, valuable insights in the fluid flow in the packing have been gained from the results of tomographic measurements [96,97]. Analysis of tomographic images provides quantitative characteristics of the gas–liquid interfacial area, mixing points of liquid flow and fluid patterns other than liquid films. By taking this information into account, the HA model can be refined accordingly. The HA method can not only be used as an alternative design and modeling technique for columns equipped with structured packing, but can also be applied for theoretical optimization of packing geometry, because of its independence of experimentally determined mass transfer coefficients.

10.4.4 Computational fluid dynamics

The most rigorous approach to the description of a process or a phenomenon is based on the classical equations of fluid dynamics. These equations are partial differential equations; they provide a local description of the transport phenomena and are supplied by the corresponding initial and boundary conditions. When solved, such models yield local velocity, pressure, temperature, and concentration fields, which can be used for the determination of all relevant process characteristics.

However, the application of fluid dynamic equations to the modeling of distillation units is not straightforward. The major difficulty lies in the very complex, two-phase flow pattern caused both by column internals geometry and by intricate phase interactions. In most cases, the real flow, above all the position of the vapor–liquid interface, cannot be captured, and, thus, the transport equations and boundary conditions cannot be determined properly [5]. In this regard, significant modeling and numerical work still has to be done.

In distillation units with a simplified film-like flow, mass and heat transport phenomena can be considered in a rather rigorous way. If the film flow is laminar and non-wavy, the velocity profiles can be obtained by solving a reduced form of the Navier–Stokes equations for simple flat or cylindrical (nonwavy) film-flow geometry. Furthermore, the governing diffusion-convection equations take a simple form and can be solved even for multicomponent systems [60,79]. This is done by using the Maxwell–Stefan equations for the description of multicomponent diffusion and a compact matrix-type form of the governing mass-transfer equations. Moreover, simultaneous heat and mass transfer during film distillation can also be resolved [60].

Fluid flow in the majority of distillation units is, however, much more complex, both because of the column internals geometry and to the intricate phase interactions. Analytical and simple numerical solutions do not apply here. Instead, two-dimensional and three-dimensional numerical simulations of coupled flow and transport phenomena have to be performed, which is usually done with the aid of modern numerical facilities and tools of CFD.

CFD simulation today represents a powerful approach to many physicochemical problems, and the availability of advanced commercial tools facilitates its application. In the area of single-phase flow, a significant progress can be recognized. But multiphase flow simulations, especially in combination with mass transfer phenomena, are still under development.

When the phase contact is very intensive, so that the phase interaction can be considered as for the *interpenetrating continua*, the so-called Euler–Euler flow modeling can be applied [98–100]. Using this approach, some groups approached direct modeling of diverse distillation units (see Ref. [101]). The major drawback of this approach is that there is no physically grounded method to determine source terms appearing in the species transfer equations (the rate of mass transfer per unit volume). Basically, such source terms should describe the interfacial mass transfer; however, attempts to solve this problem by using macroscopic (e.g., film model based) models are inconsistent (see Ref. [102]).

When the fluid phases in a distillation column flow separately, a free moving interface between them should be localized. Different strategies to handle moving interfaces are classified into moving mesh methods, fixed mesh methods, and a combination of both. In the first method, a moving mesh is used to track the interface. As the form of the interface changes, the mesh is adjusted in accordance with the change. Methods falling under this strategy are called *front tracking methods* [103]. The second approach uses a fixed (Eulerian) mesh, while the interface is tracked using different procedures, e.g., special markers or functions. Such methods are called *front capturing methods*; they include marker-and-cell, surface capturing, and volume capturing approaches [103].

The development of these methods represents an ongoing research area. Nevertheless, the simulation of large-scale distillation columns appears too difficult, mostly because of numerical difficulties and conflict of different scales in the unit model. In most cases, the position of the vapor–liquid interface is difficult to capture, and, thus, the boundary conditions cannot be determined properly [5]. Thus, the direct application of fluid dynamic equations to the modeling of distillation units remains very complicated. CFD can be applied for the determination of process and model parameters in the context of the complementary modeling concept that has recently been suggested in [5]. For instance, CFD modeling of small periodic elements in packed columns provides valuable information about the turbulent flow there and can further be used within the HA modeling. Moreover, pressure drop characteristics can be determined by direct numerical simulation, thus providing parameter correlations for the RBA modeling [104,105].

10.5 Modeling of more complex distillation processes

Classical distillation represents a basic separation operation applicable to many different systems and is usually the first choice for separating mixtures. Its significance is even higher, because it serves as a basis for more complex operations combining the distillation principle with other principles or functionalities. In particular, these complex operations comprise

- azeotropic and extractive distillation, in which an additional component is deliberately introduced into the system in order to increase the difference in volatility of the hardest-to-separate component,

- thermally coupled distillation columns, in which heat integration, pressure change, or a separation of a column onto two parts (prefractionator and main column) using a dividing wall, is applied,
- reactive distillation (RD) combining reaction and separation step in a single distillation column,
- membrane distillation (MD), where a membrane separates the vapor and liquid phases and a selective evaporation through this membrane results in the component separation,
- micro-distillation processes that take place in very small units, mostly driven by capillary forces
- some other modern technologies aiming at intensification of distillation unit performance (e.g., cyclic distillation).

The modeling of these operations largely falls within the scope of the methods discussed in this Chapter, and we will just briefly review the possible peculiarities.

10.5.1 Azeotropic and extractive distillation

In many cases, it is quite difficult or even impossible to separate a binary mixture by simple distillation; for example, when the volatilities of both mixture components are close or when these components form an azeotrope. In such cases, more complex operations involving an additional substance deliberately introduced into the initial system can help to increase the difference in volatility.

For instance, a third component, called *an entrainer*, can be added to the initial system, forming a low-boiling new azeotrope with one of the mixture components. This new azeotrope has to be easily separable from another initial mixture component [6,58]. An example of such an operation is given in [106], where the difficult separation of acetic acid from water is significantly facilitated by butyl acetate as an entrainer building a heteroazeotrope with water. This operation is often referred to as *azeotropic distillation*, and it belongs to a more general group of azeotropic distillation methods [26].

Another example is given by *extractive distillation* in which a new substance, *a solvent*, is added that is relatively nonvolatile compared to the components to be separated. This new component favorably alters the relative volatility of the original mixture components and facilitates separation. Because of its low volatility, the solvent is charged continuously near the top of the distillation unit, so that a sufficient amount is maintained throughout the column. An example of such an operation, in which the separation of toluene from isooctane is greatly enhanced by adding phenol, is considered in [58].

Certainly, adding an extra substance to a process is problematic, because of an additional separation effort and possible impurity. Moreover, for both azeotropic and extractive distillation, usually, an additional column is necessary. Therefore, these operations can only be brought in use if their overall cost is lower than that of conventional distillation. The key issue of a successful application is a proper choice of the additional component. According to Ref. [58], extractive distillation can be considered as more desirable than azeotropic distillation, for more flexibility with respect to the choice of an additional substance, and for lower solvent consumption.

The modeling of both azeotropic and extractive distillation is generally very similar to the modeling of simple distillation. This is because of the basic similarity of the underlying phenomena inside the column. In this respect, both equilibrium-based and nonequilibrium-based modeling are applicable. However, the description of the thermodynamic equilibrium in these very complex, nonideal systems is difficult, and the relevant data are scarcer.

10.5.2 Thermally coupled distillation columns

It is very well known that distillation is an extremely energy consuming separation process, and, because of its widest application, distillation operations are responsible for about 40% of the energy used in the chemical process industries. For this reason, the optimization of distillation offers a great energy saving potential, and, therefore, a significant effort has been done to improve the heat utilisation. Along these lines, several new ideas have been suggested. For instance, by combining rectifying and stripping columns in an annular arrangement to exchange heat between them under elevated pressure in the rectifying section, energy savings of nearly 50% can be achieved. This concept is called *Heat-Integrated Distillation Columns* (HIDiC) [107,108]. Other interesting opportunities are related to the use of side heat exchangers or heat pump assisted distillation [109,110]. However, the greatest optimization potential seems to be achievable when the thermally coupled (*Petlyuk*) or *dividing wall column* (DWC) configurations are applied [111−113].

In both the Petlyuk column and in the DWC, a configuration with a prefractionator column and a main column is used. The prefractionator of the Petlyuk column has neither reboiler nor condenser. The DWC concept goes even further: both columns are integrated in one single shell. Technically this is realized by implementing a dividing wall [112]. Both the Petlyuk and DWC configurations are thermodynamically equivalent; however, being a one shell column, a DWC permits additional capital cost and installation space reduction [114]. According to Schultz et al. [115], the DWC will become a standard distillation tool within the next 50 years.

An excellent review on the DWC technology, covering both the theoretical description and the patent area, was recently given by Dejanovic et al. [114]. Industrial applications and control issues are considered in [116].

In order to investigate different distillation configurations, their setup and operating parameters have to be determined. For the DWC, this is especially difficult regarding the large number of required design parameters (see Ref. [117]). In addition to conventional column variables (e.g. reflux ratio, distillate stream), further parameters, e.g., the location and height of the dividing wall, and the distribution of liquid above and vapor below the dividing wall, have to be determined.

Triantafyllou and Smith [118] presented the so-called *decomposition method*, which simplifies the design problem by replacing a complex integrated configuration by a (disintegrated) sequence of conventional single columns. The latter can be designed by applying already existing, mostly shortcut, approaches. This allows a quick design estimation under consideration of required product specifications (e.g., product concentration).

Modeling of the DWC has mostly been based on the equilibrium stage concept (see Refs [114,116]). In the meantime, a commercial software tool for the simulation of the Petlyuk configuration is available [119]. The first rate-based model was developed and validated by Mueller et al. [120] for a ternary alcohol mixture methanol−isopropanol−butanol. Here, the film model was applied to both sides of the dividing wall resulting in a kind of a "double" film model necessary for the simultaneous treatment of the prefractionator and main column. A special method had to be developed for the heat transfer modeling in order to account for the additional heat flux through the dividing wall. Furthermore, the RBA modeling method was extended to cover chemical reacting systems in the so-called *reactive dividing wall column* [121].

10.5.3 **Reactive distillation**

In reactive distillation (RD), reaction and distillative separation take place within the same zone of a distillation column. Reactants are converted to products with simultaneous separation of the products and recycle of unused reactants. The RD process can be efficient both in size and cost of capital equipment, as well as in energy required to achieve complete conversion of the reactants. Therefore, the combination of a reactor and a distillation column offers great potential for overall savings [67,122].

Chemical reactions in RD commonly take place in the liquid phase and, depending on the catalyzing mechanism, RD processes can be divided into homogeneous ones, either autocatalyzed or homogeneously catalyzed, and heterogeneous processes, in which the reaction is catalyzed by a solid catalyst. The latter is often referred to as catalytic distillation (CD).

The modeling of RD can be performed fairly well with the extended equilibrium stage model, in which reaction kinetics is considered as a source term in the balance equations. It is because common RD reactions (esterifications, transesterifications, etherifications) are slow as compared to the mass transfer, and, thus, reaction kinetics represents a limiting step in RD processes. This is especially clear in homogeneously catalyzed RD column as well as in the reaction zones of CD processes. Outside the reaction zones, mass transfer becomes important. In the case of very fast reactions, like, for instance, in systems containing electrolytes, the chemical equilibrium is achieved nearly instantly, and hence, usage of transformed compositions is advantageous according to Ung and Doherty [123], as it allows RD processes to be handled by the same methods as those available for conventional, non-RD operations.

The rate-based modeling of homogeneously catalyzed RD, with a liquid catalyst acting as a mixture component, and auto-catalyzed RD, is very similar to that of classical distillation considered in Section 10.4.2, with the only difference being that the reaction source term appears in Eqn (10.32) for the liquid phase. The modeling of CD may become much more difficult, if all relevant factors in and around solid phase, such as intrinsic kinetics and mass transport inside the porous catalyst, have to be considered explicitly (see Refs [124−126]). However, it is often assumed that all internal (inside the porous medium) and external mass transfer resistances can be lumped together [127,128]. The catalyst surface is then totally exposed to the liquid bulk conditions and can be completely described by the bulk variables. This results in the so-called *pseudo-homogeneous models*. By this name, a similarity to a simpler homogeneous bulk-phase reaction is reflected.

Detailed reviews on the RD modeling are given in Refs [67,129].

10.5.4 **Membrane distillation**

Membrane distillation (MD) is one of few nonisothermal membrane separation processes used in various applications, especially related to aqueous systems. The process is called membrane distillation because of its apparent similarity to conventional distillation, as both technologies are applied to vapor/liquid systems and both require heat to be supplied to the feed in order to achieve the separation. However, the MD behavior is determined primarily by selective diffusion through membrane pores, and, thus, the phase flow and contact is completely different resulting in different modeling.

While the properties, availability, and cost of the membrane represent critical issues in MD, its modeling is also important for the process design and optimization. Here, stage-based models are hardly applicable: above all, the treatment of mass and heat transfer in the feed and permeate channels and in the membrane play an important part [130]. The modeling approach may differ depending on the technology applied (direct contact MD, sweeping gas MD, vacuum MD, air gap MD [130]), and on the required accuracy.

A rigorous consideration of the transport phenomena inside the membrane is very difficult. In general, mass transfer occurs by the diffusive and convective transport of volatile species through the membrane pores [130]. Here, Knudsen diffusion becomes important, and the dusty gas model can be applied [131,132]. For the molecular diffusion description, the Maxwell−Stefan equations (see Eqn (10.30)) are applied in [133−136].

A membrane is often treated in a simplified way, as an arrangement of uniform and noninterconnected cylindrical pores. Recently, some pore distribution mechanisms have been implemented [137−140]. Interconnected pores with size distribution were considered based on Monte Carlo simulations [141−142]. A comprehensive review of different models applied for MD is given by Khayet [130]. He claims that most of the developed MD models are based on simplified, one-dimensional transport, and the use of at least one adjustable parameter to govern the MD permeate flux. Moreover, many semiempirical models apply heat and mass transfer correlations developed for rigid, nonporous heat exchangers, which represent a questionable approach.

There exist a few attempts to describe the coupled flow and transport phenomena in MD in a rigorous way, using partial differential equations of fluid dynamics written for both channels and for the membrane (see, e.g., Ref. [143]). However, they cannot be considered trustworthy enough, because of numerous assumptions made with respect to the membrane phenomena. On the other hand, because of the stochastic nature of porous frameworks, any application of CFD-based methods would be difficult. An HA approach may represent a reasonable compromise.

10.5.5 Microdistillation

Microdistillation technology is necessary for the implementation of classical downstream unit operations within the overall micro-scale production process. A characteristic feature of microprocess technology operations is that the contacting phases move through micro-structures. The prefix "micro" denotes structures with typical dimensions varying between submillimeters and sub-microns. Consequently, compared to conventional equipment, microstructured units reveal extremely large specific surface-to-volume ratios. Moreover, at microscale, diffusion, and heat conduction paths are shorter. Therefore, lower driving concentration and temperature differences are required for the transfer of a certain amount of mass or heat; alternatively, for fixed inlet and outlet conditions, shorter contact times and thus smaller unit volumes are necessary [144,145].

However, a stable counter-current operation in such small units represents a challenge [146]. Recent developments are based on the realization of a contact between a thin liquid film and a vapor phase [147] or on an elegant "scaling down" a sieve distillation column [148]. In both cases, the stage modeling is generally applicable, although the RBA appears preferable. More difficult is to find a proper modeling way for the microdistillation idea presented in Ref. [149], in which arrangement of gas–liquid segmented flow is suggested for the phase separation. In principle, a CFD-based description would be most relevant here. However, as discussed in Section 10.4.4, CFD modeling of complex flows with free moving interface cannot be considered mature yet.

Another concept was suggested by [150] for a distillation column with dimensions in the submillimeter range. This distillation operation was called zero-gravity distillation, because the process can be conducted in an arbitrary-oriented apparatus, ranging from the vertical position, traditional for distillation units, to the horizontal one, traditional for heat pipes [151]. The reason is that the driving force of the liquid transport is taken over by the action of capillary pressure gradient at the interface instead of gravity used in common distillation.

Thus, in zero-gravity distillation, the liquid phase flows slowly through the porous layer. Using the HA between porous and film flow, the actual liquid-phase movement along the apparatus was substituted in [152] by a simpler film-like movement. Similar as described in Section 10.4.3, this provides an opportunity to apply the matrix-form equations describing the mass transport in the column. In Ref. [152], a three-component distillation in the test system ethanol—isopropanol—water was simulated in a horizontally oriented column segment, with promising results indicating a high separation performance potential of the microdistillation columns.

10.6 Concluding remarks

Modeling of distillation is a very wide area. It has a long history and great potential. It comprises a broad spectrum of ideas, methods, and algorithms. It covers both conceptual and detailed design tasks and it is indispensable for column optimization and process intensification. Distillation can be modeled by theoretical, numerical, graphical, and experimental techniques, and combinations of those.

In this Chapter, we attempted to classify and systematically overview this diversity. We covered the entire spectrum of modeling accuracy, starting from shortcut and finishing with most rigorous CFD methods.

In addition to classical distillation, we briefly discussed some more complex processes based on the distillation principles that have attracted more and more attention in the past decades. Among these processes are extractive, azeotropic, and membrane distillation, thermally coupled distillation configurations, reactive distillation and microdistillation.

Distillation will remain a key separation technology for years, and its adequate modeling will be an important means for its design, control, and optimization. Regarding the modern trend toward process intensification, a deep understanding of general and specific features of distillation will become ever more essential, and hence the role of distillation modeling will grow.

References

[1] K.E. Porter, Why research is needed in distillation, Chem. Eng. Res. Des. 73, Part A (1995) 357—361.

[2] E.J. Henley, J.D. Seader, Equilibrium Stage Separation Operations in Chemical Engineering, Wiley, New York, 1981.

[3] J.G. Stichlmair, J.R. Fair, Distillation: Principles and Practice, Wiley-VCH, New York, 1998.

[4] J.D. Seader, The rate-based approach for modeling staged separations, Chem. Eng. Progr. 85 (1989) 41—49.

[5] E.Y. Kenig, Complementary modelling of fluid separation processes, Chem. Eng. Res. Des. 86 (2008) 1059—1072.

[6] M.F. Doherty, M.F. Malone, Conceptual Design of Distillation Systems, McGraw-Hill, New York, 2001.

[7]   F.B. Petlyuk, V.S. Avet'yan, Investigation of the rectification of three-component mixtures with infinite reflux, Theor. Found. Chem. Eng. 5 (1971) 499−507.

[8]   L.A. Serafimov, V.S. Timofeev, M.I. Balashov, Rectification of multicomponent mixtures. II. Local and general characteristics of the trajectories of rectification processes at infinite reflux ratio, Acta Chim. Acad. Sci. Hung. 75 (1973) 193−211.

[9]   N. Bekiaris, G.A. Meski, C.M. Radu, M. Morari, Multiple steady states in homogeneous azeotropic distillation, Ind. Eng. Chem. Res. 32 (1993) 2023−2038.

[10] N. Bekiaris, M. Morari, Multiple steady states in distillation: ∞/∞ predictions, extensions, and implications for design, synthesis, and simulation, Ind. Eng. Chem. Res. 35 (1996) 4264−4280.

[11] M.F. Doherty, J.D. Perkins, On the dynamics of distillation processes. I: simple distillation of multicomponent non-reacting, homogeneous liquid mixtures, Chem. Eng. Sci. 33 (1978) 281−301.

[12] L.A. Serafimov, V.T. Zharov, V.S. Timofeev, Rectification of multicomponent mixtures. I. Topological analysis of liquid-vapour phase equilibrium diagrams, Acta Chim. Acad. Sci. Hung. 69 (1971) 383−396.

[13] M.F. Doherty, J.D. Perkins, On the dynamics of distillation processes. III: the topological sturcture of ternary residue curve maps, Chem. Eng. Sci. 34 (1979) 1401−1414.

[14] V.T. Zharov, L.A. Serafimov, Physicochemical Fundamentals of Simple Distillation and Rectification, Khimiya, Leningrad, 1975 (in Russian).

[15] A. Vogelpohl, Definition of distillation lines for the separation of multi-component mixtures, Chem. Eng. Technol. 65 (1993) 515−522.

[16] V.N. Kiva, E.K. Hilmen, S. Skokestad, Azeotropic phase equilibrium diagrams: a survey, Chem. Eng. Sci. 58 (2003) 1903−1953.

[17] O. Ryll, S. Blagov, H. Hasse, ∞/∞-analysis of homogeneous distillation processes, Chem. Eng. Sci. 84 (2012) 315−332.

[18] J. Ulrich, M. Morari, Operation of homogeneous azeotropic distillation column sequences, Ind. Eng. Chem. Res. 42 (2003) 4512−4534.

[19] B.T. Safrit, A.W. Westerberg, Algorithm for generating the distillation regions for azeotropic multicomponent mixtures, Ind. Eng. Chem. Res. 36 (1997) 1827−1840.

[20] R.E. Rooks, V. Julka, M.F. Doherty, M.F. Malone, Structure of distillation regions for multicomponent azeotropic mixtures, AIChE J. 44 (1998) 1382−1391.

[21] S. Blagov, H. Hasse, Topological analysis of vapor-liquid equilibrium diagrams for distillation process design, Phys. Chem. Chem. Phys. 4 (2002) 896−908.

[22] M. Bellows, A. Lucia, Geometry Separation Boundaries: Four-component Mixtures, AIChE J. 53 (2007) 1770−1778.

[23] T. Poepken, J. Gmehling, Simple method for determining the location of distillation region boundaries in quarternary systems, Ind. Eng. Chem. Res. 43 (2004) 777−783.

[24] G. Feng, L.T. Fan, P.A. Seib, B. Bertok, L. Kalotai, F. Friedler, Graph-theoretic method for the algorithmic synthesis of azeotropic-distillation systems, Ind. Eng. Chem. Res. 42 (2003) 3602−3611.

[25] M.R. Fenske, Fractionation of straight-run Pennsylvania gasoline, Ind. Eng. Chem. 24 (1932) 482−485.

[26] R.H. Perry, D.W. Green, Perry's Chemical Engineers' Handbook, McGraw-Hill, New York, 2007.

[27] A.J.W. Underwood, Fractional distillation of multi-component mixtures, Chem. Eng. Progr. 44 (1948) 603−614.

[28] A.J.W. Underwood, Fractional distillation of ternary mictures. Part I, J. Inst. Pet. 31 (1945) 111−118.

[29] A.J.W. Underwood, Fractional distillation of ternary mixtures. Part II, J. Inst. Pet. 32 (1946) 598−613.

[30] S.G. Levy, D.B.V. Dongen, M.F. Doherty, Design and synthesis of homogeneous azeotropic distillations. 2. Minimum refluc calculations for nonideal and azeotropic columns, Ind. Eng. Chem. Fundam. 24 (1985) 463−474.

[31] V. Julka, M.F. Doherty, Geometric behavior and minimum flows for nonideal multicomponent distillation, Chem. Eng. Sci. 45 (1990) 1801−1822.

[32] V. Julka, M.F. Doherty, Geometric nonlinear analysis of multicomponent nonideal distillation − a simple computer-aided-design procedure, Chem. Eng. Sci. 48 (1993) 1367−1391.

[33] D.Y.-C. Thong, M. Jobson, Multicomponent homogeneous azeotropic distillation. 2. Column design, Chem. Eng. Sci. 56 (2001) 4393−4416.

[34] J. Bausa, R. von Watzdorf, W. Marquardt, Shortcut methods for nonideal multicomponent distillation: 1. Simple columns, AIChE J. 44 (1998) 2181−2198.

[35] S. Brüggemann, W. Marquardt, Rapid screening of design alternatives for nonideal multiproduct distillation processes, Comp. Chem. Eng. 29 (2005) 165−179.

[36] S. Brüggemann, W. Marquardt, Conceptual design of distillation processes for mixtures with distillation boundaries: 1. Computational assessment of split feasibility, AIChE J. 57 (2010) 1526−1539.

[37] N.L. Franklin, Counterflow cascades: part I, Chem. Eng. Res. Des. 64 (1986) 56−66.

[38] N.L. Franklin, Counterflow cascades: part II, Chem. Eng. Res. Des. 66 (1988) 47−64.

[39] N.L. Franklin, The theory of multicomponent countercurrent cascades, Chem. Eng. Res. Des. 66 (1988) 65−74.

[40] E. Sorel, La rectification de l'alcool, Gauthier-Villars et fils, Paris, 1894.

[41] C.D. Holland, Fundamentals of Multicomponent Distillation, McGraw-Hill, New York, 1981.

[42] T.L. Wayburn, J.D. Seader, Homotopy continuation methods for computer-aided process design, Comput. Chem. Eng. 11 (1987) 7−25.

[43] H. Hasse, B. Bessling, R. Böttcher, OPEN CHEMASIM™: breaking paradigms in process simulation, Comp. Aided Chem. Eng. 21 (2006) 255−260.

[44] W.L. McCabe, E.W. Thiele, Graphical design of fractionating columns, Ind. Eng. Chem. 17 (1925) 605−611.

[45] M. Ponchon, Etude graphique de la distillation fractionnée industrielle, Tech. Moderne 13 (1921) 20−24, 55−58.

[46] R. Savarit, Elements de distillation, Arts et Métiers 75 (1922), 65, 142, 178, 241, 266, 307.

[47] E.V. Murphree, Rectifying column calculations − with particular reference to n-component mixtures, Ind. Eng. Chem. 17 (1925) 747−750.

[48] H.Z. Kister, Distillation Design, McGraw-Hill, New York, 1992.

[49] T.K. Sherwood, R.L. Pigford, C.R. Wilke, Mass Transfer, McGraw-Hill, New York, 1975.

[50] W.E. Stewart, R. Prober, Matrix calculation of multicomponent mass transfer in isothermal systems, Ind. Eng. Chem. Fund. 3 (1964) 224−235.

[51] H.L. Toor, Solution of the linearized equations of multicomponent mass transfer, AIChE J. 10 (1964) 448−455, 460−465.

[52] A. Górak, Simulation Thermischer Trennverfahren Fluider Vielkomponenten-gemische, in: G. Schuler (Ed.), Prozeßsimulation, Wiley-VCH, Weinheim, 1995, pp. 349−408.

[53] R. Taylor, R. Krishna, Multicomponent Mass Transfer, Wiley, New York, 1993.

[54] R. Billet, Packed Towers, Wiley-VCH, Weinheim, 1995.

[55] J. Mackowiak, Fluid Dynamics of Packed Columns, Springer-Verlag, Berlin, 2010.

[56] E.Y. Kenig, Multicomponent multiphase film-like systems: a modeling approach, Comput. Chem. Eng. 21 (1997) 355−360.

[57] A. Shilkin, E.Y. Kenig, A new approach to fluid separation modelling in the columns equipped with structured packings, Chem. Eng. J. 110 (2005) 87−100.

[58] R.E. Treybal, Mass Transfer Operations, third ed., McGraw-Hill, New York, 1975.

[59] Z. Olujic, M. Jödecke, A. Shilkin, G. Schuch, B. Kaibel, Equipment improvement trends in distillation, Chem. Eng. Process. 48 (2009) 1089−1104.

[60] E.Y. Kenig, Modeling of Multicomponent Mass Transfer in Separation of Fluid Mixtures, VDI-Verlag, Düsseldorf, 2000.

[61] W.K. Lewis, W.G. Whitman, Principles of gas absorption, Ind. Eng. Chem. 16 (1924) 1215−1220.

[62] R. Higbie, The rate of absorption of a pure gas into a still liquid during short periods of exposure, Trans. Am. Inst. Chem. Engrs. 31 (1935) 365−383.

[63] P.V. Danckwerts, Significance of liquid-film coeffficients in gas absorption, Ind. Eng. Chem. 43 (1951) 1460−1467.

[64] H.L. Toor, J.M. Marchello, Film-penetration model for mass and heat transfer, AIChE J. 4 (1958) 97−101.

[65] L.E. Scriven, Flow and transfer at fluid interfaces, Chem. Eng. Educ. (1968−1969), 150−155 (Fall 1968), 26−29 (Winter 1969), 94−98 (Spring 1969).

[66] J.O. Hirschfelder, C.F. Curtiss, R.B. Bird, Molecular Theory of Gases and Liquids, Wiley, New York, 1964.

[67] E.Y. Kenig, A. Górak, Modeling of Reactive Distillation, in: F. Keil (Ed.), Modeling of Process Intensification, Wiley-VCH, Weinheim, 2007, pp. 323−363.

[68] N. Kolev, Packed Bed Columns, Elsevier B. V, Amsterdam, 2006.

[69] R. Billet, M. Schultes, Fluid dynamics and mass transfer in the total capacity range of packed columns up to the flood point, Chem. Eng. Technol. 18 (1995) 371−379.

[70] K. Onda, H. Takeuchi, Y. Okumoto, Mass transfer coefficients between gas and liquid phases in packed columns, J. Chem. Eng. Jpn. 1 (1968) 56−62.

[71] Z. Olujic, Development of a complete simulation model for predicting the hydraulic and separation performance of distillation columns equipped with structured packings, Chem. Biochem. Eng. Q. 11 (1997) 31−46.

[72] J.A. Rocha, J.L. Bravo, J.R. Fair, Distillation columns containing structured packings − 2. Mass transfer model, Ind. Eng. Chem. Res. 35 (1996) 1660−1667.

[73] G.Q. Wang, X.G. Yuan, K.T. Yu, Review of mass-transfer correlations for packed columns, Ind. Eng. Chem. Res. 44 (2005) 8715−8729.

[74] T. Kuppan, Heat Exchanger Design Handbook, Talor & Francis, New York, 2000.

[75] R.K. Shah, D.P. Sekulic, Fundamentals of Heat Exchanger Design, Wiley and Sons, Inc., Hoboken, 2003.

[76] VDI-Heat Atlas, second ed., Springer-Verlag Berlin Heidelberg, 2010.

[77] T.H. Chilton, A.P. Colburn, Mass transfer (absorption) coefficients: prediction from data on heat transfer and fluid friction, Ind. Eng. Chem. 26 (1934) 1183−1187.

[78] R.E. Tsai, A.F. Seibert, R.B. Eldridge, G.T. Rochelle, Influence of viscosity and surface tension on the effective mass transfer area of structured packings, Energy Procedia 1 (2009) 1197−1204.

[79] E.Y. Kenig, Studies into kinetics of mass and heat transfer in separation of multicomponent mixtures: Part I and II, Theor. Found. Chem. Eng. 28 (1994) 199−216, 305−325.

[80] P.V. Danckwerts, Continious flow systems − distribution of residence times, Chem. Eng. Sci. 2 (1953) 1−13.

[81] W.P.M. van Swaaij, J.C. Charpentier, J. Villermaux, Residence time distribution in the liquid phase of trickle flow in packed columns, Chem. Eng. Sci. 24 (1969) 1083−1095.

[82] C. Noeres, E.Y. Kenig, A. Górak, Modelling of reactive separation processes: reactive absorption and reactive distillation, Chem. Eng. Process. 42 (2003) 157−178.

[83] L. Chen, J.-U. Repke, G. Wozny, S. Wang, Extension of the mass transfer calculation of three-phase distillation in a packed column: nonequilibrium model based parameter estimation, Ind. Eng. Chem. Res. 48 (2009) 7289−7300.

[84] J. Choe, W.L. Luyben, Rigorous dynamic models of distillation columns, Ind. Eng. Chem. Res. 26 (1987) 2158−2161.

[85] C.C. Pantelides, The consistent initialization of differential-algebraic systems. SIAM, J. Sci. Stat. Comp. 9 (1988) 213−231.

[86] A. Shilkin, E.Y. Kenig, Z. Olujic, A hydrodynamic-analogy-based model for efficiency of structured packing columns, AIChE J. 52 (2006) 3055−3066.

[87] U. Brinkmann, E.Y. Kenig, R. Thiele, M. Haas, Modelling and simulation of a packed sulphur dioxide absorption unit using the hydrodynamic analogy approach, Chem. Eng. Trans. 18 (2009) 195−200.

[88] U. Brinkmann, T. Schildhauer, E.Y. Kenig, Hydrodynamic analogy approach for modelling of reactive stripping with structured catalyst supports, Chem. Eng. Sci. 65 (2010) 298−303.

[89] A. Shilkin, K. Heinen, C. Grossmann, A. Lautenschleger, A. Janzen, E.Y. Kenig, On the development of an energy efficient packing for vacuum distillation, in: Int. Conf. "Distillation and Absorption", Eindhoven, 2010.

[90] M. Zogg, Modifizierte Stoffübergangskoeffizienten für bilanzmäßige Stoffübergangs-berechnungen an laminaren Rieselfilmen, Chem. Ing. Techn. 44 (1972) 930−936.

[91] L. Zhao, R.L. Cerro, Experimental characterization of viscous film flows over complex surfaces, Int. J. Multiphas. Flow 18 (1992) 495−516.

[92] S. Shetty, R.L. Cerro, Fundamental liquid flow correlations for the computation of design parameters for ordered packings, Ind. Eng. Chem. Res. 36 (1997) 771−783.

[93] P. Valluri, O.K. Matar, G.F. Hewitt, M.A. Mendes, Thin film flow over structured pack-ings at moderate Reynolds numbers, Chem. Eng. Sci. 60 (2005) 1965−1975.

[94] S. Pelkonen, Multicomponent Mass Transfer in Packed Distillation Columns (Ph.D. thesis), University of Dortmund, Dortmund, 1997.

[95] I. Mueller, U. Brinkmann, E.Y. Kenig, Modeling of transport phenomena in two-phase film-flow systems: application to monolith reactors, Chem. Eng. Commun. 198 (2011) 629−651.

[96] S. Aferka, J. Steube, A. Janzen, E.Y. Kenig, M. Crine, P. Marchot, et al., Investigation of Liquid Flow Pattern inside a Structured Packing Using X-ray Tomography, 6th International Symposium on Process Tomography, Cape Town, 2012.

[97] A. Janzen, J. Steube, S. Aferka, E.Y. Kenig, M. Crine, P. Marchot, et al., Investigation of liquid flow morphology inside a structured packing using X-ray tomography, Chem. Eng. Sci. 102 (2013) 451−460.

[98] Y. Pan, M.P. Dudukovic, M. Chang, Dynamic simulation of bubbly flow in bubble columns, Chem. Eng. Sci. 54 (1999) 2481−2489.

[99] J.B. Joshi, Computational flow modelling and design of bubble column reactors, Chem. Eng. Sci. 56 (2001) 5893−5933.

[100] D. Zhang, N.G. Deen, J.A.M. Kuipers, Numerical simulation of dynamic flow behaviour in a bubble column: a study of closures for turbulence and interface forces, Chem. Eng. Sci. 61 (2006) 7593–7608.

[101] G.B. Liu, K.T. Yu, X.G. Yuan, C.J. Liu, A numerical method for predicting the performance of a randomly packed distillation column, Int. J. Heat Mass Transfer 52 (2009) 5330–5338.

[102] E.Y. Kenig, A. Shilkin, T. Atmakidis, Comments on "simulations of chemical absorption in pilot-scale and industrial-scale packed columns by computational mass transfer" by Liu et al, Chem. Eng. Sci. 63 (2008) 4239–4240.

[103] J.H. Ferziger, M. Peric, Computatiuonal Methods for Fluid Dynamics, third ed., Springer-Verlag, Berlin, 2002.

[104] M. Kloeker, E.Y. Kenig, A. Górak, On the development of new column internals for reactive separations via integration of CFD and process simulation, Catal. Today 79–80 (2003) 479–485.

[105] Y. Egorov, F. Menter, M. Kloeker, E.Y. Kenig, On the combination of CFD and rate-based modelling in the simulation of reactive separation processes, Chem. Eng. Process. 44 (2005) 631–644.

[106] D.F. Othmer, Azeotropic separation, Chem. Eng. Progr. 59 (1963) 67–78.

[107] M. Nakaiwa, K. Huang, A. Endo, T. Ohmori, T. Akiya, T. Takamatsu, Internally heat-integrated distillation columns: a review, Chem. Eng. Res. Des. 81 (2003) 162–177.

[108] Z. Olujic, F. Fakhri, A.D. Rijke, J.D. Graauw, P.J. Jansens, Internal heat integration – the key to an energy-conserving distillation column, J. Chem. Tech. Biotech. 78 (2003) 241–248.

[109] J.L. Humphrey, G.E. Keller II, Separation Process Technology, McGraw-Hill, New York, 1997.

[110] C.D. Grant, Energy management in chemical industry, in: B. Elvers, et al. (Eds.), Ullman's Encyclopedia of Industrial Chemistry, VCH Publishers, New York, 1998, B3, pp. 12.11–12.16.

[111] F.B. Petlyuk, V.M. Platonov, D.M. Slavinskii, Thermodynamically optimal method for separating multicomponent mixtures, Int. Chem. Eng. 5 (1965) 555–561.

[112] G. Kaibel, Distillation columns with vertical partitions, Chem. Eng. Technol. 10 (1987) 92–98.

[113] R. Smith, Thermally-coupled columns: distillation, in: I.D. Wilson, et al. (Eds.), Encyclopedia of Separation Science, 9, Academic Press, London, 2000, pp. 4363–4371.

[114] I. Dejanovic, L. Matijasevic, Z. Olujic, Dividing wall column – a breakthrough towards sustainable distilling, Chem. Eng. Process. 49 (2010) 559–580.

[115] M.A. Schultz, D.G. Stewart, J.M. Harris, S.P. Rosenblum, M.S. Shakur, D.E. O'Brien, Reduce costs with dividing-wall columns, Chem. Eng. Progr. 98 (2002) 64–71.

[116] O. Yildirim, A.A. Kiss, E.Y. Kenig, Dividing-wall columns in chemical process industry: a review on current activities, Separ. Purif. Technol. 80 (2011) 403–417.

[117] I. Mueller, C. Pech, D. Bhatia, E.Y. Kenig, Rate-based analysis of reactive distillation sequences with different degrees of integration, Chem. Eng. Sci. 62 (2007) 7327–7335.

[118] C. Triantafyllou, R. Smith, The design and optimization of fully thermally coupled distillation columns, Chem. Eng. Res. Des. 70 (1992) 118–132.

[119] Aspen Plus User Guide, Version 10.2, Aspen Technology, Inc., Cambridge, MA, 2000.

[120] I. Mueller, M. Kloeker, E.Y. Kenig, Rate-based modelling of dividing wall columns – a new application to reactive systems, in: PRES'2004. 7th Conference on process

integration, modelling and optimisation for energy savings and pollution reduction, Prague, 2004.

[121] I. Mueller, E.Y. Kenig, Reactive distillation in a dividing wall column: rate-based modeling and simulation, Ind. Eng. Chem. Res. 46 (2007) 3709−3719.

[122] M.F. Doherty, G. Buzad, Reactive distillation by design, Trans. IChemE 70 (1992) 448−458.

[123] S. Ung, M.F. Doherty, Vapor-liquid equilibrium in systems with multiple chemical reactions, Chem. Eng. Sci. 50 (1995) 23−48.

[124] K. Sundmacher, U. Hoffmann, Multicomponent mass and energy transport on different length scales in a packed reactive distillation column for heterogeneously catalysed fuel ether production, Chem. Eng. Sci. 49 (1994) 3077−3089.

[125] R. Krishna, J.A. Wesselingh, The Maxwell−Stefan approach to mass transfer, Chem. Eng. Sci. 52 (1997) 861−911.

[126] A. Higler, R. Krishna, R. Taylor, Nonequilibrium modeling of reactive distillation: a dusty fluid model for heterogeneously catalyzed processes, Ind. Eng. Chem. Res. 39 (2000) 1596−1607.

[127] Z. Yuxiang, X. Xien, Study on catalytic distillation processes. Part II. Simulation of catalytic distillation processes − quasi-homogenous and rate-based model, Trans. IChemE 70 (1992) 465−470.

[128] A. Górak, A. Hoffmann, Catalytic distillation in structured packings: methyl acetate synthesis, AIChE J. 47 (2001) 1067−1076.

[129] R. Taylor, R. Krishna, Modelling reactive distillation, Chem. Eng. Sci. 55 (2000) 5183−5229.

[130] M. Khayet, Membranes and theoretical modeling of membrane distillation: a review, Adv. Coll. Interf. Sci. 164 (2011) 56−88.

[131] E.A. Mason, A.P. Malinauskas, Gas Transport in Porous Media: The Dusty-Gas Model, Elsevier Scientific Pub. Co, Amsterdam, New York, 1983.

[132] K.W. Lawson, D.R. Lloyd, Membrane distillation, J. Membr. Sci. 124 (1997) 1−25.

[133] C.A. Rivier, M.C. Garcia-Payo, I.W. Marison, U. von Stockar, Separation of binary mixtures by thermostatic sweeping gas membrane distillation I. Theory and simulations, J. Membr. Sci. 201 (2002) 1−16.

[134] C. Gostoli, G.C. Sarti, Separation of liquid mixtures by membrane distillation, J. Membr. Sci. 41 (1989) 211−224.

[135] F.A. Banat, F.A. Al-Rub, R. Jumah, M. Al-Shannag, Application of Stefan−Maxwell approach to azeotropic separation by membrane distillation, Chem. Eng. J. 73 (1999) 71−75.

[136] F.A. Banat, F.A. Al-Rub, R. Jumah, M. Shannag, Theoretical investigation of membrane distillation role in breaking the formic acid−water azeotropic point: comparison between Fickian and Stefan−Maxwell-based models, Int. Commun. Heat Mass Transfer 26 (1999) 879−888.

[137] L. Martínez, F.J. Florido-Díaz, A. Hernández, P. Prádanos, Estimation of vapor transfer coefficient of hydrophobic porous membranes for applications in membrane distillation, Sep. Purif. Technol. 33 (2003) 45−55.

[138] J. Phattaranawik, R. Jiraratananon, A.G. Fane, Effect of pore sizedistribution and air flux on mass transport in direct contact membrane distillation, J. Membr. Sci. 215 (2003) 75−85.

[139] M. Khayet, T. Matsuura, Pervaporation and vacuum membrane distillation processes: modeling and experiments, AIChE J. 50 (2004) 1697−1712.

[140] M. Khayet, A. Velázquez, J.I. Mengual, Modelling mass transport through a porous partition: effect of pore size distribution, J. Non-Equilib. Thermodyn. 29 (2004) 279–299.

[141] A.O. Imdakm, T. Matsuura, A Monte Carlo simulation model for membrane distillation processes: direct contact (MD), J. Membr. Sci. 237 (2004) 51–59.

[142] M. Khayet, A.O. Imdakm, T. Matsuura, Monte Carlo simulation and experimental heat and mass transfer in direct contact membrane distillation, Int. J. Heat Mass Transfer 53 (2010) 1249–1259.

[143] K. Charfi, M. Khayet, M.J. Safi, Numerical simulation and experimental studies on heat and mass transfer using sweeping gas membrane distillation, Desalination 259 (2010) 84–96.

[144] K.F. Lam, E. Sorensen, A. Gavriilidis, Review on gas-liquid separations in microchannel devices, Chem. Eng. Res. Des. 91 (2013) 1941–1953.

[145] E.Y. Kenig, Y. Su, A. Lautenschleger, P. Chasanis, M. Grünewald, Micro-separation of fluid systems: a state-of-the-art review, Sep. Purif. Technol. 120 (2013) 245–264.

[146] P. Chasanis, J. Kern, M. Grünewald, E.Y. Kenig, Mikrotrenntechnik: Entwicklungsstand und Perspektiven, Chem. Ing. Techn. 82 (2010) 215–228.

[147] A.L. Tonkovich, W.W. Simmons, L.J. Silva, D. Qiu, S.T. Perry, T. Yuschak, Distillation Process Using Microchannel Technology, Velosys, Inc., US 7305850, B2, 2007.

[148] A. Ziogas, G. Kolb, H.-J. Kost, V. Hessel, Entwicklung einer leistungsstarken Mikrorektifikationsapparatur für analytische und präparative Anwendungen, Chem. Ing. Techn. 83 (2011) 465–478.

[149] L.R. Hartmann, R.H. Sahoo, C.B. Yen, F.K. Jensen, Distillation in microchemical systems using capillary forces and segmented flow, Lab. Chip 9 (2009) 1843–1849.

[150] D.R. Seok, S.-T. Hwang, Zero-gravity distillation utilizing the heat pipe principle (micro-distillation), AlChE J. 31 (1985) 2059–2065.

[151] E.A. Ramirez-Gonzâlez, C. Martínez, J. Alvarez, Modeling zero-gravity distillation, Ind. Eng. Chem. Res. 31 (1992) 901–908.

[152] J. Tschernjaew, E.Y. Kenig, A. Górak, Mikrodestillation von Mehrkomponentensystemen, Chem. Ing. Techn. 68 (1996) 272–276.

Appendix

Application of Underwood's equation to the parametric study for a simple distillation column analysis

The power of the Underwood method is illustrated here with the analysis of the structure of a two-dimensional parameter space. For a simple distillation column with a given feed, assuming both rectifying and stripping sections to be infinitely high, there remain two degrees of freedom. They might be, for instance, related to the reflux ratio and the distillate-to-feed flow rate ratio.

Here, Franklin's interpretation of Underwood's equation [37–39] is used. For clarity, only ternary mixtures are considered, and the column feed is assumed to be a saturated liquid. The presented methodology can, nevertheless, be extended to mixtures with an arbitrary number of components and for an arbitrary state of the column feed.

10.A1 **Basic relationships**

Under the constant molar overflow assumption, for each cross-section p of the column, a simple relationship follows from the mass balance:

$$\mathbf{y}^p - m\mathbf{x}^p = \text{const} \tag{10.A1}$$

where

$$m = L\big/V \tag{10.A2}$$

The vector γ given by

$$\gamma = \mathbf{y}^p - m\mathbf{x}^p \tag{10.A3}$$

is denoted by Franklin [37] as *a net interstage flow*—a dimensionless difference between the vapor and liquid flows. Equation (10.A1) represents the so-called *operating line equation* and it states that the net interstage flow γ is a constant characterizing the entire column section (stripping or rectifying).

The distillate and bottom product compositions are closely related to the net interstage flows in the rectifying and stripping sections, respectively. Assuming total condenser $(\mathbf{y}^d = \mathbf{x}^d)$, it holds:

$$\gamma^r = (1 - m^r) \cdot \mathbf{x}^d \tag{10.A4}$$

Similarly, assuming total reboiler $(\mathbf{y}^b = \mathbf{x}^b)$,

$$\gamma^s = (1 - m^s) \cdot \mathbf{x}^b \tag{10.A5}$$

Underwood [27] combined the operating line equation, Eqn (10.A3), with the constant relative volatility model, Eqn (10.4), and derived a relationship that is now referred to as *Underwood's equation*:

$$\sum_{j=1}^{n} \frac{\alpha_j \cdot \gamma_j}{\alpha_j - \theta} = 1 \tag{10.A6}$$

Equation (10.A6) is equivalent to a polynomial of order n with respect to the quantity θ. For a rectifying section $(m^r < 1)$, all elements of vector γ^r are positive, and there always exist exactly n different real roots θ_k^r of Eqn (10.A6) that may be ordered as:

$$\alpha_1 > \theta_1^r > \alpha_2 > \ldots > \alpha_{n-1} > \theta_{n-1}^r > \alpha_n > \theta_n^r \tag{10.A7}$$

Similarly, for a stripping section ($m^s > 1$), all elements of vector $\boldsymbol{\gamma}^s$ are negative and there always exist exactly n different real roots θ_k^s of Eqn (10.A6) that may be ordered as:

$$\theta_1^s > \alpha_1 > \theta_2^s > \alpha_2 > \ldots > \alpha_{n-1} > \theta_n^s > \alpha_n \qquad (10.A8)$$

As mentioned in Section 10.3.1.3, Eqn (10.A6) is usually used for the estimation of the minimum reflux ratio. For instance, if the distillate composition $\mathbf{x}^d$ and one of the roots θ^r of Eqn (10.A6) were known, combining Eqn (10.A4) with Eqn (10.A6) would allow a direct determination of the corresponding m^r-value, which, in turn, is related to the reflux ratio $R_f = \dfrac{m^r}{m^r - 1}$.

Finding the roots θ represents, however, a separate problem, as sections in the column are coupled at the feed point and thus require their simultaneous consideration. For a simple distillation column, Underwood [27] showed that for all *nonlimiting* separations (cf. Table 10.A1), at least one common root $\theta_k = \theta_k^r = \theta_{k+1}^s$ always exists. Such common roots can be found from the following relationship:

$$\sum_{i=1}^{n} \frac{\alpha_j \cdot x_j^f}{\alpha_j - \theta} = 1 - q \qquad (10.A9)$$

Thus, for any section of the column, Eqn (10.A9) coupled with Eqn (10.A6), together with simple component balance equations and an appropriate number of distillation process specifications, allows either to solve the so-called minimum reflux problem, if the product quality is specified, or to find the products and composition profiles, if the reflux ratio is specified (see, e.g., Refs [2,27,37,38]).

For a simple distillation column, the minimum reflux calculation based on Eqn (10.A9) is fairly straightforward, and the solution is found by a purely algebraic method [2,26]. However, application of Eqn (10.A6) to complex columns with multiple feeds and withdrawals as well as to distillation sequences requires more detailed analysis of relationships described by Underwood's equation.

Underwood [27] found that each root θ of Eqn (10.A6) defines a so-called *invariant hyper-plane* in the composition space given by the following equation:

$$\sum_{j=1}^{n} \frac{\alpha_j \cdot x_j}{\alpha_j - \theta} = 0 \qquad (10.A10)$$

The term "invariant" means here that, if the liquid composition $\mathbf{x}$ at a certain cross-section satisfies Eqn (10.A10), the entire composition profile also satisfies Eqn (10.A10). This equation is in fact the key point of Underwood's method. Whereas Eqns (10.A6) and (10.A9) only provide the means to determine the unknown quantities θ, Eqn (10.A10) reveals the very meaning of them: for the partial reflux operation, each specific root θ_k determines an appropriate hyper-plane, and,

Table 10.A1 Separation Types for an Ideal Ternary Mixture Separated in a Simple Distillation Column

Separation Type	Number of Common Roots	Common Roots[a]	Other Known Roots	Rectifying Section		Stripping Section		Limiting Separation
				Product Type	Pinch Point	Product Type	Pinch Point	
ABC–ABC	2	$\begin{cases} \theta_{AB} = \theta^r_A = \theta^S_B \\ \theta_{BC} = \theta^r_B = \theta^S_C \end{cases}$	—	ABC	C^r	ABC	A^S	No
AB–ABC	1	$\theta_{AB} = \theta^r_A = \theta^S_B$	$\theta^r_C = \alpha_C$	AB	B^r	ABC	A^S	No
ABC–BC	1	$\theta_{BC} = \theta^r_B = \theta^S_C$	$\theta^S_A = \alpha_A$	ABC	C^r	BC	B^S	No
A–ABC	0	—	$\begin{cases} \theta^r_B = \alpha_B \\ \theta^r_C = \alpha_C \end{cases}$	A	A^r	ABC	A^S	Yes
ABC–C	0	—	$\begin{cases} \theta^S_A = \alpha_A \\ \theta^S_B = \alpha_B \end{cases}$	ABC	C^r	C	C^S	Yes
AB–BC	0	—	$\begin{cases} \theta^r_C = \alpha_C \\ \theta^S_A = \alpha_A \end{cases}$	AB	B^r	BC	B^S	Yes

[a] Common roots are found from Eqn (10.A9) with an assumption of saturated liquid feed ($q = 1$) and ordered according to $\alpha_A > \theta_{AB} > \theta_B > \alpha_B > \theta_{BC} > \alpha_C$.

taken together, they build up the region in the composition space where all possible composition profiles may be located.

For this reason, Franklin [37] suggested to replace the original molar fractions **x** by the transformed **z**-concentrations given by

$$z_k = \sum_{i=1}^{n} \frac{\alpha_i \cdot x_i}{\alpha_i - \theta_k} \tag{10.A11}$$

The transformed **z**-concentrations allow distillation processes at partial reflux to be described by the same set of equations that are formulated for the total reflux, just by replacing the vector pair $(\mathbf{x}, \boldsymbol{\alpha})$ by the pair $(\mathbf{z}, \boldsymbol{\theta})$. In particular, the Fenske equation for the total reflux operation, Eqn (10.3), can be rewritten for the partial reflux operation as

$$\frac{z_{1,j}}{z_{1,k}} = \left(\frac{\theta_j}{\theta_k}\right)^N \frac{z_{N,j}}{z_{N,k}} \tag{10.A12}$$

Furthermore, the hyper-planes defined by Eqn (10.A10) build a deformed composition simplex (or **Z**-simplex) with the following properties:

$$\begin{cases} z_j \geq 0; \quad j = 1, n \\ \sum_{j=1}^{n} z_j = 1 \end{cases} \tag{10.A13}$$

The composition profile of the corresponding column section at partial reflux lies either inside of this **Z**-simplex or on its boundary. Examples of **Z**-simplexes for rectifying and stripping sections are presented in Section 10.A2.

Vertices of **Z**-simplexes at partial reflux are analogs of pure components at total reflux. They represent the points (denoted as pinch or singular points) where the separation driving force reduces to zero resulting in an infinite column height required to reach them. Singular points are extremely important when considering distillation in an infinitely high section, as the total "infinite height" of this section is "concentrated" within such points.

Vertices of **Z**-simplexes for partial reflux operation should be seen as inheritors of appropriate pure components. Therefore, they are denoted here with the same letters, with an upper index r or s indicating the appropriate column section, e.g., $\mathbf{A}^r$ for an **A**-vertex in the $\mathbf{Z}^r$-simplex of the rectifying section.

If either the reflux or the reboil ratio increases to infinity, then m-ratio for the appropriate section tends to one and the roots of Eqn (10.A6) become equal to values of the corresponding relative volatilities

$$\lim_{m \to 1} \theta_j = \alpha_j \tag{10.A14}$$

while the Z-simplex contracts to the original composition simplex.

Equations (10.A1)–(10.A14) build a full analytical model describing a distillation process at partial reflux for a column of an arbitrary height. This model can be applied both for single distillation columns and for column sequences.

10.A2 Modeling of distillation processes in infinitely high columns

For columns with infinitely high sections, there exist a number of additional useful relationships, simplifying the model application. Here we give a simple illustrative example, in which a ternary mixture $\mathbf{A}$, $\mathbf{B}$, $\mathbf{C}$ is considered, with component ordered according to their relative volatility ($\mathbf{A}$ is the lightest boiling one). In this case, the following relationships are valid:

- If a net component interstage flow $\gamma_\mathbf{I}$ of any component $\mathbf{I}$ is equal to zero (i.e., the molar fraction of component $\mathbf{I}$, either in a distillate for a rectifying section or in a bottom product for a stripping section, is equal to zero), then the appropriate root $\theta_\mathbf{I}$ is equal to the relative volatility of component $\mathbf{I}$, and, consequently, the corresponding invariant hyper-plane coincides with the appropriate edge of the composition simplex. For example, for a rectifying section, if the heaviest component $\mathbf{C}$ is absent in the distillate, then

$$\lim_{x_\mathbf{C}^d \to 0} \theta_\mathbf{C}^r = \alpha_\mathbf{C}$$
$$\lim_{x_\mathbf{C}^d \to 0} \mathbf{A}^r\mathbf{B}^r = \mathbf{AB} \tag{10.A15}$$

From Eqn (10.A15), it does not follow that $\mathbf{A}^r = \mathbf{A}$; it only states that singular points $\mathbf{A}^r, \mathbf{B}^r$ for a process at partial reflux and appropriate points $\mathbf{A}$, $\mathbf{B}$ for a process at total reflux lie on the same line.

- If a section product is a pure component, then this component is a singular point for processes at both total and partial reflux, e.g., for a rectifying section:

$$\lim_{x_\mathbf{A}^d \to 1} \mathbf{A}^r = \mathbf{A} \tag{10.A16}$$

- Similar to distillation processes at total reflux, the molar fraction of a more volatile component in the distillate cannot be equal to zero, if some less volatile component is present there. Analogously, the molar fraction of a less volatile component in the bottom product cannot be equal to zero, if some more volatile component is present there.

- If a section product contains all mixture components, this product point and the whole composition profile of the section lie inside the $\mathbf{Z}$-simplex. If the section is considered to be infinitely high, the profile contains the only singular point. This is either the terminus point $\mathbf{C}^r$ at the lowest cross-section of a rectifying section or the terminus point $\mathbf{A}^s$ at the highest cross-section of a stripping section.

- The product composition, either of distillate for a rectifying section or of bottom product for a stripping section, and the coordinates of any vertex **I** of the corresponding **Z**-simplex are related by

$$\frac{x_i^{\mathbf{I}}}{x_i^{\text{product}}} \cdot \frac{m}{1-m} = \frac{\theta_{\mathbf{I}}}{\alpha_i - \theta_{\mathbf{I}}} \tag{10.A17}$$

- Finally, there exists a relationship between the product of relative volatilities and the product of the roots of Underwood's equation

$$\frac{\displaystyle\prod_{j=\mathbf{A},\mathbf{B},\mathbf{C}} \theta_j}{\displaystyle\prod_{j=\mathbf{A},\mathbf{B},\mathbf{C}} \alpha_j} = m \tag{10.A18}$$

Since each infinitely high section contains at least one singular point, for an ideal ternary mixture, there exist six possible separation types in a simple distillation column. These types are listed in Table 10.A1. Three types (**A**−**ABC**, **AB**−**BC** and **ABC**−**C**) are denoted as *limiting*, because they can be realized at total reflux, whereas three other, *non-limiting* types (**ABC**−**ABC**, **AB**−**ABC** and **ABC**−**BC**), are met for partial reflux operation only. Figure 10.A1 illustrates the regions of different separation types using the parameter space diagram reflux ratio−reboil ratio. Alternatively, Figure 10.A2 shows the same separation

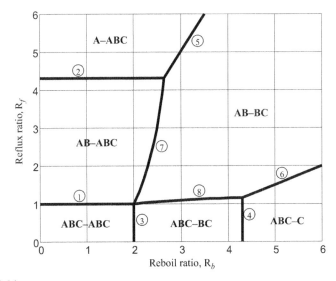

FIGURE 10.A1

Structure of the parameter space reflux ratio−reboil ratio for a simple distillation column: $x^f = [0.3, 0.3, 0.4]$, $\alpha = [3, 2, 1]$. Equations describing boundary lines one to eight are given in Table 10.A2.

type diagram in the parameter space reflux ratio—distillate-to-feed molar ratio. The first diagram is more comprehensible, whereas the second one, based on actual control parameters of a distillation process, is of higher practical value. Figures 10.A3–10.A6 illustrate column profiles and the corresponding **Z**-simplices for some of separation types listed in Table 10.A1.

The separation type **ABC–ABC** is realized at small values of reflux and reboil ratios (cf. Figure 10.A1). Neither the distillate nor the bottom product reaches the boundary of the composition triangle, and both these products lie within the triangle (cf. Figure 10.A3). There are two common roots (cf. Table 10.A1), and there exists a common singular point $C^r = A^s = \mathbf{x}^f$, which is the downward terminus point for a rectifying section and the upward terminus point for a stripping section.

For this separation type, it is well known [2] that distillate and bottom products always lie on the material balance line passing through the tie-line vector $\mathbf{y}^*(\mathbf{x}^f) - \mathbf{x}^f$ for the feed composition $\mathbf{x}^f$. The vapor leaving the feed tray is in equilibrium with the liquid feed composition $\mathbf{x}^f$ thus forming a pinch point. Similar pinch point appears in binary distillation when the minimum reflux condition is considered. It is worth noting that such behavior is not specific just to Underwood's model (constant relative volatility and constant molar overflow assumptions); rather, it is observed in any distillation process with infinitely high sections.

The distillate composition can be found here directly from one of the two mass balance equations, written either for the rectifying section

$$\mathbf{x}^d = \left(R_f + 1\right) \cdot \mathbf{y}^*\left(\mathbf{x}^f\right) - R_f \cdot \mathbf{x}^f \tag{10.A19}$$

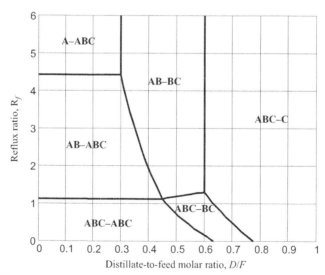

FIGURE 10.A2

Structure of the parameter space reflux ratio—distillate-to-feed ratio for a simple distillation column: $\mathbf{x}^f = [0.3, 0.3, 0.4]$, $\boldsymbol{\alpha} = [3, 2, 1]$.

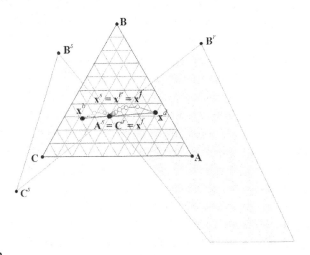

FIGURE 10.A3

Example of the **ABC–ABC** type separation: $x^f = [0.3, 0.3, 0.4]$, $\alpha = [3, 2, 1]$, $R_f = 0.7$, $R_b = 1.0$. Points x^r and x^s are liquid-phase compositions at the lowest cross-section of the rectifying section and at the top cross-section of the stripping section, respectively.

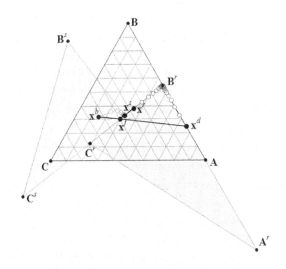

FIGURE 10.A4

Example of the **AB–ABC** type separation: $x^f = [0.3, 0.3, 0.4]$, $\alpha = [3, 2, 1]$, $R_f = 2.0$, $R_b = 1.0$. Points x^r and x^s are defined similarly, as in Figure 10.A3.

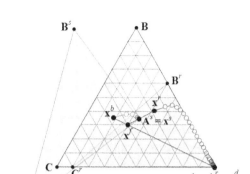

FIGURE 10.A5

Example of the **A–ABC** type separation: $x^f = [0.3, 0.3, 0.4]$, $\alpha = [3, 2, 1]$, $R_f = 5.0$, $R_b = 1.0$. Points x^r and x^s are defined similarly, as in Figure 10.A3.

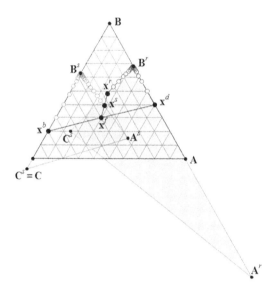

FIGURE 10.A6

Example of the **AB–BC** type separation: $x^f = [0.3, 0.3, 0.4]$, $\alpha = [3, 2, 1]$, $R_f = 2.0$, $R_b = 3.0$. Points x^r and x^s are defined similarly, as in Figure 10.A3.

or for the stripping section

$$\mathbf{x}^b = (R_b + 1) \cdot \mathbf{x}^f - R_b \cdot \mathbf{y}^*(\mathbf{x}^f) \qquad (10.A20)$$

Thus, the cumbersome computations of the roots of Underwood's equation are not necessary here. By increase of either the reflux or the reboil ratio, either the distillate or the bottom product finally reaches the boundary of the composition simplex—namely, the distillate product can achieve the **AB** edge, or the bottom product can achieve the **BC** edge of the composition simplex. On further increase of the reflux—reboil ratio, the separation type **ABC**−**ABC** cannot be realized anymore and another separation (**AB**−**ABC** or **ABC**−**CD**) has to be considered.

For the separation type **AB**−**ABC** illustrated in Figure 10.A4, compositions $\mathbf{x}^r$, $\mathbf{x}^s$ and $\mathbf{x}^f$ are not equal, and there is only one common root of Underwood's equation $\theta_{\mathbf{A}^r} = \theta_{\mathbf{B}^s} = \theta_{\mathbf{AB}}$. The composition profile in the rectifying section contains a saddle $\mathbf{B}^r$ in the middle part, and the stripping section profile has an upward terminus point $\mathbf{A}^s$. As there is no component **C** in the distillate, $\mathbf{A}^r\mathbf{B}^r = \mathbf{AB}$ and $\theta_{\mathbf{C}^r} = \alpha_{\mathbf{C}}$. For this separation type, the location of point $\mathbf{B}^r$ depends only on the feed composition and not on the reflux ratio.

With increasing reflux and reboil ratios, the distillate point $\mathbf{x}^d$ moves along the edge **AB**, while the bottom product $\mathbf{x}^b$—in the interior of the composition triangle. There exist critical values of the process parameters, at which either the distillate product achieves the **A** vertex or the bottom product achieves the **BC** edge of the simplex. On further increase of the reflux—reboil ratio, the separation type **AB**−**ABC** cannot be realized anymore and another separation (**A**−**ABC** or **AB**−**BC**) has to be considered.

For the separation type **A**−**ABC** (cf. Figure 10.A5), there is no common root of Underwood's equation, but, as the distillate consists of only pure component A (and therefore, the composition of distillate is a priori known), all process variables can easily be found. The composition profile of the rectifying section has an upward terminus point in a node $\mathbf{A}^r$, whereas the stripping section profile also has an upward terminus point $\mathbf{A}^s$. As there are no components **B** and **C** in the distillate, $\theta_{\mathbf{B}^r} = \alpha_{\mathbf{B}}$ and $\theta_{\mathbf{C}^r} = \alpha_{\mathbf{C}}$.

For this limiting separation type, further increase of the reflux ratio alone cannot improve the distillate quality, because pure **A** has already been gained. As a consequence, at constant distillate flow rate, neither product changes with varying reflux ratio, although the profile itself is affected by the reflux. With this in mind, determination of the minimal reflux ratio for fixed products becomes a feasible task targeting the appropriate state on the boundary line 2 (Figure 10.A1) that subdivides regions of the **A**−**ABC** and **AB**−**ABC** separation types. On the contrary, an increase of the distillate withdrawal causes recovery increase of **A** in the distillate, so that the bottom product may reach the edge **BC** and the separation type changes to **AB**−**BC**.

For the limiting separation type **AB**−**BC** (cf. Figure 10.A6), there is, again, no common root of Underwood's equation (cf. Table 10.A1), but, similar to the case **A**−**ABC**, distillate and bottom product compositions can be found directly from the component material balance equations. The composition profiles in both sections have a saddle type singular point in the middle, $\mathbf{B}^r$ and $\mathbf{B}^s$, respectively.

Two other separation types, **ABC**−**BC** and **ABC**−**C**, are not discussed here, because these two types are simply the inverted types **AB**−**ABC** and **A**−**ABC**, respectively. As illustrated for the separation type **A**−**ABC**, only limiting separation types allow for varying reflux ratio without changing the products quality.

Table 10.A2 Equations of Boundary Lines Subdividing the Parameter Space Reflux Ratio–Reboil Ratio into Different Separation Type Regions (cf. Figure 10.A1)

Number	Boundary Line Equation
1	$R_f = \dfrac{y_C^*(\mathbf{x}^f)}{x_C^f - y_C^*(\mathbf{x}^f)}$
2	$R_f = \dfrac{\theta_{AB}}{\alpha_A - \theta_{AB}}$
3	$R_b = \dfrac{x_A^f}{y_A^*(\mathbf{x}^f) - x_A^f}$
4	$R_b = \dfrac{\alpha_C}{\theta_{BC} - \alpha_C}$
5	$R_b = \dfrac{x_A^f}{x_B^f + x_C^f} \cdot (1 + R_f)$
6	$R_b = \dfrac{x_A^f + x_B^f}{x_C^f} \cdot (1 + R_f)$
7	$R_b = \dfrac{(1 + R_f) \cdot x_A^f}{\dfrac{\alpha_A - \theta_{AB}}{\alpha_A - \alpha_B} \cdot \left(1 + R_f - \dfrac{\alpha_B}{\theta_{AB}} \cdot R_f\right) - x_A^f}$
8	$R_f = \dfrac{R_b \cdot x_C^f}{\dfrac{\alpha_C - \theta_{BC}}{\alpha_C - \alpha_B} \cdot \left(\dfrac{\alpha_B}{\theta_{BC}} \cdot (1 + R_b) - R_b\right) - x_C^f} - 1$

Consequently, determination of the minimal reflux ratio for *fixed* product compositions is feasible only for such separations, with the aim to determine states on the boundary lines 2, 6, 7 and 8, respectively (cf. Figure 10.A1). In contrast, for the non-limiting separations, product compositions are affected by the reflux ratio, so that a feasible minimal reflux problem can only be formulated if the required products are not fully specified, e.g., through specification of concentrations for the selected key components in the products.

Generally, for nonideal mixtures, partial reflux operation can be analyzed by numerical methods only. This makes the solution of the minimal reflux problem significantly more complex. Nevertheless, it is worth noting that basic qualitative regularities illustrated here using Underwood's model remains similar: there still exist 'Z-simplices', however, now with nonlinear boundaries and, probably, with a more complex structure, if additional singular points—azeotropes—are present. For this reason, the idea of Z-simplices is used in the rectification body method [34] for the minimum reflux problem in a nonideal mixture, with the assumption that the boundaries can be linearly approximated with sufficient accuracy.

Optimization of Distillation Processes

11

José A. Caballero[1], Ignacio E. Grossmann[2]

Department of Chemical Engineering, University of Alicante, Alicante, Spain[1], Department of Chemical Engineering, Carnegie Mellon University, Pittsburgh, PA, USA[2]

CHAPTER OUTLINE

Distillation: Fundamentals and Principles. http://dx.doi.org/10.1016/B978-0-12-386547-2.00011-9

11.1 Introduction

Distillation is the most important operation for separation and purification in process industries, and this situation is unlikely to change in the near future. In order to get an idea of the importance of distillation, Humphrey [1] estimated that in the United States there are 40,000 distillation columns in operation that handle more than 90% of separations and purifications. The capital investment for these distillation systems is estimated to be 8 billion US$. Using data by Mix et al. [2], Soave and Feliu [3] estimated that distillation accounts about 3% of the total US energy consumption, which is equivalent to 2.87×10^{18} J (2.87 million TJ) per year, or to a power consumption of 91 GW, or 54 million tons of crude oil. Distillation columns use very large amounts of energy because of the evaporation steps that are involved. Typically more than half of the process heat distributed to a plant is dedicated to supply heat in the reboilers of distillation columns [4]. Unfortunately, this enormous amount of energy is introduced in the bottom of the column and approximately the same amount of energy is removed in the top, but at significantly lower temperature, which renders a very inefficient process, but also one of the most effective for the separation of mixtures.

The general separation problem was defined more than 40 years ago by Rudd and Watson [5] as the transformation of several source mixtures into several product mixtures. Interestingly, in 1983 Westerberg [6] claimed that this problem was essentially unsolved. Today we can say that this general problem has not been completely solved. We will focus on the more restricted, and much more studied, problem of separating a single source mixture into several products using only distillation columns. In general, to separate a complex mixture a sequence of columns is necessary.

Before going into the details of column sequencing it would be useful for the interested reader to review some optimization background and the corresponding methods. This information can be found in the Appendix of this chapter.

11.2 Optimization of a single distillation column

The optimization of distillation columns involves the selection of the number of trays, the feed location, and the operating conditions to minimize a performance function, usually the total investment and operating costs. Discrete decisions are related to the calculation of the number of trays and feed and product locations, and continuous decisions are related to the operation conditions. Due to the discrete—continuous nature of the problem and to the complex equations involved, it is common to use shortcut or aggregated models together with some rules of thumb that under some assumptions have proved to produce good results, at least in the first stages of design where a rigorous design is neither necessary nor convenient due to the large computational effort needed. Taking this fact into account, we first show an overview of the most common shortcut methods and then we will present the alternatives for rigorous optimization of a single column.

11.2.1 Shortcut methods

11.2.1.1 Fenske–Underwood–Gilliland method

The most used and successful method for distillation design is the Fenske–Underwood–Gilliland (FUG) method [7–9]. The FUG method assumes a constant molar overflow and constant relative volatilities in all the trays of the distillation column. Although these conditions seem too restrictive, they can be applied to a large class of mixtures (i.e. hydrocarbon separations, alcohols, etc.). This method considers two extreme ideal situations. The first is when the distillation column operates at total reflux (no feed is entering or exiting from the column), which allows calculating the minimum number of trays for a given separation of two key components. The second situation is when the column operates at pinch conditions (an infinite number of trays), which allows calculating the minimum reflux. The optimal situation lies at some point in between these two extreme cases.

If we assume a total reflux (see Figure 11.1), the equilibrium equations for the key components at the reboiler (R) are:

$$(y_{LK})_R = K_{LK}(x_{LK})_R$$
$$(y_{HK})_R = K_{HK}(x_{HK})_R$$
(11.1)

where the K-factor represents the distribution of the light key (LK) or heavy key (HK) component between the vapor phase, with composition y, and the liquid phase with composition x.

Dividing those equations results in Eqn (11.2):

$$\left(\frac{y_{LK}}{y_{HK}}\right)_R = \alpha_R \left(\frac{x_{LK}}{x_{HK}}\right)_R$$
(11.2)

where α is the relative volatility and the ratio of K-factors: $\alpha_R = K_{LK}/K_{HK}$

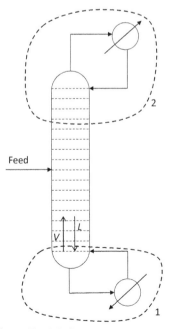

FIGURE 11.1 Column Operating at Total Reflux

In the total reflux conditions, the feed, distillate, and bottoms are all zero. An overall mass balance in the reboiler yields:

$$V = L \tag{11.3}$$

A mass balance including the reboiler and a tray N (envelope 1 in Figure 11.1) gives the following equations:

$$\left.\begin{array}{l} V(y_{LK})_R = L(x_{LK})_N \\ V(y_{HK})_R = L(x_{HK})_N \end{array}\right\} \Rightarrow \begin{array}{l} (y_{LK})_R = (x_{LK})_N \\ (y_{HK})_R = (x_{HK})_N \end{array} \tag{11.4}$$

The liquid composition of the reboiler stage, which is a given specification, is correlated with the liquid composition of the previous stage N, which can be consequently calculated.

Dividing equations in (11.4) and substituting in Eqn (11.2), we get:

$$\left(\frac{x_{LK}}{x_{HK}}\right)_N = \alpha_R \left(\frac{x_{LK}}{x_{HK}}\right)_R \tag{11.5}$$

Proceeding backwards from tray N to tray $N-1$ to $N-2$, and until we reach the composition of the distillate stage, which is known, we get:

$$\left(\frac{x_{LK}}{x_{HK}}\right)_D = \alpha_1 \cdot \alpha_2 \cdot \alpha_3 \cdots \cdots \alpha_{N-1} \cdot \alpha_N \cdot \alpha_R \left(\frac{x_{LK}}{x_{HK}}\right)_R \tag{11.6}$$

Extending the procedure to all the stages in the distillation column and assuming an average relative volatility for all the stages, we finally get the well-known Fenske equation [7] that relates the minimum number of trays with the composition of key components in distillate and bottoms:

$$\left(\frac{x_{LK}}{x_{HK}}\right)_D = \alpha^{N,min} \left(\frac{x_{LK}}{x_{HK}}\right)_R \tag{11.7}$$

The other extreme situation is when the column operates at minimum reflux conditions (an infinite number of trays). In this situation the concentration profiles reach a "pinch point" in which the concentrations do not change from one stage to another:

$$\begin{array}{l} x_{j-1} = x_j = x_{j+1} \\ y_{j-1} = y_j = y_{j+1} \end{array} \tag{11.8}$$

A mass balance around the envelope 2 (Figure 11.1) in the rectifying section for all of the i components yields:

$$V_{R,min}y_{j+1,i} = L_{R,min}x_{j,i} + Dx_{D,i} \tag{11.9}$$

where $V_{R,min}$ and $L_{R,min}$ are the minimum flow rates of vapor and liquid in the rectifying section (subscript R) and D represents the distillate flow rate.

The equilibrium conditions for component i in tray $(j+1)$ are given by:

$$y_{j+1,i} = K_i x_{j+1,i} \tag{11.10}$$

and because of the pinch point conditions (see Eqn (11.8)), this is equivalent to:

$$y_{j+1,i} = K_i x_{j,i} \tag{11.11}$$

The equilibrium constant of component i can be written in terms of relative volatility referred to component k:

$$K_i = \alpha_{i,k} K_k \tag{11.12}$$

Substituting Eqns (11.11) and (11.12) in (11.9) and taking the sum over all components:

$$\sum V_{R,min} y_{j+1,i} = \frac{\sum D x_{D,i}}{1 - \frac{L_{R,min}}{V_{R,min} \alpha_{i,k} K_k}} \tag{11.13}$$

Equation (11.13) is usually rewritten as follows:

$$V_{R,min} = D \sum \frac{\alpha_{i,k} x_{D,i}}{\alpha_{i,k} - \frac{L_{R,min}}{V_{R,min} K_k}} \tag{11.14}$$

In the same way for the stripping section (subscript S), we reach:

$$-V_{S,min} = B \sum \frac{\alpha_{i,k} x_{B,i}}{\alpha_{i,k} - \frac{L_{S,min}}{V_{S,min} K_k}} \tag{11.15}$$

In conditions of minimum reflux, Underwood proved that:

$$\phi_{R,R} = \frac{L_{R,min}}{V_{R,min} K_k} = \frac{L_{S,min}}{V_{S,min} K_k} = \phi_{R,S} \tag{11.16}$$

Equations (11.15) and (11.16) allow solving the problem. ϕ_R is the so-called Underwood root. However, because usually the feed to the system is completely specified, it is convenient to substitute one of those equations by a linear combination of both as follows: Subtracting Eqns (11.15) and (11.16) and from an overall mass balance:

$$V_{R,min} - V_{S,min} = \sum \frac{\alpha_{i,k}(D x_{D,i} + B x_{B,i})}{\alpha_{i,k} - \phi_R} = \sum \frac{\alpha_{i,k} F z_{F,i}}{\alpha_{i,k} - \phi_R} = F(1 - q_{LS}) \tag{11.17}$$

where q_{LS} is the liquid fraction in the feed stream. Values of q_{LS} greater than 1 indicate a subcooled feed stream (F). Negative values indicate a superheated vapor.

Therefore, given the feed conditions, it is possible to use Eqn (11.17) to calculate the Underwood roots φ_R. Equation (11.17) has N roots, but only $N-1$ correspond to values of ϕ_R with physical meaning and is bounded by the relative volatilities:

$$0 < \phi_{R,1} < \alpha_N < \phi_{R,2} < \alpha_{N-1} < \phi_{R,3} < \cdots < \alpha_2 < \phi_{R,N} < \alpha_1 \tag{11.18}$$

Of those $N-1$ Underwood roots, only those whose value is between the relative volatilities of the key components are active. Therefore, if the recovery of the key components is specified (i.e. >95%) assuming that all the components lighter than the light key are obtained in the distillate and all those heavier than the heavy key are obtained in the bottoms stream, it is possible to use Eqn (11.17) to calculate the active Underwood roots and then Eqn (11.15) or (11.16) to determine the distribution of the intermediate non-key components. If there are S intermediate non-key components, we have $S+1$ active Underwood roots and from (11.15) we can write $S+1$ equations where the unknowns are the S molar fractions of the distributed components plus the minimum vapor flow rate.

The minimum number of theoretical trays (N_{min}) and the minimum reflux ratio (RR_{min}) are two extreme operating conditions; the actual operation must be some place in between. The optimum is usually located in values between 1.1 and 1.5 times the minimum reflux. To optimize the column, a general shortcut method for determining the number of stages required for a multicomponent distillation at finite reflux ratios would be very useful. Unfortunately, such a method has not been developed. However, Gilliland [8] used empirical data to correlate the number of stages at finite reflux ratios with the number of stages and to the minimum reflux ratio. He presented his results in a graphical correlation using the following two parameters:

$$Y = \frac{N - N_{min}}{N + 1} \quad X = \frac{RR - RR_{min}}{RR + 1} \tag{11.19}$$

Molokanov [10] fit the Gilliland correlation to the following equation:

$$Y = 1 - \exp\left[\frac{1 + 54.4X}{11 + 117.2X}\left(\frac{X - 1}{X^{0.5}}\right)\right] \tag{11.20}$$

Implicit in the application of the Gilliland correlation is that the theoretical stages must be optimally distributed between the rectifying and stripping sections. Again there is not an equation based on first principles that allows determining such a distribution, but according to Seader and Henley [11] a reasonably good approximation is given by the Kirkbride equation [12]:

$$\frac{N_R}{N_S} = \left[\frac{z_{HK}}{z_{LK}}\left(\frac{x_{LK,B}}{x_{HK,D}}\right)^2 \frac{B}{D}\right]^{0.206} \tag{11.21}$$

Application of the Kirkbride equation requires knowledge of the distillate and bottoms composition at the specified finite reflux ratio. Seader and Henley [12] suggest that the distribution of components at finite reflux is close to that estimated by the Fenske equation at total reflux conditions.

Due to the wide application of the FUG method it has been modified to deal with multiple feeds, side draws, or complex column configurations [13–21]. Interestingly, the Underwood method can be extended to azeotropic systems [22]. The idea consists of treating azeotropes as pseudocomponents. An N component system with A azeotropes is treated as an enlarged $(N + A)$ component system. This enlarged system is divided into compartments, where each compartment behaves like a non-azeotropic distillation region formed by the singular points that appears in it.

11.2.2 Group methods

Another approach that deserves especial attention is based on group methods. Group methods (GMs) basically use approximate calculations to relate the outlet stream properties to the inlet stream specifications and number of equilibrium trays. These approximation procedures are called group methods because they provide only an overall treatment of the stages in the cascade without considering detailed changes in the temperature and composition of individual stages. However, they are much easier to solve because they involve fewer variables and constraints. They can be used to represent a cascade of trays in many countercurrent operations like absorption, stripping, distillation, leaching, or extraction [23]. Although due to their initial limitations GMs were used mainly in absorption, recent developments have reached excellent results in distillation [23].

Group methods were originally devised for simple hand calculations that were performed in an iterative manner. However, their equation-based nature allows for easy incorporation in a mathematical programming model. The specifications for the entering vapor V_{N+1} and the entering liquid L_0 are the inputs to the model. The method evaluates the properties of the outputs ($V_1; L_N$) in terms of the inputs and the characteristics of the cascade. In the following analysis we assume adiabatic operation and a known pressure drop in the cascade. The following presentation follows the lines of Kamath et al. [23].

The fundamental equations for the group contribution methods are the mass and energy balance in the cascade:

$$V_{N+1}y_{N+1,i} + L_0x_{0,i} = V_1y_{1,i} + L_Nx_{N,i} \quad i \in C$$
$$V_{N+1}h_{V,N+1} + L_0h_{L,0} = V_1h_{V,1} + L_Nh_{L,N} \quad (11.22)$$

where C refers to the set of components.

The performance equation of the cascade, derived initially by Kremser in 1930 [24], is:

$$V_1y_{1,i} = V_{N+1}y_{N+1,i}\phi_{A,i} + L_0x_{0,i}(1 - \phi_{S,i}) \quad i \in C \quad (11.23)$$

where $\phi_{A,i}, \phi_{S,i}$ denote the recovery factors for absorption and stripping sections.

There are $2(|C|+1)$ variables in the model given by (11.22) and (11.23). We have $(2|C|+1)$ independent equations—C mass balances, C performance equations, and the energy balance—and therefore we have one degree of freedom.

The recovery factors in Eqn (11.23) are given by:

$$\phi_{A,i} = \frac{A_{e,i} - 1}{A_{e,i}^{N+1} - 1}; \quad \phi_{S,i} = \frac{S_{e,i} - 1}{S_{e,i}^{N+1} - 1} \quad i \in C \quad (11.24)$$

where $A_{e,i}, S_{e,i}$ are the effective absorption and stripping factors, and represent average values for all the trays contained in the cascade. Edmister [25] proposed the following average scheme:

$$A_{e,i} = \left[A_{N,i}(A_{1,i} + 1) + 0.25\right]^{0.5} - 0.5$$
$$S_{e,i} = \left[S_{1,i}(S_{N,i} + 1) + 0.25\right]^{0.5} - 0.5 \quad (11.25)$$

Equation (11.25) uses factors at the top and bottom of the cascade. These factors are in turn calculated using the following expressions:

$$
\left.\begin{array}{ll}
A_{1,i} = \dfrac{L_1}{K_{1,i}V_1}; & A_{N,i} = \dfrac{L_N}{K_{N,i}V_N} \\[3mm]
S_{1,i} = \dfrac{1}{A_{1,i}}; & S_{N,i} = \dfrac{1}{A_{N,i}}
\end{array}\right\} \quad i \in C \qquad (11.26)
$$

Equation (11.26) introduces two new variables L_1 and V_N, that do not appear previously in the model. Therefore, the model has three degrees of freedom. Different approaches have been used with GMs that differ on how these three degrees of freedom are satisfied.

Kresmer [24] proposed the following three approximations:

$$
\begin{aligned}
L_1 &= L_0 \\[2mm]
V_N &= V_{N+1} \\[2mm]
T_N &= \frac{T_0 + T_{N+1}}{2}
\end{aligned} \qquad (11.27)
$$

Kremser did not included the energy balance and instead used the following approximation:

$$
T_1 = \frac{T_0 + T_{N+1}}{2} \qquad (11.28)
$$

Kremser assumed that there was not too much change either in the liquid in the first stage or in the vapor in the last stage. Besides, he used identical approximations for temperatures (T) of the vapor and liquid streams exiting the cascade, and they are both considered to be equal to the arithmetic mean of the temperature of entering vapor and liquid streams. Although these seem crude approximations, it is necessary to take into account that Kremser developed the model for the recovery of gasoline from natural gas, where only a small fraction is absorbed.

Edmister [26], for the case of distillation systems, proposed a different set of approximations to satisfy the three degrees of freedom. However, he proposed different equations depending on whether the cascade is an absorber or a stripper. For the absorber they are:

$$
V_N = V_{N+1}\left(\frac{V_1}{V_{N+1}}\right)^{1/N} \qquad (11.29)
$$

$$
\frac{T_N - T_1}{T_N - T_0} = \frac{V_{N+1} - V_2}{V_{N+1} - V_1} \qquad (11.30)
$$

$$
L_1 = L_0 + V_2 - V_1 \qquad (11.31)
$$

Equation (11.29) gives an approximation for V_N assuming that the vapor contraction per stage is the same percentage of the vapor flow to that stage. Equation (11.30) assumes that the temperature change of the liquid is proportional to volume of gas absorbed. Finally, Eqn (11.31) is a rigorous mole balance for L_1, but it contains the new variable V_2 that can be approximated by an analogous assumption to Eqn (11.29):

$$V_2 = V_1 \left(\frac{V_{N+1}}{V_1}\right)^{1/N} \tag{11.32}$$

For the stripping cascade, the equations are similar to those for the absorber but the dependencies are in terms of the molar flow of liquid instead of vapor:

$$L_1 = L_0 \left(\frac{L_N}{L_0}\right)^{1/N} \tag{11.33}$$

$$\frac{T_0 - T_1}{T_0 - T_N} = \frac{L_0 - L_1}{L_0 - L_N} \tag{11.34}$$

$$V_N = V_{N+1} + L_{N-1} - L_N \tag{11.35}$$

$$L_{N-1} = L_N \left(\frac{L_0}{L_N}\right)^{1/N} \tag{11.36}$$

The major limitation of the Edmister approach is clearly that in complex cascades (i.e. multiple feeds and or side streams), for some of the sections it is not clear whether such sections behave like a stripper or like an absorber. In order to overcome those limitations, Kamath et al. [23] proposed an alternate set of specifications for the degrees of freedom. The first two equations are based on the fact that since the outlet streams are coming out of the first and last trays of the cascade, they must be under vapor—liquid equilibrium. Hence, for the outlet vapor they imposed that it should be at dew point conditions:

$$\sum_{i \in C} \frac{y_{1,i}}{K_{1,i}} = 1 \tag{11.37}$$

In addition, the outlet liquid must be saturated liquid:

$$\sum_{i \in C} K_{N,i} x_{N,i} = 1 \tag{11.38}$$

Note that these two equations are not approximations and try to capture the physical behavior of the system. To satisfy the third degree of freedom, Kamath and co-workers proposed the following equation:

$$L_1 - L_N = V_1 - V_N \tag{11.39}$$

Equation (11.39) is based on an approximation of mole balance with an assumption that the decrease in vapor at the bottom is approximately equal to the increase in liquid at the top and vice versa.

11.2.3 Aggregated models

Caballero and Grossmann [27], using as base the work on heat and mass transfer networks proposed by Bagajewicz and Manousiouthakis [28], proposed an aggregated model based on mass balances and equilibrium feasibility, expressed in terms of flows, inlet concentrations, and recoveries. The energy balance can then be decoupled from the mass balance and the utilities can be calculated for each separation task. The main assumptions for this model are as follows.

Each single column is divided into two sections (two mass exchange zones). In each of the sections the molar flow rate of vapor and liquid are assumed to be constant.

The pinch point can be located only in the extreme points of the sections. If this is not the case, this behavior must be "captured" a priori in order to correctly implement the model. Feasibility of mass exchange is established when both ends of a mass exchanger's operating line lie below the equilibrium curve (and above if the equilibrium curve is based on the heavier component). Because the liquid curve is concave, thermodynamic feasibility of mass exchange can be verified by examining the end points at each stream. In a multi-component mixture, the feasibility constraints will depend on the separation that the column performs. For example, in a column with three components—say A, B, and C—in which we want to perform the separation A/BC (A is the most volatile and C the least), the following constraints must hold at the ends of the streams:

$$y_A \leq K_A x_A; \quad y_i \geq K_i x_i \quad i = B, C \tag{11.40}$$

The model for a column is as described next. It is assumed that the pinch point can be in the extreme points of each section. Therefore for a conventional distillation column there are four pinch point candidates: $S = [s \in (\text{top, mt, mb, bot}) | \text{pinch point candidates}]$ where mt is the bottom part of the top section and mb is the top of the bottom section.

Overall mass balances for each section:

$$\left. \begin{array}{l} V_{in_{i,mt}} + L_{in_{i,top}} = V_{in_{i,top}} + L_{in_{i,mt}} \\ V_{in_{i,bot}} + L_{in_{i,mb}} = V_{in_{i,mb}} + L_{in_{i,bot}} \end{array} \right\} \quad i \in \text{COM} \tag{11.41}$$

$$\left. \begin{array}{l} V_s = \sum_i V_{in_{i,s}} \\ L_s = \sum_i L_{in_{i,s}} \end{array} \right\} \quad i \in COM, \; s \in S \tag{11.42}$$

$$V_{\text{top}} = V_{\text{mt}}; \quad V_{\text{mb}} = V_{\text{bot}} \quad L_{\text{top}} = L_{\text{mt}}; \quad L_{\text{mb}} = L_{\text{bot}} \tag{11.43}$$

where V_{in}, L_{in} make reference to the flow rate of the individual components in the vapor and liquid, respectively; and L, V are the overall liquid and vapor molar flow rates, respectively.

Overall mass balance:

$$F_i = \text{pt}_i + \text{pb}_i \quad i \in \text{COM} \tag{11.44}$$

where F is the individual flow rate of the component i in the feed, and pt and pb are the individual flow rates of the top and bottom products, respectively.

Mass and energy balances in the feed section:

It is assumed that the feed is introduced at its bubble point and it mixes with the liquid stream:

$$F_i + L_{in_{i,mt}} + V_{in_{i,mb}} = L_{in_{i,mb}} + V_{in_{i,mt}} \quad i \in COM \tag{11.45}$$

$$\left. \begin{array}{l} \sum_i F_i h_{L,i} + \sum_i L_{in_{i,mt}} h_{L,i,mt} + \sum_i V_{in_{i,mb}} h_{V,i,mb} = \\ \sum_i L_{in_{i,mb}} h_{L,i,mb} + \sum_i V_{in_{i,mt}} h_{V,i,mt} \end{array} \right\} \quad i \in COM \tag{11.46}$$

where h_V, h_L correspond to the specific enthalpies of the vapor and liquid, respectively.

Mass balances in condenser and reboiler that are treated as splitters:

$$\left. \begin{array}{l} V_{in_{i,top}} = pt_i + L_{in_{i,top}} \\ pt_i = \zeta_1 V_{in_{i,top}} \\ L_{in_{i,top}} = (1 - \zeta_2) V_{in_{i,top}} \end{array} \right\} \quad i \in COM \tag{11.47}$$

$$\left. \begin{array}{l} L_{in_{i,bot}} = pb_i + V_{in_{i,bot}} \\ pb_i = \zeta_2 L_{in_{i,bot}} \\ V_{in_{i,bot}} = (1 - \zeta_2) L_{in_{i,bot}} \end{array} \right\} \quad i \in COM \tag{11.48}$$

where ζ_1, ζ_2 are split fractions to be determined and COM is the set of components.

Equilibrium equations:

The equations are not restricted to any particular equilibrium model. In general:

$$K_{i,S} = f(x_{1s}, x_{2s}, \ldots x_{ns}, , p, T_s) \quad i \in COM, \quad s \in S \tag{11.49}$$

where K is the equilibrium constant, $x_{j,s}$ ($j = 1, 2 \ldots n$) is the molar fraction of the component j in the liquid fraction at position s in the column, p is the pressure in the column, and T is the temperature in section s of the column.

It is assumed that a total condenser is used, and that the bottom product is extracted from the reboiler as liquid. Therefore, all products are saturated liquids. Of course these equations can be modified to deal with vapor products:

$$\left. \begin{array}{l} \sum_i pt_i K_{i,con} = \sum_i pt_i \\ \sum_i pb_i K_{i,reb} = \sum_i pb_i \end{array} \right\} \quad i \in COM \tag{11.50}$$

where reb and con refer to the reboiler and condenser, respectively.

The temperature increases from the top to the bottom of the column:

$$T_{con} \leq T_{top} \leq T_{mt} \leq T_{mb} \leq T_{bot} \leq T_{reb} \tag{11.51}$$

The feasibility pinch constraints can be generalized as follows. These constraints have two functions. First, they represent the pinch constraints, and second they distribute the non-key components:

$$\frac{V_{\text{in}_{i,s}}}{V_s} \leq K_{i,s}\frac{L_{\text{in}_{i,s}}}{L_s} \quad i \in \text{COM}, \quad s \in S \tag{11.52}$$

if the product i is mostly present in the top of the column; or:

$$\frac{V_{\text{in}_{j,s}}}{V_s} \geq K_{j,s}\frac{L_{\text{in}_{j,s}}}{L_s} \quad j \in \text{COM}, \quad s \in S \tag{11.53}$$

if the product j is mostly present in the bottom of the column.

A recovery factor (ϕ_i) can be fixed for each component:

$$F_i\phi_i \leq \text{pt}_i \quad or \quad F_i\phi_i \leq \text{pb}_i \quad i \in \text{COM} \tag{11.54}$$

depending on whether the product is obtained as a top or bottom product.

In the original work, the authors used the vapor flow rate as an objective function. Because the column has two sections, they minimize the maximum of those two flows in the column:

$$\text{Min Max}\left(V_{\text{top}}, V_{\text{bot}}\right) \tag{11.55}$$

Defining a new variable α, it is possible to transform the min−max problem to a regular minimization problem as follows:

$$\begin{aligned} &\text{Min } \alpha \\ &s.t. \; \alpha \geq V_{\text{top}} \\ &\quad\;\; \alpha \geq V_{\text{bot}} \end{aligned} \tag{11.56}$$

Note that the previous model given by Eqns (11.43)−(11.58) only includes mass balances, and an energy balance in the feed section.

It is worth noting that Eqns (11.43)−(11.56) represent the aggregation of the equations of a tray- by-tray model. In particular, the mass balance Eqns (11.43)−(11.47) and (11.49), (11.50) represent a linear combination of component mass balance in each tray with the assumption of equimolar flow. The enthalpy balances are relaxed because they are removed, except for the feed tray in Eqn (11.48). Finally the equilibrium equations are relaxed by two inequalities (11.54) and (11.55), which are imposed at the extremes of each section. Thus, if the same thermodynamic model is used, the aggregated model will yield a lower bound in the vapor flows with respect to a rigorous tray-by-tray model with equimolar flows. Furthermore, if the heat of vaporization decreases with relative volatility, the model also predicts a lower bound of the utilities (in this case energy balances are added to the reboiler and condenser). This is due to the relaxation of the equilibrium equations, which in turn will overpredict recoveries of lighter-than-key components.

Introducing heat balances in the reboiler and condenser allows calculating heat duties and temperatures that are useful for heat integration or to use another objective function including specific costs for utilities.

One of the keys to the success of the FUG approach, and to a lesser extent the GM and aggregated models, is that it is possible to include all the equations in a mathematical programming model and determine the optimal operating conditions and investment costs. This approach is commonly used either for the preliminary design of a single column or for determining the best or more promising sequences of distillation columns in the separation of multicomponent mixtures, as will be shown in next sections.

Some other methods that had acquired importance are described in the remainder of this section.

The *Boundary Value Method (BVM)*, proposed by Levy et al. [29] and extended with different works over the last 25 years [30−35], can be used to determine the minimum reflux ratio and feasible design parameters for a column separating a ternary homogeneous mixture. This BVM requires fully specified product compositions, the feed composition and the thermal condition of the feed. Once these specifications have been made, only one degree of freedom remains between the reflux and boil-up ratios. Specifying the reflux (or boil-up) ratio, the rectifying and stripping profiles can be calculated starting from the fully specified products. If these two profiles intersect, the separation is feasible. The number of trays, composition profiles, etc. are then obtained. The optimal operating conditions can be obtained by iterative calculations. Julka and Doherty [33] extended the BVM to multicomponent mixtures. In this case, a split is feasible if two stages that lie on the composition profile of two different sections have the same liquid composition.

The *Rectification Body Method (RBM)*, proposed by Bausa et al. [36], can be used for the determination of minimum energy requirements of a specified split. For a given product, branches of the pinch point curves can be found. Rectification bodies can be constructed by joining points on the branches with straight lines. For either section of a column, a rectification body can be constructed. The intersection of the rectification bodies of two sections of a column indicates its feasibility. The RBM can be used to calculate the minimum reflux ratio and minimum energy cost and to test the feasibility of a split. Because faces on rectification bodies are linearly approximated by joining branches of pinch point curves using straight lines, this method cannot guarantee accurate results. No information about column design (number of stages and operating reflux ratio) is obtained. The calculation of pinch point curves is, furthermore, computationally intensive [22].

The *Reversible Distillation Model (RDM)* developed by Koehler et al. [37] assumes that heat can be transferred to and from a column at zero temperature difference and that no contact of non-equilibrium liquid and vapor streams is allowed. Reversible distillation path equations are derived by rearranging the column material balances as well as the equilibrium relationships for the most and least volatile components. The solution of this reduced set of equations requires that the flow rates of the most and least volatile components be specified at the feed plate. Numerical methods based on any reversible distillation model require knowledge of the products that can be achieved by the distillation before starting the computations for finding the minimum reflux.

The *Driving Force Method* proposed by Gani and Bek-Pedersen [38] is a simple graphical method based on driving force for separation. Here the separation driving force is defined as $F_{Di} = |y_i - x_i|$, where the subscript $i = $ LK denotes the light key component. Gani and Bek-Pedersen proved that the minimum or near minimum energy requirements generally correspond to a maximum in the driving force. The proposed method is quite simple and applies to two product distillations with N stages.

The *Shortest Stripping Line* approach was developed by Lucia and Taylor [39] and extended by Lucia et al. [40,41]. The authors showed that exact separation boundaries for ternary mixtures are given by the set of locally longest residue curves (or distillation lines at infinite reflux) from any given unstable node to any reachable stable node. They also showed that the longest residue curve is related with the highest energy consumption for a given separation. Then the shortest curve should produce the minimum energy required for the same separation. The concept of shortest stripping lines can be extended to find minimum energy requirements in reactive distillation, hybrid separation processes, and reaction/separation/recycle systems regardless of the underlying thermodynamic models.

Although some of previous methods have been automated, not all of them can be easily included within a deterministic optimization algorithm. But in this context they are valuable tools for getting precise initial values and reliable bounds on the main variables for the rigorous design of distillation columns.

11.2.4 Rigorous tray-by-tray optimization models

As commented above, the economic optimization of a distillation column involves the selection of the number of trays and feed location as well as the operating conditions to minimize the total investment and operating cost. Continuous decisions are related to the operational conditions and energy involved in the separation, while discrete decisions are related to the total number of trays, the tray positions of each feed, and product streams. A major challenge is to perform the optimization using tray-by-tray models that assume phase equilibrium.

11.2.4.1 Mixed Integer Nonlinear Programming model

The simplest type of distillation design problem is the one where there is a fixed number of trays, and the goal is to select the optimal feed tray location [42]. Figure 11.2 shows that a superstructure that can be postulated is one where the feed is simply split into as many as there are trays, excluding condenser and reboiler. Of course the candidate trays can be constrained to a given set of trays according to the knowledge that the designer has about the physical behavior of the column. This is in essence the superstructure that was proposed by Sargent and Gaminibandara in 1976 [43]. The model can be easily written as a Mixed-Integer Nonlinear Programming (MINLP) model by considering all the mass and enthalpy balances, phase equilibrium equations, and that molar fraction

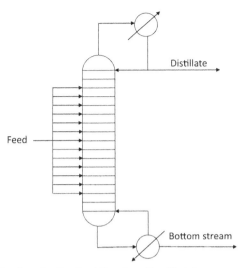

FIGURE 11.2 Superstructure for the Feed Tray Location Model

summation equals one in each phase (MESH equations). In addition, the following mixed-integer constraints must be added:

Let $z_i : i \in$ FLOC denote the binary variable associated with the selection of i as the feed tray, FLOC denote the set of trays in which the feed can enter the column, and $F_i \cdot i \in$ FLOC denote the amount of feed entering tray i:

$$
\begin{aligned}
&\sum_{i \in \text{FLOC}} F_i = F \\
&\sum_{i \in \text{FLOC}} z_i = 1 \\
&F_i - F z_i \leq 0 \quad i \in \text{FLOC} \\
&z_i \in \{0, 1\}; F_i \geq 0 \quad i \in \text{FLOC}
\end{aligned}
\tag{11.57}
$$

The second and third constraints in (11.57) assure that the feed is entering in a single tray; this follows from the fact that only one tray can be selected (second constraint in (11.57)) and that if the tray $i \in$ FLOC is selected as the feed tray, the amount of feed entering other locations is zero because if $z_j = 0, j \neq i$ then the third equation in (11.57) forces the flow $F_j \leq 0 \, j \neq i$.

An interesting property of the MINLP model for a fixed number of trays is that computational experience has shown that this problem is frequently solved as a relaxed Nonlinear Programming (NLP) model. The physical explanation is that one can expect the optimal distribution to be one where the feed is all directed into a single tray where the composition matches closely the composition of the feed [42,44–46].

Besides the MESH equations and the constraints in (11.57), specification on purity, whether recovery of some components is distillate or bottoms, etc., must be added to completely define the MINLP model.

When the objective is to optimize not only the feed tray position but also the number of trays, the complexity of the model greatly increases, as shown in the model by Viswanathan and Grossmann in 1993 [45]. These authors proposed a superstructure that involves a variable reflux location as depicted in Figure 11.3. The basic idea was to consider a fixed feed tray with an upper bound of trays specified above and below the feed. The reflux is then returned to all trays above, and the reboil returned to all trays below the feed. Basically, this model determines which are the optimal tray locations for the reflux and reboil streams. The model relies on the MESH equations in each tray; specification on recoveries, purities, etc. The variable reflux/reboil return can be modeled as described below.

Defining the following sets:

$$T = \{t \mid is\ a\ tray\ in\ the\ column\}$$
$$RF_T = \{t \mid Candidate\ tray\ for\ reflux\ return\}$$
$$RB_T = \{t \mid Candidate\ tray\ for\ reboil\ return\}$$

Let Ld, Vr be the reflux and reboil flow rate returned to the column, respectively; and let r_t $t \in RF_t$; b_t $t \in RB_t$ be binaries that take the value 1 if the reflux/reboil is returned to the tray t:

$$Ld = \sum_{t \in RF_t} ref_t$$

$$ref_t \leq Ld^{Up} r_t \quad t \in RF_t$$

$$Vr = \sum_{t \in RB_t} reb_t$$

$$reb_t \leq Vr^{Up} b_t \quad t \in RB_t \tag{11.58}$$

$$Ld \geq 0, Vr \geq 0$$

$$ref_t \geq 0, r_t \in \{0, 1\} \quad \forall t \in RF_t$$

$$reb_t \geq 0, b_t \in \{0, 1\} \quad \forall t \in RB_t$$

Viswanathan and Grossmann [45] also extended the model to include more than a single feed. The model is a combination of the two presented above; the different feeds are able to go to any tray in the column (or a subset of trays previously selected) and the reflux and reboil streams are postulated to return to a subset of different trays.

Although in principle this model is suitable for optimizing the feed tray location and the number of trays, a major difficulty is related to the non-existing trays. In these trays, there is a zero liquid flow (rectifying section) or a zero vapor flow (stripping section), which can produce numerical problems due to the convergence of equilibrium equations with a zero value in the flow of one of the phases. In other words, the vapor–liquid equilibrium equations must be satisfied in trays where no mass transfer takes place.

Despite the increase of the computational time of the model and convergence problems, the model of Viswanathan and Grossmann has been successfully applied by

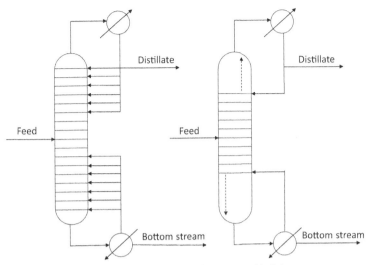

FIGURE 11.3 Superstructure of Viswanathan and Grossmann Model

different research groups. For example, Ciric and Gu [47] used the MINLP approach for the synthesis of ethylene glycol via ethylene oxide in a kinetic controlled reactive distillation column. Bauer and Stichlmair [48] applied the MINLP approach to the synthesis of sequences of azeotropic columns. Dünnebier and Pantelides [49] used the model to generate sequences of thermally coupled distillation columns.

The superstructure presented by Viswanathan and Grossmann (see Figure 11.3) is not the only possible alternative for the simultaneous determination of the feed tray position and the total number of trays. Barttfeld et al. [44] studied the impact of different representations and models that can be used for the optimization of a single distillation column. Figure 11.4 shows three representations that are different to the original by Viswanathan and Grossmann that achieve the same objective. First, in Figure 11.4(a), a condenser and a reboiler are placed in all candidate trays for exchanging energy. This means that a variable reflux/reboil stream is considered by moving the condenser/reboiler. Otherwise, in the representation of the variable reflux location in Figure 11.3, the condenser and reboiler are fixed equipment. These two alternatives are the same if one piece of fixed equipment is considered at each column end. However, when variable heat exchange locations are modeled as a part of the optimization procedure, some differences arise. In one case the problem consists of finding the optimal location for the energy exchanged, while in the other the optimal location for a secondary feed stream (reflux/reboil) is considered. The variable heat exchange has an important advantage; the energy can be exchanged at intermediate trays temperatures, possibly leading to more energy-efficient designs [50]. The results have shown that the most energy-efficient MINLP representation involves the variable reboiler and feed tray location shown in Figure 11.4(b).

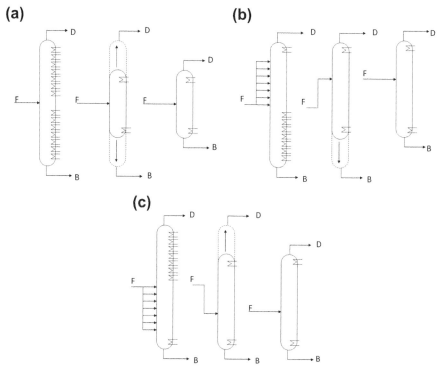

FIGURE 11.4 MINLP Distillation Column Representations

(a) Variable reboiler and condenser location. (b) Variable reboiler location. (c) Variable condenser location.

11.2.5 Generalized disjunctive programming models

Yeomans and Grossmann [51] proposed a Generalized Disjunctive Programming (GDP) model that overcomes the numerical difficulties of the MINLP models. The basic idea consist of dividing the trays in the distillation column in permanent trays (they exist in all the cases) and conditional trays (they can exist or not, depending on the optimal solution). For each existing tray, the mass and energy transfers are taking into account and modeled using the MESH equations: component mass balances, tray energy balance, equilibrium equations, and the summation of liquid and vapor mole fractions equal to 1. For a non-existing or inactive tray, the model considers a simple bypass of liquid and vapor streams without mass or energy transfer, which gives rise to trivial mass and energy balance equations (inlet and outlet flows and enthalpies are equal for both liquid and vapor phases). Because the MESH equations include the solution for trivial mass and energy balances, the only difference between existing and non-existing trays is the application of the equilibrium equations. As for the permanent trays, all the equations for an existing tray apply. Figure 11.5 shows a superstructure for this approach.

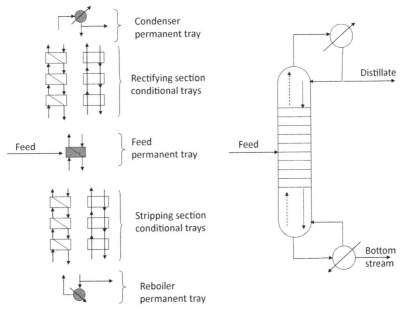

FIGURE 11.5 Superstructure for GDP Optimization

Conceptually the GDP model for the design of a single distillation column can be written as follows:

$$min : TAC = Total\ Annual\ Cost$$
$$s.t.$$

$$MESH\ equations\ for\ permanent\ trays$$
$$Mass/Energy\ balances\ for\ conditional\ trays \qquad (11.59)$$
$$purity, recovery...contraints$$

$$\begin{bmatrix} Y_t \\ Equilibrium\ equation \end{bmatrix} \vee \begin{bmatrix} \neg Y_t \\ Bypass\ equations \\ (Input - Output\ relationships) \end{bmatrix} \quad t \in CondTrays$$

$$\Omega(Y) = True.$$

The logical relationships in Eqn (11.59) are necessary to avoid the degeneracy due to equivalent solutions, i.e. in a given distillation section two solutions with the same number of trays but different distribution. This problem can be solved by forcing all existing trays to be consecutive. For example, assuming that the trays are numbered from the top to the bottom of the column:

$$\begin{aligned} Y_t &\Rightarrow Y_{t+1} \quad t \in REC \\ Y_t &\Rightarrow Y_{t-1} \quad t \in STR \end{aligned} \qquad (11.60)$$

where REC, STR make reference to the set of conditional trays in the rectifying and stripping sections, respectively. In that way all the existing trays will be around the permanent feed tray.

As in the case of MINLP models, Barttfeld et al. [44] considered different representations for the GDP model with fixed and variable feeds, as shown in Figure 11.6. The computational results showed that the most effective structure is the one with a fixed feed, which was the original representation used by Yeomans and Grossmann [51].

As mentioned above, GDP formulations provide better numerical behavior than MINLP models, but because of the nonlinearities and non-convexities inherent to the distillation models, both MINLP and GDP formulations require good initial values and bounds to converge. Getting good initial values is not straightforward. Barttfeld et al. [52] proposed a preprocessing phase to generate good initial estimates. The

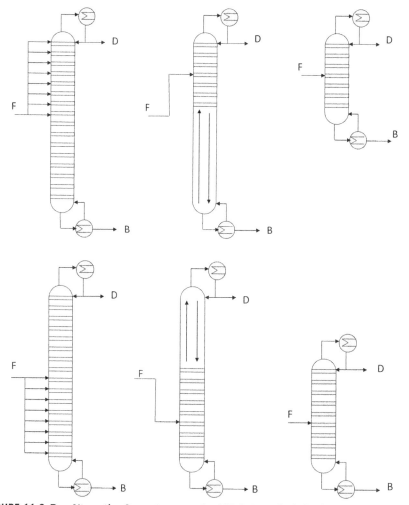

FIGURE 11.6 Two Alternative Superstructures for GDP Column Optimization

column topology in this phase corresponds to the one used for the economic optimization, except that the number of trays is fixed to the maximum specified. This means that the same upper bound on the number of trays has to be employed as well as the potential feed and product location. The initial design considered is the one that involves the minimum reflux conditions as well as minimum entropy production. This reversible separation provides a feasible design, and hence a good initial guess to the economic optimization. For the case of zeotropic columns, overall mass and energy balances are formulated as an NLP problem to compute the reversible products. This formulation is a well-behaved problem that provides initial values and bounds for the rigorous tray-by-tray optimization problem.

Another option is to start with a simpler representation of the column through some shortcut method, and successively increase the complexity of the model using the results of previous steps to initialize the following, both at the level of model or even in the solution algorithm. For example, Harwardt and Marquardt [53] used a multistep approach for the design of internally heat-integrated distillation columns (HIDiC) and vapor recompression (VRC). The results of a shortcut step, such us the minimum energy demand and the concentration profile estimated based on pinch points, were used to initialize the optimization. Based on these results, a simplified model that comprises only component mole balances and equilibrium relations, but no energy balances, is solved. In subsequent solution steps, the energy balance was included again and the model resolved. Two extra interesting modifications were added to the model. First, the vapor—liquid equilibrium calculations were performed as an external user-defined function; in other words, they were dropped from the equation-based environment and solved as an implicit external function. This approach reduces the size of the optimization problem and enhances the flexibility to choose more complex thermodynamic models. Second, to solve the problem they use the so-called Successive Relaxed MINLP (SR-MINLP) proposed by Kraemer et al. [54]. They proposed to reformulate the MINLP or GDP problems as pure continuous problems with tailored big-M constraints, where all discrete decisions are represented by continuous variables. The discrete decisions are enforced by non-convex constraints that force the continuous variables to take discrete values. In this form the GDP problem is reformulated as:

$$\min Z = f(x) + \sum_{k \in K} c_k$$

$$s.t. \ g(x) \leq 0$$

$$r_{i,k}(x) \leq M_{i,k} \left(\sum_{j \in D_k \setminus \{i\}} y_{j,k} \right)$$

$$-M_{i,k} \left(\sum_{j \in D_k \setminus \{i\}} y_{j,k} \right) \leq c_k - \gamma_{i,k} \leq M_{i,k} \left(\sum_{j \in D_k \setminus \{i\}} y_{j,k} \right) \qquad (\text{R} - \text{GDP})$$

$$Ay \leq b$$

$$\varphi_{\text{FB}} \left(y_{i,k}, \sum_{j \in D_k \setminus \{i\}} y_{j,k} \right) = 0 \quad i \in D_k, \ k \in K$$

Equation φ_{FB} is the so called Fischer–Burmeister function that constitutes the special constraints that force the integer decisions, in which at most one $y_{i,k}$ must be 1.

$$y_{i,k} + \sum_{j \in D_k \setminus \{i\}} y_{j,k} - \sqrt{y_{i,k}^2 + \left(\sum_{j \in D_k \setminus \{i\}} y_{j,k} \right)^2} = 0 \qquad (11.61)$$

Due to the nonconvex nature of (11.61), this continuous reformulation suffers from the drawback that the quality of the local optimal solution is highly dependent of the specific initial values to start the solution procedure. To counter this drawback of the continuous reformulation, these authors relax the Fischer–Burmeinsteir function according to:

$$y_{i,k} + \sum_{j \in D_k \setminus \{i\}} y_{j,k} - \sqrt{y_{i,k}^2 + \left(\sum_{j \in D_k \setminus \{i\}} y_{j,k} \right)^2} - M_{\text{FB}} \leq 0$$

The resulting SR-MINLP is solved in a sequential solving procedure where the problem is tightened with each step by reducing the value of the Big-M parameters.

Even with all these difficulties, complex problems have been successfully solved, including reactive distillation [47,55], azeotropic sequences [48,56,57], and hybrid membrane/distillation systems [58], among others.

Although the results reported in this work have shown that there has been significant progress in the optimal design of complex distillation columns, it is clear that significant research is still needed in this area. For instance, the generation of a superstructure to azeotropic systems of more than three components remains an open question. The integration of these rigorous synthesis models as a part of a flowsheet superstructure has not been accomplished. At this point this has only been performed with shortcut models. Finally, a major challenge that remains is the rigorous global optimization.

11.3 Synthesis of distillation sequences

As mentioned in the introduction, the general separation problem was defined more than 40 years ago by Rudd and Watson [5] as the transformation of several source mixtures into several product mixtures. In this section we will focus on the more restricted, and much more studied, problem of separating a single source mixture into several products using only distillation columns. Focusing even more, we look in particular at two kinds of problems: (1) when the product sets contain nonoverlapping species with one another—sharp separations; and (2) when there are overlapping species—non-sharp separations. The nature of these two problems requires different solution approaches. In the case of sharp separations, we can differentiate two cases: when each distillation column performs a sharp separation between consecutive keys, and when non-sharp separations are allowed in some columns—nonconsecutive keys. Historically, sharp separation sequences were assumed to be performed by conventional columns having one feed and producing two products, and including a reboiler and a condenser. Here, we will follow this approach. Later we will show that this case arises naturally as a particular case of the more general thermally coupled distillation.

11.3.1 Sharp separation: only conventional columns

The problem receiving the most and earliest attention has been the sharp separation of a single source mixture using conventional columns. The problem of enumerating the sequences without heat integration is straightforward [6]. However, the selection of the best alternative in terms of total cost and/or energy consumption is not so easy due to the large number of feasible alternatives when the number of components to be separated increases. The earliest attempts were based on case studies in order to develop heuristics with the objective of selecting the preferred structure [59—61]. Sets of heuristics are due to Rudd, Powers, and Siirola [62] and Seader and Westerberg [63].

The first approaches using optimization algorithms used the tree search of alternatives. Thomson and King [64] used a heuristic pseudoalgorithm search that was almost a branch and bound search. It could fail by cycling, but when it worked it was very fast [6]. Hendry and Hughes [65] proposed a dynamic programming algorithm. Additional important papers of these first works can be found in other references [66—68].

11.3.2 Superstructures

According to Grossmann et al. [69], in the application of mathematical programming techniques to design and synthesis problems it is always necessary to postulate a superstructure of alternatives. This is true whether one uses a high-level aggregated model or a fairly detailed model. Although in some cases this is more or less straightforward, this is not true in the general case. The alternative representations of MINLP or GDP structures for a single column presented above shows that even in simple cases the representation is not unique. There are two major issues that arise in postulating a superstructure. The first is to determine, given a set of alternatives to be analyzed, what are the major types of representations that can be used, and what are the implications for the modeling. The second is for a given representation that is selected, what are all the feasible alternatives that must be included to guarantee that the global optimum is not overlooked.

As for types of superstructures, Yeomans and Grossmann [70] have characterized two major types of representation using the concepts of Tasks, States, and Equipment. A State is the minimum set of physical and chemical properties needed to characterize a stream in a given context. They can be quantitative, like pressure or temperature, or qualitative, i.e. a mixture of BCD indicating that we have a stream formed by the compounds of BCD inside some specifications, which does not exclude the presence or other compounds. A Task is the chemical or physical transformation that relates two or more States. The Equipment is the physical device in which a Task is performed.

The first major representation is the State Task Network (STN), which is motivated by the work in scheduling by Kondili, Pantelides, and Sargent [71]. The basic idea here is to use a representation that uses only two types of nodes: States and Tasks (see Figure 11.7). The assignment of Equipment is dealt with implicitly through the model. Both the case of One-Task One-Equipment (OTOE), in which a given Task is assigned a single piece of Equipment; or the Variable Task Equipment (VTE), in which a given Task can be performed by different pieces of Equipment, were considered. The second representation is the State Equipment Network (SEN) which was motivated by the work of Smith [72]. In this case the superstructure uses two nodes, States and Equipment, which assumes an a priori assignment of the different Tasks to Equipment based on the knowledge of the designer about the process (see Figure 11.8).

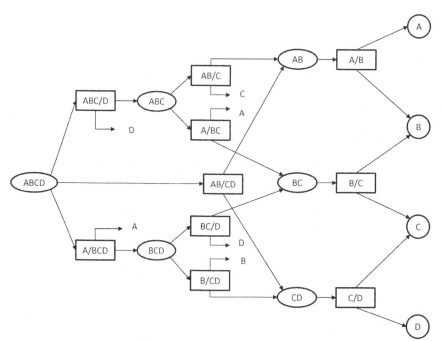

FIGURE 11.7 STN Superstructure for the Sharp Separation of a Four-Component Zeotropic Mixture

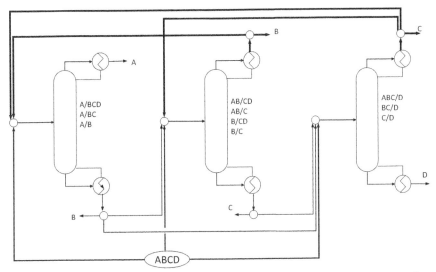

FIGURE 11.8 SEN Superstructure for the Sharp Separation of a Four-Component Zeotropic Mixture

11.3.3 Linear models for sharp split columns

One of the first approaches to synthesize distillation sequences using MILP methods is due to Andrecovich and Westerberg [73]. The following presentation, although with some modifications, is based on their work.

If there is a fixed pressure and reflux ratio, then by performing shortcut calculations with any of the methods previously presented, it is possible to obtain linear mass balance relationships in terms of the feed flow rates as given by the following equation:

$$D_i = \gamma_i F_i$$
$$B_i = (1 - \gamma_i) F_i \tag{11.62}$$

where D_i and B_i represent the mass flow rates of components in the distillate and bottom streams, and γ_i are the corresponding recovery fractions that are typically obtained from the mass balance in the shortcut model for a selected feed composition. By assuming the fractions Ψ_i to be constant, it is clear that Eqn (11.62) reduces to a linear expression. It is possible to consider a further simplification without significantly increasing the error, that consists of assuming 100% recoveries of key components in each column. It is possible to determine a priori for each column the composition and total flow (or the component molar flow) entering the column.

From the above assumptions, in 1985 Andrecovich and Westerberg [73] proposed to model the heat duties of the condenser and reboiler and the capital cost in terms of the total flow rate entering each column. Assuming the same loads in condenser and reboiler, the heat duties for column $k(\dot{Q}_k)$ can be expressed as the linear functions:

$$\dot{Q}_k = K_k F_k \tag{11.63}$$

where K_k is a constant derived from a shortcut calculation. Finally, the annualized cost of the column, that includes the fixed charge cost model for investment and the utility cost will be given by:

$$C_k = C_{\text{fix},k} y_k + \beta_k F_k + (C_H + C_C)\dot{Q}_k \tag{11.64}$$

where $C_{\text{fix},k}$ is the annualized fixed charge cost in terms of the $0-1$ binary variable y_k, β_k is the size factor for the column in terms of the total flow entering that column, and C_H, C_C are the unit costs for the heating and cooling in the reboiler and condenser, respectively.

It is worth noting that instead of using the K_k factors, or assuming the same loads in condenser and reboiler, or even assuming a linear size factor with the flow, it is possible to perform a rigorous optimization of each separation and exactly obtain the heat loads and optimal sizes of each possible distillation column, and therefore obtain the optimal separation sequence with the only approximation of 100% recovery.

Based on previous considerations, Andrecovich and Westerberg postulated the superstructure shown in Figure 11.9. Note that this superstructure corresponds to a State Task Network (STN) with an a priori assignment of Tasks to Equipment (One-Task One-Equipment, OTOE) according to the Yeomans and Grossmann classification [70].

The model, a modification of the original proposed by Andrecovich and Westerberg, can be written as follows:

Index sets

COL	[k|k is a column]
S	[m|m is a mixture] (i.e. ABCD, ABC, AB, BC, A...)
COMP	[i|i is a component]
IP(m)	[m|m is an intermediate mixture] (i.e. AB, ABC, BCD...)
TN(m)	[m|m is a terminal mixture] (i.e. A, B, C, D...)
SD(m,k)	[Distillate of column k goes to mixture m]
SB(m,k)	[Bottom stream of column k goes to mixture m]
SF(m,k)	[mixture m is the feed of column k]
Init(k)	[columns k that have as feed the initial mixture]

Variables

$F_{k,i}, D_{k,i}, B_{k,i}$	Individual molar flow rates of feed, distillate, and bottoms of column k
$\dot{Q}_k$	Heat load in column k

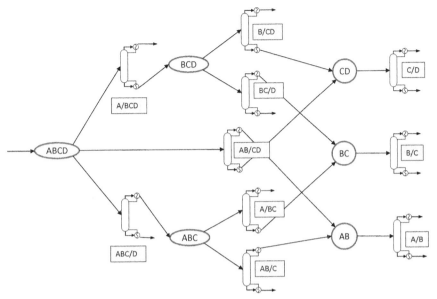

FIGURE 11.9 Superstructure Proposed by Andrecovich and Westerberg for the Sharp Separation of a Four-Component Mixture using Sharp Split and Consecutive Key Components

$$\min : Total\ Cost = \sum_{k \in COL} C_{\mathrm{fix},k} y_k + \beta_k F_k + (C_H + C_C)\dot{Q}_k$$

$s.t.$

$$F_0 z_i = \sum_{k \in Init} F_{k,i} \quad \forall i \in COMP$$

$$\sum_{k \in SD(m,k)} D_{k,i} + \sum_{k \in SB(m,k)} B_{k,i} = \sum_{k \in SF(m,k)} F_{k,i} \quad i \in COMP;\ m \in IP$$

$$\sum_{k \in SD(m,k)} D_{k,i} + \sum_{k \in SB(m,k)} B_{k,i} = F_0 z_i \quad i \in COMP;\ m \in TN \qquad (A-W)$$

$$\dot{Q}_k - K_k \sum_{i \in COMP} F_{k,i} = 0 \quad k \in COL$$

$$F_{k,i} = D_{k,i} + B_{k,i} \quad k \in COL;\ i \in COMP$$

$$D_{k,i} = F_{k,i} \quad k \in COL;\ i \in COMP/i \leq light\ key$$

$$F_{k,i} \leq M y_k \quad k \in COL;\ i \in COMP$$

The first three constraints in the Andrecovich and Westerberg model correspond to the mass balance in the initial node, in intermediate nodes, and in terminal nodes, respectively. The fourth constraint represents the relation between the total flow and the heat load for a given column, equivalent to Eqn (11.63). The fifth and sixth constraints are the mass balances in a given column including the total sharp separation of keys (that must be consecutive). Finally, the last constraints force the variables to be zero if the column is not selected.

11.3.4 Nonlinear models for sharp split columns

In some situations the assumptions made for linear models can introduce significant errors. For instance, the feed entering at each possible column cannot be calculated a priori because the assumption of 100% recoveries of key components does not hold true. In this case, the calculation of a given column in the sequence and the determination of the optimal column sequence must be performed simultaneously.

Due to the mathematical complexity associated with a rigorous distillation column, the optimal determination of column sequences has been generally carried out using shortcut methods. But even with those shortcut methods, there is an intrinsic relationship between the superstructure, the model complexity, and the associated numerical performance. Although there are other alternatives that will be briefly commented upon at the end of this section, we will focus here on the models related to the two superstructures described above: STN and SEN.

Although the general problem consists of separating an N component mixture in M groups of compounds ($M \leq N$) in such a way that any of this groups contains a key component that must be sharp separated from the rest (i.e. separate C3-C4-C5-C6 hydrocarbons), for the sake of simplicity and without loss of generality, we can assume that we want to sharp separate an N component mixture into its N pure constituents using conventional columns and consecutive keys. Under these conditions, generating an STN superstructure is straightforward: we only need to identify the States, the possible Tasks, and simply join the Tasks with the States. For example, in a zeotropic four-component mixture (ABCD) ordered by decreasing volatilities, the States correspond to each of the possible mixtures. ABCD, ABC, BCD, AB, BC, CD. The possible Tasks are the following:

From	ABCD:	A/BCD	AB/CD	ABC/D
From	ABC:	A/BC	AB/C	
From	BCD:	B/CD	BC/D	
From	AB:	A/B		
From	BC:	B/C		
From	CD:	C/D		

Then the resulting superstructure is the one shown in Figure 11.8. It is possible to go a step further and assign a distillation column to each of the Tasks. Then the STN approach reduces to the superstructure proposed by Andrecovich and Westerberg [73]; see Figure 11.10.

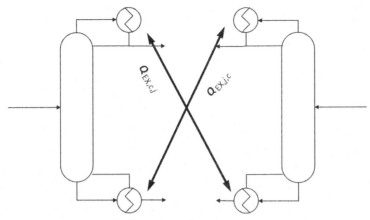

FIGURE 11.10 Superstructure for Heat Integration

All possible matches between condensers and reboilers are considered.

Using the Underwood shortcut model, the STN formulation can be written as follows. Define the following index sets:

IP	[m|m is an intermediate State (i.e. ABC; BCD; AB...)]
IF	[m|m is a final State (i.e. A, B, C...)]
COL	[K|k is a column (Task) in the superstructure]
FS_F	[Columns k whose feed is the initial mixture]
FS_m	[Columns k whose feed is an intermediate State m]
DS_m	[Columns k that produce State m as a distillate]
BS_m	[Columns k that produce State m as a bottom stream]
DP_m	[Columns k that produce final product m as a distillate]
BP_m	[Columns k that produce final product m as a bottom stream]
COM	[i|i is a component in the mixture]

The variables of the problem are as follows:

$FT_k; F_{i,k}$	Molar flow, total and of component i, entering the column k
$DT_k; D_{i,k}$	Distillate molar flows in column k
$BT_k; B_{i,k}$	Bottoms molar flows in column k
$V_{R,k}, L_{R,k}$	Molar flows of vapor and liquid in rectifying section of column k
$V_{S,k}, L_{S,k}$	Molar flows of vapor and liquid in stripping section of column k
Y_k	Boolean variable. True if the column k is selected

Data:

F_{0i}	Component molar flow entering the system
rec_i	Component recoveries.

The GDP model can be written as follows:

$$\min : \sum_{k \in \text{COL}} (Total\ Cost)_k$$

s.t.

$$\sum_{k \in \text{FS}_F} FT_k = F_0; \qquad \sum_{k \in \text{FS}_F} F_{i,k} = F_{0i}$$

$$\sum_{k \in \text{DS}_m} D_{i,k} + \sum_{k \in \text{BS}_m} B_{i,k} = \sum_{k \in \text{FS}_m} F_{i,k} \quad \forall i \in \text{COM}; \quad m \in \text{IP}$$

$$\sum_{k \in \text{DP}_m} D_{i,k} + \sum_{k \in \text{BP}_m} B_{i,k} \geq \text{rec}_i F_{0i} \quad i \in \text{COM}$$

$$
\begin{bmatrix}
Y_k \\
\\
F_{i,k} = D_{i,k} + B_{i,k} \quad i \in \text{COM} \\
\\
V_{R,k} = DT_k + L_{R,k} \\
\\
L_{S,k} = BT_k + V_{S,k} \\
\\
FT_k + V_{S,k} + L_{R,k} = V_{R,k} + L_{S,k} \\
\\
FT_k = \sum_{i \in \text{COL}} F_{i,k} \\
\\
DT_k = \sum_{i \in \text{COL}} D_{i,k} \\
\\
BT_k = \sum_{i \in \text{COL}} B_{i,k} \\
\\
\sum_{i \in \text{COM}} \dfrac{\alpha_i F_{i,k}}{\alpha_i - \phi_R} = V_{R,k} - V_{S,k} \\
\\
\sum_{i \in \text{COL}} \dfrac{\alpha_i F_{i,k}}{\alpha_i - \phi_R} = V_{R,k} \\
\\
Total\ cost_k = f(V_R, V_S, L_R, L_S \ldots)
\end{bmatrix}
\lor
\begin{bmatrix}
\neg Y_k \\
F_{i,k} = 0 \\
D_{i,k} = 0 \\
B_{i,k} = 0 \\
V_{R,k} = 0 \\
V_{S,k} = 0 \\
L_{R,k} = 0 \\
L_{S,k} = 0 \\
FT_k = 0 \\
DT_k = 0 \\
BT_k = 0
\end{bmatrix}
(\text{M} - \text{STN})
$$

$$\Omega(Y) = True$$

The first three constraints are mass balances in the initial feed node, in the intermediate States, and in the final States (products), respectively. The disjunctions include all the equations to be solved if a given column is selected. The logical relationships are basically connectivity equations that can be obtained from the superstructure. For a four-component mixture, the logical relationships are the following:

$$
\begin{aligned}
&Y_{\mathrm{A/BCD}} \ \vee \ Y_{\mathrm{AB/CD}} \ \vee \ Y_{\mathrm{ABC/D}}\\
&Y_{\mathrm{A/BCD}} \Rightarrow Y_{\mathrm{B/CD}} \vee Y_{\mathrm{BC/D}}\\
&Y_{\mathrm{ABC/D}} \Rightarrow Y_{\mathrm{A/BC}} \vee Y_{\mathrm{AB/C}}\\
&Y_{\mathrm{AB/CD}} \Rightarrow Y_{\mathrm{A/B}} \wedge Y_{\mathrm{C/D}}\\
&Y_{\mathrm{A/BC}} \Rightarrow Y_{\mathrm{ABC/D}} \wedge Y_{\mathrm{B/C}}\\
&Y_{\mathrm{AB/C}} \Rightarrow Y_{\mathrm{ABC/D}} \wedge Y_{\mathrm{A/B}}\\
&Y_{\mathrm{BC/D}} \Rightarrow Y_{\mathrm{A/BCD}} \wedge Y_{\mathrm{B/C}}\\
&Y_{\mathrm{B/CD}} \Rightarrow Y_{\mathrm{A/BCD}} \wedge Y_{\mathrm{C/D}}\\
&Y_{\mathrm{A/B}} \Rightarrow Y_{\mathrm{AB/CD}} \vee Y_{\mathrm{AB/C}}\\
&Y_{\mathrm{B/C}} \Rightarrow Y_{\mathrm{BC/D}} \vee Y_{\mathrm{A/BC}}\\
&Y_{\mathrm{C/D}} \Rightarrow Y_{\mathrm{AB/CD}} \vee Y_{\mathrm{B/CD}}
\end{aligned}
\tag{11.65}
$$

In general, generating a SEN superstructure is not so straightforward, because the designer has to decide a priori which Equipment to use. However, in this particular case, we know that we need exactly $N-1$ distillation columns (N is the number of components). The problem is then to assign a possible set of Tasks to each distillation column in such a way that all the feasible alternatives are included. Figure 11.8 shows one possibility for a four-component mixture, although the assignment of Tasks to columns is not unique. It is interesting to note that Novak et al. [74] used this superstructure before the formalization by Yeomans and Grossmann [70].

In the SEN superstructure, the logic of the problem is transferred to the selection of the particular Task that a given column must perform and then to the streams connecting the different columns, but all the particular equations for a given column becomes permanent equations. Like in the STN approach, using the Underwood equations, the model can be written as follows. Define the following index sets:

COL	[k|k is a column]
Task	[t|t is a separation Task]
CT_{kt}	[t|task t is assigned to column k]
ST	[s|s is a stream]
DF_{kk}	[Indicates that there is a distillate stream from column k to column kk]
BF_{kk}	[Indicates that there is a bottoms stream from column k to column kk]
$PD_{i,k}$	[Pure product i is produced as distillate in column k]
$PB_{i,k}$	[Pure product i is produced as a bottom stream in column k]
SZ_t	[Stream s that does not exist (zero flow) if Task t is selected]

The variables of the problem are:

FT_k; $F_{i,k}$	Molar flow, total and of component i, entering the column k
DT_k; $D_{i,k}$	Distillate molar flows in column k
BT_k; $B_{i,k}$	Bottoms molar flows in column k
$V_{R,k}$, $L_{R,k}$	Molar flows of vapor and liquid in rectifying section of column k
$V_{S,k}$, $L_{S,k}$	Molar flows of vapor and liquid in stripping section of column k
ST_k; $S_{i,k}$	Molar flow, total and of component i, of external feed flow that goes to the mixer at the inlet of column k

$$\min : \sum_{k \in \text{COL}} (Total\ cost)_k$$

$$s.t.F_{0,i} = \sum_{k \in \text{COL}} S_{i,k}$$

$$S_{i,k} + \sum_{kk \in \text{DF}_k} D_{i,kk} + \sum_{kk \in \text{BF}_k} B_{i,kk} = F_{i,k} \quad i \in \text{COMP}; k \in \text{COL}$$

$$\sum_{k \in \text{PD}_i} D_{i,k} + \sum_{k \in \text{PB}_i} B_{i,k} \geq \text{rec}_i F_{0,i} \quad i \in \text{COMP}$$

$$F_{i,k} = D_{i,k} + B_{i,k} \quad i \in \text{COM}$$

$$V_{\text{R},k} = DT_k + L_{\text{R},k}$$

$$L_{\text{S},k} = BT_k + V_{\text{S},k}$$

$$FT_k + V_{\text{S},k} + L_{\text{R},k} = V_{\text{R},k} + L_{\text{S},k}$$

$$FT_k = \sum_{i \in \text{COL}} F_{i,k}$$

$$DT_k = \sum_{i \in \text{COL}} D_{i,k} \qquad\qquad \text{(M − SEN)}$$

$$BT_k = \sum_{i \in \text{COL}} B_{i,k}$$

$$\sum_{i \in \text{COM}} \frac{\alpha_i F_{i,k}}{\alpha_i - \phi_{\text{R},k}} = V_{\text{R},k} - V_{\text{S},k}$$

$$\sum_{i \in \text{COM}} \frac{\alpha_i F_{i,k}}{\alpha_i - \phi_{\text{R},k}} = V_{\text{R},k}$$

$$Total\ cost_k = f(V_{\text{R}}, V_{\text{S}}, L_{\text{R}}, L_{\text{S}}...)$$

$$\underset{t \in \text{CT}_k}{\vee} \begin{bmatrix} Y_{t,k} \\ \alpha_i \leq \varphi_{\text{R},k} \leq \alpha_j \quad i = \text{HK}; j = \text{LK} \\ \left. \begin{array}{l} D_{i,k} = 0 \\ B_{i,k} = 0 \end{array} \right\} \quad k \in \text{SZ}_t \end{bmatrix}$$

$$\Omega(Y) = True$$

The first constraint is the mass balance in the node for the external feed. The second one represents mass balances in the mixers entering the column. The third is a recovery constraint. The rest before the disjunctions are the equations that define each distillation column. Inside the disjunctions only the specific equations that define the Task assigned to each column are included. The logical relationships are the same as in the STN model.

When both models are reformulated as MINLPs, the SEN produces models with fewer numbers of equations and in general is more robust from a numerical point of view [27,75]. However, the reformulation to MINLP is usually easier with an STN model.

Finally, it is worth noting that the SEN or STN approaches are not the only options for superstructure optimization. Between these two extreme alternatives there are a large number of superstructures with intermediate characteristics that can be generated by aggregation of some Tasks or by an initial partial assignment of Equipment; see Ref. [27] for an example. Other alternatives include the work by Bagajewicz and Manousiouthakis [28]. These authors developed a model in which a distillation column was considered as a composite heat and mass exchanger operation. Assuming constant countercurrent molar flow rate, the mass exchange inside the distillation columns can be treated as a pure mass transfer operation. Therefore, in this case a distillation network could be treated as a separable heat/mass exchange network. Papalexandri and Pistikopoulos [76] introduced a multipurpose mass/heat transfer module as a building block of a systematic representation of conventional and nonconventional process units and process structures.

11.3.5 Heat-integrated distillation sequences

In this section we will assume that the heat integration will only be considered between different separation tasks. Thus the possibility of synthesizing multieffect columns will be excluded. The interested reader can find information, for example, in the works by Andrecovich and Westerberg [73,77]. In the rest of the section, for the sake of simplicity, we assume a single hot and a single cold utility (e.g. steam and cooling water), although the extension to multiple utilities is straightforward. The following model is based on the works by Paules and Floudas [78,79] and Raman and Grossmann [80]. The starting point is a model that allows calculating a sequence of columns (shortcut, aggregated, rigorous) including specifically the heat loads and temperatures in condenser and reboiler, and their dependence with the column pressure. The model assumes that a given condenser could eventually exchange heat with any reboiler (and vice versa) (see Figure 11.10).

If $T_{\text{reb},c}$ and $T_{\text{con},c}$ are the reboiler and condenser temperatures of column c, $T_{\text{min},e,a}$ is the minimum exchanger approach temperature, and T_{st} and T_{cool} are the temperatures of steam and cooling utility, the two following constraints apply:

$$\left.\begin{array}{l} T_{\text{reb},c} \leq T_{\text{st}} - T_{\text{min},ea} \\ T_{\text{con},c} \leq T_{\text{cool}} + T_{\text{min},ea} \end{array}\right\} c \in \text{COL} \qquad (11.66)$$

To consider the potential exchanges of heat, the variable $Q_{EX,k,j}$ is introduced, which is the amount of heat exchanged between the condenser of column k and the reboiler of column j. We also define the binary variable $w_{k,j}$ which is equal to 1 if the condenser of column k supplies heat to the reboiler of column j and zero otherwise. Thus, the following conditional constraints apply:

$$\left.\begin{array}{l} Q_{EX,k,j} \leq M_k w_{k,j} \\ T_{con,k} \geq T_{reb,j} + T_{min,ea} - M_{k,j}(1 - w_{k,j}) \end{array}\right\} \quad k,j \in \text{COl} \; k \neq j \qquad (11.67)$$

If $w_{k,j} = 1$ the temperature of the condenser in column k must be larger than the temperature in the reboiler of column j. If $w_{k,j} = 0$, the heat exchanged between the condenser of column k and the reboiler of column j is forced to be zero.

Heat balances must hold for cooling and heating utilities:

$$\left.\begin{array}{l} \sum_{j \in \text{COL}} Q_{EX,c,j} + Q_{cool,c} = Q_{reb,c} \\ \sum_{j \in \text{COL}} Q_{EX,j,c} + Q_{heat,c} = Q_{con,c} \end{array}\right\} \quad c \in \text{COL} \qquad (11.68)$$

Finally, the logic constraints can be added that relate the existence of columns with the selection of matches:

If a column is not selected, the corresponding matches to that column cannot take place:

$$\left.\begin{array}{l} w_{j,k} \leq y_k \\ w_{k,j} \leq y_k \end{array}\right\} \quad k,j \in \text{COL}; k \neq j \qquad (11.69)$$

Either column j supplies heat to column k, or vice versa:

$$w_{j,k} + w_{k,j} \leq 1 \quad k,j \in \text{COL}; k \neq j \qquad (11.70)$$

One interesting characteristic of the model in Eqns (11.66)–(11.70) is that it can be "added" to any existing process model. The only modification will be in the objective function to take into account the reduction in utilities consumption.

11.3.6 Synthesis of complex distillation sequences (Thermally coupled distillation)

From the point of view of energy requirements, separation sequences using conventional columns (a single feed with two product streams, a condenser, and a reboiler) suffer from an inherent inefficiency produced by the thermodynamic irreversibility during the mixing of streams at the feed, top, and bottom of the column [81]. This remixing is inherent to any separation that involves an intermediate boiling component and can be generalized to an N-component mixture. Theoretical studies developed by Petluyk and coworkers [81] showed that this inefficiency can be improved by removing some heat exchangers and introducing thermal coupling between columns (see Figure 11.11). If a heat exchanger is removed, the liquid reflux (or vapor load) is provided by a new stream that is

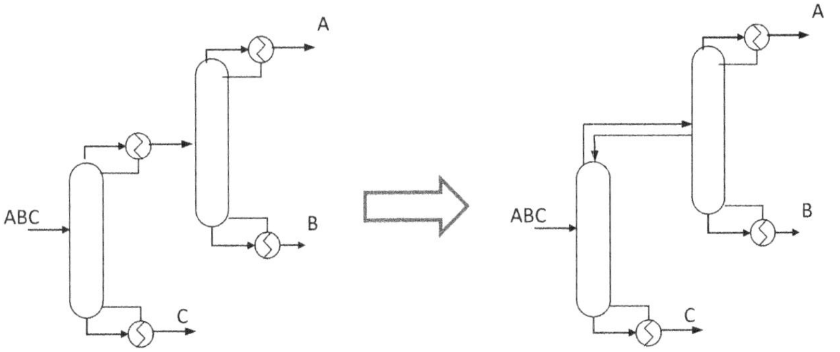

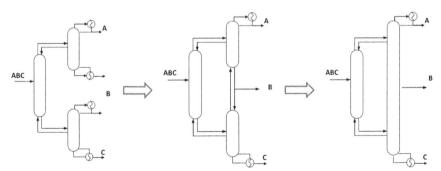

FIGURE 11.11 Introduction of a Thermal Couple by Removing the Intermediate Condenser

withdrawn from another column. In this way, it is possible to reduce the energy consumption, and in some circumstances also the capital costs. A fully thermally coupled (FTC) configuration is reached when the entire vapor load is provided by a single reboiler and all the reflux by a single condenser (see Figure 11.12). Different researchers have shown that thermally coupled configurations are capable of typically achieving 30% energy reduction when compared to conventional systems [82–84]. Halvorsen and Skogestad [85–87] proved that the minimum energy consumption for an N-component mixture is always obtained for the FTC configuration.

The first known thermally coupled system dates from 1949, due to Wright [88], now known as a divided wall column. While the final detailed theoretical study was developed by Petlyuk et al. [81], the lack of reliable design methods and concerns about the operation and control of these columns have prevented their application. The discovery of the concept of a "thermodynamically equivalent configuration" [89,90], the further development of this new concept [91,92], and improved control

FIGURE 11.12 Fully Thermally Coupled Configuration (Petluyk Configuration) for a Three-Component Mixture

Only one condenser and one reboiler for the entire system.

strategies [93–97] have made complex columns a realistic alternative to conventional designs.

Despite the reduction in energy consumption, there is a price to be paid when using thermally coupled systems:

1. The energy must be supplied under the worst conditions, at the highest temperature in the reboiler and removed at the lowest temperature in the condenser, preventing, in some cases, the use of utilities such as medium- or low-pressure steam.
2. When using conventional columns, it is common to constrain the alternatives to sharp separations. For example, in a three-component mixture (ABC), sorted by volatilities, we postulate initially separations A from BC (A/BC) and AB from C (AB/C). However, in FTC distillation, we could introduce sloppy separations—an intermediate product is allowed to distribute along the column—and therefore we increase the number of column sections. This increase does not imply an increase in the number of columns, but usually an increase in the total number of trays. A detailed discussion on the number of column sections needed for a given separation can be found in Agrawal [98] and Caballero and Grossmann [99–102].
3. In FTC systems, the minimum vapor flow is that of the most difficult separation, and therefore some column sections will have larger diameters.
4. Operation is more difficult due to the large number of interconnections between columns. Therefore it cannot be concluded that complex configurations are always superior, compared to sequences of simple columns. Instead, the optimum configuration will be dependent on the specific mixture and feed conditions.

Therefore the objective should be to find the optimal sequence of columns in a search space that includes all of the alternatives from systems with only conventional columns and a sharp split of consecutive keys to fully thermally coupled systems in which the vapor load for the entire system is supplied by a single reboiler and the reflux provided by a single reboiler going through all of the intermediate alternatives.

The first task therefore consists of determining the characteristics that a feasible, eventually optimal, sequence must have. Depending on the number of distillation columns used to separate an N-component mixture into N product streams, the sequence can be classified as more than $N-1$ column configurations, exactly $N-1$ columns, or less than $N-1$ column configurations.

In the case of zeotropic mixtures, sequences with exactly $N-1$ columns, named basic configurations by Agrawal [103], are characterized by the following three features:

1. Mixtures (or states) with the same components are transferred only once from one distillation column to another.
2. A final product is obtained in a single location of the sequence.

3. The feed stream and all the intermediate mixtures are split into exactly two product streams by two columns sections.

Configurations that violate the first two features and obey the third produce sequences with more than $N-1$ columns. These configurations, also referred as nonbasic, have higher operating cost than the best basic configuration [102,104,105]. Nonbasic configurations also tend to have higher capital cost due to the additional distillation columns, and therefore nonbasic configurations can be removed from the search space.

Configurations that violate the third feature have higher operating costs than the best basic configuration due to increased heat duty, especially for getting high-purity products. However, the reduced number of columns in some situations could compensate for the extra energy consumption. In the literature some of these cases can be found, for example, in Brugma [106], Kaibel [90], Kim et al. [107], or Errico and Rong [108]. However in this work, we will focus only on basic configurations.

It is important to make some remarks on previous features and classification.

1. Sequences obeying the three features can always be arranged in $N-1$ columns, although the total number of separation tasks can be larger. Consider the following example. We want to separate a four-component mixture, ABCD, where the components are sorted by decreasing volatilities (A is the most volatile component, D the least). Say we perform the following separation tasks: AB/BCD, which means separate A from CD letting the B component be optimally distributed between distillate and bottoms; A/B; BC/CD; B/C; and C/D. In this example we can identify five separation tasks, and this sequence can be performed using five distillation columns, but this sequence can be easily arranged in three distillation columns (see Figure 11.13). In fact it can be rearranged in 16 thermodynamically equivalent configurations using three distillation columns [92] (see next point).

2. When a thermal couple appears, the arrangement of separation tasks in the actual columns is not unique. Using the two flows of the thermal couple it is possible to move a column section from an actual column to another and obtain different arrangement of tasks in actual columns. Figure 11.14 shows an example. All the configurations obtained moving column sections using a thermal coupled are said to be thermodynamically equivalent. From the point of view of simulation, all of the temperatures, flows, concentrations, etc. are the same. From a practical point of view there are some differences due to different pressure losses and some practical considerations in transfer, mainly of vapor streams. But at the preliminary design we can consider these configurations to be equivalent. A detailed discussion on thermodynamically equivalent configurations can be found in references [89,91,92,109].

3. The total number of thermodynamically equivalent configurations can be very large. For example, in a five-component mixture there are 203 basic configurations but more than 8000 arrangements in actual columns—most of them with very similar performance in terms of total cost. To avoid this degeneracy it is

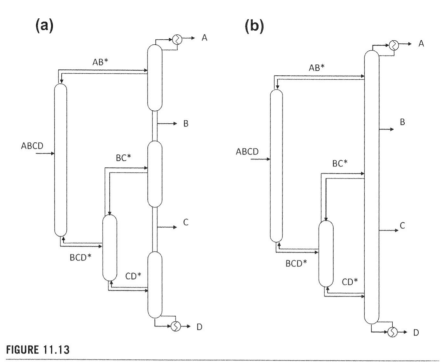

FIGURE 11.13

The basic sequence with 5 separation tasks (see text) can be arranged from (a) in three columns shown in (b).

convenient to represent a sequence in terms of the separation tasks involved instead of the particular equipment used to perform the separation. In other words, all the thermodynamically equivalent configurations perform the same sequence of separation tasks.

As can be seen from the previous paragraphs, some major considerations must be taken into account in order to develop an MINLP or GDP model to select the best column sequence:

1. A superstructure based on equipment (i.e. SEN) is neither possible nor convenient if we want to avoid the degeneracy due to thermodynamically equivalent configurations.
2. A procedure is needed for estimating the cost of the tasks without assuming a specific column configuration.
3. A practical set of logical relationships is needed that allows constraining the search to basic configurations.

Related to superstructure representations, the STN approach can be easily adapted to any complex zeotropic system. The procedure is equivalent to generating an STN superstructure for sharp split and consecutive keys but now including all the

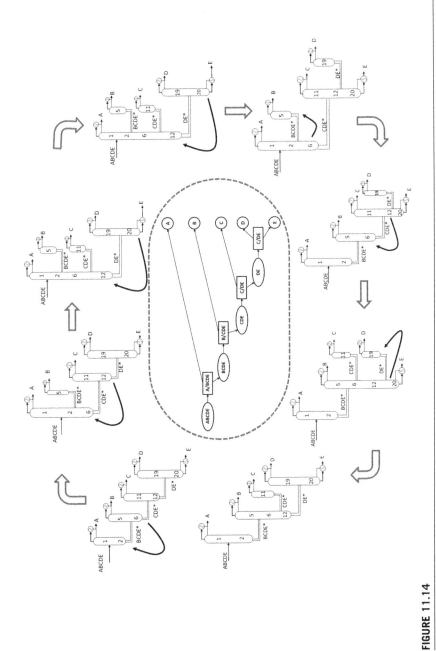

FIGURE 11.14

State task representation (center) and its eight thermodynamic equivalent configurations in actual columns.

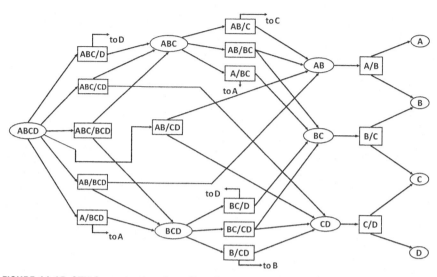

FIGURE 11.15 STN Superstructure for a Four-Component Mixture

Sharp split of not necessarily consecutive key components.

possible sharp splits between two non-necessarily consecutive keys. Figure 11.15 shows this superstructure for a four-component mixture.

In the STN superstructure the number of tasks rapidly grows with the number of components. In order to use a more compact representation, a superstructure obtained by the aggregation in a single "super-task" of some task is used (see Figure 11.16). This superstructure is similar to that proposed by Sargent and

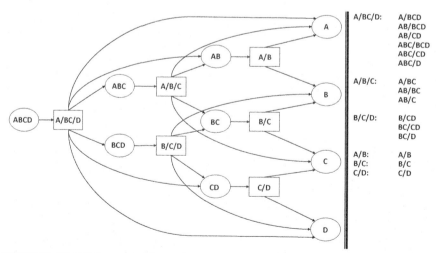

FIGURE 11.16 STN-Aggregated Superstructure

Gaminibandara [43] in 1976, although here the bypasses are explicitly included to account for all the alternatives, and it is a task-based approach instead of a column-based approach.

Even though no particular configuration in actual columns is assumed, a given separation task is formed by two column sections: a rectifying and a stripping section (although maybe in the final arrangement of tasks in actual columns these two sections are placed in different columns). In this sense it is possible to consider a separation task like a pseudocolumn, which conceptually facilitates the modeling; i.e. it is possible to use any of the shortcut, aggregated, or even rigorous models presented before to any of these pseudocolumns. With this approach it is also possible to calculate the cost of each column section. It is possible that the final cost of each column section slightly increases in the final arrangement in columns (i.e. by increasing the diameter of some sections in order to build a single-diameter column). Taking also into account that the total number of actual columns is always $N-1$ and that operating costs do not depend on the particular arrangement in columns of a set of tasks, it is possible to accurately estimate the total annual cost of a sequence or at least to get a tight lower bound to the final cost.

As discussed in the previous paragraphs, a sequence of distillation columns that includes columns ranging from conventional to fully thermally coupled columns must include the space of all the basic configurations.

The first rule-based algorithm for generating the full set of basic configurations was proposed by Agrawal [103]. Following this line, Ivakpour and Kasiri [110] later proposed a formulation in which distillation configurations are represented mathematically as upper triangular matrices. Independently Shah and Agrawal [111] presented an alternate matrix formulation in which distillation configurations are generated by exploring all possible instances of the presence or absence of transfer streams. Particularly, this last approach is very efficient for generating the full set of basic configurations. All of these rule-based approaches have proven to be effective when generating alternatives, but the enumeration of all the alternatives is, except in the case of a reduced number of alternatives, an inefficient strategy. These approaches can be easily adapted to metaheuristic optimization methods (i.e. genetic algorithms, particle swarm optimization, etc.). However, formulating them within a deterministic mathematical programming framework is not obvious.

Based mainly on the observations in the seminal paper by Agrawal [98], Caballero and Grossmann [99,101,102] proposed a complete set of logical rules in terms of Boolean variables that implicitly include all the basic column configurations. These logical equations can be transformed into algebraic linear equations in terms of binary variables and integrated within a mathematical programming environment. The objective in those works was not to generate explicitly all the basic configurations, but to develop a set of logic equations that ensure a strong relaxation when solving the resulting MINLPs that include all the performance equations of the distillation columns, and trying to extract the best configuration without an explicit enumeration of all the alternatives. It is interesting to point out that Shah and Agrawal [111] proposed a valid alternative set of equations in terms also of binary variables that could also be integrated in a mathematical programming environment. However, their focus was on checking quickly if a given alternative is a basic one (with excellent performance). It was not on the performance when those equations are integrated with the model of the columns in a mathematical programming environment. In fact, some of their equations can be obtained from the aggregation of some of the logic relations presented by Caballero and Grossmann [101,102], and therefore a worse relaxation can be expected.

For the model and logical relationships it is necessary first to define the following index sets:

TASK = [t|t is a given task]
   e.g. TASK = [(ABC/BCD), (AB/BCD), (ABC/CD), (AB/BC), (AB/C), (B/CD), (BC/CD), (A/B), (B/C), (C/D)]
STATES = [s|s is a state]
   e.g. STATES = [(ABCD), (ABC), (BCD), (AB), (BC), (CD), (A), (B), (C), (D)]
IM$_S$ = [s|s is an intermediate state. All but initial and final products]
   e.g. IMS = [(ABC), (BCD), (AB), (BC), (CD)]
COMP = [i|i is a component to be separated in the mixture]
   e.g. COMP = [A, B, C, D]
FS$_T$ = [t|t is a possible initial task; Task that receives the external feed]
   e.g. TASK [(A/BCD), (AB/BCD), (AB/CD), (ABC/BCD), (ABC/CD), (ABC/D)]
TS$_s$ = [tasks t that the state s is able to produce]
   e.g. TS$_{ABCD}$ = [(AB/BCD), (ABC/BCD), (ABC/CD)]
      TS$_{ABC}$ = [(AB/BC), (AB/C)]
      TS$_{BCD}$ = [(B/CD), (BC/CD)]
ST$_s$ = [tasks t that are able to produce state s]
   e.g. ST$_{ABC}$ = [(ABC/CD), (ABC/BCD)]
      ST$_{BCD}$ = [(AB/BCD), (ABC/BCD)]
      ST$_{AB}$ = [(AB/BCD), (AB/BC), (AB/C)]
      ST$_{BC}$ = [(AB/BC), (BC/CD)]
      ST$_{CD}$ = [(ABC/CD), (B/CD), (BC/CD)]
RECTs = [task t that produces state s by a rectifying section]
   e.g. RECT$_{ABC}$ = [(ABC/CD), (ABC/BCD)]
      RECT$_{AB}$ = [(AB/BCD), (AB/BC), (AB/C)]
      RECT$_{BC}$ = [(BC/CD)]
STRIPs = [task t that produces state s by a stripping section]
   e.g. STRIP$_{BCD}$ = [(AB/BCD), (ABC/BCD)]
      STRIP$_{BC}$ = [(AB/BC)]
      STRIP$_{CD}$ = [(ABC/CD), (B/CD), (BC/CD)]
FPs = [s|s is a final state (pure products)]
   e.g. FP = [(A), (B), (C), (D)] do not confuse with components, although the name is the same.

P_REC$_s$ = [tasks t that produce final product s through a rectifying section]
  e.g. PRE$_A$ = [(A/B)]
      PRE$_B$ = [(B/CD), (B/C)]
      PRE$_C$ = [(C/D)]
P_STR$_s$ = [tasks t that produce final product s through a stripping section]
  e.g. PST$_B$ = [(A/B)]
      PST$_C$ = [(AB/C), (B/C)]
      PST$_D$ = [(C/D)]

and the following Boolean variables:

Y_t	True if the separation task t exists. False, otherwise
Zs	True if the state s exists. False, otherwise
Ws	True if the heat exchanger associated to the state s exists. False, otherwise

1. A given state s can give rise to at most one task:

$$\underset{t \in TS_t}{\vee} Y_t \underline{\vee} R; \quad s \in COL \tag{L1}$$

where R is a dummy Boolean variable that means "do not choose any of the previous options."

2. A given state can be produced at most by two tasks; one must come from the rectifying section of a task and the other from the stripping section of a task:

$$\left. \begin{array}{c} \underset{t \in RECT_s}{\vee} Y_t \underline{\vee} R \\[1em] \underset{t \in STRIP_s}{\vee} Y_t \underline{\vee} R \end{array} \right\} \quad s \in STATES \tag{L2}$$

where R has the same meaning as in Eqn (L1). Note that if we want only systems with the minimum number of column sections at a given state, except products, this should be produced at most by one contribution. Note also that when at least a state is produced by two contributions, the number of separation tasks is not the minimum.

3. All of the products must be produced at least by one task:

$$\underset{t \in (P\_REC_s \cup P\_STR_s)}{\vee} Y_t; \quad s \in FP \tag{L3}$$

4. If a given final product stream is produced only by one task, the heat exchanger associated with this state (product stream) must be selected. A given final product must always exist, produced by a rectifying section, a stripping section, or both. Therefore an equivalent form of expressing this logical relationship is that if a final product is not produced by any rectifying (stripping) section, the heat exchanger related to that product must exist:

$$\left.\begin{array}{c} \neg\left(\underset{t\in \text{P\_REC}_s}{\vee} Y_t\right) \Rightarrow W_s \\[2ex] \neg\left(\underset{t\in \text{P\_STR}_s}{\vee} Y_t\right) \Rightarrow W_s \end{array}\right\} \quad s\in \text{FP} \qquad \text{(L4)}$$

5. If a given state is produced by two tasks (a contribution coming from a rectifying section and the other from a stripping section of a task), then there is not a heat exchanger associated with that state (stream):

$$\left(Y_t\wedge Y_k\right) \Rightarrow \neg W_s \quad \left\{\begin{array}{l} t\in \text{RECT}_s \\ k\in \text{STRIP}_s \\ s\in \text{STATES} \end{array}\right. \qquad \text{(L5)}$$

6. Connectivity relationships between tasks in the superstructure are as follows:

$$\left.\begin{array}{ll} Y_t \Rightarrow \underset{k\in \text{TS}_s}{\vee} Y_k; & t\in \text{ST}_s \\[2ex] Y_t \Rightarrow \underset{k\in \text{ST}_s}{\vee} Y_k; & t\in \text{TS}_s \end{array}\right\} \quad s\in \text{STATES} \qquad \text{(L6)}$$

7. If a heat exchanger associated with any state is selected, then a task that generates that state must also be selected:

$$W_s \Rightarrow \underset{\text{ST}_s}{\vee} Y_t; \quad s\in \text{STATES} \qquad \text{(L7)}$$

8. If a separation task t produces a state s by a rectifying section, and that state has a heat exchanger associated, then it must be a condenser. If the state is produced by a stripping section then it must be a reboiler:

$$\begin{array}{ll} Y_t\wedge W_s \Rightarrow WC_s & t\in \text{RECT}_s \\ Y_t\wedge W_s \Rightarrow WR_s & t\in \text{STRIP}_s \end{array} \qquad \text{(L8)}$$

It is convenient to complete the previous rule by adding that:

9. If a given state does not have a heat exchanger, then both WC and WR associated with that state must be false:

$$\neg W_s \Rightarrow \neg WC_s\wedge\neg WR_s \quad s\in \text{STATES} \qquad \text{(L9)}$$

It is important to note that if the problem is solved as an MI(N)LP or GDP, the variables WC and WR do not need to be declared as binary and they can be considered as continuous with values between 0 and 1. Equations (L8) and (L9) force WC and WR to take integer values when Y and W are integer. Therefore, the variables WR and WC do not increase the combinatorial complexity of the problem.

10. It is worth mentioning that the set of logical rules previously presented in terms of separation tasks can be easily rewritten in terms only of states: *"There is a*

one to one correspondence between the sequence of tasks and the sequence of states and vice-versa." The relationship between tasks and states is as follows:

$$Y_t \Rightarrow Z_s; \quad t \in ST_S \tag{L10}$$

$$Z_s \Rightarrow \bigvee_{t \in TS_s} Y_t \tag{L11}$$

Equation (L10) can be read as: "If the task t, that belongs to the set of task produced by the state s, exists, then the state s must exist." And Eqn (L11) can be read as: "If the state s exists, then at least one of the tasks that the state s is able to produce must exist."

We should note that if the problem is solved as an MI(N)LP, it is only necessary to declare as binary either Y_t or Z_s, but not both. Whether or not Y_t is declared as binary, Z_s can be declared as continuous between 0 and 1, and vice-versa.

The previous equations ensure that any sequence of tasks and the selected heat exchangers is a feasible separation that can be arranged in $N-1$ distillation columns.

A detailed description of the model is too extensive to include here. The interested reader is referred to the original works [99,101,102]. However, a conceptual model showing the most important points in the model is presented here. The following is referred to a pure STN superstructure.

The objective function is any performance measure of the system, i.e. total annualized cost:

$$\min : \text{TAC} = \frac{r(1+r)^L}{(1+r)^L - 1}[Capital\ costs] + Operating\ costs \tag{11.71}$$

where the capital cost are annualized assuming a depreciation interval of L and an interest rate r (typical values are $r = 0.1$; $L = 10$ year).

The disjunctions related with the existence of a given task:

$$\begin{bmatrix} Y_t \\ Equations\ of\ the\ task \\ (shortcut, aggregated, rigorous...) \\ Task\ cost_t = f(V_R, V_S, L_R, L_S, ...) \end{bmatrix} \vee \begin{bmatrix} \neg Y_t \\ x_t = 0 \\ Task\ cost_t = 0 \end{bmatrix} \quad (D-1)$$

A graphical conceptual representation of this disjunction is shown in Figure 11.17.

If an intermediate state (Z_S) exists, this state could have associated a heat exchanger (W_S) or form a thermal couple. If there is a heat exchanger, this can be a condenser or a reboiler. If the state does not exist, then all flows related to that state must be zero. The conceptual graphical representation in Figure 11.18 illustrates this situation.

In the conceptual representation of Figure 11.18, the variable Z_S takes the value "true" if the state s exists, and takes the value "false" otherwise. It is important to recall that there is a one-to-one relationship between a sequence of tasks and a sequence of states. Therefore, the introduction of the new Boolean variable Z_S

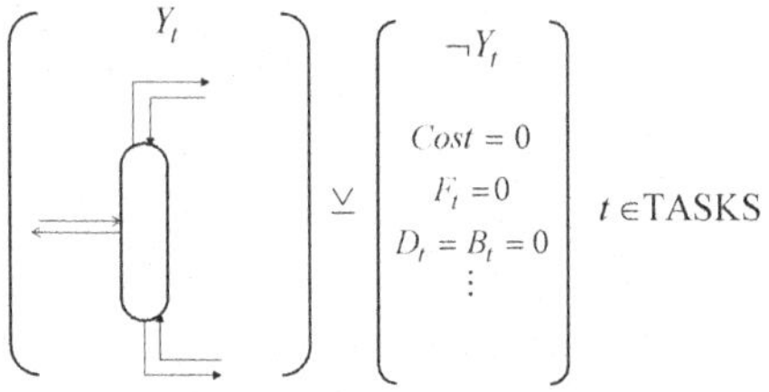

FIGURE 11.17 Conceptual Representation of Disjunction (D-1)

does not increase the combinatorial complexity of the model. Even more, if the problem is solved using an MINLP reformulation, it is necessary to define as binary variables those that are either related to tasks, Y_t, or those that are related to states, Z_S (the other variables can be defined as a continuous variables bounded between 0 and 1). The logical relationships will force the other set of variables to take the correct integer values.

The second term in the main disjunction (when the Boolean Z_S takes the value of false), is introduced for the sake of completeness, but it is redundant. Note that if a given state does not exist, the logical relationships will force that all the tasks that could be generated by the state, and all the tasks that could generate the state, do not exist as well. Therefore, the second term in disjunction D-1 also forces all the variables related to those tasks to be zero.

The disjunction inside the first term in Figure 11.18 is related to the existence or not of a heat exchanger in a state (if W_S is true, a heat exchanger is selected). Again, it is worth noting that there is a one-to-one relationship between assigning heat exchangers to states or to tasks. The logical relationships force that if one is selected (i.e. tasks), the corresponding correct state is selected, and vice versa. If the heat exchanger is selected, the equations are different depending if the heat exchanger is a condenser or a reboiler. The innermost disjunction (related to WC_S, true if the heat exchanger is a condenser; or WR_S, true if the heat exchanger is a reboiler) includes the energy balance in the condenser or reboiler and the cost equations.

If the heat exchanger does not exist ($\neg W_s$), but the state exists, then we have a thermal couple. The equations inside this term of the disjunction are simply mass balances to ensure the correct liquid and vapor flow transfer between columns.

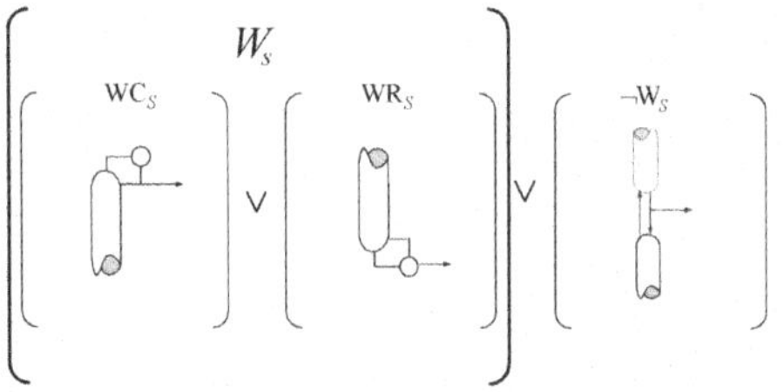

FIGURE 11.18 Disjunctive Representation of Alternatives if State _s_ Exists

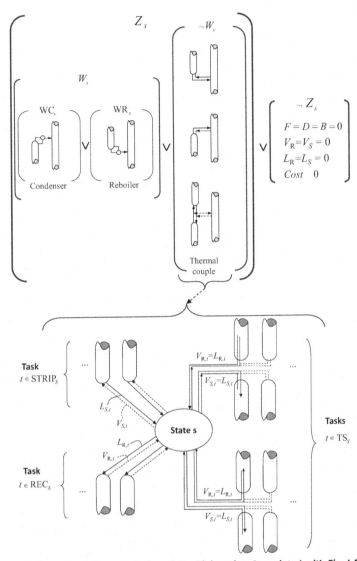

FIGURE 11.19 Conceptual Representation of the Disjunction Associated with Final States (Pure Products)

A final state is a state related with a pure product, or in general with a stream that leaves the system (the sequence of columns), and then these states must always exist. The most volatile product will always have a condenser and the heaviest a reboiler. However, the rest could have a condenser or a reboiler if it is produced by a single contribution, or no heat exchanger at all if produced by two contributions. In this last case the internal liquid and vapor flows of at least one of the tasks that generate the state must be adjusted to satisfy the mass balances. Figure 11.19 shows this situation.

The only remaining equations are mass balances in the initial mixture and desired recoveries of each final product.

References

[1] J. Humphrey, Separation processes: playing a critical role, Chem. Eng. Prog. 91 (10) (1995) 43–54.

[2] T. Mix, et al., Energy conservation in distillation, Chem. Eng. Prog. 74 (4) (1978) 49.

[3] G. Soave, J.A. Feliu, Saving energy in distillation towers by feed splitting, Appl. Therm. Eng. 22 (8) (2002) 889.

[4] J. Kunesh, et al., Distillation: still towering over other options, Chem. Eng. Prog. 91 (10) (1995) 43.

[5] D.F. Rudd, C.C. Watson, Strategy of Process Engineering, Wiley, New York, 1968.

[6] A.W. Westerberg, A Review of the Synthesis of Distillation Based Separation Systems. Department of Electrical and Computer Engineering, Paper 100, 1983. http://repository.cmu.edu/ece/100.

[7] M.R. Fenske, Fractionation of straight-run Pennsylvania gasoline, Ind. Eng. Chem. 24 (5) (1932) 482–485.

[8] E.R. Gilliland, Multicomponent rectification estimation of the number of theoretical plates as a function of the reflux ratio, Ind. Eng. Chem. 32 (9) (1940) 1220–1223.

[9] A. Underwood, Fractional distillation of multicomponent mixtures, Chem. Eng. Prog. 44 (1948) 603–614.

[10] Y.K. Molokanov, et al., An approximation method for calculating the basic parameters of multicomponent fraction, Int. Chem. Eng. 12 (2) (1972) 209.

[11] J.D. Seader, E.J. Henley, in: second ed.Separation Process Principles, John Willey & Sons, Inc., New Jersey, 2006.

[12] G.G. Kirkbride, Process design procedure for multicomponent fractionators, Pet. Refin. 23 (1944) 87.

[13] K.N. Glinos, M.F. Malone, Design of sidestream distillation-columns, Ind. Eng. Chem. Process Des. Dev. 24 (3) (1985) 822–828.

[14] K.N. Glinos, I.P. Nikolaides, M.F. Malone, New complex column arrangements for ideal distillation, Ind. Eng. Chem. Process Des. Dev. 25 (3) (1986) 694–699.

[15] I.P. Nikolaides, M.F. Malone, Approximate design of multiple feed/side draw distillation systems, Ind. Eng. Chem. Res. 26 (9) (1987) 1839–1845.

[16] T.E. Short, J.H. Erbar, Minimum reflux for complex fractionators, Pet. Chem. Eng. 11 (1963) 180–184.

[17] H. Sugie, B.C.Y. Lu, On the determination of minimum reflux ratio for a multicomponent distillation column with any number of side-cut streams, Chem. Eng. Sci. 25 (1970) 1837–1846.

[18] J.A. Wachter, T.K.T. Ko, R.P. Andres, Minimum reflux behavior of complex distillation columns, AIChE J. 34 (1988) 1164–1184.

[19] F.J. Barnes, D.N. Hanson, C.J. King, Calculation of minimum reflux for distillation columns with multiple feeds, Ind. Eng. Chem. Process Des. Dev. 11 (1972) 136–140.

[20] N.A. Carlberg, A.W. Westerberg, Temperature heat diagrams for complex columns. 2. *Underwoods method for side strippers and enrichers*, Ind. Eng. Chem. Res. 28 (9) (1989) 1379−1386.

[21] N.A. Carlberg, A.W. Westerberg, Temperature heat diagrams for complex columns. 3. *Underwoods method for the Petlyuk configuration*, Ind. Eng. Chem. Res. 28 (9) (1989) 1386−1397.

[22] L. Guilian, et al., Shortcut design for columns separating azeotropic mixtures, Ind. Eng. Chem. Res. 43 (14) (2004) 3908−3923.

[23] R.S. Kamath, I.E. Grossmann, L.T. Biegler, Aggregate models based on improved group methods for simulation and optimization of distillation systems, Comput. Chem. Eng. 34 (8) (2010) 1312−1319.

[24] A. Kremser, Theoretical analysis of absorption process, Natl. Pet. News 22 (1930) 43−49.

[25] W.C. Edmister, Design of hydrocarbon absorption and stripping, Ind. Eng. Chem. 35 (1943) 837−839.

[26] W.C. Edmister, Absorption and stripping factor functions for distillation calculation by manual and digital computer methods, AIChE J 3 (1957) 165−171.

[27] J.A. Caballero, I.E. Grossmann, Aggregated models for integrated distillation systems, Ind. Eng. Chem. Res. 38 (6) (1999) 2330−2344.

[28] M.J. Bagajewicz, V. Manousiouthakis, Mass/heat-exchange network representation of distillation networks, AIChE J. 38 (11) (1992) 1769−1800.

[29] S.G. Levy, D.B. Van Dongen, M.F. Doherty, Design and synthesis of homogeneous azeotropic distillations. 2. *Minimum reflux calculations for nonideal and azeotropic columns*, Ind. Eng. Chem. Fundam. 24 (4) (1985) 463−474.

[30] D. Barbosa, M.F. Doherty, Design and minimum-reflux calculations for single-feed multicomponent reactive distillation columns, Chem. Eng. Sci. 43 (7) (1988) 1523−1537.

[31] Z.T. Fidkowski, M.F. Doherty, M.F. Malone, Feasibility of separations for distillation of nonideal ternary mixtures, AIChE J. 39 (8) (1993) 1303−1321.

[32] Z.T. Fidkowski, M.F. Malone, M.F. Doherty, Nonideal multicomponent distillation: use of bifurcation theory for design, AIChE J. 37 (12) (1991) 1761−1779.

[33] V. Julka, M.F. Doherty, Geometric behavior and minimum flows for nonideal multicomponent distillation, Chem. Eng. Sci. 45 (7) (1990) 1801−1822.

[34] S.G. Levy, M.F. Doherty, Design and synthesis of homogeneous azeotropic distillations. 4. *Minimum reflux calculations for multiple-feed columns*, Ind. Eng. Chem. Fundam. 25 (2) (1986) 269−279.

[35] H.N. Pham, P.J. Ryan, M.F. Doherty, Design and minimum reflux for heterogeneous azeotropic distillation columns, AIChE J. 35 (10) (1989) 1585−1591.

[36] J. Bausa, R. von Watzdorf, W. Marquardt, Shortcut methods for nonideal multicomponent distillation: 1. *Simple columns*, AIChE J. 44 (10) (1998) 2181−2198.

[37] J. Koehler, P. Aguirre, E. Blass, Minimum reflux calculations for nonideal mixtures using the reversible distillation model, Chem. Eng. Sci. 46 (12) (1991) 3007−3021.

[38] R. Gani, E. Bek-Pedersen, Simple new algorithm for distillation column design, AIChE J. 46 (6) (2000) 1271−1274.

[39] A. Lucia, R. Taylor, The geometry of separation boundaries: I. *Basic theory and numerical support*, AIChE J. 52 (2) (2006) 582−594.

[40] A. Lucia, A. Amale, R. Taylor, Distillation pinch points and more, Comput. Chem. Eng. 32 (6) (2008) 1342−1364.

[41] A. Lucia, B.R. McCallum, Energy targeting and minimum energy distillation column sequences, Comput. Chem. Eng. 34 (2010) 931−942.

[42] J. Viswanathan, I.E. Grossmann, A combined penalty function and outer-approximation method for MINLP optimization, Comput. Chem. Eng. 14 (7) (1990) 769−782.

[43] R.W.H. Sargent, K. Gaminibandara, Optimum design of plate distillation columns, in: L.W.C. Dixon (Ed.), Optimization in Action, Academic Press, London, 1976, p. 267.

[44] M. Barttfeld, P.A. Aguirre, I.E. Grossmann, Alternative representations and formulations for the economic optimization of multicomponent distillation columns, Comput. Chem. Eng. 27 (3) (2003) 363−383.

[45] J. Viswanathan, I.E. Grossmann, Optimal feed locations and number of trays for distillation-columns with multiple feeds, Ind. Eng. Chem. Res. 32 (11) (1993) 2942−2949.

[46] J. Viswanathan, I.E. Grossmann, An alternate MINLP model for finding the number of trays required for a specified separation objective, Comput. Chem. Eng. 17 (9) (1993) 949−955.

[47] A.R. Ciric, D. Gu, Synthesis of nonequilibrium reactive distillation processes by MINLP optimization, AIChE J. 40 (9) (1994) 1479−1487.

[48] M.H. Bauer, J. Stichlmair, Design and economic optimization of azeotropic distillation processes using mixed-integer nonlinear programming, Comput. Chem. Eng. 22 (9) (1998) 1271−1286.

[49] G. Dunnebier, C.C. Pantelides, Optimal design of thermally coupled distillation columns, Ind. Eng. Chem. Res. 38 (1) (1999) 162−176.

[50] I.E. Grossmann, P.A. Aguirre, M. Barttfeld, Optimal synthesis of complex distillation columns using rigorous models, Comput. Chem. Eng. 29 (6) (2005) 1203−1215.

[51] H. Yeomans, I.E. Grossmann, Optimal design of complex distillation columns using rigorous tray-by-tray disjunctive programming models, Ind. Eng. Chem. Res. 39 (11) (2000) 4326−4335.

[52] M. Barttfeld, P.A. Aguirre, Optimal synthesis of multicomponent zeotropic distillation processes. 1. *Preprocessing Phase and rigorous optimization for a single unit*, Ind. Eng. Chem. Res. 41 (21) (2002) 5298−5307.

[53] A. Harwardt, W. Marquardt, Heat-integrated distillation columns: vapor recompression or internal heat integration? AIChE Journal (58) (2012) 3740.

[54] K. Kraemer, S. Kossack, W. Marquardt, An efficient solution method for the MINLP optimization of chemical processes, Comput. Aided Chem. Eng. (2007) 105−110.

[55] J.R. Jackson, I.E. Grossmann, A disjunctive programming approach for the optimal design of reactive distillation columns, Comput. Chem. Eng. 25 (11−12) (2001) 1661−1673.

[56] M. Barttfeld, P.o.A. Aguirre, I.E. Grossmann, A decomposition method for synthesizing complex column configurations using tray-by-tray GDP models, Comput. Chem. Eng. 28 (11) (2004) 2165−2188.

[57] M.H. Bauer, J. Stichlmair, Superstructures for the mixed integer optimization of nonideal and azeotropic distillation processes, Comput. Chem. Eng. 20 (1996) S25−S30.

[58] I.K. Kookos, Optimal design of membrane/distillation column hybrid processes, Ind. Eng. Chem. Res. 42 (2003) 1731−1738.

[59] F.J. Lockhart, Multi-column distillation of natural gasoline, Pet. Refin. (26) (1947) 104.

[60] V.D. Harbert, Which tower goes where? Pet. Refin. 36 (3) (1957) 169.

[61] D.L. Heaven, Optimum Sequence of Distillation Columns in Multicomponent Fractionation, University of California, Berkeley, 1969.

[62] D.F. Rudd, G.J. Powers, J.J. Siirola, Process Synthesis, Prentice Hall, New York, 1973.

[63] J.D. Seader, A.W. Westerberg, A combined heuristic and evolutionary strategy for synthesis of simple separation sequences, AIChE J. 23 (1977) 951.

[64] R.W. Thomson, C.J. King, Systematic synthesis of separation schemes, AIChE J. (1972) 941.

[65] J.E. Hendry, R.E. Hughes, Generating process separations flowsheets, Chera. Eng. Prog. 68 (1972) 69.

[66] A.W. Westerberg, G. Stephanopoulos, Studies in process synthesis − I. *Branch and bound strategy with list techniques for the synthesis of separation schemes*, Chem. Eng. Sci. 30 (1975) 963.

[67] B.F.R. Rodrigo, J.D. Seader, Synthesis of separation sequences by ordered branch search, AIChE J. 21 (1975) 885.

[68] M.A. Gomez, J.D. Seader, Separator sequence synthesis by a predictor based ordered search, AIChE J. (1976) 970.

[69] I.E. Grossmann, J.A. Caballero, H. Yeomans, Mathematical programming approaches to the synthesis of chemical process systems, Korean J. Chem. Eng. 16 (4) (1999) 407−426.

[70] H. Yeomans, I.E. Grossmann, A systematic modeling framework of superstructure optimization in process synthesis, Comput. Chem. Eng. 23 (6) (1999) 709−731.

[71] E. Kondili, C.C. Pantelides, R.W.H. Sargent, A general algorithm for short-term scheduling of batch-operations .1. *Milp formulation*, Comput. Chem. Eng. 17 (2) (1993) 211−227.

[72] E.M. Smith, C. Pantelides, Design of reaction/separation networks using detailed models, Comput. Chem. Eng. S83 (1995) 19.

[73] M.J. Andrecovich, A.W. Westerberg, An MILP formulation for heat-integrated distillation sequence synthesis, AIChE J. 31 (9) (1985) 1461−1474.

[74] Z. Novak, Z. Kravanja, I.E. Grossmann, Simultaneous synthesis of distillation sequences in overall process schemes using an improved MINLP approach, Comput. Chem. Eng. 20 (12) (1996) 1425−1440.

[75] H. Yeomans, I.E. Grossmann, Disjunctive programming models for the optimal design of distillation columns and separation sequences, Ind. Eng. Chem. Res. 39 (6) (2000) 1637−1648.

[76] K.P. Papalexandri, E.N. Pistikopoulos, Generalized modular representation framework for process synthesis, AIChE J. 42 (4) (1996) 1010−1032.

[77] M.J. Andrecovich, A.W. Westerberg, A simple synthesis method based on utility bounding for heat-integrated distillation sequences, AIChE J. 31 (3) (1985) 363−375.

[78] G.E. Paules, C.A. Floudas, Stochastic-programming in process synthesis − a 2-Stage model with MINLP recourse for multiperiod heat-integrated distillation sequences, Comput. Chem. Eng. 16 (3) (1992) 189−210.

[79] C.A. Floudas, G.E. Paules, A mixed-integer nonlinear programming formulation for the synthesis of heat-integrated distillation sequences, Comput. Chem. Eng. 12 (6) (1988) 531−546.

[80] R. Raman, I.E. Grossmann, Symbolic-integration of logic in mixed-integer linear-programming techniques for process synthesis, Comput. Chem. Eng. 17 (9) (1993) 909−927.

[81] F.B. Petlyuk, V.M. Platonov, D.M. Slavinsk, Thermodynamically optimal method for separating multicomponent mixtures, Int. Chem. Eng. 5 (3) (1965) 555.

[82] Z.T. Fidkowski, R. Agrawal, Multicomponent thermally coupled systems of distillation columns at minimum reflux, AIChE J. 47 (12) (2001) 2713−2724.

[83] H. Rudd, Thermal coupling for energy efficiency, Chem. Engineer-London (525) (1992) S14−S15.

[84] C. Triantafyllou, R. Smith, The design and optimization of fully thermally coupled distillation-columns, Chem. Eng. Res. Des. 70 (2) (1992) 118−132.

[85] I.J. Halvorsen, S. Skogestad, Minimum energy consumption in multicomponent distillation. 1. *V-min diagram for a two-product column*, Ind. Eng. Chem. Res. 42 (3) (2003) 596−604.

[86] I.J. Halvorsen, S. Skogestad, Minimum energy consumption in multicomponent distillation. 2. *Three-product Petlyuk arrangements*, Ind. Eng. Chem. Res. 42 (3) (2003) 605−615.

[87] I.J. Halvorsen, S. Skogestad, Minimum energy consumption in multicomponent distillation. 3. *More than three products and generalized Petlyuk arrangements*, Ind. Eng. Chem. Res. 42 (3) (2003) 616−629.

[88] R.O. Wright, Fractionation Apparatus. U.S. Patent, Editor, United States, 1949.

[89] R. Agrawal, Z.T. Fidkowski, More operable arrangements of fully thermally coupled distillation columns, AIChE J. 44 (11) (1998) 2565−2568.

[90] G. Kaibel, Distillation columns with vertical partitions, Chem. Eng. Tech. 10 (1987) 92.

[91] R. Agrawal, A method to draw fully thermally coupled distillation column configurations for multicomponent distillation, Chem. Eng. Res. Des. 78 (A3) (2000) 454−464.

[92] J.A. Caballero, I.E. Grossmann, Thermodynamically equivalent configurations for thermally coupled distillation, AIChE J. 49 (11) (2003) 2864−2884.

[93] M. Serra, A. Espuna, L. Puigjaner, Control and optimization of the divided wall column, Chem. Eng. Process. 38 (4−6) (1999) 549−562.

[94] M. Serra, A. Espuna, L. Puigjaner, Controllability of different multicomponent distillation arrangements, Ind. Eng. Chem. Res. 42 (8) (2003) 1773−1782.

[95] M. Serra, et al., Study of the divided wall column controllability: influence of design and operation, Comput. Chem. Eng. 24 (2−7) (2000) 901−907.

[96] M. Serra, et al., Analysis of different control possibilities for the divided wall column: feedback diagonal and dynamic matrix control, Comput. Chem. Eng. 25 (4−6) (2001) 859−866.

[97] E.A. Wolff, S. Skogestad, Operation of integrated 3-Product (Petlyuk) distillation-columns, Ind. Eng. Chem. Res. 34 (6) (1995) 2094−2103.

[98] R. Agrawal, Synthesis of distillation column configurations for a multicomponent separation, Ind. Eng. Chem. Res. 35 (4) (1996) 1059−1071.

[99] J.A. Caballero, I.E. Grossmann, Generalized disjunctive programming model for the optimal synthesis of thermally linked distillation columns, Ind. Eng. Chem. Res. 40 (10) (2001) 2260−2274.

[100] J.A. Caballero, I.E. Grossmann, Logic based methods for generating and optimizing thermally coupled distillation systems, in: European Symposium on Computer Aided Process Engineering-12, Elsevier, The Hague, The Netherlands, 2002.

[101] J.A. Caballero, I.E. Grossmann, Design of distillation sequences: from conventional to fully thermally coupled distillation systems, Comput. Chem. Eng. 28 (11) (2004) 2307–2329.

[102] J.A. Caballero, I.E. Grossmann, Structural considerations and modeling in the synthesis of heat-integrated-thermally coupled distillation sequences, Ind. Eng. Chem. Res. 45 (25) (2006) 8454–8474.

[103] R. Agrawal, Synthesis of multicomponent distillation column configurations, AIChE J. 49 (2) (2003) 379–401.

[104] A. Giridhar, R. Agrawal, Synthesis of distillation configurations: I. *Characteristics of a good search space*, Comput. Chem. Eng. 34 (1) (2010) 73.

[105] A. Giridhar, R. Agrawal, Synthesis of distillation configurations. *II. A search formulation for basic configurations*, Comput. Chem. Eng. 34 (1) (2010) 84.

[106] A.J. Brugma, Process and Device for Fractional Distillation of Liquid Mixtures. U.S. Patent 2.295.256, 1942.

[107] J.K. Kim, P.C. Wankat, Quaternary distillation systems with less than N-1 columns, Ind. Eng. Chem. Res. 43 (14) (2004) 3838–3846.

[108] M. Errico, B.-G. Rong, Modified simple column configurations for quaternary distillations, Comput. Chem. Eng. 36 (0) (2012) 160–173.

[109] R. Agrawal, More operable fully thermally coupled distribution column configurations for multicomponent distillation, Chem. Eng. Res. Des. 77 (A6) (1999) 543–553.

[110] J. Ivakpour, N. Kasiri, Synthesis of distillation column sequences for nonsharp separations, Ind. Eng. Chem. Res. 48 (18) (2009) 8635–8649.

[111] V.H. Shah, R. Agrawal, A matrix method for multicomponent distillation sequences, AIChE J. 56 (7) (2010) 1759–1775.

Appendix

Optimization background

The economic optimization of a distillation column involves the selection of the number of trays and feed location, as well as the operating conditions to minimize the total investment and operating cost. Continuous decisions are related to the operating conditions and energy involved in the separation, while discrete decisions are related to the total number of trays and the tray positions of each feed and product streams. If we also consider the order in which the separation is performed, it is possible to find significant differences in total cost and energy consumption between a good sequence and a bad one [1]. As the number of components increases, the number of possible alternatives can be enormous, and the selection of the correct sequence becomes a major challenge.

Using an equation-based environment there are two major formulations for the mathematical representation of problems involving discrete and continuous variables: Mixed-Integer Nonlinear Programming (MINLP) and Generalized Disjunctive Programming (GDP) [2]. Both approaches have been employed in the literature to model distillation columns.

MINLP methods

The most common form of MINLP problems is the special case in which 0−1 variables are linear while the continuous variables are nonlinear:

$$\begin{aligned}
&\min : Z = \mathbf{c}^T \mathbf{y} + f(\mathbf{x}) \\
&s.t.\ \mathbf{Dy} + \mathbf{h}(\mathbf{x}) = \mathbf{0} \\
&\quad\ \mathbf{By} + \mathbf{g}(\mathbf{x}) \leq \mathbf{0} \qquad \text{(MINLP)} \\
&\quad\ \mathbf{x} \in X, \mathbf{y} \in \{0, 1\}^m
\end{aligned}$$

Major methods for solving MINLP problems include the Branch and Bound method [3−5], which is an extension of the linear case, except that NLP subproblems are solved at each node. The Generalized Benders Decomposition (GBD) [6,7] and Outer Approximation (OA) methods [8−11] are iterative methods that solve a sequence of alternate NLP subproblems and MILP master problems that predict lower bounds and new values for the 0−1 variables. The difference between GBD and OA methods lies in the definition of the MILP master problem. The OA method uses accumulated linearizations of the objective function and the constraints, while GBD uses accumulated Lagrangean functions parametric in 0−1 variables. The LP/NLP-based Branch and Bound method [12,13] essentially integrates both subproblems within one tree search. The Extended Cutting Plane (ECP) method [14,15] does not solve NLP problems and relies only on successive linearizations. All of these algorithms can be classified in terms of the following basic subproblems [2] that are involved in each of these methods.

NLP subproblems

1. NLP relaxation:

$$\begin{aligned}
&\min : Z_{\text{LB}} = \mathbf{c}^T \mathbf{y} + f(\mathbf{x}) \\
&s.t.\ \mathbf{Dy} + \mathbf{h}(\mathbf{x}) = \mathbf{0} \\
&\quad\ \mathbf{By} + \mathbf{g}(\mathbf{x}) \leq \mathbf{0} \\
&\quad\ \mathbf{x} \in X, \qquad\qquad\qquad \text{(NLP} - \text{R)} \\
&\quad\ 0 \leq y_j \leq 1 \quad j \in REL \\
&\quad\ y_i \in \{0, 1\}^{m - |\text{REL}|}
\end{aligned}$$

Where the subset of binary variables REL is relaxed to continuous bounded by their extreme values (0−1). When the dynamic subset REL includes all the binary variables, NLP-R corresponds to the continuous relaxation and provides an absolute lower bound to the MINLP problem.

2. NLP subproblem for fixed y^k:

$$\begin{aligned}
&\min : Z_{\text{UB}}^k = \mathbf{c}^T \mathbf{y}^k + f(\mathbf{x}) \\
&s.t.\ \mathbf{Dy}^k + \mathbf{h}(\mathbf{x}) = \mathbf{0} \qquad \text{(NLP)} \\
&\quad\ \mathbf{By}^k + \mathbf{g}(\mathbf{x}) \leq \mathbf{0} \\
&\quad\ \mathbf{x} \in X
\end{aligned}$$

Which yields an upper bound to the MINLP problem, provided that this NLP problem has a feasible solution. If this is not the case, the following feasibility subproblem must be solved.

3. Feasibility problems for fixed y^k:

$$
\begin{aligned}
\min : \; & \beta \\
s.t. \; & \beta \geq \mathbf{s}^1 \\
& \beta \geq \mathbf{s}^2 \\
& \beta \geq \mathbf{u} \\
& \mathbf{D}\mathbf{y}^k + \mathbf{h}(\mathbf{x}) = \mathbf{s}^1 - \mathbf{s}^2 \qquad \text{(NLP – F)} \\
& \mathbf{B}\mathbf{y}^k + \mathbf{g}(\mathbf{x}) \leq \mathbf{u} \\
& \mathbf{s}^1 \geq 0, \mathbf{s}^2 \geq 0, \mathbf{u} \geq 0 \\
& \mathbf{x} \in X; \quad \beta \in \Re^1
\end{aligned}
$$

This can be interpreted as the minimization of the infinity norm as a measure of the infeasibility of the corresponding NLP subproblem. It should be noted that for an infeasible subproblem the solution of the NLP-F yields a positive value of the scalar beta.

Master (MILP) cutting plane

The convexity of the nonlinear functions is exploited by replacing them with supporting hyperplanes, that are generally, but not necessarily, derived at the solution of the NLP subproblem. In particular, the new binary values (y^{k+1}) are obtained from an MILP cutting plane problem that is based on K points (x^k), $k = 1,2,\dots$ (K is generated at the K previous steps):

$$
\left.
\begin{aligned}
\min : \; & Z_L^k = \alpha \\
s.t. \; & \alpha \geq \mathbf{c}^T \mathbf{y} + f\left(\mathbf{x}^k\right) + \nabla f\left(\mathbf{x}^k\right)^T \left(\mathbf{x} - \mathbf{x}^k\right) \\
& sign\left(\lambda_j\right)\left[\mathbf{d}_j^T \mathbf{y} + h_j\left(\mathbf{x}^k\right) + \nabla h_j(\mathbf{x}^k)^T\left(\mathbf{x} - \mathbf{x}^k\right)\right] \leq 0 \quad j \in J \\
& \mathbf{b}_i^T \mathbf{y} + g_i\left(\mathbf{x}^k\right) + \nabla g_i(\mathbf{x}^k)^T\left(\mathbf{x} - \mathbf{x}^k\right) \quad i \in I \\
& \mathbf{x} \in X; \quad \mathbf{y} \in \{0,1\}^m
\end{aligned}
\right\} k = 1,2,3\dots K \quad \text{(M – MILP)}
$$

where λ_j are the multipliers of the corresponding equation $j \in J$ in the NLP problem. The index j makes reference to the equations in J and the index i to the inequalities in I. The solution of M-MILP yields a valid lower bound to the original MINLP problem, which is nondecreasing with the number of linearization points K.

The different methods can be classified according to their use of the subproblems (NLP-R; NLP, NLP-F) and the specific specialization of the M-MILP, as seen in Figure 11.A1.

Generalized disjunctive programming

An alternative approach for representing discrete–continuous optimization problems is by using models consisting of algebraic constraints, logic disjunctions, and logic propositions [16–20] This approach not only facilitates the development

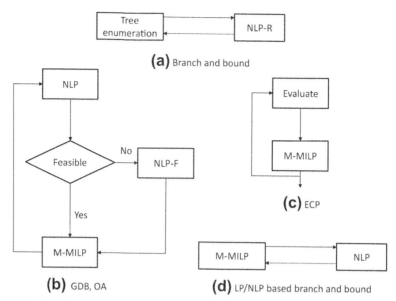

FIGURE 11.A1 Major Steps Involved in the MINLP Algorithms

of the models by making the formulation intuitive, but it also keeps in the model the underlying logic structure of the problem that can be exploited to find the solution more efficiently. The general structure of a GDP can be represented as follows [21]:

$$
\begin{aligned}
\min Z &= f(x) + \sum_{k \in K} c_k \\
s.t. \quad & g(x) \leq 0 \\
& \bigvee_{i \in D_k}
\begin{bmatrix}
Y_{i,k} \\
r_{i,k}(x) \leq 0 \\
c_k = \gamma_{i,k}
\end{bmatrix}
\quad k \in K \qquad \text{(GDP)} \\
& \Omega(Y) = True \\
& x^{Lo} \leq x \leq x^{Up} \\
& x \in \Re^n, \quad c_k \in \Re^1, \quad Y_{i,k} \in \{True, False\}
\end{aligned}
$$

where $f: R^n \rightarrow R^1$ is a function of the continuous variables x in the objective function. $g: R^n \rightarrow R^1$ belongs to the set of global constraints, and the disjunctions $k \in K$ are composed of a number or terms $i \in D_k$ that are connected by the OR operator. In each term there is a Boolean variable $Y_{i,k}$, a set of inequalities $r_{i,k}: R^n \rightarrow R^m$, and a cost variable c_k. If $Y_{i,k}$ is True, then $r_{i,k} \leq 0$ and $c_{i,k} = \gamma_{i,k}$ are enforced, otherwise they are ignored. The $\Omega(Y) = True$ are logic propositions for the Boolean variables expressed in the conjunctive normal form:

$$\Omega(Y) = \underset{t=1,2...T}{\wedge} \left[\underset{Y_{i,k} \in R_t}{\vee} \left(Y_{i,k} \right) \underset{Y_{i,k} \in Q_t}{\vee} \left(\neg Y_{i,k} \right) \right] \quad \text{(CNF)}$$

where for each clause t, $t = 1, 2, 3...T$, R_t is the subset of Boolean variables that are non-negated, and Q_t is the subset of Boolean variables that are negated. It is assumed that the logic constraints $\underset{i \in D_k}{\vee} Y_{i,k}$ are included in the general equation $\Omega(Y) = True$.

In order to take advantage of the existing MINLP solvers, GDPs are often reformulated as an MINLP problem and solved using the standard solvers. In order to do so, two main transformations can be used in which the disjunctive constraints are expressed in terms of algebraic equations and the propositional logic is expressed in terms of linear equations.

The disjunctive constraints in (GDP) can be transformed by using either the big-M (BM) [22] or the convex hull reformulation (CH) [18].

The BM reformulation is as follows:

$$\begin{aligned}
&r_{i,k}(x) \leq M(1 - y_{i,k}) \quad i \in D_k, \quad k \in K \\
&\sum_{i \in D_k} y_{i,k} = 1 \quad k \in K \\
&x \in R^n \\
&y_{i,k} \in \{0, 1\} \quad i \in D_k, \quad k \in K
\end{aligned} \quad \text{(BMR)}$$

Where the variable $y_{i,k}$ has a one-to-one correspondence with the Boolean variable $Y_{i,k}$. If the binary variable takes a value of one the inequality constraint is enforced; otherwise, if the parameter M is large enough the constraint becomes redundant.

The CH reformulation can be written as follows:

$$x = \sum_{i \in D_k} v^{i,k} \quad k \in K$$

$$y_{i,k} r\left(v^{i,k} / y_{i,k} \right) \leq 0 \quad i \in D_k; \quad k \in K$$

$$y_{i,k} x^{\text{Lo}} \leq v^{i,k} \leq y_{i,k} x^{\text{Up}} \quad i \in D_k; \quad k \in K \quad \text{(CHR)}$$

$$\sum_{i \in D_k} y_{i,k} = 1 \quad k \in K$$

$$x \in R^n, \quad v^{i,k} \in R^n, \quad y_{i,k} \in \{0, 1\} \quad i \in D_k; \quad k \in K$$

There is also a one-to-one correspondence between disjunctions in GDP and CH. The size of the problem is increased by introducing a new set of disaggregated variables $v^{i,k}$ as well as new constraints. On the other hand, as proved in Grossmann and Lee [23] and extensively discussed by Vecchietti et al. [24], the convex hull reformulation is at least as tight and generally tighter than the BM when the discrete domain is relaxed, which can impact the efficiency of MINLP solvers since they rely heavily on the quality of those relaxations.

It is worth remarking that the term $y_{i,k} r\left(v^{i,k}/y_{i,k}\right) \leq 0$ is convex if $r_{i,k}(x)$ is a convex function, but requires an adequate approximation to avoid singularities. Sawaya and Grossmann [25] proposed the following reformulation, which yields an exact approximation for values of binaries equal to 1 or 0, for any value of $\varepsilon \in [0,1]$ in where the feasibility and convexity are maintained:

$$y_{i,k} r\left(v^{i,k}/y_{i,k}\right) \approx \left[(1-\varepsilon)y_{i,k} + \varepsilon\right] r\left(v^{i,k}/\left[(1-\varepsilon)y_{i,k} + \varepsilon\right]\right) - \varepsilon r_{i,k}(0)\left(1 - y_{i,k}\right)$$

(A11.1)

It should be note that the approximation in (A11.1) assumes that $r_{i,k}(x)$ is defined in $x = 0$ and that the inequality $y_{i,k}x^{Lo} \leq v^{i,k} \leq y_{i,k}x^{Up}$ is enforced.

The propositional logic in terms of Boolean variables, conjunctions (AND operator), disjunctions (OR operator), negations, implications, equivalences (double implications), and exclusive disjunctions (XOR operator) is transformed in a set of linear algebraic constraints in terms only of binary variables [19,26,27] (again there is a one-to-one relationship between binary and Boolean variables). This set of new linear equations is only feasible if the original set of logical propositions is true. This transformation can be done systematically through a set of three recursive steps to get the logic in its conjunctive normal form. Once the logic is expressed in its conjunctive normal form, the transformation is straightforward. Details of the procedure can be found, for example in the text book by Biegler et al. [27]. The final result is a set of linear equations than can be added either in the BMR or in the CHR problems:

$$\mathbf{A}\mathbf{y} \leq \mathbf{b}$$
$$\mathbf{y} \in \{0, 1\}^m$$

(A11.2)

In order to fully exploit the logic structure of the GDP problems, two other solution methods have been proposed for the case of convex nonlinear GDP problems: the Disjunctive Branch and Bound (DBB) [18] and the Logic Based Outer Approximation (LBOA) methods [28].

The basic idea of the DBB method is to directly branch on the constraints corresponding to particular terms in the disjunctions, while considering the convex hull of the remaining disjunctions. Although the tightness of the relaxation at each node is comparable with the obtained when solving the CH reformulation, the size of the problems solved are smaller and the numerical robustness improved.

For the case of the LBOA method, the idea is similar to the OA for MINLP problems. The main idea is to solve iteratively a master problem given by a linear GDP problem, which will give a lower bound of the solution and an NLP subproblem that will give an upper bound. The fixed values of the Boolean variables determine which equations must be included in each NLP subproblem:

$$\min : Z = f(x) + \sum_{k \in K} c_k$$

$$s.t. \ g(x) \leq 0$$

$$\left. \begin{array}{l} r_{i,k}(x) \leq 0 \\ c_k = \gamma_{i,k} \end{array} \right\} for \ Y_{i,k} = True; \ i \in D_k, \ k \in K \quad (S-NLP)$$

$$x^{Lo} \leq x \leq x^{Up}$$

$$x \in R^n, \ c_k \in R^1$$

As only the constraints that belong to the active terms in the disjunction are imposed, the result is a reduced-size S-NLP compared to the direct application of OA in the MINLP reformulation. The initialization requires that all the terms in the disjunctions appear at least once in any NLP, so it is initially necessary to solve a set of S-NLP subproblems to accomplish this requirement. The master linear GDP can then be written as follows:

$$\min Z = \alpha + \sum_{k \in K} c_k$$

$$\left. \begin{array}{l} s.t. \ \alpha \geq f\left(x^l\right) + \nabla f\left(x^l\right)^T \left(x - x^l\right) \\ g\left(x^l\right) + \nabla g\left(x^l\right)^T \left(x - x^l\right) \leq 0 \end{array} \right\} l = 1, 2, 3 \ldots L$$

$$\bigvee_{i \in D_k} \left[\begin{array}{c} Y_{i,k} \\ r_{i,k}\left(x^l\right) + \nabla r_{i,k}\left(x^l\right)^T \left(x - x^l\right) \leq 0 \quad l \in L_{i,k} \\ c_k = \gamma_{i,k} \end{array} \right] k \in K \quad (M-LGDP)$$

$$\Omega(Y) = True$$

$$x^{Lo} \leq x \leq x^{Up}$$

$$\alpha \in R^1; \ x \in R^n; \ c_k \in R^1; \ Y_{i,k} \in \{True, False\}; \ i \in D_k, \ k \in K$$

References

[1] A.W. Westerberg, A Review of the Synthesis of Distillation Based Separation Systems. Department of Electrical and Computer Engineering, Paper 100, 1983. http://repository. cmu.edu/ece/100.

[2] I.E. Grossmann, Review of nonlinear mixed-integer and disjunctive programming techniques, Optim. Eng. (3) (2002) 227–252.

[3] B. Borchers, J.E. Mitchell, An improved branch and bound algorithm for mixed integer nonlinear programs, Comput. Oper. Res. 21 (4) (1994) 359–367.

[4] O.K. Gupta, V. Ravindran, Branch and bound experiments in convex nonlinear integer programming, Manage. Sci. 31 (12) (1985) 1533.

[5] R.A. Stubbs, S. Mehrotra, A branch-and-cut method for 0-1 mixed convex programming, Math. Program. 86 (3) (1999) 515–532.

[6] J.F. Benders, Partitioning procedures for solving mixed-variables programming problems, Numerische Mathematik 4 (1) (1962) 238–252.

[7] A.M. Geofrion, Generalized benders decomposition, J. Optim. Theory Appl. 10 (4) (1972) 237–259.

[8] M.A. Duran, I.E. Grossmann, An outer-approximation algorithm for a class of mixed-integer nonlinear programs, Math. Program. 36 (3) (1986) 307–339.

[9] R. Fletcher, S. Leyffer, Solving mixed integer nonlinear programs by outer approximation, Math. Program. 66 (1) (1994) 327–349.

[10] G.R. Kocis, I.E. Grossmann, Relaxation strategy for the structural optimization of process flow sheets, Ind. Eng. Chem. Res. 26 (9) (1987) 1869–1880.

[11] X. Yuan, et al., Une Methode d'optimisation Nonlineare en Variables Mixtes pour la Conception de Procedes, RAIRO 22 (1988) 331.

[12] P. Bonami, et al., An algorithmic framework for convex mixed integer nonlinear programs, Discrete Optim. 5 (2) (2008) 186–204.

[13] I. Quesada, I.E. Grossmann, An LP/NLP based branch and bound algorithm for convex MINLP optimization problems, Comput. Chem. Eng. 16 (10–11) (1992) 937–947.

[14] T. Westerlund, R. Pörn, Solving pseudo-convex mixed integer optimization problems by cutting plane techniques, Optim. Eng. 3 (3) (2002) 253–280.

[15] T. Westerlund, F. Pettersson, An extended cutting plane method for solving convex MINLP problems, Comput. Chem. Eng. 19 (Suppl. 1(0)) (1995) 131–136.

[16] N. Beaumont, An algorithm for disjunctive programs, Eur. J. Oper. Res. 48 (3) (1990) 362–371.

[17] J.N. Hooker, M.A. Osorio, Mixed logical-linear programming, Discrete Appl. Math. 97 (1999) 395–442.

[18] S. Lee, I.E. Grossmann, New algorithms for nonlinear generalized disjunctive programming, Comput. Chem. Eng. 24 (9–10) (2000) 2125–2141.

[19] R. Raman, I.E. Grossmann, Modeling and computational techniques for logic-based integer programming, Comput. Chem. Eng. 18 (7) (1994) 563–578.

[20] M. Turkay, I.E. Grossmann, Disjunctive programming techniques for the optimization of process systems with discontinuous investment costs multiple size regions, Ind. Eng. Chem. Res. 35 (8) (1996) 2611–2623.

[21] J.P. Ruiz, et al., Generalized Disjunctive Programming: Solution Strategies Algebraic Modeling Systems, in: J. Kallrath (Ed.), Springer Berlin Heidelberg, 2012, p. 57–75.

[22] G. Nemhauser, L. Wolsey, Integer and Combinatorial Optimization, John Wiley & Sons, 1999.

[23] I.E. Grossmann, S. Lee, Generalized convex disjunctive programming: nonlinear convex hull relaxation, Comput. Optim. Appl. 26 (1) (2003) 83–100.

[24] A. Vecchietti, S. Lee, I.E. Grossmann, Modeling of discrete/continuous optimization problems: characterization and formulation of disjunctions and their relaxations, Comput. Chem. Eng. 27 (3) (2003) 433–448.

[25] N.W. Sawaya, I.E. Grossmann, Computational implementation of non-linear convex hull reformulation, Comput. Chem. Eng. 31 (7) (2007) 856–866.

[26] H.P. Williams, Model Building in Mathematical Programming, Wiley, Chichester England and New York, 1990.

[27] L.T. Biegler, I.E. Grossmann, A.W. Westerberg, Systematic Methods of Chemical Process Design, P.H.I.S.i.t.P.a.C.E. Sciences, Prentice Hall, New Jersey, 1997.

[28] M. Turkay, I.E. Grossmann, Logic-based MINLP algorithms for the optimal synthesis of process networks, Comput. Chem. Eng. 20 (8) (1996) 959–978.

Index

Note: Page numbers followed by f indicate figures; t, tables; b, boxes.

CPSIA information can be obtained at www.ICGtesting.com
Printed in the USA
LVOW03*2110190215

427619LV00013B/226/P